Maximum Entropy and Bayesian Methods
Garching, Germany 1998

Fundamental Theories of Physics

An International Book Series on The Fundamental Theories of Physics: Their Clarification, Development and Application

Volume 105

Maximum Entropy and Bayesian Methods

Garching, Germany 1998

Proceedings of the 18th International Workshop on Maximum Entropy and Bayesian Methods of Statistical Analysis

edited by

Wolfgang von der Linden

Institute for Theoretical Physics,
Technical University of Graz,
Graz, Austria

Volker Dose

Rainer Fischer

and

Roland Preuss

Max Planck Institute for Plasma Physics,
Garching, Germany

SPRINGER-SCIENCE+BUSINESS MEDIA, B.V.

Library of Congress Cataloging-in-Publication Data

International Workshop on Maximum Entropy and Bayesian Methods of Statistical Analysis (18th : 1998 : Garching bei München, Germany)
Maximum entropy and Bayesian methods, Garching, Germany 1998 : proceedings of the 18th International Workshop on Maximum Entropy and Bayesian Methods of Statistical Analysis / edited by Wolfgang von der Linden ... [et al.].
p. cm. -- (Fundamental theories of physics ; v. 105)
Includes index.
ISBN 978-94-010-5982-4 ISBN 978-94-011-4710-1 (eBook)
DOI 10.1007/978-94-011-4710-1
1. Maximum entropy method Congresses. 2. Bayesian statistical decision theory Congresses. 3. Bayesian statistical decision theory--Industrial applications Congresses. I. Linden, Wolfgang von der. II. Title. III. Series.
Q370.I575 1998
530'.07'27--dc21 99-27057

ISBN 978-94-010-5982-4

Printed on acid-free paper

DEDICATED TO: EDWIN THOMPSON JAYNES

July 5, 1922 to April 30, 1998

CONTENTS

PREFACE xi

EDWIN THOMPSON JAYNES
G.L. Bretthorst 1

A TRIBUTE TO EDWIN T. JAYNES
M. Tribus 11

THE NEAR-IRRELEVANCE OF SAMPLING FREQUENCY DISTRIBUTIONS
G. Larry Bretthorst 21

OUTLIER TOLERANT PARAMETER ESTIMATION
V. Dose and W. von der Linden 47

UNIVERSALITY, PRIOR INFORMATION AND MAXIMUM ENTROPY IN FOAMS
N. Rivier, B. Dubertret, T. Aste, and H. Ohlenbusch 57

TEXTURE RECONSTRUCTION IN NOISY IMAGES
M. Walessa and M. Datcu 65

BAYESIAN 3D X-RAY RECONSTRUCTION FROM INCOMPLETE NOISY DATA
V. Vengrinovich, Yu. Denkevich, and G.-R. Tillack 73

A BAYESIAN REFLECTION ON SURFACES
David R. Wolf 83

QUANTUM ENTROPY REGULARIZATION
R. N. Silver 91

BAYESIAN RECONSTRUCTION OF ELECTRON ENERGY DISTRIBUTIONS FROM EMISSION LINE INTENSITIES
R. Fischer, W. Jacob, W. von der Linden, and V. Dose 99

DEPTH PROFILE RECONSTRUCTION FROM RUTHERFORD BACKSCATTERING DATA
U. v. Toussaint, K. Krieger, R. Fischer, V. Dose 107

BAYESIAN IMAGE SEGMENTATION USING A DYNAMIC PYRAMIDAL STRUCTURE
H. Rehrauer, K. Seidel, and M. Datcu 115

ADAPTIVE IMAGE RECONSTRUCTION WITH PREDICTIVE MODELS
V. M. Artemiev, A. O. Naumov, and G.-R. Tillack 123

ARE WE CRUISING A HYPOTHESIS SPACE?
C. C. Rodriguez 131

FIDUCIAL INFERENCE
Diemer Salomé 141

POSTERIOR MOMENTS OF THE CAUCHY DISTRIBUTION
David R. Wolf 149

BAYESIAN REASONING VERSUS CONVENTIONAL STATISTICS IN HIGH ENERGY PHYSICS
G. D'Agostini 157

MODEL COMPARISON IN PLASMA ENERGY CONFINEMENT SCALING
R. Preuss, V. Dose, and W. von der linden 171

STRAIGHT LINES
Do Kester 179

BAYESIAN ANALYSIS WITH APPLICATION TO THE TIMING RESIDUALS OF A PULSAR
Gilles Zalamansky Christian P. Robert 189

BAYESIAN LABELING OF REMOTE SENSING IMAGE CONTENT
M. Schröder, K. Seidel, and M. Datcu 199

SOME RECENT DEVELOPMENTS IN BAYESIAN STATISTICS AND ECONOMETRICS
Arnold Zellner 207

CONVERGENT BAYESIAN FORMULATIONS OF BLIND SOURCE SEPARATION AND ELECTROMAGNETIC SOURCE ESTIMATION
K. H. Knuth and H. G. Vaughan, jr. 217

A MAXIMUM ENTROPY APPROACH FOR ESTIMATING NONLINEAR DYNAMIC MODELS
Amos Golan 227

PROBABILITY AND ENTROPY IN QUANTUM THEORY
Ariel Caticha 237

UNREAL PROBABILITIES:*Partial Truth with Clifford Numbers*
C. C. Rodriguez 247

YET ANOTHER DERIVATION OF THE PRINCIPLE OF MAXIMUM ENTROPY
Anthony J.M. Garrett 271

EFFICIENT ESTIMATION OF MISSING INFORMATION IN MULTIVALUED SINGLY CONNECTED NETWORKS USING MAXIMUM ENTROPY
D. Holmes 289

OVERVIEW AND BAYESIAN PERSPECTIVE OF PHASE UNWRAPPING
G. Nico, L. Guerriero, G. Palubinskas, and M. Datcu 301

OPERATION OF THE BAYES INFERENCE ENGINE
K.M. Hanson and G.S. Cunningham 309

THE PRIOR-PREDICTIVE VALUE
W. von der Linden, R. Preuss, and V. Dose 319

AN UNSUPERVISED CLUSTERING METHOD BY ENTROPY MINIMIZATION
G. Palubinskas 327

APPLICATIONS OF THE MAXIMUM ENTROPY METHOD IN CRYSTALLOGRAPHY
Martin Schneider and Sander van Smaalen 335

TRACE METAL DETECTION IN NATURAL WATER
A. Taglauer, R. Niessner, W. von der Linden, V. Dose, and E. Taglauer 341

OSCILLATOR STABILITY ANALYSIS USING A BAYESIAN APPROACH
Gilles Zalamansky and Francois Vernotte 345

INDEX 353

LIST OF PARTICIPANTS 357

PREFACE

In 1978 Edwin T. Jaynes and Myron Tribus initiated a series of workshops to exchange ideas and recent developments in technical aspects and applications of Bayesian probability theory. The first workshop was held at the University of Wyoming in 1981 organized by C.R. Smith and W.T. Grandy. Due to its success, the workshop was held annually during the last 18 years. Over the years, the emphasis of the workshop shifted gradually from fundamental concepts of Bayesian probability theory to increasingly realistic and challenging applications.

The **18th international workshop on Maximum Entropy and Bayesian Methods** was held in Garching / Munich (Germany) (27-31. July 1998). Opening lectures by G. Larry Bretthorst and by Myron Tribus were dedicated to one of the pioneers of Bayesian probability theory who died on the 30th of April 1998: Edwin Thompson Jaynes. Jaynes revealed and advocated the correct meaning of 'probability' as the state of knowledge rather than a physical property. This interpretation allowed him to unravel longstanding mysteries and paradoxes. Bayesian probability theory, "the logic of science" - as E.T. Jaynes called it - provides the framework to make the best possible scientific inference given all available experimental and theoretical information.

We gratefully acknowledge the efforts of Tribus and Bretthorst in commemorating the outstanding contributions of E.T. Jaynes to the development of probability theory.

Following the general scope of the annual workshop the talks were mostly informal presentations of recent development in technical aspects and applications covering a wide variety of research areas, such as physics (high energy physics, plasma physics, astro-physics, foundation of quantum mechanics), biology, medicine, econometrics, urban planing, image reconstruction, communication theory. In most fields the applications are changing from academic studies to challenging real-world problems with significant success. The number of participants varied between 60 and 90, a decent number for active discussions during and between the sessions. More information about the conference can be found on the conference-homepage: http://www.ipp.mpg.de/OP/maxent98/.

This volume contains a collection of papers presented at the workshop, which testifies the far-reaching applicability and importance of Bayesian probability theory. According to Larry Bretthorst, E.T. Jaynes had little tolerance for incompetence. We are convinced that the papers in these proceedings would have found his approval.

Finally, we wish to acknowledge financial support by the Deutsche Physikalische Gesellschaft and the Max-Planck-Institut für Plasmaphysik.

Wolfgang von der Linden
Institut für Theoretische Physik
Technische Universität Graz
Austria

Volker Dose, Rainer Fischer, Roland Preuss
Max-Planck-Institut für Plasmaphysik
Garching / Munich
Germany

CONFERENCE PHOTO

Volker Dose, his wife Luise and Ray Smith at the wellcome reception.

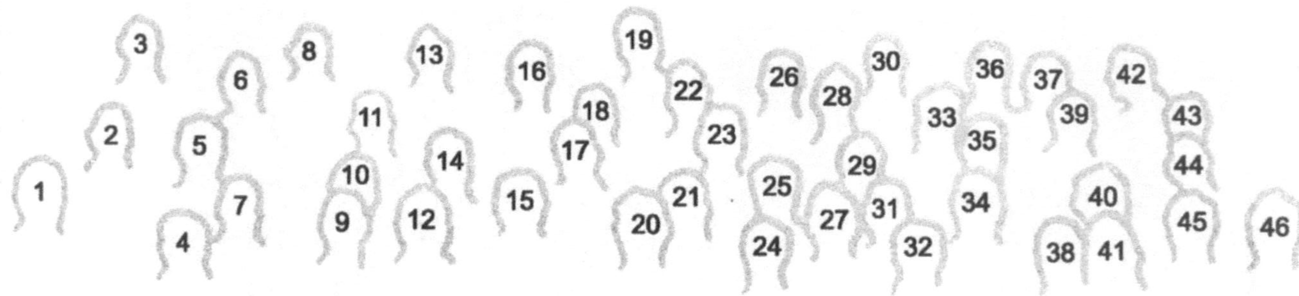

Legend for group photograph: 1. Tj. Romke Bontekoe, 2. Amos Golan, 3. Roland Preuss, 4. Kevin H. Knuth, 5. Arnold Zellner, 6. Karl Ertl, 7. Valentin M. Artemiev, 8. Ray Smith, 9. Do Kester, 10. Udo von Toussaint, 11. Peter Coles, 12. Rainer Fischer, 13. John Skilling, 14. Larry Bretthorst, 15. Gary J. Erickson, 16. Casper J. Albers, 17. Anthony Garrett, 18. Peter C. Cheeseman, 19. Sander Van Smaalen, 20. Volker Dose, 21. Kenneth M. Hanson, 22. Diemer Salomé, 23. Wolfgang von der Linden, 24. Sabina Figge, 25. Reinhard Drube, 26. Michael Schröder, 27. Peter Goldstrass, 28. Hubert Rehrauer, 29. Carlos C. Rodriguez, 30. Robert L. Fry, 31. Myron Tribus, 32. Gintautas Palubinskas,Marc Walessa, 33. David Hestenes, 34. Martin Schneider, 35. Richard N. Silver, 36. Gerhard Venus, 37. Klaus Engelhardt, 38. Ariel Caticha, 39. Gilles Zalamansky, 40. Ingeborg Zeising, 41. Amara Graps, 42. Mathias Anton, 43. Caio Görner, 44. Giulio D'Agostini, 45. Ales Gottvald, 46. Josh Rychert.

Conference secretaries: Ingeborg Zeising, Gerhard Venus, and Sabina von Polenz.

EDWIN THOMPSON JAYNES

July 5, 1922 to April 30, 1998

G. LARRY BRETTHORST
Dept. of Chemistry
Washington University
St. Louis Mo 63130

On July 5th, 1922 Edwin Thompson Jaynes Jr. was born in Waterloo, Iowa to Ethyl and Edwin Jaynes. At the time the Jayneses lived near Cedar Falls, Iowa. Ed had two sisters, and an older brother. His father, a surgeon, practiced in Waterloo. By 1922 standards his family was fairly well off. However, his father died when Ed was very young, leaving his mother to care for four small children. Consequently, the Jaynes family moved to Parkersburg, Iowa where Ethyl's parents lived.

Ed grew up around Parkersburg. His first word was "kitty" which he uttered when he was seven months old. He took his first step at 13 months. He called his mother "Ma." His family did have some fine furniture, linen, silverware, a Knabe grand piano, a classical music collection, and a large library, including the Harvard Classics, housed in elegant glass-front mahogany bookcases – things left over from when his father was alive – but no money. Their only source of income was from renting half of the house. They had to grow most of their own food. Jaynes noted in [5] that his grandfather must have had to help them out financially many times.

The presence of the piano and library must have exerted a great influence on Ed, because he considered a career as a concert pianist. At his death he had a Bosendorfer grand piano in his home valued at approximately $100,000 dollars and hundreds of tapes of himself playing compositions from various classical composers. His library consisted of more than a thousand books, including statistics, physics, music, chemistry, biology, history, and philosophy.

Ed left Parkersburg in 1938 to enter Cornell College in Mount Vernon, Iowa. He returned to Parkersburg occasionally to visit his mother and grandmother. Interestingly, his mother kept his grades from Cornell College. In his first semester he received a B in English, an A in mathematics, a B in German, and a B in chemistry. By the second semester he was doing better, receiving all A's except for a B in mathematics. By the end of his second year his grades were straight A's. That continued for the third and fourth year, with one exception. He received a C in philosophy. He finished his undergraduate schooling in 1942 receiving a B. A. in physics.

During the time he was attending Cornell College he supported himself by working and by loans from the Louise Foote Foundation. He noted with pride that

W. von der Linden et al. (eds.), Maximum Entropy and Bayesian Methods, 1–10.

these loans were repaid on time and that "my entire college education cost my mother a total of $25: one winter she bought me an overcoat" [5].

Between his junior and senior year, he worked at the Warner Institute with Drs. Gustav Martin and Marvin R. Thompson. After receiving his B. A. he had intended to return to the Warner Institute for another summer, but his plans changed because of the war. From 1942 to 1944 he worked for the Sperry Gyroscope Company on Long Island helping to develop Doppler radar.

At the end of 1944, he became Ensign Jaynes and worked at the Anacostia Naval Research Lab in Washington D. C. developing microwave systems. During his stay in the Navy he spent some time on Guam. When he was discharged in 1946, he was a lieutenant (j.g.). There are two surviving documents written by Ensign Jaynes: the first is a series of 9 lectures on solving circuit problems using Laplace and Fourier transforms [1]; the second, is titled "Theory of Microwave Coupling Systems" [2]. These two documents constitute the earliest known professional writings of Ed Jaynes.

Jaynes left the Navy in 1946 and headed for California. In the summer of 1946 he worked in the W. W. Hansen Laboratories of Physics at Stanford on the design of the first linear electron accelerator. In "Disturbing The Memory" [4] Jaynes notes

> I first met Julian Schwinger, Robert Dicke, and Donald Hamilton during the War when we were all engaged in developing microwave theory, measurement techniques, and applications to pulsed and Doppler radar; Schwinger and Dicke at the MIT Radiation Laboratory, Hamilton and I at the Sperry Gyroscope Laboratories on Long Island. Bill Hansen (for whom the W. W. Hansen Laboratories at Stanford are now named) was running back and forth weekly, giving lectures at MIT and bringing us back the notes on the Schwinger lectures as they appeared, and I accompanied him on a few of those trips.
>
> I first met Edward Teller when he visited Stanford in the Summer of 1946 and Hansen, Teller, and I discussed the design of the first Stanford LINAC, then underway. After some months of correspondence I first met J. R. Oppenheimer in September 1946, when I arrived at Berkeley as a beginning graduate student, to learn quantum theory from him – the result of Bill Hansen having recommended us strongly to each other. When in the Summer of 1947 Oppy moved to Princeton to take over the Institute for Advanced Study, I was one of four students that he took along. The plan was that we would enroll as graduate students at Princeton University, finish our theses under Oppy although he was not officially a Princeton University faculty member; and turn them in to Princeton (which had agreed to this somewhat unusual arrangement in view of the somewhat unusual circumstances). My thesis was to be on Quantum Electrodynamics.
>
> But, as this writer learned from attending a year of Oppy's lectures (1946–47) at Berkeley, and eagerly studying his printed and spoken words for several years thereafter, Oppy would never countenance any retreat from the Copenhagen position, of the kind advocated by Schrödinger and Einstein. He derived some great emotional satisfaction from just those elements of mysticism that

> Schrödinger and Einstein had deplored, and always wanted to make the world still more mystical, and less rational.
>
> This desire was expressed strongly in his 1955 BBC Reith lectures (of which I still have some cherished tape recordings which recall his style of delivery at its best). Some have seen this as a fine humanist trait. I saw it increasingly as an anomaly – a basically anti–scientific attitude in a person posing as a scientist – that explains so much of the contradictions in his character.
>
> As a more practical matter, it presented me with a problem in carrying out my plan to write a thesis under Oppy's supervision, quite aside from the fact that his travel and other activities made it so hard to see him. Mathematically, the Feynman electromagnetic propagator made no use of those superfluous degrees of freedom; it was equally well a Green's function for an unquantized EM field. So I wanted to reformulate electrodynamics from the ground up without using field quantization. The physical picture would be very different; but since the successful Feynman rules used so little of that physical picture anyway, I did not think that the physical predictions would be appreciably different; at least, if the idea was wrong, I wanted to understand in detail why it was wrong.
>
> If this meant standing in contradiction with the Copenhagen interpretation, so be it; I would be delighted to see it gone anyway, for the same reason that Einstein and Schrödinger would. But I sensed that Oppy would never tolerate a grain of this; he would crush me like an eggshell if I dared to express a word of such subversive ideas. I could do a thesis with Oppy only if it was his thesis, not mine.

Eugene Wigner became Ed's thesis advisor in 1948.

His dissertation was a calculation of the electrical and magnetic properties of ferroelectric materials. Ferroelectric materials are crystalline substances which have a permanent electric polarization (an electric dipole moment per unit volume) that can be reversed by an electric field. His dissertation "Ferroelectricity" was finished in 1950 and he received his Ph.D. in physics. He published his first paper in 1950 while still at Princeton. It was titled "The Displacement of Oxygen in BaTiO3" [1]. This paper is essentially a one page summary of some of his thesis results. The paper is so short that it does not begin to hint at the amount of work and original thought that went into his thesis calculations. Joel Snow, one of Jaynes' early graduate students, refers to "Ferroelectricity" as a "*tour de force*".[1] Jaynes' thesis was extensively modified and later published by the Princeton University Press in 1953 [3] in the series *Investigations In Physics.* It was volume one of this prestigious series. *Investigations In Physics* has featured many famous authors, including John Von Neumann, Eugene Wigner, and Eugene Feenberg.

After finishing his degree, Jaynes returned to Stanford in 1950. He stayed through 1960. In his early work Jaynes was both theoretician and experimentalist. For example, his fourth paper was on the observation of a paramagnetic resonance

[1] Snow, Joel A., "Recollections of an Independent Thinker," in *Probability & Physics: Essays in Honor of Edwin T. Jaynes,* W. T. Grandy, Jr. and P. W. Millonni (Eds.), Cambridge University Press, Great Britain, 1993.

in a single crystal of barium titanate [4], essentially an experimental paper. His second paper, on the concept and measurement of the impedance in periodically loaded wave guides [2], had both theoretical and experimental aspects. Jaynes continued to maintain an active research laboratory well into the 70's (many years after moving to St. Louis). Indeed, among his papers there was a copy of the Sunday magazine supplement of the St. Louis Globe-Democrat dated May 30, 1967 containing a photograph of Ed Jaynes in his laboratory in Crow Hall on the Washington University campus. The photograph shows him working with a high energy laser. During this period his students were testing some of the predictions of neoclassical and quantum theory.

While he was at Stanford he also supported himself consulting with Varian Associates, the U. S. Army Corp of Engineers, and the University of California at Livermore. While consulting he wrote a number of reports for both Varian and the U. S. Army. Many of the U. S. Army reports still survive, but are not available for general release; a condition that will change shortly. Two of the reports done for Varian still survive, but are only available from the Varian corporate library. Varian, at that time, was a young upstart company that could not afford to pay Jaynes in cash, so they paid him in stock. Additionally, Jaynes' records indicate that he continued to buy Varian stock throughout most of this period. At the time of his death this stock constituted about one fourth of Jaynes' total wealth.

Prior to 1957 Jaynes published a total of 6 articles [[1],[2],[4]–[7]]. These articles essentially grew out of his thesis work. However, his interest were varied as illustrated by the fact that these 6 papers covered such diverse research areas as solid state, classical electrodynamics, electron spin resonance, and nuclear magnetic resonance. Yet these papers are all related, they are all applications of classical electrodynamics to real physical problems. Jaynes had essentially four different areas of research: his first could be called applied classical electrodynamics; his second, information theory (entropy as a measure of information); his third, probability theory; and finally, semiclassical and neoclassical radiation theory.

During the years preceding 1957, Jaynes was preparing a set of lecture notes on probability theory. This material eventually was presented to the Field Research Laboratory of the Socony-Mobil Oil Company. Socony-Mobil in turn published, at least internally, a collection of five of these lecture notes [3]. Jaynes did try to publish the first of these lectures, "How Does The Brain Do Plausible Reasoning," in 1960. However, this work was rejected by the referee and Jaynes eventually gave up on publishing it. It was later rediscovered in the Stanford Microwave Laboratory library and, with Jaynes's permission, it was published in 1988 [62]; some 28 years after Jaynes first tried to publish it.

In 1957 Jaynes published his first articles in information theory, "Information Theory and Statistical Mechanics," [8,9]. In these two articles Jaynes reformulated statistical mechanics in terms of probability distributions derived by the use of the principle of maximum entropy. This reformulation of the theory simplified the mathematics, allowed for fundamental extensions of the theory, and reinterpreted statistical mechanics as inference based on incomplete information. These articles were published over the objection of a reviewer. (Jaynes comments on this review in "Where do we Stand on Maximum Entropy," [37]). Jaynes kept that review framed and hanging on the wall of his office until the day he died.

The two 1957 articles, by themselves, would have been a career for most scientists; but Jaynes was far from finished. In the three years he remained at Stanford he published articles on wave guides [10,13], relativity [11], information theory [12], masers [14] and 50 others after moving to Washington University in St. Louis.

In the years immediately preceding his departure from Stanford (1960) he was becoming increasingly dissatisfied with the *publish* or *perish* mentality plaguing Stanford, a condition he talked about in "Backward Look to the Future" [73]. So in 1960 he packed his belongings, sold his house, and moved to St. Louis, Missouri where he joined the physics faculty of Washington University.

Around the time Jaynes moved to St. Louis, he set out on his remaining research interest, reformulating quantum electrodynamics to avoid quantization of the electromagnetic field. Jaynes published his first paper on this subject in 1963 with Fred Cummings. It was titled "Comparison of Quantum and Semiclassical Radiation Theory with Application to the Beam Maser" [17]. This work, while not know to workers in probability and information theory, is probably Jaynes' most cited and notorious article. Jaynes continued to publish articles on both semiclassical and neoclassical radiation theory more or less continuously until he retired [17,25,26,31,34,35,39]. However, much of this research is in the theses of his graduate students and has never been published in the open literature.

While his graduate students were working on semiclassical and neoclassical radiation theory, he was continuing with his other research interests. For example, in statistical mechanics he published 14 articles from 1960 onward. In probability theory, especially during the 80's, he published 21 articles. These articles addressed fundamental questions within these theories and often extended them to new application areas. Many of these articles are published in proceedings volumes – it was commonplace for mainstream journals to reject his manuscripts. Consequently, he often had to wait many years to respond in print to a critic.

In 1982 Jaynes took a two year sabbatical. He spent the first year at the University of Wyoming as an Adjunct Professor. While there he taught statistics, gave a few colloquia, generally renewed old friendships, enjoyed himself, thought, and wrote. Ray Smith told me that he, his son, and Gary Erickson took Jaynes camping in the Snowy mountains. Apparently Jaynes enjoyed this experience so much that when he returned to St. Louis he went out and bought a tent and sleeping bag in the hopes of going camping again some day.

The second year of his sabbatical was spent as a Fellow at St. John's College in Cambridge, England. The time Jaynes spent at St. John's college was the highlight of his life. Had it been possible, I think he would have stayed. During his time at Cambridge he attended the weekly meetings John Skilling held with his graduate students. He wrote a number of papers related to the discussions that were going on at these meetings; "Monkeys Kangaroos and N" being the most notable of these [57]. He wandered the campus thinking about the history and magnitude of human accomplishments that are associated with Cambridge. He mentioned spending parts of several weeks trying to find the tomb of Rev. Thomas Bayes. He succeeded at this, because in his possessions were a number of pictures of Bayes' tomb.

Jaynes retired in 1992 after a long and productive career. Jaynes' contributions to science were of the highest caliber. His work in reformulating statistical mechanics has illuminated the foundations of that theory and enabled extensions to non-equilibrium systems. His dedication to rooting out contradictions in quantum mechanics is legendary. He may have single-handedly sparked more debate in quantum mechanics than any other person in the last 40 years. The verdict on his neoclassical radiation theory is still not in, and may not be for many more years. Neoclassical radiation theory may yet prove to be a better description of nature than quantum electrodynamics. He also helped take an interpretation of probability theory from being virtually unknown to a healthy research area that is being applied daily in economics, biology, physics, nuclear magnetic resonance and many other disciplines. His writing helped to clarify the foundations of probability theory in a way never achieved before. He wrote profusely, in a warm and friendly way that enabled one to see complex points as if they were intuitively obvious. He spoke as he wrote. When he criticized someone's work, he always stuck to the facts; he never reverted to name calling. His friendship was hard to earn, and hard to keep, for he had little tolerance for incompetence – and he would undoubtedly be uncomfortable with all of the attention being lavished on him now that he is dead.

Bibliography

1. Jaynes, E. T., 1950, "Displacement of Oxygen in $BaTiO_3$," *Phys. Rev.,* **79,** 1008.
2. Jaynes, E. T., 1952, "The Concept and Measurement of Impedance in Periodically Loaded Wave Guides," *J. Appl. Phys.* **23,** 1077.
3. Jaynes, E. T., 1953, "Ferroelectricity," Princeton Univ. Press, London.
4. Hornig, A. W., E. T. Jaynes, and H. E. Weaver, 1954, "Observation of Paramagnetic Resonances in Single Crystals of Barium Titanate," *Phys. Rev.* **96,** 1703.
5. Jaynes, E. T., 1955, "Matrix Treatment of Nuclear Induction," *Phys. Rev.* **98,** 1099.
6. Jaynes, E. T., 1955, "Nonlinear Dielectric Materials," *Proc. IRE.,* **43,** 1733.
7. Vartanian, P. H., and E. T. Jaynes, 1956, "Propagation in Ferrite-Filled Transversely Magnetized Waveguide," *IRE Trans. on Microwave Theory and Techniques,* **MTT-4,** 140.
8. Jaynes, E. T., 1957, "Information Theory and Statistical Mechanics," *Phys. Rev.,* **106,** 620.
9. Jaynes, E. T., 1957, "Information Theory and Statistical Mechanics II," *Phys. Rev.,* **108,** 171.
10. Jaynes, E. T., 1958, "Ghost Modes in Imperfect Waveguides," *Proc. IRE.,* **46,** 416.
11. Jaynes, E. T., 1958, "Relativistic Clock Experiments," *Am. J. Phys.,* **26,** 197.
12. Jaynes, E. T., 1959, "Note on Unique Decipherability," *IRE Trans. on Information Theory,* **IT-5,** 98.
13. Forrer, M. P., and E. T. Jaynes, 1960, "Resonant Modes in Waveguide Windows," *IRE Trans. on Microwave Theory and Techniques,* **MTT-8,** 147.
14. Jaynes, E. T., 1960, "The Maser as a Parametric Amplifier," in *Quantum Electronics,* C. H. Townes (ed.), Columbia Univ. Press, New York, p. 237.
15. S. Heims, and E. T. Jaynes, 1962, "Theory of Gyromagnetic Effects and Some Related Magnetic Phenomena," *Rev. Mod. Phys.,* **34,** 143.
16. Jaynes, E. T., 1963. "New Engineering Applications of Information Theory," in *Engineering Uses of Random Function Theory and Probability,* J.L. Bogdanoff and F. Kozin (eds.), Wiley, New York, p. 163.
17. Jaynes, E. T., and F. W. Cummings, 1963, "Comparison of Quantum and Semiclassical

Radiation Theory with Application to the Beam Maser," *Proc. IEEE.* **51,** 89.
18. Jaynes, E. T., 1963, "Information Theory and Statistical Mechanics," in *Statistical Physics,* K. Ford (ed.), Benjamin, New York, p. 181.
19. Jaynes, E. T., 1963, "Comments on an article by Ulric Neisser," *Science,* **140,** 216.
20. Jaynes, E. T., 1965, "Gibbs vs Boltzmann Entropies," *Am. J. Phys.,* **33,** 391.
21. Jaynes, E. T., 1966, "Is QED Necessary," in *Proceedings of Second Rochester Conference on Coherence and Quantum Optics,* L. Mandel and E. Wolf (eds.), Plenum, New Your, p. 21.
22. Jaynes, E. T., 1967, "Foundations of Probability Theory and Statistical Mechanics," in *Delaware Seminar in the Foundations of Physics,* M. Bunge (ed.), Springer-Verlag, Berlin, p. 77.
23. Jaynes, E. T., 1968, "You CAN Parallel Storage Batteries," in *Popular Electronics,* **33,** 86.
24. Jaynes, E. T., 1968, "Prior Probabilities," *IEEE Trans. on Systems Science and Cybernetics,* **SSC-4,** 227.
25. Crisp, M. D., and E. T. Jaynes, 1969, "Radiative Effects in Semiclassical Theory," *Phys. Rev.,* **179,** 1253.
26. Stroud, C. R., Jr., and E. T. Jaynes, 1970, "Long-Term Solutions in Semiclassical Radiation Theory," *Phys. Rev. A.,* **1,** 106.
27. Jaynes, E. T., 1970, "Reply to Leiter's Comments," *Phys. Rev. A.,* **2,** 260.
28. Jaynes, E. T., 1971, "Violation of Boltzmann's H-Theorem in Real Gases," *Phys. Rev.,* **A4,** 747.
29. Jaynes, E. T., and F. W. Cummings, 1972, "Comparison of Quantum and Semiclassical Radiation Theories with Application to the Beam Maser," in *Laser Theory,* F. S. Barnes (ed.), IEEE Press, pp. 173-203.
30. Jaynes, E. T., 1973, "The Well-Posed Problem," in *Found. Phys.,* **3,** 477.
31. Jaynes, E. T., 1973, "Survey of the Present Status of Neoclassical Radiation Theory," in *Coherence and Quantum Optics,* L. Mandel and E. Wolf (eds.), Plenum, New York, p. 35.
32. Jaynes, E. T., 1976. "Confidence Intervals vs Bayesian Intervals," in *Foundations of Probability Theory, Statistical Inference, and Statistical Theories of Science,* W. L. Harper and C. A. Hooker (eds.), D. Reidel, Dordrecht, p. 175.
33. Jaynes, E. T., 1976, "Reply to Kempthornes' Comments," in *Foundations of Probability Theory, Statistical Inference and Statistical Theories of Science,* W. L. Harper and C. A. Hooker (eds.), D. Reidel, Dordrecht, p. 229.
34. Jaynes, E. T., 1978, "Ancient History of Free Electron Devices," in *Novel Sources of Coherent Radiation,* S. F. Jacob, M. Scully, and M. Sargent III (eds.), Addison-Wesley, Reading, MA, p. 1.
35. Jaynes, E. T., 1978, "Electrodynamics Today," in *Coherence and Quantum Optics IV,* L. Mandel and E. Wolf (eds.), Plenum Press, New York, p. 495.
36. O'Donnell, M., E. T. Jaynes, and J. G. Miller, 1978, "General Relationships between Ultrasonic Attenuation and Dispersion," *J. Acoust. Soc. Am.,* **63,** 1935.
37. Jaynes, E. T., 1979, "Where do we Stand on Maximum Entropy?" in *The Maximum Entropy Formalism,* R. D. Levine and M. Tribus (eds.), M. I. T. Press, Cambridge, MA, p. 15.
38. Jaynes, E. T., 1979, "Concentration of Distributions at Entropy Maxima," in *E. T. Jaynes: Papers on Probability, Statistics and Statistical Physics*, R. D. Rosenkrantz (ed.), D. Reidel, Dordrecht, p. 315.
39. Jaynes, E. T., 1980, "Quantum Beats," in *Foundations of Radiation Theory and Quantum Electrodynamics,* A. O. Barut (ed.), Plenum Press, New York, p. 37.
40. Jaynes, E. T., 1980, "Marginalization and Prior Probabilities," in *Bayesian Analysis in Econometrics and Statistics,* A. Zellner (ed.), North-Holland, Amsterdam, p. 43.
41. Jaynes, E. T., 1980, "Reply to Dawid, Stone, and Zidek," in *Bayesian Analysis in Econometrics and Statistics,* A. Zellner (ed.), North-Holland Amsterdam, p. 83.
42. Jaynes, E. T., 1980, "What is the Question?," in *Bayesian Statistics,* J. M. Bernardo, M. H. deGroot, D. V. Lindly, and A. F. M. Smith (eds.), Valencia Univ. Press, Valencia.
43. Jaynes, E. T., 1980, "The Minimum Entropy Production Principle," *Ann. Rev. Phys. Chem.* **31,** 579.
44. Matthys, D. R., and E. T. Jaynes, 1980, "Phase-Sensitive Optical Amplifier," *J. Opt. Soc. Am.,* **70,** 263.

45. O'Donnell, M., E. T. Jaynes, and J. G. Miller, 1980, "Kramers-Kronig Relationship between Ultrasonic Attenuation and Phase Velocity," *J. Acoust. Soc. Am.*, **69**, 696.
46. Jaynes, E. T., 1982, "On the Rationale of Maximum-Entropy Methods," *Proc. IEEE.*, **70**, 939.
47. Jaynes, E. T., 1984, "The Intuitive Inadequacy of Classical Statistics," *Epistemologia,* **VII**, 43.
48. Jaynes, E. T., 1984, "Prior Information and Ambiguity in Inverse Problems," *Inverse Problems,* D. W. McLaughlin (ed.), Am. Math. Soc., Providence, RI, [SIAM-AMS Proceedings **14**, 151].
49. Jaynes, E. T., 1985, "Where do we go from Here?," in *Maximum-Entropy and Bayesian Methods in Inverse Problems,* C. R. Smith and W. T. Grandy, Jr. (eds.), D. Reidel, Dordrecht, p. 21.
50. Jaynes, E. T., 1985, "Generalized Scattering," in *Maximum-Entropy and Bayesian Methods in Inverse Problems,* C. R. Smith and W. T. Grandy, Jr. (eds.), D. Reidel, Dordrecht, p. 377.
51. Jaynes, E. T., 1985, "Entropy and Search-Theory," in *Maximum-Entropy and Bayesian Methods in Inverse Problems,* C. R. Smith and W. T. Grandy, Jr. (eds.), D. Reidel, Dordrecht, p. 443.
52. Jaynes, E. T., 1985, "Highly Informative Priors," in *Bayesian Statistics 2,* J. M. Bernardo, M. H. deGroot, D. V. Lindley, and A. F. M. Smith (eds.), Elsevier, Amsterdam, p. 329.
53. Jaynes, E. T., 1985, "Macroscopic Prediction," in *Complex Systems – Operational Approaches,* H. Haken (ed.), Springer-Verlag, Berlin, p. 254.
54. Jaynes, E. T., 1985, "Some Random Observations," *Synthese,* **63**, 115.
55. Jaynes, E. T., 1986, "Predictive Statistical Mechanics," in *Frontiers of Nonequilibrium Statistical Physics,* G. T. Moore and M. O. Scully (eds.), Plenum Press, New York, p. 33.
56. Jaynes, E. T., 1986, "Bayesian Methods: General Background," in Maximum-Entropy and Bayesian Methods in Applied Statistics, J. H. Justice (ed.), Cambridge Univ. Press, Cambridge, p. 1.
57. Jaynes, E. T., 1986, "Monkeys, Kangaroos and N," in *Maximum-Entropy and Bayesian Methods in Applied Statistics,* J. H. Justice (ed.), Cambridge Univ. Press, Cambridge, p. 26.
58. Jaynes, E. T., 1986, "Some Applications and Extensions of the de Finetti Representation Theorem," in *Bayesian Inference and Decision Techniques with Applications: Essays in Honor of Bruno de Finetti,* P. K. Goel and A. Zellner (eds.), North-Holland, Amsterdam, p. 31.
59. Muckenheim, W., G. Ludwig, C. Dewdney, P. R. Holland, A. Kyprianidis, J. P. Vigier, N. C. Petroni, M. S. Bartlett, and E. T. Jaynes, 1986, "A Review of Extended Probability," *Phys. Repts.,* **133**, 337.
60. Jaynes, E. T., 1987, "Bayesian Spectrum and Chirp Analysis," in *Maximum Entropy and Bayesian Spectral Analysis and Estimation Problems,* C. R. Smith and G. J. Erickson (eds.), D. Reidel, Dordrecht, p. 1.
61. Jaynes, E. T., 1987, "Comments on a Review by P.W. Atkins," *Contemp. Phys.,* **28**, 501.
62. Jaynes, E. T., 1988, "How Does the Brain Do Plausible Reasoning?" in *Maximum-Entropy and Bayesian Methods in Science and Engineering,* **1**, G. J. Erickson and C. R. Smith (eds.), Kluwer, Dordrecht, p. 1. [This first appeared as a Stanford Microwave Laboratory Report in 1957]
63. Jaynes, E. T., 1988, "The Relation of Bayesian and Maximum Entropy Methods," in *Maximum-Entropy and Bayesian Methods in Science and Engineering,* **1**, G. J. Erickson and C. R. Smith (eds.), Kluwer, Dordrecht, p. 25.
64. Jaynes, E. T., 1988, "Detection of Extra-Solar-System Planets," in *Maximum-Entropy and Bayesian Methods in Science and Engineering,* **1**, G. J. Erickson and C. R. Smith (eds.), Kluwer, Dordrecht, p. 147.
65. Jaynes, E. T., 1988, "The Evolution of Carnot's Principle," in *Maximum-Entropy and Bayesian Methods in Science and Engineering,* **1**, G. J. Erickson and C. R. Smith (eds.), Kluwer, Dordrecht, p. 267.
66. Jaynes, E. T., 1989, "Clearing up Mysteries – The Original Goal," in *Maximum-Entropy and Bayesian Methods,* J. Skilling (ed.), Kluwer, Dordrecht, p. 1.
67. Jaynes, E. T., 1990, "Probability in Quantum Theory," in *Complexity, Entropy, and the Physics of Information,* W. H. Zurek (ed.), Addison-Wesley, Redwood City, CA, p. 381.

68. Jaynes, E. T., 1990, "Probability Theory as Logic," in *Maximum-Entropy and Bayesian Methods,* P. F. Fougère (ed.), Kluwer, Dordrecht, p. 1.
69. Jaynes, E. T., 1991, "Notes On Present Status And Future Prospects," in *Maximum Entropy and Bayesian Methods,* W. T. Grandy, Jr. and L. H. Schick (eds.), Kluwer, Dordrecht, p. 1.
70. Jaynes, E. T., 1991, "Scattering of Light by Free Electrons," in *The Electron,* D. Hestenes and A. Weingartshofer (eds.), Kluwer, Dordrecht, p. 1.
71. Jaynes, E. T., 1991, "Commentary on Two Articles by C.A. Los," in Computers and Mathematics with Applications, **3** of special issue of "On System-Theoretic Methods in Economic Modeling", S. Mittnik (ed.), Pergamon Press, New York.
72. Jaynes, E. T., 1992, "The Gibbs Paradox," in *Maximum-Entropy and Bayesian Methods,* G. Erickson, P. Neudorfer, and C. R. Smith (eds.), Kluwer, Dordrecht.
73. Jaynes, E. T., 1993, "A Backward Look to the Future," in *Physics and Probability,* W. T. Grandy, Jr. and P. W. Milonni, Cambridge Univ. Press, Cambridge, England.

Unpublished Works:

1. Jaynes, E. T., 1945, "Circuit Analysis by Laplace and Fourier Transforms," a series of nine lectures given at the Combined Research Group, Naval Research Laboratory, Anacostia Station, Washington, D. C.
2. Jaynes, E. T., 1945, "Theory of Microwave Coupling Systems," Combined Research Group, Report No. 84, Naval Research Laboratory, Anacostia Station, Washington, D. C.
3. Jaynes, E. T., 1959, "Probability Theory in Science and Engineering," No. 4 in "Colloquium Lectures in Pure and Applied Science," Socony-Mobil Oil Co. USA.
4. Jaynes, E. T. 1984, "Disturbing The Memory," an unpublished and incomplete manuscript available either by anonymously FTP or on the world wide web from "bayes.wustl.edu".
5. Jaynes, E. T., 1990, "Recollections and Mementos of G. A. Pfeiffer," an unpublished manuscript available either by anonymously FTP or on the world wide web from "bayes.wustl.edu".

Book Reviews:

1. Jaynes, E. T., 1952, "Advanced Theory of Waveguides," by L. Lewin, *Rev. Sci. Inst.* **23**, 373.
2. Jaynes, E. T., 1953, "Theory of Electric Polarization," by C. J. F. Böttcher, *Electronics Magazine.*
3. Jaynes, E. T., 1955, "Laplace Transforms for Electrical Engineers," by B. J. Starkey, *Proc. IRE.* **43**, 898.
4. Jaynes, E. T., 1956, "Spheroidal Wave Functions," by J. A. Stratton, *et al, Proc. IRE.* **44**, 951.
5. Jaynes, E. T., 1958, "Principles of Electricity (3rd ed.)," L. Page and N. I. Adams Jr., *Proc. IRE* **46**, 1664.
6. Jaynes, E. T., 1958, "Television Engineering, Vol. IV: General Circuit Techniques," by S. W. Amos and D. C. Birkinshaw, *Proc. IRE* **46**, 1974.
7. Jaynes, E. T., 1959, "Principles of Circuit Synthesis," by E. S. Kuh and D. O. Pederson, *Proc. IRE* **47**, 1676.
8. Jaynes, E. T., 1961, "Statistical Theory of Communication," by Y. W. Lee, *Am. J. Phys.* **29**, 276.
9. Jaynes, E. T., 1962, "The Fermi Surface," by W. A. Harrison and M. B. Webb (eds.), *Am. J. Phys.* **30**, 231.
10. Jaynes, E. T., 1962, "Atomic Theory and the Description of Nature," by N. Bohr, *Am. J. Phys.* **30**, 658.
11. Jaynes, E. T., 1962, "Wave Mechanics of Crystalline Solids," by R. A. Smith, *Am. J. Phys.* **30**, 846.
12. Jaynes, E. T., 1963, "The Algebra of Probable Inference," by R. T. Cox, *Am. J. Phys.* **31**, 66.

13. Jaynes, E. T., 1963, "Semiconductor Device Physics," by A. Nussbaum, *Am. J. Phys.* **31**, 220.
14. Jaynes, E. T., 1963, "New Perspectives in Physics," by L. DeBroglie, *Am. J. Phys.* **31**, 225.
15. Jaynes, E. T., 1963, "Lasers," by B. A. Lengyel, *Am. J. Phys.* **31**, 739.
16. Jaynes, E. T., 1963, "Noise and Fluctuations," by D. K. C. MacDonald, *Am. J. Phys.* **31**, 946.
17. Jaynes, E. T., 1964, "An Introduction to Transport Theory," by G. M. Wing, *Am. J. Phys.* **32**, 235.
18. Jaynes, E. T., 1968, "Principles of Statistical Mechanics – The Information Theory Approach," by A. Katz, *IEEE Trans. on Information Theory,* **IT-14**, 611.
19. Jaynes, E. T., 1979, "Inference, Method, and Decision: Towards a Bayesian Philosophy of Science," by R. D. Rosenkrantz, *J. Am. Stat. Assn.*, **74**, 740.
20. Jaynes, E. T., 1981, "Works on the Foundations of Statistical Physics," by N. S. Krylov, *J. Am. Stat. Assn.*, **76**, 742.
21. Jaynes, E. T., 1982, "Methods of Statistical Physics," by A. I. Akhiezer and S. V. Peletminskii, *Physics Today,* **35** (8), 57.

A TRIBUTE TO EDWIN T. JAYNES

M. TRIBUS
350 Britto Terrace, Fremont
CA 94539 USA[1]

This is going to be a very difficult talk for me. My emotional ties to the memory of Ed Jaynes are very strong. He changed my life in many ways. I shall always be grateful for the chance I had to know and to work with him.

I met Ed Jaynes as the result of a decade of study and research into the foundations of thermodynamics. When I was examined for my doctoral degree, at UCLA, I was asked to explain the connection between the entropy defined by Claude Shannon and the entropy defined by Clausius a century earlier. Shannon's work was then very new. Neither I nor my committee knew the answer. I was not at all satisfied with the answer I gave. That was in 1948 and I continued to fret about it for the next ten years. I read everything I could find that promised to explain the connection between the entropy of Clausius and the entropy of Shannon. I got nowhere. I felt in my bones there had to be a connection; I couldn't see it.

I wanted the answer to that question for a very specific reason. In 1941 as an undergraduate student in chemistry, I had been introduced to elementary statistical mechanics by the Nobelist Professor William F. Giauque. In his course in thermodynamics he showed us how to use statistical mechanical methods to compute the specific heat of nitrogen from spectral lines with an accuracy greater than could be measured in any heat experiment. I was so thrilled by the beauty of the calculations that when, after WW II, I became an instructor, I wanted my students to know about it too. But I could not find a way to make statistical mechanics logically satisfying to myself, and therefore, explicable to undergraduate students of engineering. In those days, around 1946 to the mid 50's, I found the education of engineers to be stultifying and wanted to enrich it with more intellectually demanding material. But I did not know how to bridge the gap. Every treatment of statistical mechanics I found was convoluted in its logic. Every treatment, that is, until I met Ed Jaynes.

One day, in 1958 a student of mine, Zyssimo Typaldos (now deceased) replied to my question about the connection between Shannon and Clausius by saying, "Oh, that's already been solved." He referred me to Ed Jaynes famous paper in Phys Rev. [1]. I read it the next day and was inspired. Here was my Rosetta Stone! I went home and worked with that paper for a week, almost without sleep. All of my

[1]Email: mtribus@compuserve.com

W. von der Linden et al. (eds.), Maximum Entropy and Bayesian Methods, 11–20.

studies for a decade suddenly fell in place. A straightforward connection between Shannon's Entropy and the entropy of Clausius poured onto the paper, almost without my hand.

I was ecstatic to discover that Ed Jaynes was then at Stanford. I took an overnight train and showed up in his office, the next morning, surely acting like a crazed man. I remember going to the blackboard in his office and pouring out the derivations of the laws of thermodynamics and many of their consequences. Ed listened to my presentation, encouraged me, and that started our friendship. I was disappointed that he did not share my glow of excitement at having cracked the code on thermodynamics; but then I did not realize that he was after much bigger issues and that in the scheme of things, this elucidation of thermodynamics, though it was important to me, was only a small by product of what he was about to do.

The next summer I invited Ed to come to UCLA and participate in a two week summer course on information theory and thermodynamics. In that seminar Ed taught us the material that had been prepared for the Mobil Oil company. He told us that those notes were to be the basis for a book on probability and its applications. I must admit that when we started to teach this course together I had a much more limited objective. I merely wanted to show a new way to think about thermodynamics. But as Ed began his lectures and taught us more and more about the foundations of inductive logic, it became clear. He was teaching us that we were not just discussing statistical physics. We were working on the foundations of inductive logic. We were working on the question of what to do when we know something but do not know everything. We were working on how to acknowledge and build upon our uncertainty. He was not just teaching us to think. He was teaching us how to think about thinking!

Ed repeated the summer courses at UCLA in 1959 and 1960. Each experience with Ed taught me more. When I moved from Los Angeles to New Hampshire to become the Dean of Engineering at Dartmouth, I had just finished writing my book, "Thermostatics and Thermodynamics, An Introduction to Energy, Information and States of Matter". This book was based squarely on Ed's work. Writing the book posed a dilemma, for Ed had not yet published his own book. The fundamental ideas on which my book was based were taken from his work; without his contribution, I had nothing to which to refer for justification of many points. I couldn't just copy his material without writing his book, too. But his book was not yet in print. I could only cite his book as a reference, saying it was "in print". That was 36 years ago and the book has still not issued!

In the first week of April, 1961, I gave a seminar at MIT on the new way to derive thermodynamics. There a critical audience of Joe Keenan's students tried to rip it apart. At one point Benoit Mandelbrot attacked Maxent, saying, "Everyone knows that Shannon's derivation is in error." It happened that Claude Shannon was in residence at MIT that week, so I went to see him. He was immediately able to dispel Mandelbrot's criticism, but then he went on to lecture to me on his misgivings about using his definition of entropy for applications beyond communication channels. I was not able to convince him then of the significance of what Ed had done.

In 1962, Ed joined me in repeating the summer course at Dartmouth. I still think fondly of those warm Summer days in Hanover, New Hampshire, when Ed was at the board, developing one application after another. In those days a kind of tension developed between us, for Ed, quite properly objected to the way I presented his materials. He felt that there were still many ideas to be proven and that I took too many of them on faith. More than once he would take over from my presentation and teach me more of what I obviously did not know.

I have had a lifelong experience with the introduction of new ideas. Once, when I was working with Irving Langmuir on the introduction of his ideas on weather modification through cloud seeding, and there was great opposition to what he had to say, he told me: "There is nothing harder to see than a new *good* idea. If it is new, people will not understand it. If it is good, they will be forced to adopt it. They don't want to." He then went on tell me about his experiences with the grid in the vacuum tube, with the tungsten filament, with coated lenses and a host of other inventions for which he was famous.

I once discussed with Ed the problems inherent in the introduction of new ideas to replace old ones. He told me that his strategy was to hold up publication of his proofs and criticisms until he had them so solidly in place that no one could answer them. That, too, is one of the hallmarks of Ed's work. His proofs are irrefutable. But they did not often change minds.

I found Langmuir's observations true also. After my book on Thermodynamics appeared I sat back and waited for a call from the Nobel Committee. They never wrote. But other people did write and it was not at all what I wanted to hear. My attempts at getting engineering professors to adopt the new way to look at thermodynamics were a complete failure. Not only that, but I was attacked as someone who had taken intellectual shortcuts. Once, during a presentation at IIT in Chicago, I showed how Ed Jaynes' methods developed, in four lines, the equations Fowler and Guggenheim required hundreds of pages to reach. Not once did we mention ergodicity. The presentation did not convince most of the audience. Their reaction: Surely we had cheated somewhere.

It became apparent that Ed was not likely to write his book very soon and I felt that people needed to know more about what his methods were if they were to accept my book on thermodynamics. So I wrote a second book, "Rational Descriptions, Decisions and Designs." I discussed this book with Ed at every opportunity and he gave me encouragement. That book did not sell very well. I persuaded the publisher to publish it in three languages. In three languages it didn't sell either. A few years ago the publisher said he would be pulping the remaining copies! So the need for Ed's book remains.

From 1960 to 1974 I did not see Ed very often. He moved from Stanford to St. Louis. I moved from UCLA to New Hampshire, then to Washington, DC, then to the Xerox Corporation in Rochester, New York and finally, in 1974, to MIT. His influence continued, however, as I moved from one place to another.

In Washington I introduced Bayesian inference into Federal Government in a variety of ways: The National Bureau of Standards (as it was then called, now it is National Institute of Standards and Technology), taught a class on Ed's methods in 1970. (Who says an Assistant Secretary of Commerce has no clout!) I used decision analysis in the administration of the Flammable Fabrics Act to affect the standards on children's nightwear. Ron Howard of Stanford University developed a Bayesian Analysis for me, of the decision to seed or not to seed a hurricane. At the Xerox corporation various strategies were analyzed by Bayesian methods to find the best way to introduce a new product.

When I moved to MIT in 1974, I called Ed and asked him if he would be interested in another conference. In 1978 we held a two day conference in Cambridge, Massachusetts, on "The Maximum Entropy Formalism." Fortythree people came. This was the first time that the subject had been formally introduced to MIT. I had a hidden agenda. I hoped MaxEnt would catch on at MIT. Well, it didn't, but enough people from elsewhere became interested that that first conference became the launching pad for the conference we are at today. These conferences provided Ed with a keenly interested audience and a place to present his current thinking. I took the occasion of this lecture to review the proceedings of these conferences and though I knew he covered many subjects, I did not realize just how broad his interests were until I made this review.

What stands out in Ed's work is the extraordinary breadth of his interests and knowledge. He not only pioneered the development of the MaxEnt formalism, he demonstrated it in a wide range of applications.

At this first conference, at MIT, Ed presented a paper "Where Do We Stand on Maximum Entropy?" which reviewed the history of the Principle of Maximum Entropy. Ed demonstrated that the *methods* used by Shannon were exactly the same as those used by Gibbs. I wish I had known that when I first discussed Ed's work with Shannon and had been rebuffed. He also introduced his analysis of the data of Wolf, the Swiss astronomer who 'tested' concepts of randomness from 18501890 by tossing a die 20,000 times and recording the values! (In a later paper Ed reports of a visit to the laboratory where the dice were still available. He had accurate measurements made of the dice and showed they were unsymmetrical in just the way he had calculated in this paper!) Ed also discussed the potential of applications of MaxEnt in the fields of turbulence and biology. He also dealt with irreversible processes, general methods of prediction and epistemology. This paper is so characteristic of Ed's work. He not only discusses a method, but he offers conjectures on how to use the method in a wide variety of topics. In those cases where he touches on things I know a little bit about, I am always amazed at how much of an expert he is.

The first workshop in this present series was held at the University of Wyoming, June 8 to 10, 1981. At that workshop Ed presented a paper, "Where do we go from Here?" In that paper Ed reviewed applications of MaxEnt to communication, kinetic theory and image reconstruction.

The second workshop was also held at the University of Wyoming, August 9 to 11, 1982, just after Ed Jaynes' 60th birthday. At that workshop Ed presented a paper on "Generalized Scattering". Here he introduced the notion that the scattering of waves was equivalent to the scattering of information. This treatment was especially interesting to me because in it he begins by developing a generalized 'first law' this way:

Start with the definition of expectation:

$$\langle A \rangle = \sum_i p_i A_i \ .$$

Differentiate:

$$\delta \langle A \rangle = \sum_i \delta p_i A_i + \sum_i p_i \delta A_i \ .$$

Note that the last sum is the expectation of the deviation of A:

$$\delta \langle A \rangle - \langle \delta A \rangle = \sum_i \delta p_i A_i \ .$$

Note that the term on the right in equation (3) is the only term in which the probability changes. Therefore, this term must be intimately associated with changes in entropy. Therefore, let: $\delta Q = \delta p_i A_i$ to obtain a 'generalized first law'

$$\delta \langle A \rangle - \langle \delta A \rangle = \delta Q \ .$$

which looks more familiar when written:

$$\delta U - \delta W = \delta Q \ .$$

In one fell swoop, Ed has produced a generalization of both work (δW) and heat (δQ). In thermodynamics this result has an obvious interpretation but Ed showed that it had meaning in many other contexts. In presenting this point we find Ed making one of those remarks that are the hallmark of his work:

> *The first law of thermodynamics (or at least a relation that in textbooks is often called the 'first law') is seen here as a special case of a general rule: A small change in the predicted value of any quantitywhatever its physical meaningmay be resolved into parts arising from a change in the distribution (the 'generalized heat') and from a change in the physical quantity (the 'generalized work').*

In a few lines Ed introduces a generalization applicable to a wide variety of problems and at the same time recognizable as a familiar application. (It is a little more complicated than that since there is nothing in equation (3) that requires that the extremizing function be the entropy defined as in MaxEnt. Any extremizing function that depends only on the set of probabilities p_i will produce the same result) .

At the 1983 conference Ed presented his paper, "Bayesian Spectrum and Chirp Analysis", in which he examined the problem of doing signal analysis in the presence of noise. This paper, alone, would be worthy of a Nobel Prize, in my opinion, for it laid out, clearly, how conventional Fourier analysis treat noise as though it were a part of the signal and how the MaxEnt method picks out the signal. In this paper, as he does in many others, Ed not only lays out the correct solution, he analyzes other, nonBayesian and nonMaxEnt solutions, and shows how they are inconsistent with themselves.

I spoke earlier of a kind of tension between us back in the 60's and it was about this issue. He felt that I was too quick to present a solution without having worked out in fine detail what was wrong with the existing methods. The difficulty is, such an approach requires abilities that Ed had, but that few of the rest of us possess.

The 1984 conference was held in Alberta, Canada where Jaynes presented an introductory tutorial, "Bayesian Methods: General Background". In this paper Ed not only described the various contributions to reasoning under uncertainty; he examined them critically, from Herodotus (500 BC) through Bernoulli, Bayes, Laplace, Gibbs, Jeffreys, Cox, Shannon and modern writers. He shows how each contributor provided a clarification of thinking that was not and could not be understood by the existing 'authorities'. In rereading this paper, today, 14 years later, I found in it perspectives and analysis which I overlooked then. I urge each of you to look at that paper again, for it sets so many contributions in a context that only MaxEnt could provide and only Ed could explain.

This paper brings out another of Ed's traits. He was intensely interested in the history of thought. Time and again he reminds us of the origins of our current approaches.

At this same 1984 conference Ed undertook an examination of image construction with a paper under the title, "Monkeys, Kangaroos and N". In this paper he discusses the different methods people have used to decide that one answer was better than another. For example, most books on statistical mechanics will consider that entropy is defined by $S = k \ln W$ and not by $S = -k \sum_i p_i \ln p_i$. Most of us have known that in the limit of large numbers, these two definitions become the same. Ed, however, is not content with this answer. He considers the nature of the differences in these two definitions by looking at the "Kangaroo Problem" in which the size of N makes the difference. Which definition of entropy to use depends upon what problem is being solved. That paper contains a sentence that should be on the wall of every reviewer of statistical papers: "...we must learn to define the problem much more carefully than in the past..." I recall Ed once saying "Whenever I solve a problem, I never get criticized for what I did but for not solving a different problem."

At the 1986 conference Ed presented "Detection of Extra Solar System Planets", which was based on the thesis of his student and friend, Larry Bretthorst. The paper postulates a rather simple 'spread function' for the optical system and shows how to use the maximum entropy principle to decide if there really are two sources in the image, and if so, how far apart they probably are and of what brightness. Ed uses that example to illustrate how MaxEnt verified predictions made by Laplace, using the same spirit of inquiry that Laplace used, but now in a way that could be explained more clearly.

At the 1987 conference Ed discussed "The Evolution of Carnot's Principle". I was particularly interested in this paper because it covered much the same ground as in my own writings, but in a much more compact and elegant way. He began with the logic of Sadi Carnot's discussion of the impossibility of perpetual motion of the second kind. What I found interesting was how he formulated the treatment so that it looked, mathematically, very much like his 1985 paper on plausible reasoning. He followed the line of development from Carnot to Kelvin to Clausius to Gibbs and to Boltzman, each time illustrating the role of probable conjecture in the developments. This paper ties it all together in Jaynes' inimitable way, blending history, philosophy, epistemology, mathematics and logical proof.

At the 1988 conference Ed returned to his main theme under the title, "Clearing Up Mysteries The Original Goal". He began with an acknowledgment of the importance of Harold Jeffreys' great work. He had had a chance to speak with Sir Harold and Lady Jeffreys about it and reported to us that they were pleased that, at long last, Sir Harold's work was being followed up by a new generation. I have no doubt that Sir Harold was delighted to meet Ed and that he saw in Ed the perpetuation of his own contributions. Ed knew this as well and it gave them both pleasure. This was another special Jaynes characteristic. He was a leader because he "walked the talk"; he was a role model, going out of his way to respect what has gone before, all the while extending the work.

This particular paper contains several important contributions. One of them was a new derivation of the diffusion coefficient, rederiving Einstein's result in a new and much more informative way. Again, he is not content merely to present us with the solution, he helps us to understand its relevance to larger questions. He wrote: "*The above example may indicate the price that kinetic theory has paid for its failure to comprehend and use the Bayesian methods that Harold Jeffreys gave us 50 years ago, and how many other puzzles need to be reexamined from that viewpoint.*" This remark reminds me of a comment made by Max Planck in his Treatise on Thermodynamics [2]. In the preface to the first edition Planck discusses three possible ways to look at the subject. These are 1) Kinetic Theory, 2) Statistical Mechanics and 3) Classical macroscopic (phenomenological) Thermodynamics. Planck commented that the first method had "Obstacles, at present insurmountable..." Regarding the second method, he wrote, "It confines itself to the most important hypothesis of the mechanical theory of heat, that heat is due to motion, but refuses on principle to specialize as to the character of this motion." I had read this material before, but until Ed Jaynes pointed it out, had not realized how the underlying reasoning was equivalent to MaxEnt.

Ed also has pointed out to us an important distinction which I know I have often forgotten. He wrote, on the difference between inference and prediction: "*In physical prediction we are trying to describe the real world; in inference we are describing only our state of knowledge about the world.*"

Ed summed up his position on the meaning of the word probability in the following paragraph:

> *In our system, a probability is a theoretical construct, on the epistemological level, which we assign in order to represent a state of knowledge, or that we calculate from other probabilities according to the rules of probability theory. A frequency is a property of the real world, on the ontological level, that we measure or estimate. So for us, probability theory is not an Oracle telling how the world must be: it is a tool for learning (1) Is our state of knowledge adequate to describe the world? or (2) For which aspects of the world is our information adequate to make predictions?*

In the same paper Ed also presented an analysis of the efficiency of muscles. This is another pioneering work which has not yet been adequately disseminated. I recently received a manuscript from Thomas Schneider of the Frederick Cancer Research and Development Center in Maryland, in which he makes use of some of Ed's reasoning, so the diffusion has begun. Schneider presented a related paper at an earlier conference [3]. From Ed's analysis, it would appear that muscles act like Maxwell Demons, picking out the necessary 'hot spots' in the molecules which provide the energy.

In the 1989 conference, held at Dartmouth, Ed returned to his favorite theme, "The Mind Projection Fallacy", the confusion between states of knowledge and states of things. He writes:

> *For one who has no comprehension of physical law, but is aware of his own consciousness and volition, the natural question to ask is not : 'What is causing it?', but rather : 'Who is causing it?'*

This comment has meaning far beyond physics. I have been working for the last 18 years in the field of quality management. Managers, today, generally do not know that they are dealing with complex systems of people, machines, materials, energy, information and energy. Following the lead of W. Edwards Deming, I have been trying to teach them that when something goes wrong they should look first to fix the system, not the blame. But because the concept of a system is strange to them, the immediately try to find out, "Who caused it?"

Ed sums up the "Mind Projection Fallacy" by describing the error as due to two complementary forms:

$$\begin{array}{lcl} \text{A) (My own imagination)} & \longrightarrow & \text{(Real property of Nature)} \\ \text{B) (My own ignorance)} & \longrightarrow & \text{(Nature is indeterminate)} \end{array}$$

In a beautiful set of examples, Ed demonstrates how the "Mind Projection Fallacy" has created confusion in a wide variety of statistical problems. There is no reason to repeat the arguments here; Ed does it so much better than an interpreter can. I would only say in passing that his critique of Quantum Mechanics has many supporters and I hope that in my lifetime I shall see the resolution of the "mysteries" Ed decries.

The 1990 conference at Laramie, Wyoming, was the last of these conference s Ed attended. After that he became too ill to travel. In his paper "Notes on Present Status and Future Prospects" Ed took on the tone of the elder statesman of MaxEnt, which he assuredly was, and gave us a prediction for the future. As we should have expected, it was based on a study of the past. He recounted the familiar story of what happens to new ideas. As Sigmund Freud said:

> *Every new idea in science must pass through three phases. In Phase 1 everybody says the man is crazy and the idea is all wrong. In Phase 2 they say that the idea is correct, but of no importance. In Phase 3 they say that it is correct and important, but we knew it all along.*

Ed repeated his critiques of conventional statistics, artificial intelligence and fuzzy logic and ended up reminding us of our obligations to the future in these words.

> *...unless we ourselves take the education of the next generation very seriously, and see to it that the needed textbooks and courses of instruction come into existence in our libraries and universities, we may be overwhelmed by the sheer number of barbarians being bred today.*

We are all aware that Ed left behind an unfinished manuscript. Some of us have had a chance to read parts of it. I hope that a way will be found to complete and publish it, for in that manuscript Ed has put the heart and soul of his work. The result is magnificent.

All of us here today who have been touched and changed by Edwin T. Jaynes' contributions now have a solemn obligation to continue his work. Ed left a legacy that is worth preserving and extending. He, almost single handedly, established the methods of maximum entropy on a solid footing. He clarified the role of statistical inference in a way that consolidates more than two centuries of inquiry and debate. He has laid to rest countless paradoxes of inference. He did it all on his own time; he did not rely on governmental subsidies. He demonstrated the value of what he had done in a wide variety of fields, and in so doing demonstrated that they were linked by a common logic.

We are sorry to have lost him; we are grateful for what he did.

References

1. E. Jaynes, "Information theory and statistical mechanics," *Phy.Rev.*, **106**, p. 620, 1957.
2. M. Planck, Dover Publications, 1945. This Third Edition was originally published in 1910. The first edition was published in 1897.
3. T. Schneider, "Information and entropy of patterns in genetic switches," in *Maximum Entropy and Bayesian Methods in Science and Engineering*, G. Erickson and C. Smith, eds., vol. 2, pp. 147–154, Kluwer Academic Publishers, Dordrecht, 1988.

Ken Hanson and Richard Silver enjoying the boat-trip and the bavarian beer

"THE NEAR-IRRELEVANCE OF SAMPLING FREQUENCY DISTRIBUTIONS[1]"

G. LARRY BRETTHORST
Washington University
Dept. of Chemistry
St. Louis Mo 63130

Abstract. Jaynes, in his unfinished work on probability theory [2], discusses the Gaussian distribution extensively. In that discussion he demonstrates that when using a Gaussian distribution to estimate a location parameter, the only properties of the noise that enter the calculation are the first and second moments of the true noise; the underlying ensemble sampling distribution for the noise, the distribution from which the noise was actually sampled, completely cancels out of the inference. Exactly the same inferences will be made regardless of the underlying ensemble sampling distribution for the noise, provided the first and second moments of the true noise are the same. In this paper we review Jaynes' calculation, demonstrate it explicitly, generalize it to more complex models, and show that for regression style models the underlying ensemble sampling distribution for the noise is irrelevant to our inferences provided the mean and mean-square projection of the model onto the true noise are the same.

1. Introduction

The problem of assigning the joint probability for the noise, effectively the likelihood, has plagued probability theory since the time of Laplace. The outstanding question has always been, what probability density function should be assigned? It seems intuitively obvious that if one knows the underlying ensemble sampling distribution for the noise then one should assign this function as the likelihood. However, this is typically not done; rather one usually assigns a Gaussian distribution for the likelihood, and then proceeds to perform the calculation. Some authors have noted that this cavalier approach is unusually effective in practice [6]. The problem with the use of a Gaussian distribution for the likelihood is simply why are the estimates so good? Jaynes, in his unfinished work on probability theory [2], gives an answer to this question: namely that the Gaussian does not represent the

[1] In Chapter 7 of Jaynes' unfinished work on probability theory [2], there is a section by this title. In that section Jaynes gives convincing reasons why the Gaussian represents a state of knowledge about the true noise.

W. von der Linden et al. (eds.), Maximum Entropy and Bayesian Methods, 21–46.

underlying ensemble sampling distribution of the noise at all, rather it represents a particularly uninformed state of knowledge, a state of knowledge that accurately reflects what the experimenter knows about the true noise in data.

To understand Jaynes' conclusion and how to resolve the problem of assigning the joint probability for the noise, one needs to go back to first principles and to understand what we are actually doing when we apply probability theory. This discussion, Section 2, will be done using an example taken from Nuclear Magnetic Resonance (NMR), but the example is unimportant. Virtually any scientific experiment could be substituted for the NMR example and it would not change the discussion. The rules of probability theory will be applied just enough to reach the point where the joint probability for the noise must be supplied, and we will show that it is the probability for the *noise actually realized in our data sample* that must be assigned. In Section 3, the principal of maximum entropy is used to derive the Gaussian probability density function, and there it is shown that the only information this probability density function incorporates is knowledge of the first and second moments of the *actual noise in the sample at hand.* In Section 4, Jaynes' example of estimating a location parameter [2] will be used to show that a Gaussian distribution uses only the first and second moment of the actual noise in the data, and that all other characteristics of the noise have been made irrelevant to our inferences. At the end of this section, we illustrate the near-irrelevance of ensemble sampling distributions for the noise with a numerical example. In Section 5, we will generalize the calculations to much more complex models, and show that for regression style models the ensemble sampling distribution for the noise is irrelevant provided the mean-square of the actual noise and the mean projection of the model onto the actual noise remains the same across different noise samples. When models contain nonlinear parameters we show that for any fixed value of the nonlinear parameters an additional condition is needed for invariance of the estimated location parameters: that the mean-square of the data must be the same across noise samples. If these conditions are met, then again, the underlying ensemble sampling distribution for the noise will cancel from the calculation, and we shall obtain the same results regardless of the underlying ensemble sampling distribution for the noise. At the end of this section, we again illustrate the calculation with a numerical example.

2. What Are We Doing When We Use Probability Theory?

When we are solving a problem using probability theory there is always some hypothesis about which we wish to make inferences. This hypothesis might be as simple as estimating a location parameter, or as complicated as determining the number of sinusoidal signals in NMR data. Hypotheses of this nature contain two implicit assumptions: we have some facts, usually data, and these facts can be used to measure the quantity of interest, *i.e.*, they are relevant. Measuring is the process of relating some aspect of reality to an appropriate standard. Mathematics is the science that quantifies the measuring process, *i.e.*, mathematics is the science of measurement. Parameter estimation using probability theory is the science of measurement when the facts do not uniquely determine the measurement.

As noted, measuring a quantity presupposes we have some data, and a mathematical relationship between these data and the quantity to be measured. *Without this mathematical relationship there can be no quantification or estimation.* This mathematical relationship is the process by which the standard of measurement is brought into the problem. The data are simply the facts on which the measurement is based. The relationship between the data and the measured quantity, the model, might include many different parameters about which we have no interest. These parameters are nonetheless part of the problem. For example, when estimating the number of resonances in an NMR experiment, the data, d_i, and number of resonances, $\hat{m}$, are related by

$$d_i = \sum_{j=1}^{\hat{m}} \hat{A}_j \cos(\hat{\omega}_j t_i + \hat{\theta}) \exp\{-\hat{\alpha}_j t_i\} + n_i, \tag{1}$$

where $\hat{A}_j$, $\hat{\omega}_j$, $\hat{\alpha}_j$ are the true amplitude, frequency, and decay rate constant of the jth resonance. The common phase is represented by $\hat{\theta}$, and n_i is the *the actual* noise in the ith data value. In this paper, the word "noise" will be used to refer to the true noise in a data sample, while the word "error" will be used to refer to a running index, a hypotheses about which we must infer, that appears in our probability theory calculations. This model, Eq. (1), contains many unknown quantities. In the estimation process these unknown quantities become hypotheses of the form "the amplitude of the jth sinusoid was A_j," where A_j indexes an entire series of hypotheses about which we must infer. In this example we designated the hypotheses as A_j and not $\hat{A}_j$ to distinguish the running index, A_j, from the true but unknown value of the amplitude $\hat{A}_j$. This distinction will become increasingly important in the calculations that follow. In estimating the number of resonances m all of the parameters, $A_j, \omega_j, \alpha_j, \theta$, are uninteresting and must eliminated from the probability for m using the rules for probability theory.

Note that in Eq. (1) we have assumed additive noise. This additivity assumption is not necessary. For example, the noise might be multiplicative, or it might contaminate the data in some other more complex way. The exact functional form is not important, what is important is that the functional relationship can be supposed known. If the exact way the noise enter the calculation is unknown, then many different assumptions about how the noise enter may be postulated and tested using probability theory. Each different assumption would constitute a different model or hypothesis that is to be tested.

Before we apply the rules of probability theory to determine what calculation is to be performed, we must state exactly what is known about the various parameters appearing in the model. We know the data, $D \equiv \{d_1 \cdots d_N\}$, and we know something about the amplitudes, the frequencies, decay rate constants and phase of the sinusoids. This is obviously true because we are analyzing a real physical experiment. The data did not come into existence in a vacuum. Indeed in NMR, the signals have certain general strengths that are known in advance – if this were not so then the spectrometer could not have been designed. Additionally, the sinusoids were sampled at some rate; this presupposes that the frequencies could not have been higher than a certain value, else we would have digitized them incorrectly. We also know something about the overall decay rate constants because the signal was sampled for only a finite amount of time, and this presupposes that we sampled the signal over the time interval that the signal was large. Indeed if the signal sampling rate or the signal acquisition time were inappropriate, the experimenter would simply reset these parameters and reacquire the data.

Last, the size of the noise is known at least approximately because a digitizer was used to acquire the data and the physical characteristics of the digitizer prohibit it from responding to signals smaller than a certain known value; nor can it sample signals larger than some known value because these signals would overflow the digitizer. If either of these conditions were to occur, the experimenter would detect this condition and correct it before any data were actually saved. So the general size of the noise is also known. We are going to designate the information about the noise as I_σ and any parameters associated with this information as $\{\sigma\}$, although, at this time, we are not going to specify either the information, I_σ, or the parameters, $\{\sigma\}$. The other general background information, *i.e.*, information about the frequencies, decay rate constants, amplitudes, etc., will be designated as I.

Next, we are going to apply the rules of probability and see where they lead us. We have supposed the parameter of interest to be the number of sinusoids m, all others being nuisance parameters. The probability for the number of parameters may be computed from the joint probability for all of the parameters:

$$P(m|DI_\sigma I) = \int d\{A\}d\{\omega\}d\{\alpha\}d\theta d\{\sigma\}P(\{A\}\{\omega\}\{\alpha\}\theta\{\sigma\}m|DI_\sigma I), \qquad (2)$$

where we have removed all of the uninteresting or nuisance parameters using the sum rule of probability theory. The notation $\{x\}$ means the collection of all of the parameters of the x type, so $\{A\} \equiv \{A_1 \cdots A_m\}$. Bayes' theorem [3] may be used to factor the joint probability for the parameters, to obtain

$$\begin{aligned} P(m|DI_\sigma I) \quad \propto \quad & \int d\{A\}d\{\omega\}d\{\alpha\}d\theta d\{\sigma\} \\ & \times P(\{A\}\{\omega\}\{\alpha\}\theta\{\sigma\}m|I_\sigma I) \\ & \times P(D|\{A\}\{\omega\}\{\alpha\}\theta\{\sigma\}mI_\sigma I), \end{aligned} \qquad (3)$$

where we have dropped the normalization constant. It is at this point that one must look carefully at this equation to see exactly what including the prior information about the noise has accomplished.

In probability theory as logic, probabilities are indicated by the notation $P(X|I)$. The hypotheses appearing on the left-hand side of the vertical bar "|" are the hypotheses about which we are making inferences; while the hypotheses appearing on the right are given as true, *i.e.*, they specify the facts on which this probability is based. The notation $P(X)$ is meaningless because the information on which the truth of the hypothesis X is to be assessed has not been specified. With this in mind, look at $P(D|\{A\}\{\omega\}\{\alpha\}\theta\{\sigma\}mI_\sigma I)$. This term is the direct probability for the data given the truth of all of the parameters in the model. But if the true parameters are given, then

$$d_i - \sum_{j=1}^{m} A_j \cos(\omega_j t_i + \theta) \exp\{-\alpha_j t_i\} = e_i, \tag{4}$$

where, as noted above, we have introduced a notation without the hats to indicate that this equation is using the given values. Similarly, we have adopted the notation e_i for the given error value. This error value is a hypotheses about which we must infer. This hypotheses is of the form "the true noise value was e_i," where e_i is a running index and its numerical value would range over all valid noise amplitudes. Equation (4) is used in probability theory by introducing the joint probability for the data and the errors, and using the product and sum rules of probability theory to remove the dependence on the unknown error values:

$$\begin{aligned} P(D|\{A\}\{\omega\}\{\alpha\}\theta\{\sigma\}mI_\sigma I) &= \int de_1 \cdots de_N P(D\{e\}|\{A\}\{\omega\}\{\alpha\}\theta\{\sigma\}mI_\sigma I) \\ &= \int de_1 \cdots de_N P(D|\{e\}\{A\}\{\omega\}\{\alpha\}\theta mI) \\ &\quad \times P(e_1 \cdots e_N|\{\sigma\}I_\sigma), \end{aligned} \tag{5}$$

where $P(D|\{e\}\{A\}\{\omega\}\{\alpha\}\theta mI)$, the direct probability for the data given the errors and the parameters, is a delta function derived from Eq. (4). In this probability we dropped the dependence on the noise prior information and parameters because knowing the true error values, *i.e.*, knowing the noise values, determines things like the standard deviation for the noise and so renders knowledge of any statistics about the noise redundant. The last term, $P(e_1 \cdots e_N|\{\sigma\}I_\sigma)$, is the joint probability for the *actual noise in the data* given the noise parameters $\{\sigma\}$ and the information about the noise I_σ. In this probability we dropped the dependence on the model parameters because by assumption I_σ was all of the information about the noise. Last, the dependence on whether the noise was additive, multiplicative, or entered in some other more complex way was removed from this joint probability when we applied the rules of probability theory in Eq. (5). Indeed the joint probability for the noise, $P(e_1 \cdots e_N|\{\sigma\}I_\sigma)$, only depends on the existence of a model equation, the explicit inclusion of the prior information about the noise values, and the application of the rules of probability theory.

3. Assigning Distributions Using The Principle Of Maximum Entropy

The problem of assigning the direct probability for the data, the likelihood, has been reduced to the problem of assigning the joint probability for the actual noise in the data given whatever we know about the noise. The question that must be faced is how does one assign a joint probability density function that relates to the noise sample actually realized in the data? The easiest way to do this is state explicitly what is known about the noise and then use the principle of maximum entropy to assign this probability density function. Here, we will briefly review the principal of maximum entropy and then show in the following sections that probability density functions assigned using the principal of maximum entropy use only the information that is actually put into the assignment of these probabilities; and that, as a consequence, all noise samples having the same mean-square and the same projection onto the model will give identical location parameter estimates, irrespective of the underlying ensemble sampling distribution for which the noise sample was drawn.

Suppose one must assign a probability distribution for the ith value of a discrete parameter given the prior information I. This probability is denoted $P(i|I)$ $(1 \leq i \leq m)$. The Shannon entropy, defined as

$$H \equiv -\sum_{i=1}^{m} P(i|I) \log P(i|I), \tag{6}$$

is a measure of the amount of ignorance (uncertainty) in this probability distribution [4]. Shannon's entropy is based on a qualitative requirement: the entropy should be monotonically increasing for increasing ignorance, plus the requirement that the measure be consistent. The principle of maximum entropy then states that if one has some information I, one can assign a probability distribution, $P(i|I)$, that contains only the information I by maximizing H subject to the constraints represented by I. Because H measures the amount of ignorance in this probability distribution, assigning a probability distribution that has maximum entropy yields a distribution that is least informative (maximally ignorant) while remaining consistent with the information I: *i.e.*, $P(i|I)$ contains only the information I [4,2].

For the purposes of this discussion, we are going to pass to the limit of continuous parameters and ignore any subtleties that concern the use of probability density functions. As long as one maintains finite, normalizable probability density functions there are almost no circumstances where the use of continuous parameters will cause problems. When we do this we will assume that the appropriate prior probability for a location parameter is a uniform prior probability and we will use this uniform prior probability as the measure function in the entropy [5].

The probability that must be assigned is the joint probability for noise given that one knows the information I_σ and the parameters associated with this information $\{\sigma\}$. This probability, $P(e_1 \cdots e_N|\{\sigma\}I_\sigma)$, was derived in Eq. (5), where e_i is both a shorthand notation and an index: it stands for a hypothesis indexed by e_i of the form "the true value of the noise in the ith data value was e_i." As an index, e_i ranges over all valid values of the true noise; while the joint probability for the noise assigns a reasonable degree of belief to a particular set of hypotheses specified by the error indices.

To use the principle of maximum entropy to assign this probability there are two problems that must be solved. First, to use the principle of maximum entropy we must know what constraints we are going to use, and second we must know the numerical values of the parameters associated with these constraints. The second problem has been disposed of because when we included I_σ and $\{\sigma\}$ in the probability theory calculation we arrive at $P(e_1 \cdots e_N|\{\sigma\}I_\sigma)$ as the probability to be assigned. The parameters $\{\sigma\}$ and the prior information I_σ are given and so allow us to utilize the principle of maximum entropy. However, the first problem, has not yet been solved because we have not yet specified either I_σ or $\{\sigma\}$.

So the problem of assigning the joint probability for the noise has been reduced to specifying what is known about the noise, *i.e.*, specifying the constraints to be used in the calculation, and then applying the principal of maximum entropy. But what constraints should be used? As neither I_σ nor $\{\sigma\}$ have been specified, it would appear that we have an unlimited range of possible constraints. For example, should a constraint on correlations be included? If so, which of the many different types of correlations should be included? There are second order correlations of the form

$$\rho_s = \frac{1}{N-s}\sum_{i=1}^{N-s} n_i n_{i+s}, \tag{7}$$

where ρ_s is the known correlation coefficient and s is a measure of the correlation distance, as well as third, fourth, and higher order correlations. In addition to correlations, should a constraint on the moments of the noise be included? If so, on which moments should the joint probability for the noise depend? There are many different types of moments. There are power law moments of the form

$$\sigma^s = \frac{1}{N}\sum_{i=1}^{N} n_i^s, \tag{8}$$

as well as moments of arbitrary functions, and a host of others.

Given that we have the information necessary to incorporate any or all of these constraints into the maximum entropy calculation, the question remains which constraints should be used? If there is information that suggests that the true noise is correlated, then a correlation constraint should be used. If there is information that suggests the higher moments can deviate significantly from what one would expect from constraints on the first few moments, then again a constraint on the higher moments should be included. But if all one knows is the general magnitude and scale of the noise, then *one is always better off to leave out constraints on higher moments and correlation coefficients*, because the resulting probability density function will have higher entropy. Higher entropy distributions are by definition less informative and therefore make more conservative estimates of the parameters.

This suggests that if we wish to be conservative, *i.e.*, we do not wish to make gratuitous assumptions about things we really do not know, and we do not wish to make excessive claims for the accuracy of our parameter estimates, then we should constrain the maximum entropy distribution as little as possible, while still reflecting what is actually known about the noise. If, for example, all one knows is something about the general magnitude of the noise, then a constraint on the absolute value of the errors would be appropriate; this would result in a Laplacian assignment for the joint probability for the noise. Additionally, if one also knows something about the general scale of the noise, for example they fluctuate around zero with some standard deviation, then, as we shall see, this would result in a Gaussian assignment. Both of these assignments would give very reasonable results.

The main difference between them is really one of functionality. Because of the many properties of the Gaussian [2], the Gaussian is easier and more convenient to use than the Laplacian; although with the advent of modern computers, the difficulties associated with the use of the Laplacian are disappearing.

In the next few paragraphs the principle of maximum entropy will be used to derive the joint probability for the noise. In this derivation we will use the first and second moments of the true noise as constraints in the maximum entropy calculation. Earlier, we gave a definition of the power law moments, Eq. (8), using the true noise. In the problem we now face we do not know the true noise and so the calculation shown in Eq. (8) cannot be used; all we know is that the joint probability for the noise is to be consistent with the given power law moments. So what does it mean for a probability density to be consistent with power law moments? The answer is simple; for the joint probability for the noise to be consistent with power law moments it must have the property that

$$\sigma^s = \langle e^s \rangle \equiv \frac{1}{N}\sum_{i=1}^{N} \int de_1 \cdots de_N \; e_i^s P(e_1 \cdots e_N|\{\sigma\}I_\sigma) \tag{9}$$

where σ^s is the known value of a particular moment, and the notation $\langle e^s \rangle$ means: compute the average of the expected value of the e_i^s. Consistent means that the expected moments of the joint probability density for the noise should be equal to the observed moments of the true noise, *i.e.*, the probabilities we assign should reflect what is actually known about the noise.

We are going to place constraints on the first two moments of the joint probability for the noise. If we designate the mean of the noise as μ and the standard deviation as σ, then the constraint on the first moment is given by

$$\mu = \frac{1}{N}\sum_{i=1}^{N} \int de_1 \cdots de_N \; e_i P(e_1 \cdots e_N|\{\sigma\}I_\sigma), \tag{10}$$

and the constraint on the second moment is given by

$$\sigma^2 + \mu^2 = \frac{1}{N}\sum_{i=1}^{N} \int de_1 \cdots de_N \; e_i^2 P(e_1 \cdots e_N|\{\sigma\}I_\sigma), \tag{11}$$

where the second moment has been written as $\sigma^2 + \mu^2$ to make the resulting Gaussian come out in standard notation.

We seek the joint probability density function for the noise that has highest entropy for the given sample mean and standard deviation. To find this distribution Eqs. (10) and (11) are rewritten so they sum to zero:

$$\mu - \frac{1}{N}\sum_{i=1}^{N} \int de_1 \cdots de_N \; e_i P(e_1 \cdots e_N|\{\sigma\}I_\sigma) = 0, \tag{12}$$

and

$$\sigma^2 + \mu^2 - \frac{1}{N}\sum_{i=1}^{N} \int de_1 \cdots de_N \; e_i^2 P(e_1 \cdots e_N|\{\sigma\}I_\sigma) = 0. \tag{13}$$

Additionally, the probability for finding the true noise values somewhere in the valid range of values is one, so

$$1-\int de_1\cdots de_N P(e_1\cdots e_N|\{\sigma\}I_\sigma)=0. \tag{14}$$

Because Eqs. (12)–(14) are each zero, each may be multiplied by a constant and added to the entropy of the joint probability density function without changing the entropy of that distribution. One obtains

$$\begin{aligned} H \;=\; & -\int de_1\cdots de_N P(e_1\cdots e_N|\{\sigma\}I_\sigma)\log P(e_1\cdots e_N|\{\sigma\}I_\sigma) \\ & +\beta\left[1-\int de_1\cdots de_N P(e_1\cdots e_N|\{\sigma\}I_\sigma)\right] \\ & +\delta\left[\mu-\frac{1}{N}\sum_{i=1}^{N}\int de_1\cdots de_N\; e_i P(e_1\cdots e_N|\{\sigma\}I_\sigma)\right] \\ & +\lambda\left[\sigma^2+\mu^2-\frac{1}{N}\sum_{i=1}^{N}\int de_1\cdots de_N\; e_i^2 P(e_1\cdots e_N|\{\sigma\}I_\sigma)\right], \end{aligned} \tag{15}$$

where β, δ, and λ are called Lagrange multipliers.

To find the distribution that maximum entropy will assign for the joint probability for the noise, this expression is maximized with respect to $P(e_1'\cdots e_N'|\{\sigma\}I_\sigma)$. After some algebra, one obtains

$$P(e_1\cdots e_N|\mu\sigma)=(2\pi\sigma^2)^{-\frac{N}{2}}\exp\left\{-\sum_{i=1}^{N}\frac{(e_i-\mu)^2}{2\sigma^2}\right\}, \tag{16}$$

where

$$\lambda=\frac{N}{2\sigma^2},\qquad \delta=-\frac{N\mu}{\sigma^2},\quad \text{and}\quad \beta=\frac{N}{2}\left[\log(2\pi\sigma^2)+\frac{\mu^2}{\sigma^2}\right]-1, \tag{17}$$

and I_σ and $\{\sigma\}$ have been replaced by μ and σ in $P(e_1\cdots e_N|\{\sigma\}I_\sigma)$.

There are several interesting points to note about this probability density function. First, this is a Gaussian distribution. However, the fact that the joint probability for the noise has been assigned to be a Gaussian makes no statement about the ensemble sampling distribution of the noise; rather, it says only that for the given mean and mean-square noise value, the joint probability density function for the noise should be maximally uninformative and that maximally uninformative distribution happens to be a Gaussian. Second, the joint probability for the noise does not contain correlations. The reason for this is that a constraint on correlations must lower the entropy. By definition a probability assignment with lower entropy is more informative, and so must make more precise estimates of the parameters. Instead of saying the joint probability for the noise does not contain correlations, it would be more correct to say that this probability density function makes allowances for *every possible correlation* that could be present and so is less informative than correlated distributions. Third, if one computes the expected moments of this Gaussian, one obtains

$$\langle e^s \rangle = \exp\left\{-\frac{\mu^2}{2\sigma^2}\right\}\sigma^{2s}\frac{\partial^s}{\partial\mu^s}\exp\left\{\frac{\mu^2}{2\sigma^2}\right\} \quad (s \geq 0) \tag{18}$$

which reduces to

$$\langle e^0 \rangle = 1, \qquad \langle e^1 \rangle = \mu, \quad \text{and} \quad \langle e^2 \rangle = \sigma^2 + \mu^2 \tag{19}$$

for $s = 0$, $s = 1$, and $s = 2$, just the given mean and mean-square noise values used to assign the probability density function. So as noted previously, the maximum entropy distributions makes no other assumptions about the noise above those already implied by the constraints. Fourth, for a given value of the sample mean and mean-square the joint probability for the noise has highest entropy. Consequently, when any operation is performed on an arbitrary distribution that the preserves mean and mean-square while discarding other information, *i.e.*, the operation increases the entropy of the distribution, then that distribution necessarily will move closer and closer to a Gaussian distribution regardless of the initial assignment. The Central Limit Theorem is one special case of this phenomenon – see Jaynes [2].

4. The Jaynes Example – Estimating A Location Parameter

Maximum entropy distributions are the only distributions that have sufficient statistics. Sufficient statistics are functions of the data, and therefore the noise, that summarize all of the information in the data relevant to estimating a location parameter. We would like to demonstrate how the sufficient statistics render the underlying ensemble sampling distribution for the noise irrelevant. We will illustrate this using a Gaussian distribution for the joint probability for the noise with the understanding that *any* maximum entropy distribution will have similar properties.

Suppose the true value of a location parameter is μ_0 and one has a data set D with data elements d_i such that

$$d_i = \mu_0 + n_i. \tag{20}$$

Now because μ_0 is the true value of the mean, the n_i are the true noise samples. The problem we face is to estimate μ_0. The hypothesis about which inferences are to be made is of the form "the true value of the location parameter is μ given the data D." The parameter μ is an index that specifies the various hypotheses; while the probability $P(\mu|D\sigma I_\sigma I)$ assigns a reasonable degree of belief to these hypotheses. Following what was done in Section 2, the posterior probability for μ is given by

$$P(\mu|\sigma D I_\sigma I) = P(\mu|I)P(D|\sigma\mu I_\sigma I) \tag{21}$$

where we have assumed that the value of the noise standard deviation, σ, is known for now. $P(\mu|I)$ is the prior probability for μ, and $P(D|\mu\sigma I_\sigma I)$ is the direct probability for the data given the model parameters. Because we are interested in the consequences of assigning a Gaussian distribution for the joint probability for the noise, we will take the prior for μ to be a constant.

In this example we are going to assume that the noise could take on both positive and negative values, and that when assigning the Gaussian we had no prior information that would lead us to favor either a positive or negative noise samples. Consequently, we will assign the joint probability for the noise to be a zero-mean Gaussian, namely

$$P(e_1 \cdots e_N|\sigma) = (2\pi\sigma^2)^{-\frac{N}{2}} \exp\left\{-\sum_{i=1}^{N} \frac{e_i^2}{2\sigma^2}\right\}. \tag{22}$$

From which one obtains

$$P(\mu|D\sigma I_\sigma I) \propto P(D|\mu\sigma I_\sigma I) = (2\pi\sigma^2)^{-\frac{N}{2}} \exp\left\{-\frac{1}{2\sigma^2}\sum_{i=1}^{N}(d_i - \mu)^2\right\} \tag{23}$$

as the posterior probability for μ. This equation may be rewritten as

$$P(\mu|D\sigma I_\sigma I) \propto (2\pi\sigma^2)^{-\frac{N}{2}} \exp\left\{-\frac{N}{2\sigma^2}\left[(\overline{d} - \mu)^2 + s^2\right]\right\} \tag{24}$$

where the sufficient statistics are $\overline{d}$, the mean data value, and s^2 is the mean-square data value less the squared mean-data value, and these are just the constraints used in deriving the Gaussian distribution. Using Eq. (20), the mean data value is given by

$$\overline{d} \equiv \frac{1}{N}\sum_{i=1}^{N} d_i \tag{25}$$

$$= \frac{1}{N}\sum_{i=1}^{N} (\mu_0 + n_i) \tag{26}$$

$$= \mu_0 + \overline{n}, \tag{27}$$

with

$$\overline{n} \equiv \frac{1}{N}\sum_{i=1}^{N} n_i. \tag{28}$$

So the mean data value is equal to the true mean μ_0 plus the mean noise value $\overline{n}$. Similarly, s^2 is given by:

$$s^2 \equiv \overline{d^2} - (\overline{d})^2 \tag{29}$$

$$= \frac{1}{N}\sum_{i=1}^{N} d_i^2 - \left(\frac{1}{N}\sum_{i=1}^{N} d_i\right)^2 \tag{30}$$

$$= \frac{1}{N}\sum_{i=1}^{N} (\mu_0 + n_i)^2 - \left(\frac{1}{N}\sum_{i=1}^{N}[\mu_0 + n_i]\right)^2 \tag{31}$$

$$= \overline{n^2} - (\overline{n})^2, \tag{32}$$

with

$$\overline{n^2} = \frac{1}{N}\sum_{i=1}^{N} n_i^2 \tag{33}$$

where $\overline{n^2}$ is the mean-square noise value and $(\overline{n})^2$ is the square of the mean-noise value. Therefore, s^2 is a function of only the true noise.

Using the mean $\pm$ standard deviation as an estimate of μ one obtains

$$(\mu)_{est} = \begin{cases} \overline{d} \pm \sigma/\sqrt{N} & \sigma \text{ known} \\ \overline{d} \pm s/\sqrt{N-3} & \sigma \text{ unknown} \end{cases} . \tag{34}$$

The actual error, Δ, is given by

$$\Delta = \overline{d} - \mu_0 = \overline{n} \tag{35}$$

which depends only on the *mean of the true noise values*, while our accuracy estimate depends only on σ if the standard deviation of the noise is known, and *only on the mean and mean-square* of the true noise when the standard deviation of the noise is not known. The underlying ensemble sampling distribution for the noise has canceled out of the calculation, and the only property of the noise that survives is the actual mean and mean-square of the true noise. *All other properties of the noise are irrelevant!* Exactly the same parameter estimates will result if the underlying ensemble sampling distribution of the noise changes, provided the mean and mean-square of the new noise sample is the same as in the previous sample – just the properties needed to represent what is actually known about the noise and to render what is *not* known about them irrelevant.

From this discussion it is clear that the Gaussian represents a particularly uninformed state of knowledge. And that the underlying ensemble sampling distribution for the noise has been made irrelevant in the sense that any set of noise drawn from any ensemble sampling distribution will result in the exact same posterior probability being computed, provided the mean and mean-square noise values for the different noise samples are same. However, we would like to explicitly demonstrate this using numerical examples. According to the results that have been derived so far, the only properties of the true noise that enter the calculations are the mean and mean-square. Consequently, if we analyze a new data set containing a new noise sample generated from an entirely different ensemble sampling distribution while preserving these two quantities, the resulting probability density functions will be identical.

To illustrate this point we have prepared three data sets. These are shown in Fig. 1. Each of these three data sets contain the same constant signal, with noise having the same mean and mean-square. However, the noise in each data set were generated from three very different ensemble sampling distributions. The first data set, panel (A), contains Gaussian white noise. The second, panel (B), was generated using a Gaussian random number generator, but the results from the random number generator were rounded to integer values. Last, the third data set, panel (C), does not contain noise at all; rather, this data set is a constant plus a deterministic sinusoidal signal that is not in our model. We could have ordered the sinusoid randomly to camouflage this unmodeled signal, but by making it obvious that we have an unaccounted for signal it will better make the point that everything that has been said in the paper is equally applicable to models that only account for part of the deterministic signal present in the data. Indeed the noise itself is a deterministic signal which we have not modeled in any regular fashion, so it should not be surprising that if our estimates are invariant when we change one deterministic signal, the noise, they would also be invariant when we change another deterministic signal, in this case an unmodeled sinusoid, provided the mean and mean-square of these residual components are the same. This last point will be illustrated with Panel (C).

To ensure that each of these data sets has exactly the same mean and mean-square, we first generated the three set of numbers that we are calling noise in Fig. 1. For each set we then computed the average "noise" value, and subtracted this average from each noise value. We then computed the mean-square noise value and used that mean-square value to scale each noise value within that set so that each noise set had the same second moment. Finally, we added the constant signal to each noise set to obtain the data shown in Fig. 1.

Next, the posterior probability for μ was computed for each of these three data sets and is displayed in Fig. 2. In the three panels, the solid line is $P(\mu|\sigma DI)$, and the dotted line is $P(\mu|DI)$. On the scale shown, these probability density functions are identical. If one examines the actual numbers used to generate these plots they are indeed identical. Please note that because we know that the noise in panel (B) were generated from a Gaussian random number generator that was rounded to the nearest integer it would always be possible to include this information in the calculation, and so make a better estimate of the parameter μ. Similarly, one look at the data in Fig. 1(C) and one would know that there is a signal present in the data that the model does not account for. If we included that sinusoid in the model we would be able to estimate the value of the location parameter down to the rounding errors of the procedure used to generate the data. However, the point is not that we can do better, the point is that these three noise samples, sampled from three very different ensemble sampling distributions, give identical estimates for the location parameter. The underlying ensemble sampling distribution for the noise is simply not relevant to the questions being answered by probability theory. And it does not matter if that ensemble sampling distribution is Gaussian white noise, unaccounted for signal, or any other type of noise one can imagine. The only properties of the noise utilized by probability theory were the mean and mean-square of the actual noise in our data data.

Figure: 1 Three Data Sets Having The Same Mean and Mean-Square Noise

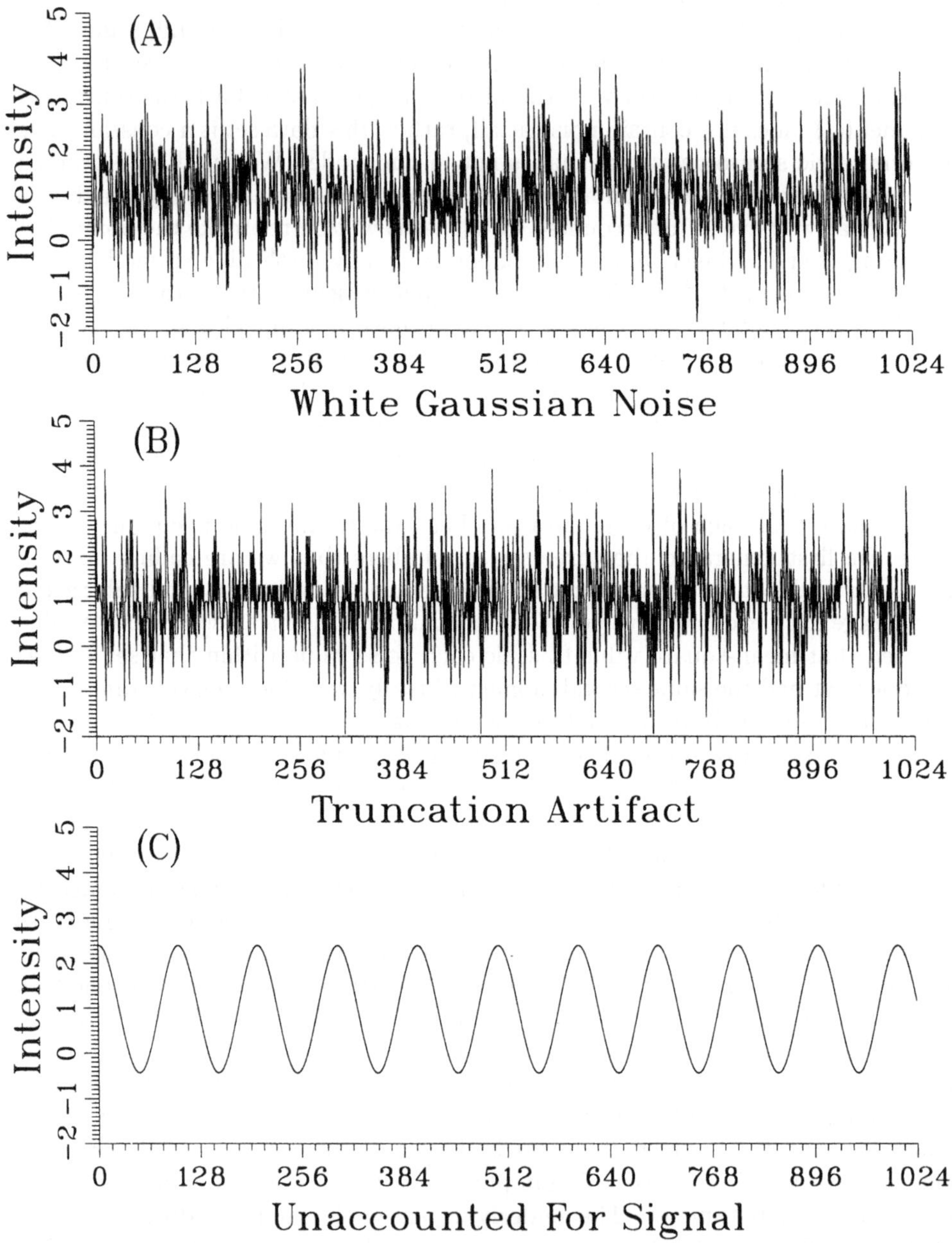

Fig. 1. Panel (A) is a constant plus Gaussian white noise, (B) a constant plus truncated Gaussian white noise, and (C) a constant plus a sinusoid. If you subtract the constant signal from each of these three data sets, the residuals have the same mean and mean-square. According to probability theory these three data sets should produce identical posterior probabilities for μ.

Figure: 2 The Posterior Probability For μ

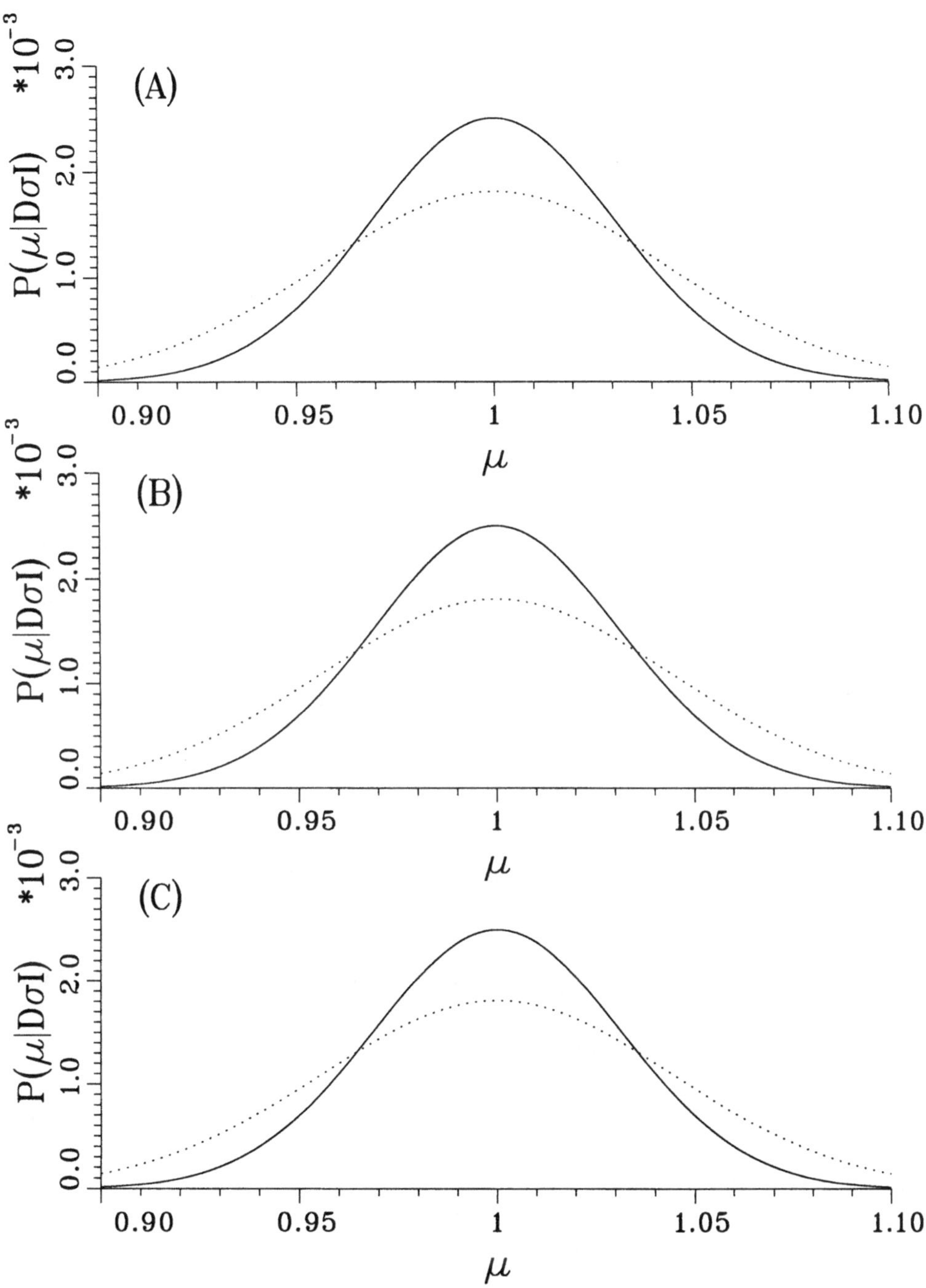

Fig. 2. The three panels are the posterior probabilities $P(\mu|\sigma DI)$, solid lines, and $P(\mu|DI)$, dotted lines, computed for the three data sets shown in Fig. 1. According to probability theory these three estimates for μ should be identical.

5. A General Nonlinear Model

The calculations given in the previous section are valid and correct as far as they go. They illustrate the near-irrelevance of ensemble sampling distributions for a particularly simple model. The question we would like to address in this section is: what happens when we have more complicated models? Do we still get the same invariance when the ensemble sampling distribution for the noise changes, and if not, how must we modify what was said in the previous section to account for these more complex models? In particular, what happens when the models contain both location and scale parameters?

The most general model that will be addressed is of the form [1]:

$$d_i = \sum_{j=1}^{m} \hat{B}_j G_j(\{\hat{\theta}\}, t_i) + n_i, \tag{36}$$

where $\hat{B}_j$ is the true amplitude of the jth model function $G_j(\{\hat{\theta}\}, t_i)$, and the $\{\hat{\theta}\}$ are the true values of a collection of parameters that appear in the model in a nonlinear fashion. Each model function is assumed to be evaluated at the same time that the data were taken. We have labeled this time as t_i, but whether this is time, space, or some other labeling is irrelevant. Additionally, we have indicated the data are one dimensional; however, whether or not the data are multidimensional is also irrelevant. Multidimensional data may always be labeled with a single index by an appropriate change of the indices. Last, n_i represents the true noise in the measurement.

What we have to say about the model is much easier to understand if we introduce an orthogonal model by the following change of parameters and functions:

$$\hat{B}_k = \sum_{j=1}^{m} \frac{\hat{A}_j \xi_{jk}}{\sqrt{\lambda_j}}, \tag{37}$$

where λ_j is the jth eigenvalue of the g_{jk} matrix, Eq. (38) below, and ξ_{jk} is the kth component of the jth eigenvector of the matrix

$$g_{jk} = \sum_{i=1}^{N} G_j(\{\hat{\theta}\}, t_i) G_k(\{\hat{\theta}\}, t_i). \tag{38}$$

In this new notation the model equation becomes

$$d_i = \sum_{j=1}^{m} \hat{A}_j H_j(\{\hat{\theta}\}, t_i) + n_i \tag{39}$$

with

$$H_j(\{\hat{\theta}\}, t_i) = \frac{1}{\sqrt{\lambda_j}} \sum_{k=1}^{m} \xi_{jk} G_k(\{\hat{\theta}\}, t_i), \tag{40}$$

and

$$\hat{A}_k = \sqrt{\lambda_k} \sum_{j=1}^{m} \hat{B}_j \xi_{kj}. \tag{41}$$

We call this model "orthogonal" because the model functions, $H_j(\{\hat{\theta}\}, t_i)$, have the property that

$$\sum_{i=1}^{N} H_j(\{\hat{\theta}\}, t_i) H_k(\{\hat{\theta}\}, t_i) = \delta_{jk} \tag{42}$$

where δ_{jk} is zero if $j \neq k$, and one otherwise. Note that introducing this change of variables and change of functions has not changed the form of the model function, *i.e.*, model Eq. (36) is formally identical to Eq. (39). Additionally, the transformation did not effect either the data, d_i, or the noise n_i. These are same in both equations.

We are going to repeat the calculations done in the previous section using this orthogonal model. The aim is to understand the effects of introducing a probability that has sufficient statistics and in particular we will concentrate on the Gaussian distribution. The object of this calculation is to understand the conditions under which the parameter estimates obtained using a Gaussian distribution are invariant when the underlying ensemble sampling distribution for the noise changes. To do this we must carefully differentiate between the true parameter values and the given parameter values in our calculations. For the amplitudes we will write A_j for the running index in our calculations. Similarly, $\{\theta\}$ will be the running index that specifies the collection of nonlinear parameters, and finally, e_i will be the given value of the error at time t_i.

In the first part of this calculation we are going to examine the estimates of the location parameters, the $\{A\}$. Consequently, the probability we will be interested in is $P(\{A\}|D\{\theta\}\sigma I)$ where, for the time being, we are going to assume that the nonlinear parameters are given. Last, because we are interested in the consequences of assigning a Gaussian for the joint probability for the noise we are going to assign a uniform prior probability for the amplitudes. With these assumptions and using Eq. (39), the posterior probability for the amplitudes is given by:

$$P(\{A\}|D\{\theta\}\sigma I) = (2\pi\sigma^2)^{-\frac{N}{2}} \exp\left\{-\frac{Q}{2\sigma^2}\right\} \tag{43}$$

where Q is defined as

$$Q \equiv \sum_{j=1}^{m} \hat{A}_j^2 + 2N \sum_{j=1}^{m} \hat{A}_j \overline{H_j(\{\hat{\theta}\})n} + N\overline{n^2} - m\overline{h^2} + \sum_{j=1}^{m} (A_j - h_j)^2 \tag{44}$$

with

$$\overline{h^2} = \frac{1}{m} \sum_{j=1}^{m} h_j^2, \tag{45}$$

$$h_j = N\left(\overline{H_j(\{\theta\})n} + \sum_{k=1}^{m} \hat{A}_k \overline{H_k(\{\hat{\theta}\}) H_j(\{\theta\})}\right), \tag{46}$$

$$\overline{H_k(\{\hat{\theta}\}) H_j(\{\theta\})} = \frac{1}{N} \sum_{i=1}^{N} H_k(\{\hat{\theta}\}, t_i) H_j(\{\theta\}, t_i), \tag{47}$$

and

$$\overline{H_j(\cdot)n} \quad = \quad \frac{1}{N}\sum_{i=1}^{N} H_j(\cdot, t_i)n_i \tag{48}$$

where we have left the argument of $\overline{H_j(\cdot)n}$ unspecified because of the two different uses in the above equations.

First, let us examine the estimates for the amplitudes. If we consider the case when the nonlinear parameters are given, then the probability for the amplitudes is given by

$$P(\{A\}|D\{\theta\}\sigma I) \propto \exp\left\{-\frac{1}{2\sigma^2}\sum_{j=1}^{m}(A_j - h_j)^2\right\} \tag{49}$$

and the expected value for each amplitude is

$$\langle A_j\rangle = h_j = N\left(\overline{H_j(\{\theta\})n} + \sum_{k=1}^{m}\hat{A}_k\overline{H_k(\{\hat{\theta}\})H_j(\{\theta\})}\right). \tag{50}$$

The expected amplitude is just the projection of the orthogonal model function onto the data. However, the data consists of two terms: the signal and the noise. The total projection of model $H_j(\{\hat{\theta}\})$ onto the signal, the second term in Eq. (50), does not change if the ensemble sampling distribution for the noise changes. The only term that changes is the total projection of the model onto the noise, the first term in Eq. (50). Consequently, for the amplitude estimates to remain invariant when the ensemble sampling distribution for the noise changes, the projections $\overline{H_j(\{\theta\})n}$ must be the same from one noise sample to another.

Think of the data as an N dimensional vector space. The model constitutes an m dimensional subspace of the full vector space. The estimated amplitudes are invariant under change in ensemble sampling distribution for the noise if *the projection of the noise onto the m dimensional subspace is the same.* In the example given by Jaynes, there is only a single constant model function, $H_1 = 1/N$, and the projection of this model onto the noise is the mean noise value, so the estimated value of μ is invariant under change in noise sample if the mean value of the noise is the same between the two samples, just the condition derived earlier.

The posterior probability for the amplitudes was derived given the nonlinear parameters, including the variance of the noise. From this posterior distribution the expected values of the amplitudes were then computed. These expected amplitudes do not change if the standard deviation for the noise is not known. However, there is a complication that must be considered when σ is unknown. Probability theory will naturally lead one to use Student's t-distribution to estimate the amplitudes. If Student's t-distribution is used our uncertainty in the estimated amplitudes will be related to the total squared-data value. But the total squared-data value depends on the projection of the true signal onto the true noise, Eq. (44), and if the ensemble sampling distribution for the noise changes this total projection could change. Consequently, changing the total squared-data value would result in a change in our estimate of how uncertain we are of the amplitudes. To ensure this does not happen, there are two additional conditions that must be met: the mean-square of the true noise, and the projection of the signal onto the noise must remain the same when the ensemble sampling distribution for the noise changes. This statement is equivalent to saying that the total squared-data value must be the same, if the estimated amplitudes are to remain the same when the ensemble sampling distribution for the noise changes.

Before we discuss what is happening with the nonlinear parameters we will illustrate the near-irrelevance of ensemble sampling distributions for these more complex models with a numerical example. The exact functional form of the model used in this example is irrelevant provided the model is of the form shown in Eq. (36). In this example we will use

$$\text{Signal} = B_1 + B_2 t + B_3 \cos(\omega t) \exp(-\beta t) \tag{51}$$

with $B_1 = -123$, $B_2 = 1$, $B_3 = 100$, $\beta = 4/N$, $\omega = 1$, $N = 256$ and dimensionless time units are used, *i.e.*, $t_i = 0, 1, \cdots, N-1$. This signal is shown in Fig. 3 The signal is an exponentially decaying sinusoid that is decaying about a line.

To illustrate that the estimates for the amplitudes are invariant when the sampling distribution for the noise changes, provided we have the same projection of the model function onto the true noise, we have prepared three noise samples – see Fig. 4. In Fig. 4 panel (A) is Gaussian white noise, (B) rounded Gaussian noise, and (C) an unmodeled signal. To prepare these noise samples, we first fixed the nonlinear parameters – here these are ω and β. For this demonstration, we fixed these parameters to their true values. We did this simply to avoid the additional complication of ensuring that the projection of the true signal onto the noise was invariant across the noise samples. After fixing the nonlinear parameters, we computed the orthogonal models and then computed the projection of the orthogonal models onto the noise. These projections times the corresponding model vector were then subtracted from each noise set. This has the effect of making the noise orthogonal to the model. Last, we took the original projections from the first noise set, the Gaussian white noise, and added them to each of the three noise sets. This ensured that the projection of the orthogonal model onto each of the noise samples is exactly the same.

The posterior probability, $P(B_3|DI)$, was computed for the signal shown in Fig. 3 plus the noise shown in the three panels in Fig. 4. These three computed probabilities are shown in Fig. 5. We could have illustrated the invariance of the amplitude estimates using any of the three amplitudes in the model. However, we choose to show only the posterior probability for the amplitude of the sinusoid. Note that on the scale of the plots shown in Fig. 5, the three probabilities are identical and a Unix difference on the files used to generate these plots verifies that they are identical.

For regression models, the underlying ensemble sampling distribution for the noise is irrelevant, because the results one obtains are invariant when the sampling distribution for the noise changes, provided the projection of the model onto the noise and the mean-square noise value are the same. Thus the underlying ensemble sampling distribution for the noise completely cancels out of the calculation. For models containing nonlinear parameters, we again have a similar result. In these

Figure 3: The Signal $= B_1 + B_2 t + B_3 \cos(\omega t)\exp(-\beta t)$

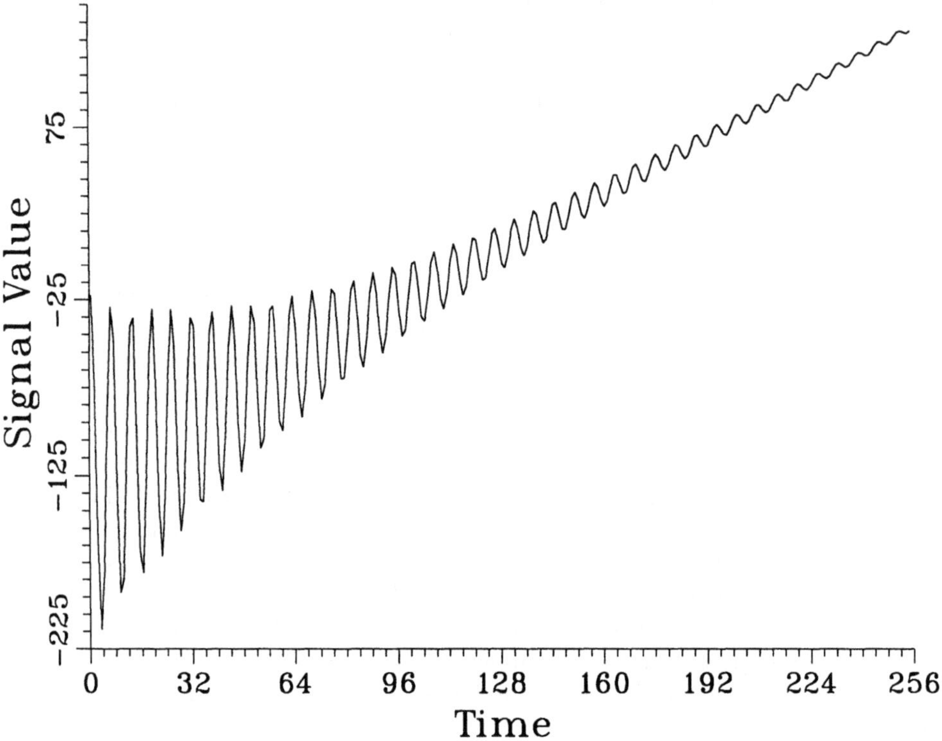

Fig. 3. This signal has parameters $B_1 = -123$, $B_2 = 1$, $B_3 = 100$, $\beta = 4/N$, $\omega = 1$, and $N = 256$. The data shown contain no noise. In order for the probability density functions for the location parameters to be invariant when the ensemble sampling distribution for the noise changes, the projection of the model onto the noise must be the same for each model vector.

models, the probability density functions for the amplitudes, given a fixed value of the nonlinear parameters, are invariant when the ensemble sampling distribution for the noise changes provided the projection of the orthogonal model onto the noise is the same in each noise sample. So again, the underlying noise sampling distribution is irrelevant for any given value of the nonlinear parameters. However, if our estimate of the uncertainty in these parameters is also to remain invariant when the ensemble sampling distribution for the noise changes, then the projection of the signal onto the noise and the mean-square noise value must also remain the same across noise samples. This last requirement is just the statement that the

Figure 4: Three Different Noise Sets

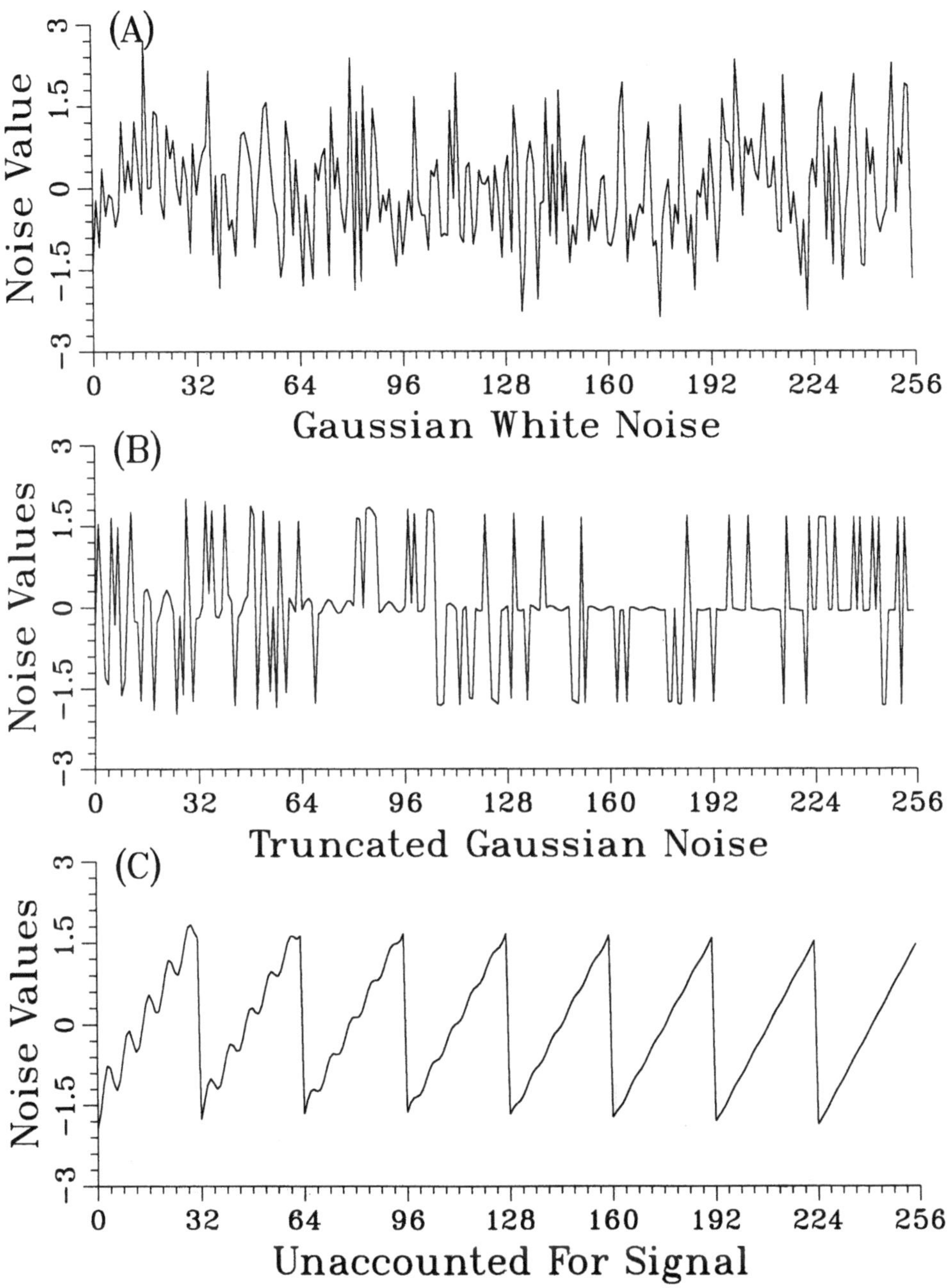

Fig. 4. Three noise sets were prepared: (A) Gaussian white noise, (B) rounded Gaussian white noise, and (C) deterministic signal. In preparing the noise the projection of the model onto the noise had to be removed for each noise set – see text for details.

Figure 5: The Posterior Probability For The Amplitude Of The Sinusoid

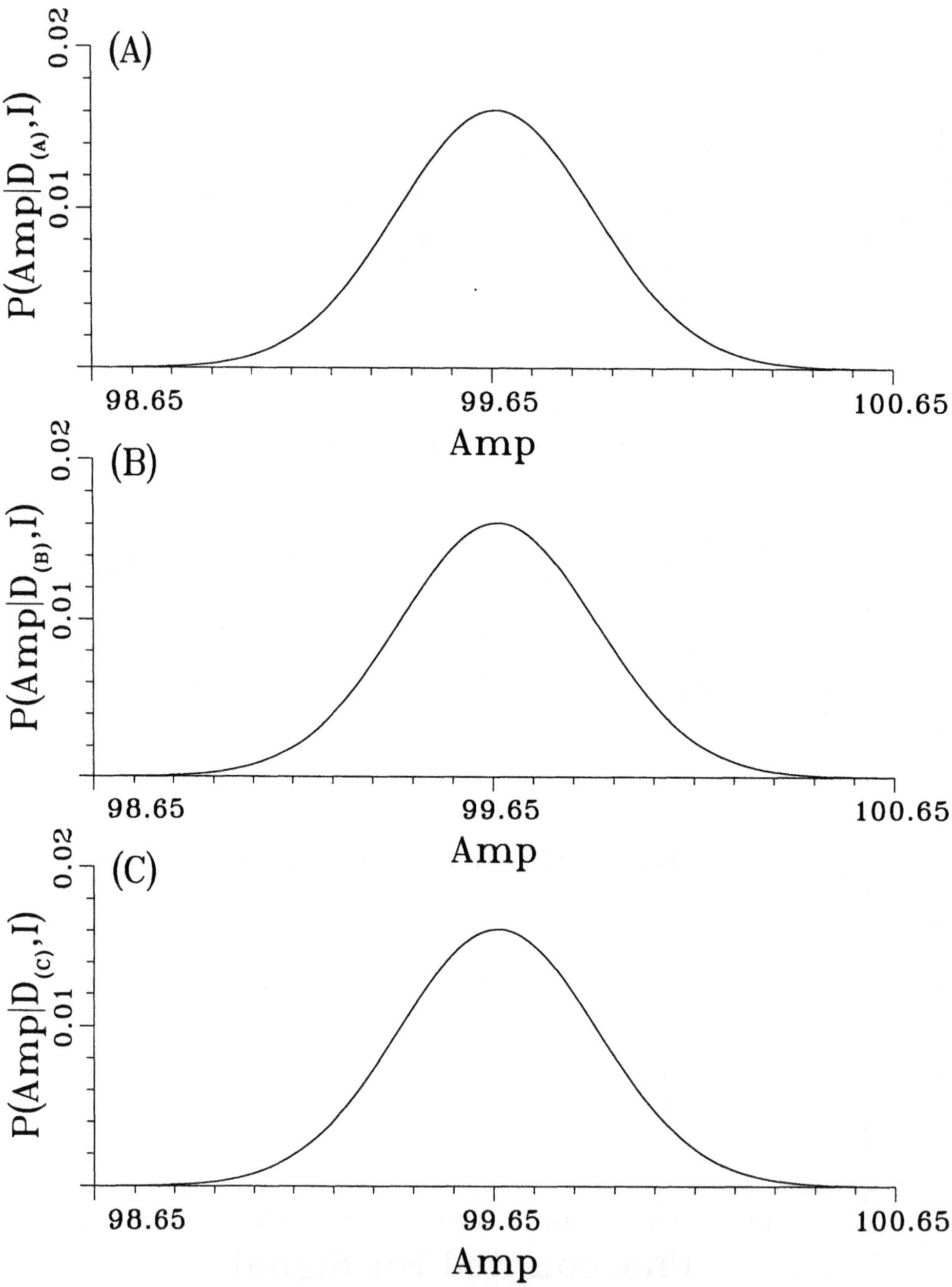

Fig. 5. The signal shown in Fig. 3 was added to the noise shown in Fig. 4 and the posterior probability for the amplitude of the sinusoid was computed, panels (A), (B), and (C) respectively. To the eye, these three probability density functions appear to be identical.

Figure 6: $P(\omega|DI)$

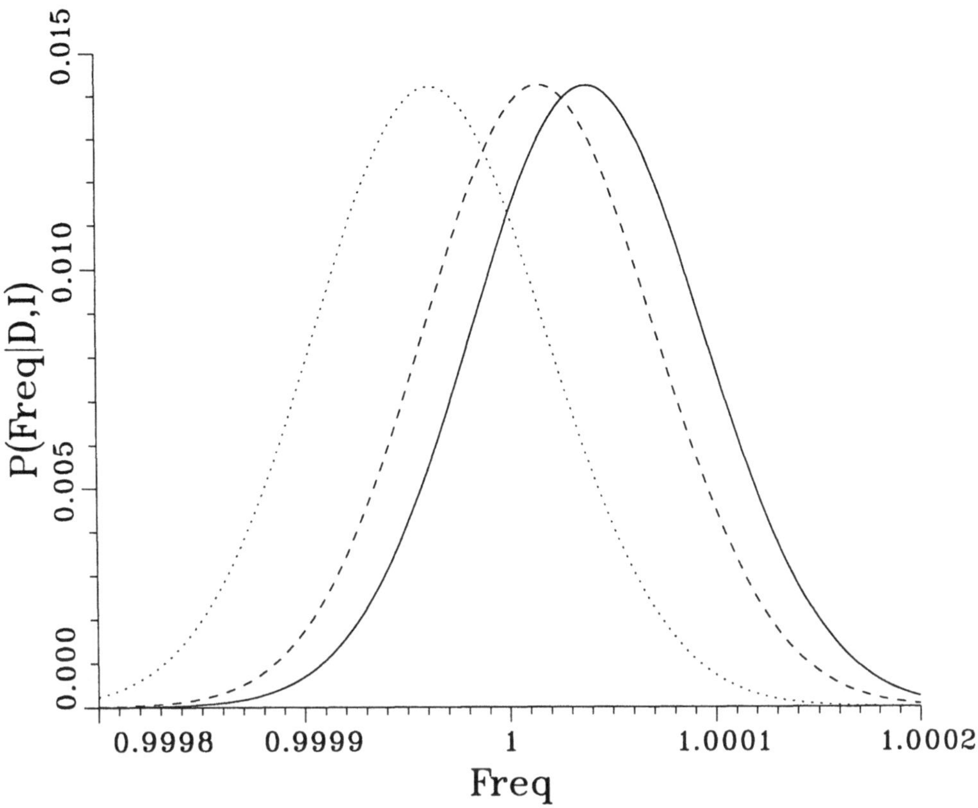

Fig. 6. The probability for the frequency was computed for the signal shown in Fig. 3 and the three noise sets shown in Fig. 4. These have been displayed here. The solid line used the Gaussian white noise, the dotted line used the rounded Gaussian noise, and the dashed line used the deterministic noise.

total squared-data value must be the same across noise samples.

The model we are using contains both location parameters, amplitudes, and parameters that appear in a nonlinear fashion. The question we would like to address is: "Are the estimates for the nonlinear parameters invariant when the ensemble sampling distribution for the noise changes?" The short answer to this is "no," and we have illustrated this by plotting the posterior probability for the frequency, $P(\omega|D\alpha I)$ for the data shown in Fig. 3 using the noise shown in Fig. 4. These plots are shown in Fig. 6. The solid line is the posterior probability for the frequency using the signal shown in Fig. 3 plus the Gaussian white noise shown in Fig. 4(A). The dotted line used the rounded Gaussian noise, and the dashed line used the deterministic noise. From this plot it is obvious that the probability for the nonlinear parameters depends on the noise sample. The real questions that should be addressed are "Why do the estimates for the nonlinear parameters depend on the noise sample?" and "Why does assigning a Gaussian distribution for the joint probability for the noise usually give reasonable estimates for these nonlinear parameters?"

First, we will derive the dependence of the joint probability for the nonlinear parameters on the noise, and then we will discuss why assigning a Gaussian distributions for the joint probability for the noise usually gives reasonable results for these nonlinear parameters. The posterior probability for the $\{\theta\}$ parameters is given by

$$\begin{aligned} P(\{\theta\}|\sigma DI) &= \int dA_1 \cdots dA_m P(\{\theta\}\{A\}|DI) \\ &= \int dA_1 \cdots dA_m P(\{\theta\}\{A\}|I) P(D|\{\theta\}\{A\}I). \end{aligned} \tag{52}$$

Because we are interested in the consequences of assigning a Gaussian distribution for the joint probability for the noise, we will take the prior $P(\{\theta\}\{A\}|I)$ to be uniform. Equation (52) is an integral over the likelihood:

$$P(\{\theta\}|\sigma DI) = \int dA_1 \cdots dA_m (2\pi\sigma^2)^{-\frac{N}{2}} \exp\left\{-\frac{Q}{2\sigma^2}\right\}, \tag{53}$$

where Q was defined earlier, Eq. (44). Each of the integrals over the amplitudes is an uncoupled Gaussian quadrature integral and is trivial to evaluate. One obtains

$$P(\{\theta\}|\sigma DI) = (2\pi\sigma^2)^{-\frac{N-m}{2}} \exp\left\{-\frac{Q'}{2\sigma^2}\right\} \tag{54}$$

as the posterior probability for the nonlinear parameters, where Q' is

$$Q' = \sum_{j=1}^{m} \hat{A}_j^2 + 2N \sum_{j=1}^{m} \hat{A}_j \overline{H_j(\{\hat{\theta}\})n} + N\overline{n^2} - m\overline{h^2}. \tag{55}$$

The first three terms in the definition of Q' are the total squared-data value, and are the same as what was derived earlier in Eq. (44). The last term, $m\overline{h^2}$, is the total squared-projection of the data onto the model, Eq. (45), and is a function of the nonlinear parameters:

$$\begin{aligned} \overline{h^2} &= \frac{1}{m}\sum_{j=1}^{m} h_j^2 \\ &= \frac{N^2}{m}\sum_{j=1}^{m}\left(\overline{H_j(\{\theta\})n} + \sum_{k=1}^{m} \hat{A}_k \overline{H_k(\{\hat{\theta}\})H_j(\{\theta\})}\right)^2. \end{aligned} \tag{56}$$

From this equation we see why the estimates for the nonlinear parameters are not invariant when the ensemble sampling distribution for the noise changes. The reason is simply that the projection of the jth model function onto the true noise is a function of the $\{\theta\}$ parameters. The only way the posterior probability for the nonlinear parameters could be invariant when the ensemble sampling distribution for the noise changes is if the noise is orthogonal to the model for *all* values of the nonlinear parameters, a condition that could not possibly exist.

However, having said this, it is now obvious why the estimated value of the nonlinear parameters are typically the correct values. It is obvious from the form of Eq. (56) that if the noise goes to zero, the parameter values that maximize the posterior probability are the true parameter value. The confounding term is $\overline{H_j(\{\theta\})n}$. If we evaluate $\overline{h^2}$ at the true parameters, then

$$\overline{h^2} = \frac{1}{m}\sum_{j=1}^{m}\left(N\overline{H_j(\{\hat{\theta}\})n} + \hat{A}_j\right)^2 \tag{57}$$

and the condition that must be satisfied for the probability to have a peak near the true parameter values is:

$$|N\overline{H_j(\{\hat{\theta}\})n}| \ll |\hat{A}_j|; \tag{58}$$

that is, the total projection of the jth model function onto the noise must be much smaller than the true amplitude of the jth model function. This condition can be fulfilled in several different ways. For example, the signal-to-noise ratio might be high, or the noise might be made up of many positive and negative values which when projected onto the signal tend to cancel and thus sum to a small number.

Finally, substituting Eq. (57) into the definition of Q', Eq. (55), a remarkable simplification takes place:

$$Q' = N\overline{n^2} - \sum_{j=1}^{m}\left(N\overline{H_j(\{\hat{\theta}\})n}\,\right)^2. \tag{59}$$

Thus, Q' is just the total squared-true error less the total squared-signal functions projected onto the noise. If the signal functions and the noise are almost orthogonal, the term on the right is essentially zero and the parameter estimates will be good. However, if the signal functions have a significant projection onto the noise, then one may or may not obtain good parameter estimates, depending on the model. The more the signal functions look like the noise, the larger the confounding term, and the parameter estimates one obtains may not be very precise.

6. Summary And Conclusion

When one explicitly includes the parameters, $\{\sigma\}$, and information about the noise, I_σ, in the probability theory calculation, probability theory naturally leads one to assign the joint probability for the noise given both the parameters and the information, Eq. (5). In deriving this equation, it becomes clear that one must assign a joint probability for the noise given what is actually *known* about the true noise.

In this paper we concentrated on the problem of assigning the joint probability for the noise when we did not have very specific information about the noise. In particular we concentrated on the case where the general order and scale, the first two moments, of the noise were supposed known. The principle of maximum entropy was then used to assign the joint probability for the noise. Use of the principle of maximum entropy, with knowledge of the first two moments of the noise results in the assignment of a Gaussian for the joint probability for the noise. Similarly, if we had used the total absolute value of the noise, a Laplacian would have been assigned. In assigning the joint probability for the noise, one must state explicitly what constraints are to be used in the calculation. Because of the way Eq. (5) was derived, any consistent set of constraints could have been incorporated into the assignment. However, in the case where little is known about the noise one is always better off to leave out constraints because the resulting maximum entropy probability density functions have higher entropy. Higher entropy distributions are by their very nature less informative then lower entropy distributions, and so are more conservative; they make allowances for every possible situation that could occur.

When the principle of maximum entropy is used to assign the joint probability for the noise, the resulting probability density functions have sufficient statistics. Sufficient statistics are functions of the data, and therefore the noise, that summarize all of the information in the data relevant to the problem being solved. As illustrated in the examples, the sufficient statistics are just the constraints used in assigning the joint probability for the noise. In the case of estimating a location parameter, this means that the parameter estimates are invariant when the ensemble sampling distribution for the noise changes, provided the mean and mean-square of the different noise samples are the same. For regression models, the estimated amplitudes remain the same when the ensemble sampling distribution for the noise changes, provided the total projection of the model onto the noise is the across noise sample. Similarly, for models containing nonlinear parameters one obtains the same result for a fixed or given value of the nonlinear parameters. Thus for any given value of the nonlinear parameters the estimated amplitudes one obtains are invariant when the ensemble sampling distribution for the noise changes.

References

1. G. Larry Bretthorst. *Bayesian Spectrum Analysis and Parameter Estimation*, volume 48. Springer-Verlag, New York, New York, 1988.
2. E. T. Jaynes. *Probability Theory – The Logic of Science.* Copies of this manuscript are available by either anonymous FTP or WWW brouser from "bayes.wustl.edu", 1993.
3. P. S. Laplace. *A Philosophical Essay on Probabilities.* Dover Publications, Inc., New York, 1951, original publication date 1814. Unabridged and unaltered reprint of Truscott and Emory translation.
4. C. E. Shannon. A mathematical theory of communication. *Bell Syst. Tech. J.*, 27:379–423, 1948.
5. J. E. Shore and R. W. Johnson. Axiomatic derivation of the principle of maximum entropy and the principle of minimum cross–entropy. *IEEE Trans. Information Theory IT*, IT-26:26–37, 1980.
6. S. M. Stigler. *Annals of Statistics*, 5:1055–1098, 1977.

OUTLIER TOLERANT PARAMETER ESTIMATION

V. DOSE AND W. VON DER LINDEN
Max-Planck-Institut für Plasmaphysik, EURATOM Association
D-85740 Garching b. München, Germany ¶

Abstract. Real world does not only provide noisy instead of perfect data. Every experimentalist has now and then to deal with outliers. The situation is simple if isolated points stick out of the general trend by a large amount. Arguments can then usually be found why such a point should be disregarded. The situation becomes critical if the outliers are not that obvious. This is usually the case for parameter space dimensions ≥ 3. We present a Bayesian solution to the outlier problem which assumes that the uncertainties assigned to the experimental data are only estimates of the true error variances.

Key words: Outlier, parameter estimation, duff data

1. Introduction

It is common experience to all experimental scientists that repeated measurements of supposedly one and the same quantity result occasionally in data which are in striking disagreement with all others. There are several conceivable reasons for such outliers. The apparatus may have performed differently due to some unstable component. If this instability has escaped the attention of the experimenter then there is no good reason to delete the outlying data point. The second possibility is that everything went well and the seemingly discordant data point has occurred as a highly unlikely event compatible with the sampling statistics. Finally, the discordant data point may signal a new and unexpected effect. This case is the most difficult and uninteresting in the context of the present paper since only repeated measurements will gradually fix the new phenomenon. By exactly this effort also the first two cases may be healed. So where is the problem? The problem is that only a small number of measurements is sufficiently fast, cheap and easy to perform such that the outlier problem can be overcome just by increased measurement efforts. Everybody who has chosen this way at some time in his career knows what increased efforts means depending on the extent to which the outlying data point deviates from the mainstream.

¶e-mail: vod@ipp-garching.mpg.de

W. von der Linden et al. (eds.), Maximum Entropy and Bayesian Methods, 47–56.

The situation which we have considered so far is simple: a single outlier and at least in principle augmentable data. Real life presents the opposite situation: a small number of discordant data which cannot be augmented with reasonable effort by new measurements. This situation is none of our imagination and characterizes not only research with B(illion) D(ollar) M(achines). It applies also to the quite different problem of the evaluation of physical constants and conversion factors from sophisticated high precision experiments carried out at different laboratories scattered all over the world. Let us illustrate by citation how the latter people proceed:'*After a thorough analysis using a number of least squares algorithms, the initial group of 38 items of stochastic input data was reduced to 22 by deleting those that were either highly inconsistent with the remaining data or had assigned uncertainties so large that they carried negligible weight.*' [1] In particular the second reason given for deleting data is curious to say the least. The selection of the 22 "good" data from the initial 38 was done in this case by CODATA, an interdisciplinary committee of the International Council of Scientific Unions. Scientific truth, however, is not a question of majority opinion. Half a page later in the same paper the authors comment on differences between the 1973 evaluation and the (still valid) 1986 results: '*The large change in K_V (a constant relating the SI unit of volt to a calibration standard) and hence in many other quantities between 1973 and 1986 would have been avoided if two determinations of F (the Faraday number) which seemed to be discrepant with the remaining data had not been deleted in the 1973 adjustment*' [1]. It is quite interesting that while deleting seemingly discrepant data in 1973 had turned out to be unacceptable, it has not prevented the committee to proceed in the same way in 1986.

The problem of outliers is so ubiquitous that it has also found its proper place in textbooks [2,3]. The proposed methods can more or less be characterized as testing the stability of a derived quantity against deletion of single data or groups of data in sophisticated but poorly justified manners. It is the purpose of this paper to reformulate and solve the problem in a consistent probabilistic framework. The present paper was stimulated by related ideas of Sivia [4] and Press [5].

2. Bayesian Analysis

For definiteness we consider in this section the problem of finding an appropriate arithmetic mean μ from a given set of data and associated experimental uncertainties. Our model equation reads

$$\mu = d_i + \xi_i \quad , \tag{1}$$

where ξ_i are the experimental errors characterized by variances σ_i^2. Of course we assume that $\langle\sigma_i^2\rangle$ is testable information. According to the principle of maximum entropy the likelihood function for the model (1) becomes

$$p(\vec{d}|\mu,\vec{\sigma},I) = \prod_i \frac{1}{\sigma_i\sqrt{2\pi}} \exp(-\frac{1}{2\sigma_i^2}(d_i-\mu)^2) \quad , \tag{2}$$

where the product over i stems from the assumption of uncorrelated errors $\{d_i - \mu\}$. The quantity we need in order to calculate $\langle\mu\rangle$ and $\langle\Delta\mu^2\rangle$ from the experimental data $\{d_i, s_i\}$ is $p(\mu|\vec{d}, \vec{s}, I)$ which we obtain from Bayes theorem as

$$p(\mu|\vec{d}, \vec{s}, I) = \frac{p(\mu|\vec{s}, I)}{p(\vec{d}|\vec{s}, I)} p(\vec{d}|\mu, \vec{s}, I) \quad . \tag{3}$$

Here s_i stands for the experimentally estimated value of σ_i. Let us assume no prior knowledge on μ. We code this in a flat improper prior. This is not harmful in the case of parameter estimation and since $p(\vec{d}|\vec{s}, I)$ is a normalizing constant in this context we have $p(\mu|\vec{d}, \vec{s}, I) \propto p(\vec{d}|\mu, \vec{s}, I)$. The problem has thus reduced to the calculation of the marginal likelihood $p(\vec{d}|\mu, \vec{s}, I)$. In order to calculate this probability from (2) we have to code our knowledge about $\vec{\sigma}$ in terms of $\vec{s}$. For completeness we first assume a set of good-natured data in which case the simplest assumption for the probability distribution of $\vec{\sigma}$, $p(\vec{\sigma}|\vec{s}, I)$ is

$$p(\vec{\sigma}|\vec{s}, I) = \prod_i \delta(\sigma_i - s_i) \tag{4}$$

from which we obtain by application of the sum rule

$$p(\vec{d}|\mu, \vec{s}, I) \propto \exp(-\frac{1}{2} \sum_i (d_i - \mu)^2 / s_i^2) \quad . \tag{5}$$

This is a gaussian with average value $\overline{\mu}$ and variance Σ_μ^2 given by

$$\overline{\mu} = \Sigma_\mu^2 \sum_i d_i / s_i^2 \quad , \; \Sigma_\mu^{-2} = \sum_i 1/s_i^2 \quad . \tag{6}$$

This is the familiar weighted mean taught already in freshmen physics laboratory courses. If these formulae fail to produce acceptable results then it is quite obvious where to modify our assumptions. It may in particular be difficult to estimate experimental uncertainties to a precision such that (4) is justified. In such a case we would generalize (4) to

$$p(\vec{\sigma}|\vec{s}, \vec{\omega}, I) = \prod_i \delta(\sigma_i - s_i/\sqrt{\omega_i}) \quad , \tag{7}$$

where we have introduced scale variables ω_i whose magnitude determine to what extent $\langle\sigma_i\rangle$ deviates from s_i. The particular parameterization with $\sqrt{\omega_i}$ has been chosen for later calculational convenience. Integrating out the true errors from (2) using (7) yields

$$p(\vec{d}|\mu, \vec{\sigma}, \vec{\omega}, I) \propto \prod_i \sqrt{\omega_i} \exp(-\frac{\omega_i}{2s_i^2}(d_i - \mu)^2) \quad . \tag{8}$$

In order to get rid of the newly introduced parameter $\vec{\omega}$ we have to specify its probability distribution. This specification codes our belief of how well the experimentally determined $\vec{s}$ approximate $\vec{\sigma}$. We should certainly expect $\langle\sigma_i\rangle = s_i$ and therefore $\langle\omega_i\rangle = 1$, otherwise it means that the determination of $\vec{s}$ suffers from a known bias. In addition we consider $\Delta^2\omega$ as testable information. Along with the fact $\omega > 0$ an appropriate prior is the Γ-distribution

$$\Gamma(\omega|a) = \frac{a^a}{\Gamma(a)}\omega^{a-1}\exp(-a\omega) \quad , \tag{9}$$

for which $\langle\Delta\omega^2\rangle = 1/a$. The parameter a specifies therefore our knowledge about the width of $p(\omega|\langle\omega\rangle, a, I)$ and therefore somehow the confidence interval of the experimentally determined quantities $\{s_i\}$. Note that (9) approaches an exponential distribution for $a \to 1$ and Jeffreys' prior for $a \to 0$. We employ (9) to marginalize over ω in (8) and obtain

$$p(\vec{d}|\mu, \vec{s}, I) \propto \prod_i \left\{a + (d_i - \mu)^2/2s_i^2\right\}^{-(a+1/2)} \quad , \tag{10}$$

from which we can now calculate $\langle\mu\rangle$ and $\langle\Delta\mu^2\rangle$. The decisive difference between (5) and (10) is obviously that (5) satisfies $\prod f(x_i) = f(\sum x_i)$ while (10) does not. The product of an arbitrary number of gaussians is again a gaussian and is strictly unimodal. (10) may well be multi-modal and will certainly be so in the case of discrepant data. For the special case $a = 1/2$ (10) reduces to a product of Lorentzians (Cauchy distributions). While the mean value μ obtained from (5) is an average weighted only by the individual $\{s_i\}$ regardless of the position of the data points, both error variance $\{s_i\}$ and position of $\{d_i\}$ enter the calculation of $\langle\mu\rangle$ and $\langle\Delta\mu^2\rangle$ from (10).

The analysis developed so far shall now be applied to two sets of precision measurements reported in [1]: The Quantum Hall(QHE) resistance as an example for a rather concordant set of data and the gyromagnetic ratio of the proton as a highly discrepant set of measurements. The former quantity has become the primary standard of electrical resistance with numerical value $h/e^2 = 25812.8056(12)\Omega$. h is Planck's constant and e the electron charge. The numbers in brackets indicate the uncertainty in the last digits. The second quantity $\gamma_p = 26752.2128(81)[10^4 s^{-1}T^{-1}]$ is the ratio of the protons magnetic moment divided by its angular momentum. By appropriate reformulation we find

$$\gamma_p = \frac{\mu_p}{I_p} = \frac{\mu_p}{\mu_n}\frac{F}{M_p} \quad , \tag{11}$$

where μ_p and μ_n are the proton and the nuclear magnetic moment respectively, F is the Faraday number and M_p the proton molar mass. Regarding M_p as an available auxiliary constant the measurement of γ_p serves therefore primarily for a better determination of the Faraday number F. The left panel of fig. 1 shows the QHE data with their error bar. The dotted line is the distribution (5), that is the posterior probability for μ on the basis of weighted least squares. The horizontal scale variable is $\mu - \mu_0$ where μ_0 is the straightforward unweighted average of the data points. It has no significance other than an appropriate shift of the μ axis for graphical presentation of the data. The dashed line is the posterior probability for μ according to (10) for $a = 1/2$. Its maximum occurs at nearly the least squares value of μ. However, we observe that this posterior is slightly wider. This reflects the fact that the error in μ according to (10) depends unlike the case of (5) not only on the quoted error variances of the contributing measurements but also on the distribution of $\{d_i\}$ themselves. The right-hand panel of fig. 1 shows data on

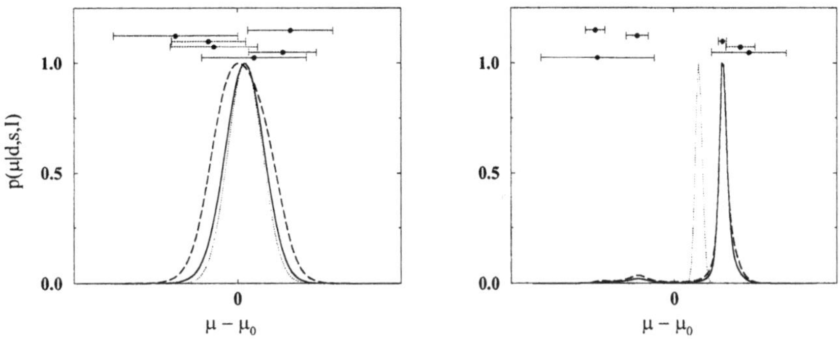

Figure 1. Posterior probability $p(\mu|\vec{d},\vec{s},I$ for the QHE (left panel) and the γ_p (right panel) data. Dotted line (5), dashed line (10) and full curve (24)

the proton gyromagnetic ratio(γ_p). This data set is of course very interesting. The dotted line indicates again the posterior probability for μ according to (5). From (6) we know already that its width is smaller than the error margin of the most precise of the contributing data. Together with the fact that its average lies in a region which does not overlap with any of the contributing data renders these results unacceptable. In fact the authors of [1] realized this fact themselves: '*Of all of the γ_p data, the most glaring discord comes from the NPL (UK) low field value (the leftmost data point in fig. 1). The measurements of the proton resonance frequency were completed in December 1975 after which the coil dimensions were measured, but no verification was made (by repeating the frequency measurement), that the measurement process did not affect the coils. Because this result is so discrepant, and because the measurements were forced to terminate prematurely we consider it to be an incomplete effort which should not be included in the final adjustment*'. Concerning deletion of data we refer the reader back to the introduction. The arbitrariness of deleting the most discrepant data point becomes quite obvious when we realize that even the truncated data set remains highly discrepant as also noted by the authors of [1]. While according to (6) they obtain a $\chi^2/N = 33$ for the full data set, the truncated set results in $\chi^2/N = 12.7$ instead of one.

The dashed curve in fig. 1 is the posterior probability for μ according to (10). It shows immediately the dilemma: The distribution is bimodal and a summary in terms of $\langle\mu\rangle$ and $\langle\Delta\mu^2\rangle$ is of very limited value. Note, however, that the prominent maximum lies in the range of the most precise of the contributing data points in accord with common sense. This result in not only numerically superior but it also points out that the data do not furnish a unique answer and that the data set might need a second thought. The straight forward application of the orthodox formula (5) could easily be deceiving.

For the present one-dimensional examples it is more or less intuitively clear upon display which data points should be regarded as an outlier and to what extent. This intuitive perception vanishes if we consider a multi-dimensional case that is a situation where μ represents a function of several variables. In particular in such cases it may be desirable to calculate $\langle\sigma_k\rangle$, the expectation value of the error variance of datum d_k in the light of the quoted error variance s_k and all the other data $\{d_i, s_i\}$. To this end we need $p(\vec{\sigma}|\vec{d}, \vec{s}, I)$ which is obtained from Bayes theorem as

$$p(\vec{\sigma}|\vec{d}, \vec{s}, I) = \frac{p(\vec{\sigma}|\vec{s}, I)}{p(\vec{d}|\vec{s}, I)} p(\vec{d}|\vec{\sigma}, I) \tag{12}$$

and

$$\langle\sigma_k\rangle = \frac{1}{p(\vec{d}|\vec{s}, I)} \int \sigma_k \, p(\vec{\sigma}|\vec{s}, I) p(\vec{d}|\vec{\sigma}, I) d\vec{\sigma} \quad . \tag{13}$$

$p(\vec{\sigma}|\vec{s}, I)$ is derived by application of the sum rule from (7) and (9) as

$$p(\vec{\sigma}|\vec{s}, I) = \int p(\vec{\omega}|I) p(\vec{\sigma}|\vec{s}, \vec{\omega}, I) d\vec{\omega} \quad . \tag{14}$$

Similarly we obtain $p(\vec{d}|\vec{\sigma}, I)$ from

$$p(\vec{d}|\vec{\sigma}, I) = \int p(\mu|I) p(\vec{d}|\mu, \vec{\sigma}, I) d\mu \quad . \tag{15}$$

Putting the various terms together we end up with

$$\langle\sigma_k\rangle / s_k = \frac{\Gamma(a)}{\Gamma(a + 1/2)} \int \left\{a + (d_k - \mu)^2 / 2s_k^2\right\}^{1/2} \rho_a(\mu, d_i) d\mu / Z \quad , \tag{16}$$

with

$$Z = \int \rho_a(\mu, d_i) d\mu \quad \text{and} \quad \rho_a(\mu, d_i) = \prod_i (a + (d_i - \mu)^2 / 2s_i^2)^{-(a+1/2)} \quad . \tag{17}$$

(16) allows to identify data point k as an outlier at a significance level δ if the right-hand side of (16) exceeds δ. The prior probability which we have used so far is (9) with $a = 1/2$. This represents the case where we assume that the expectation value of the true error variances σ_i is equal to the error s_i with a mean square deviation $\pm s_i\sqrt{2}$. It seems in order to exploit the influence of the choice of a on the final results. Fig. 2 shows such a scan of a together with mean value and posterior rms error again for the QHE data in the left panel and for the γ_p data in the right panel. Dashed curves apply in both cases to the above analysis. Since the smaller a the larger the range of the true error variance which we allow given the quoted variance, the behavior of the $\langle\Delta\mu^2\rangle^{1/2}$ is in accord with expectation. We would even accept a considerable variation of μ as a function of a in the case of discordant data as in the case of γ_p. The variation of $\langle\mu\rangle$ in the case of the QHE

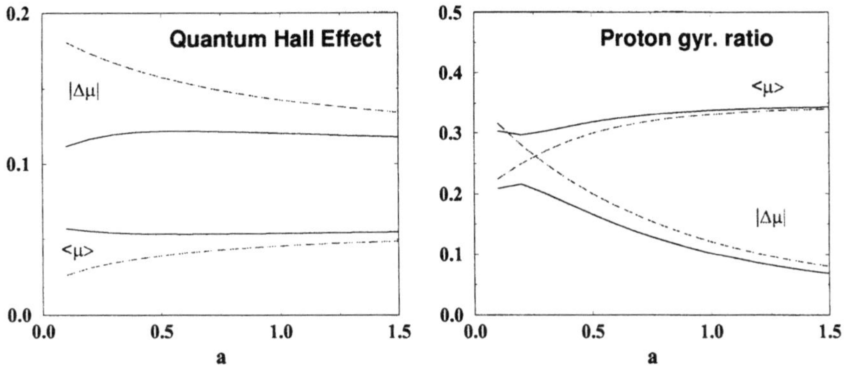

Figure 2. Dependence of $\langle\mu\rangle$ and $|\Delta\mu|$ on the variance $1/a$ of (9)

data is however less acceptable and calls for a modification of the analysis which was developed so far.

The assumption (7) that the quoted errors on every data point are uncertain according to the distribution (9) may be too strong. In fact, in real life we would expect that some of the quoted errors s_i should be treated with (4) while only a priori unknown outliers should be treated according to (7). Accordingly we modify the prior probability $p(\vec{\sigma}|\vec{s},\vec{\omega},I)$ introducing a new variable β with $0 \leq \beta \leq 1$ to

$$p(\vec{\sigma}|\vec{s},\vec{\omega},\beta,I) = \prod_i \{\beta\delta(\sigma_i - s_i/\sqrt{\omega_i}) + (1-\beta)\delta(\sigma_i - s_i)\} \quad , \tag{18}$$

where β is the probability that data point i is an outlier and the quoted error differs from the expectation value of the true error while $(1-\beta)$ is the probability that data point i is regular and the quoted error s_i is a good estimate for the expectation value $\langle\sigma_i\rangle$. For β we shall assume a flat prior in the interval [0,1]. The marginal probability $p(\vec{d}|\mu,\vec{s},I)$ is now obtained using (2), (18), (9) as

$$\begin{aligned} p(\vec{d}|\mu,\vec{s},a,I) \;\propto\; & \int d\beta \prod_i \int d\omega_i \omega_i^{a-1} e^{-a\omega_i} \frac{a^a}{\Gamma(a)} \times \\ & \int d\sigma_i \{\beta\delta(\sigma_i - s_i/\sqrt{\omega_i}) + (1-\beta)\delta(\sigma_i - s_i)\} \frac{1}{\sigma_i} e^{-\frac{1}{2\sigma_i^2}(d_i-\mu)^2} \end{aligned} \tag{19}$$

The σ_i integration is straightforward and results in a sum of two terms: the first has a structure similar to (8) multiplied by β and the second is a gaussian multiplied by $(1-\beta)$ with σ_i replaced by s_i. It is also straightforward to perform the ω_i integration. The second term in (19) does not contain ω_i and since $p(\omega_i|a)$ is normalized the ω_i integration results in a factor of 1. The second term is then according to σ_i and ω_i integrations

$$B_i = \exp[-(d_i-\mu)^2/2s_i^2]/s_i \quad , \tag{20}$$

the first term becomes

$$A_i = \frac{a^a \Gamma(a+1/2)}{s_i \Gamma(a)} \left\{ a + \frac{(d_i - \mu)^2}{2 s_i^2} \right\}^{-(a+1/2)} \quad . \tag{21}$$

We may further suppress the common factor $1/s_i$ and are left with the β-integration

$$p(\vec{d}|\mu, \vec{s}, a, I) \propto \int d\beta \prod_i^N \{\beta A_i + (1-\beta) B_i\} \quad . \tag{22}$$

The product over the N factors is a polynomial of degree N in β

$$P_N = \sum_{i=0}^{N} \binom{N}{i} \beta^i (1-\beta)^{N-i} C_i^{(N)} \quad , \tag{23}$$

with so far unknown coefficients $C_i^{(N)}$. The integral in (23) over β of P_N in $0 \leq \beta \leq 1$ is the Beta function. After simplifying the binomial coefficients, (23) becomes

$$p(\vec{d}|\mu, \vec{s}, a, I) \propto \frac{1}{N+1} \sum_{i=0}^{N} C_i^{(N)}(\mu) \quad . \tag{24}$$

Our final task, the determination of the coefficients $C_i^{(N)}$, is accomplished by $P_N = P_{N-1}(A_N \beta + B_N(1-\beta))$, where we exploited the different representations (22) and (23) of P_N. Note that the unknown coefficients in P_N are $C_i^{(N)}$ while in P_{N-1} we have $C_i^{(N-1)}$. Inserting the explicit expressions (23) for the polynomials and equating terms with equal powers in $\beta^i(1-\beta)^j$ we obtain the recurrence relation

$$C_i^{(N)} = \frac{i}{N} A_N C_{i-1}^{(N-1)} + \frac{N-1}{N} B_N C_i^{(N-1)} \quad . \tag{25}$$

The recurrence is started from $P_1 = \beta A_1 + (1-\beta) B_1 = C_0^{(1)}(1-\beta) + C_1^{(1)} \beta$, which gives $B_1 = C_0^{(1)}$ and $A_1 = C_1^{(1)}$. This completes the calculation of the marginal likelihood $p(\vec{d}|\mu, \vec{s}, a, I)$. The solid solid line in fig. 1 and fig. 2 corresponds to the results of this last piece of analysis when applied to the two previous data sets. In the case of the QHE the posterior probability (24) falls close to the least squares posterior (5). The width of (24) is slightly larger than that of (5) which signals that the QHE data are nearly but not exactly compatible with each other. The posterior (10) reflects clearly the more diffuse prior for the exact errors in a noticeably larger width. An entirely different behavior is observed in the case of the γ_p data which are highly discordant. Here the posterior (24) coincides nearly with (10) while the least squares result differs grossly from the latter approaches. (24) has thus the extremely beneficial property that it interpolates between (5) and (10) depending on the condition of the data. This is further corroborated by fig. 2 which shows as solid lines $\langle\mu\rangle$ and $\sqrt{\langle\Delta\mu^2\rangle}$ for the QHE data in the left panel and for the γ_p data in the right panel. While the posterior $\langle\mu\rangle$ and $\langle\Delta\mu^2\rangle$

calculated from (24) do not longer depend on the particular choice of a in the distribution (9) in a wide range of a-values for the good-natured QHE data, there is only a minor difference in $\langle\mu\rangle$ and $\sqrt{\langle\Delta\mu^2\rangle}$ for the discordant γ_p data when calculated either from (24) or from (10). We conclude that (18) constitutes a very flexible prior distribution to code the frequent situation of unprecise knowledge about the errors associated with experimental data.

In the prior probability (18) we have introduced a new variable β. Its numerical value is the probability that a data point is an outlier. We are now in the position to calculate the posterior probability for β $p(\beta|\vec{d},\vec{s},a,I)$. From Bayes theorem, we have $p(\beta|\vec{d},\vec{s},a,I) \propto p(\vec{d}|\beta,\vec{s},a,I)$ and using the sum rule

$$p(\beta|\vec{d},\vec{s},a,I) \propto \int p(\vec{d}|\mu,\beta,\vec{s},a,I)d\mu \quad . \tag{26}$$

The prior for μ, assumed flat, has been suppressed here. Formally (26) is equivalent to (19) with β-integration exchanged by a μ-integration. In terms of the polynomial coefficients $C_i^{(N)}$ the posterior probability for β becomes

$$p(\beta|\vec{d},\vec{s},a,I) \propto \int d\mu \sum_{i=0}^{N} \binom{N}{i} C_i^{(N)}(\mu)\beta^i(1-\beta)^{N-i} \quad . \tag{27}$$

Fig. 3 shows the posterior probability distribution of β according to (27) for the two test data sets again for $a = 1/2$. For the QHE data the function (27) is

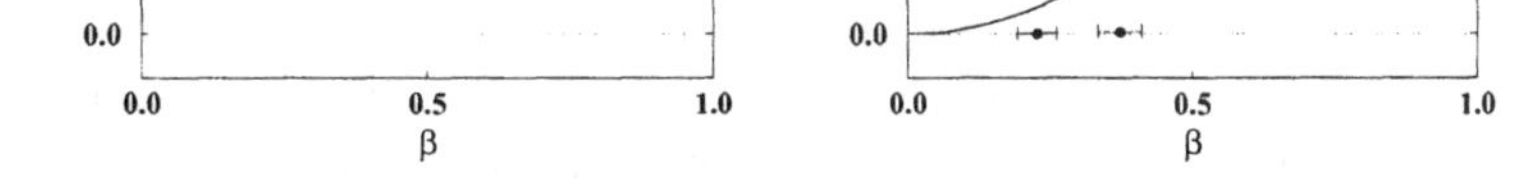

Figure 3. Average outlier probability (continuous curves) and individual outlier probability for the QHE (left) and the γ_p data (right).

monotonous and peaks at $\beta = 0$ thus indicating a low average outlier probability $\langle\beta\rangle$. For γ_p on the other hand the probability peaks at $\beta = 1$ and thus reflects that the γ_p data are highly discordant. The posterior probability for β answers therefore the interesting question as to what extent we deal with a good or a bad data set on the average.

We shall finally answer the question what is the probability that a single point d_k is either a good or a bad data point. Let us introduce the proposition Q_k: *d_k is* *an outlier.* We then need to calculate $P(Q_k|\vec{d},\vec{s},a,I)$. This probability is expressed by Bayes theorem as

$$P(Q_k|\vec{d},\vec{s},a,I) \propto \int P(Q_k|a,\beta,I)p(\vec{d}|Q_k,\vec{s},a,\beta,I)\,d\beta \quad , \tag{28}$$

where we have dropped the evidence denominator since it serves only as a normalization constant here. The marginal likelihood $p(\vec{d}|Q_k,\vec{s},a,\beta,I)$ is obtained from $p(\vec{d},\mu|Q_k,\vec{s},a,\beta,I)$ by marginalization over μ, again as before with a flat prior in μ. The prior probability $P(Q_k|a,\beta,I)$ is by definition just equal to β. The outlier probability is finally

$$P(Q_k|\vec{d},\vec{s},a,I) \propto \int d\beta d\mu\; \beta\; A_k \prod_{i\neq k}(\beta A_i + (1-\beta)B_i) \quad . \tag{29}$$

Since we have abandoned normalization we must also calculate the probability for the proposition $\overline{Q_k}$: d_k *is not an outlier.* We obtain $P(\overline{Q_k}|\vec{d},\vec{s},a,I)$ from (29) by exchange of βA_k with $(1-\beta)B_k$. This completes our analysis for the generalized arithmetic mean in the case of partly discordant data. Fig. 3 shows that according to the above analysis the probability of the QHE data for being well conditioned is above 50% for all values. The γ_p data on the other hand fall all below 50%. Interestingly enough our analysis renders not only the NPL value entirely improbable but also the VNIIM (USSR) value which is at the position of the secondary maximum in fig. 1. The selective removal of the NPL value from the data evaluation process in [1] appears therefore even more doubtful.

3. Conclusions

Bayesian analysis has been applied to the frequent situation in experimental sciences that error bars on data can only inaccurately be specified. The analysis considers this imprecise knowledge and allows further for the possibility that data may originate from the assumed sampling distribution but that also outliers discordant with the rest of the data are present. The formulation in terms of "good" and "bad" data appears to be a sensible interpolation between the two limiting cases of entirely concordant and entirely discordant data. It should (and will) be put to further tests in more complicated cases of parameter estimation.

References

1. E. Cohen and B. Taylor *Reviews of Modern Physics*, **59**, p. 1121, 1987.
2. P. Rousseeuw and A. Leroy, *Robust Regression and Outlier Detection*, John Wiley and Sons, New York,Chichester, Brisbane, Toronto, Singapore, 1987.
3. J. Taylor, *An Introduction to Error Analysis*, University Science Books, Sausalito, California, 1997.
4. D. Sivia, "Dealing with duff data," in *MAXENT96 - Proceedings of the Maximum Entropy Conference 1996*, M. Sears, V. Nedeljkovic, N. E. Pendock, and S. Sibisi, eds., p. 131, NMB Printers, Port Elizabeth, South Africa, 1996.
5. W. Press in *Proceedings of Conference in Honor of John Bakcall*, J. Ostriker, ed., p. in press, Princeton University Press, Princeton, 1998.

UNIVERSALITY, PRIOR INFORMATION AND MAXIMUM ENTROPY IN FOAMS

N. RIVIER, B. DUBERTRET, T. ASTE AND H. OHLENBUSCH
Laboratoire de Dynamique des Fluides Complexes
Université Louis Pasteur
3, Rue de l'Université, 67084 STRASBOURG, France†

Abstract. A foam is a random partition of space into cells. In two dimensions, it is a planar, regular graph, consisting of trivalent vertices, edges and cells. The number n of neighbours of a cell (which is equal to the number of its edges) is the only, local random topological variable of the problem. One looks for the most probable distribution p_n, subject to some information of the system. This information is of two types:

(i) Constraints (e.g. $\langle n \rangle = 6$), pertaining to filling (topological) space. These are inevitable, but can be made redundant. The redundancy is expressed as correlations, equations of state, between random variables on the same or different cells. These equations of state have been observed experimentally, and are known as the laws of Aboav-Weaire, Lewis, Peshkin, etc...

(ii) Prior knowledge on the system, which can be the prior probability distribution q_n, but also some symmetry or transition probability of the system. It must, however, be a local, and context-independent description of the state of the foam. We show, by simulations, that a foam is in statistical equilibrium, *i.e.* that it has the maxent, most probable distribution p_n. We will give examples of priors, and also of restrictions which are context-dependent and lead to foams off statistical equilibrium [1,2].

We will also demonstrate analytically by maxent, including selection of least informative priors, a different kind of equation of state, Lemaitre's law [3], thereby justifying its surprising universality. Lemaitre et al. noticed that a relation between two different measures ("order parameters") of disorder, was universal in two-dimensional foams. Universality of the relation may be thought, at first sight, to express their equivalence. This is not so: Lemaitre's law is the virial equation of state for foams. It is non-linear, thus non-trivial. In a gas, non-triviality reflects the specific interactions between the atoms or molecules of the particular gas. In foams, it is universal [4].

†Email: nick@ldfc.u- strasbg.fr

W. von der Linden et al. (eds.), Maximum Entropy and Bayesian Methods, 57–64.

1. Maximum ignorance as a way to wisdom

In this paper, we will extol disorder and use maximum ignorance (this is indeed what maxent inference is), not as an excuse, but to obtain a scientific result.

Our subject materials are froths or foams (soap froth, innermost layer of the epidermis, Bénard-Marangoni convection pattern, geological cracking pattern like the Giant's Causeway, etc.), in two dimensions. A froth is a space-filling, random cellular network, beautiful and interesting in spite of its disorder.

Maximum ignorance is a state of equilibrium., independent of unknown facts, unstated assumptions or outside interference. Only a system in statistical equilibrium (the state of maximum entropy) is able to respond locally and efficiently to any outside disturbance offsetting this equilibrium, like the epidermis to a wound.

This text is part of a chapter by N.R. of a forthcoming collective book (*Foams and Emulsions*, edited by J.F. Sadoc and N. Rivier, Kluwer 1998). For details and references , see the reviews [5–7]. For the other results mentioned in the abstract, see the published references [1,2].

2. Two-dimensional foams

A foam is a random cellular network, a planar, regular graph consisting of C cells (a large number), E edges and V vertices. Edges are incident on vertices and bounded by vertices. Cells are incident on edges and bounded by (a cycle of) edges. Their numbers are related by incidence relations,

$$3V = 2E = \langle n \rangle C, \tag{1}$$

and by the Euler relation

$$C - E + V = 1. \tag{2}$$

(The cell at infinity is not included in C). The number n of edges bounding a cell is the only local random variable in the foam (all other incidence and bounding numbers are fixed; in particular, randomness imposes 3 edges and cells incident on a vertex). It is shuffled by the local, elementary topological transformations, which are, in foams, $T1$ (neighbour switch), $T2$ (disappearance of a triangular cell), or mitosis (cellular division) in biological tissues. Thus, a foam is a topological pattern ("gas" of cells) in statistical equilibrium, the local, microscopic variable n attached to an individual cell being shuffled by topological collisions, like the momentum of an atom in a gas.

The Euler and incidence relations combine to yield $\langle n \rangle = 6$. (Corrections for a finite foam are of order $1/C$, and thus negligible in the thermodynamic limit). A non-hexagonal cell can be regarded as a source of curvature, carrying a topological charge $6 - n$, and for which there is a topological equivalent of Gauss theorem [8]. The elastic energy of the foam is proportional to the fluctuations of the local random variable, $\mu_2 = \sum_n p_n(n-6)^2$ (see §3). The evolution of the foam is curvature-driven: An n-sided cell shrinks at a rate proportional to $6 - n$ (von Neumann's law).

Statistical equilibrium is a thermodynamic state of maximum entropy, characterized by the distribution p_n of the cells (the Maxwell-Boltzmann distribution of velocities in a gas), and by equations of state relating the various macroscopic, thermodynamic variables, C, the total area V to fill, and the variance μ_2 of the distribution (in a gas, this variance, the average kinetic energy of an atom, is proportional to the temperature, and $U = (3/2)NkT, PV = NkT$ are the equations of state).

It has been established over the last fifteen years that foams are in statistical equilibrium, described by a state of maximum entropy restricted by constraints, linear relations between the p_n. We shall give the form of the equation of state (which is called Lemaitre law, the virial expansion in gas, of which $P/kT = N/V = \rho$ is the first term), and show that it is universal and non-trivial for all foams [4], (unless they are excessively disordered ($p_6 < 0.2, \mu_2 > 4$) [9]).

In foams, the unavoidable constraints are $\langle n \rangle = 6$, and normalization of p_n ($\langle 1 \rangle = 1$). Maximum entropy infers that, in the absence of information to the contrary, all other geometrical constraints (topological correlations, space-filling) can be expressed as linear combinations of the two unavoidable ones, and made redundant. This redundancy leads to the linear (in n) relations of Aboav (- Weaire) and Lewis, observed and discovered experimentally and/or in simulations. [6,7,1]

3. Disorder control. Lemaitre equation of state

Foams may be disordered, but they are certainly not disorganized. Disorder is measured by the thermodynamic variable

$$\mu_2 = \sum_n p_n (n-6)^2. \tag{3}$$

It manifests itself locally by curvature fluctuations, i.e. by the presence of non-hexagonal cells or by their generation. The average curvature is zero for a foam on a plane, because $\langle n \rangle = 6$ (a statement of the theorem of Gauss-Bonnet,

$$\int_S K(\mathbf{x})\mathbf{dx} = \mathbf{2}\pi\chi, \tag{4}$$

where $K(\mathbf{x})$ is the Gaussian curvature and χ is the Euler-Poincaré characteristic of the surface S). If left uncontrolled, these fluctuations destroy statistical equilibrium in the foam. For examples:

- Systematic decoration of vertices (inverse $T2$) leads to a fractal Sierpinski foam [10] with $\mu_2 = \infty$.
- Non-hexagonal cells are produced by cellular division. If our cells were to divide one after the other in a row or on a circle, our skin would develop long scars of softer tissue or pimples [7], and disorder would be concentrated on isolated defects, visible and felt far away, instead of being spread out throughout the tissue.
- A foam constructed shell-wise from a central cell [11] has an effective negative curvature due to disorder [8]. Concentric shells have more cells and are wobblier than in an hexagonal foam. To keep the foam uniform on a plane, defects (isolated cells which do not belong to a shell) are intercalated between the wobbly shells.

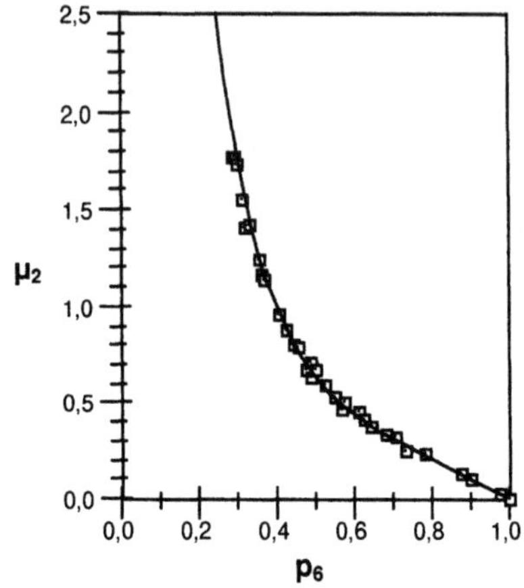

Figure 1. Lemaitre law: μ_2 as a function of p_6. Full line: Maxent theory (Mayer coefficients (11)) [4]. Squares: Experimental and numerical data [3].

Disorder is controlled by the topological energy of the foam. Cells have correlation and self energies [6,7,1].

$$\begin{aligned} U(\mu_2) &= \varepsilon \sum_{n,k} p_n p_k (n-6) A_1(n,k)(k-6) + \eta \sum_n p_n (n-6)^2 \qquad (\varepsilon, \eta > 0) \\ &= (\varepsilon \sigma \mu_2 + \eta)\mu_2 \qquad (5) \end{aligned}$$

(The first term is the average correlation energy of two neighbouring cells with topological charges $(6-n)$ and $(6-k)$. [$A_1(n,k)$ is the average number of pairs of neighbouring $n-$ and $k-$sided cells, given that the two cells exist]. The second is the average self-energy of this topological charge). The correlation energy is highest for uncorrelated foams ($\sigma = 1/6$) and vanishes for hexagonal foams ($\mu_2 = 0$). The self-energy, proportional to the fluctuations $\langle K^2 \rangle$ of the Gaussian curvature, ensures that U is positive, thus $\eta > a\varepsilon$. (The Aboav-Weaire parameter a = $-\sigma\mu_2$ equals 1 in foams locally balanced under $T2$ or mitosis [12,6]).

We will maximize the entropy, keeping the energy U, i.e. μ_2 constant. This is an additional constraint on the distribution p_n, besides the two independent, inevitable, topological ones. The new constraint cannot be made redundant since it is quadratic in n. It is therefore an additional restriction, but to the dynamics rather than to the geometry. The result is a relation (equation of state) between the thermodynamic variables μ_2 and p_6, found by Lemaitre et al. [3].

Lemaitre et al. noticed that this relation between μ_2 and $p_6 = 1 - \sum_{n \neq 6} p_n$ was universal in two-dimensional foams. Now, μ_2 and $1 - p_6$, even moments in $(n-6)$ of the distribution p_n, are both measures ("order parameters") of disorder. Universality of the relation $\mu_2(p_6)$ may be thought, at first sight, to express their equivalence. This is not so: Lemaitre's law is the virial equation of state for foams.

It is non-linear (non-trivial virial expansion). In a gas, non-triviality reflects the specific interactions between the atoms or molecules of the particular gas. In foams, the full equation of state, for $0.34 < p_6 \leq 1$, is universal. (Figure 1). For $p_6 > 0.67$, the foam has only 5-, 6- and 7-sided cells, with $(1 - p_6)/2 = p_5 = p_7$, thus $\mu_2 = 1 - p_6$. The foam has trivial virial expansion and corresponds to an ideal gas.

4. Lemaitre's law as the virial equation of state for foams

The following procedure is standard maximum entropy inference [4]. One maximises the entropy

$$S = -\sum_n p_n \ln(\frac{p_n}{q_n}) \tag{6}$$

(where q_n is the *a priori* probability of having an n-sided cell, unknown so far), restricted by the independent constraints. The latter fall into two classes and are either even or odd in $(n - 6)$.

- Inevitable, topological

$$\sum_n p_n = 1 \qquad \text{(even)}$$

$$\sum_n p_n(n-6) = 0 \qquad \text{(odd)}$$

- Metric. The foam is at constant energy if the variance μ_2 of the distribution is constrained,

$$\sum_n p_n(n-6)^2 = \mu_2 \qquad \text{(even)} \tag{7}$$

The entropy is maximized by the Boltzmann distribution

$$p_n = q_n \exp(-\beta(n-6) - \gamma(n-6)^2) \tag{8}$$

where β and γ are Lagrange multipliers enforcing the constraints. (q_n includes the normalization factor.)

Since we are interested in μ_2 and p_6, i.e. in the even moments, let us introduce a fugacity $z = \exp(-\gamma)$ and define an even moment generating function,

$$Z = \sum_{s=1}^{\infty} a_s z^{s^2} = \frac{1 - p_6}{p_6}, \tag{9}$$

with $s = |n - 6|$ and coefficients $a_s = [q_{6+s}\exp(-\beta s) + q_{6-s}\exp(\beta s)]/q_6 > 0$, dependent on the prior probabilities q_n. This is known in statistical physics as a Mayer series in the fugacity z ($Z = \rho = N/V$ for a gas). μ_2/p_6 is given by a second Mayer series, derived from the first (P/kT for a gas),

$$z\frac{\partial Z}{\partial z} = \sum_{s=1}^{\infty} a_s s^2 z^{s^2} = \frac{\mu_2}{p_6}. \tag{10}$$

Reversing the first Mayer series[1] and putting it into the second yields the virial expansion μ_2/p_6 as a series in $(1-p_6)/p_6$ (P/kT as a series in ρ for a gas). For small z, $(1-p_6)/p_6 = a_1 z = \mu_2/p_6$, the virial expansion is limited to the first term (ideal gas law $P/kT = \rho$) regardless of the Mayer coefficients a_s or the priors q_n. At higher z, the Mayer series involves several a_s and the virial expansion should accordingly also involve the priors. It does not.

This is the universality found by Lemaitre et al.

5. Universality of Lemaitre's law. Priors.

We must explain universality arising from series whose coefficients a_s are, at first sight, specific functions of the prior probabilities q_n. The dependence on n of the priors q_n is unknown and arbitrary (your guess is as good as mine), except for two, essential items of information:

i) The a_s deal with even moments in $(n-6)$. Thus, a_1 bins together priors q_5 and q_7, etc., and the odd moment constraint $\langle n \rangle = 6$ is considered as prior information in this context.

ii) $n \geq 3$.

We can now, for the third time in this paper, use maxent inference. (We used it to make constraints redundant and in the derivation of the Boltzmann distribution). Ignorance is a symmetry (whether one takes your guess or mine doesn't affect the physical relation between μ_2 and p_6, or any other macroscopic variable), so that, in the absence of any further information, every coefficient as should have equal weight *a priori*, except for its multiplicity (degeneracy):

$$\begin{aligned} a_1 &= [q_7 \exp(-\beta) + q_5 \exp(\beta)]/q_6 &&= 2 \\ a_2 &= [q_8 \exp(-2\beta) + q_4 \exp(2\beta)]/q_6 &&= 2 \\ a_3 &= [q_9 \exp(-3\beta) + q_5 \exp(3\beta)]/q_6 &&= 2 \\ a_4 &= q_{10} \exp(-4\beta)/q_6 = a_5 = \ldots &&= 1 \end{aligned} \tag{11}$$

(In the statistical mechanics of gases, the phase space (of the mechanical state $(\mathbf{x}, \mathbf{p})$ of an atom) is partitioned into equal volumes $(\Delta x . \Delta p)^3 = h^3$, all equally probable *a priori*, according to Gibbs's postulate).

The generating function Z would be a Gaussian sum (Jacobi's theta function)

$$\sum z^{s^2},$$

except that $a_{s \geq 4} = 1$ instead of 2 (because cells have at least 3 sides, the prior information (ii)) Contributions of $s \geq 4$ to the sum are utterly negligible for z small, γ large. Thus,

$$\frac{1}{2p_6} = \sum_s z^{s^2} + \frac{1}{2}$$

[1] A series $y(x)$ (y as an expansion in powers of x) is reversed, $x(y)$, when it is expressed as x in powers of y

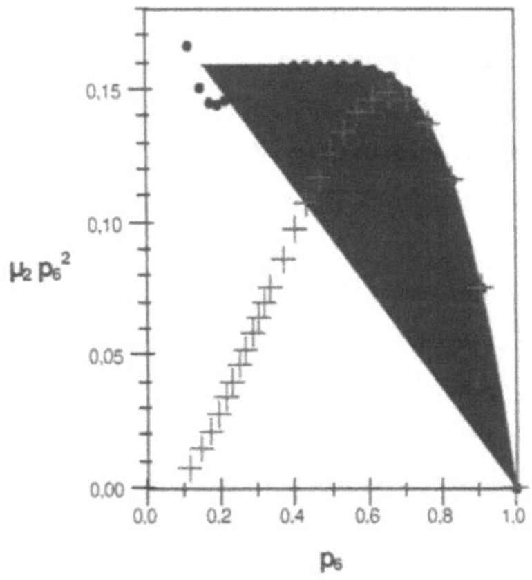

Figure 2. Lemaitre law: $\mu_2 p_6^2$ as a function of p_6. Full points: Maxent theory ($a_{s\leq 3} = 2$, $a_{s\geq 4} = 1$). Line: Jacobi theta function ($a_s = 2$). Crosses: Trivial virial expansion . Lemaitre law (virial equation of state) is given by the Jacobi theta function for $p_6 > 0.34$. It is trivial (ideal gas) for $p_6 > 0.75$. The Mayer series have been computed through $s = 10$.

$$\frac{\mu_2}{2p_6} = \sum_s s^2 z^{s^2} \tag{12}$$

for $0 < z < 0.7$ (to $O(z^{16})$). The Poisson summation formula

$$\sum_{t=-\infty}^{\infty} \exp(-\pi\tau t^2) = \frac{1}{\sqrt{\tau}} \sum_{t=-\infty}^{\infty} \exp(-\pi t^2/\tau), \tag{13}$$

gives an analytic expression for the Gaussian sums (the second is the derivative by $(\pi\tau)$ of the first), now valid for z large, $\gamma = \pi\tau$ small,

$$\begin{aligned} 2(\sum_s z^{s^2} + 1/2) &= \tau^{-\frac{1}{2}} \\ 2(\sum_s s^2 z^{s^2}) &= (\frac{1}{2\pi})\tau^{-\frac{3}{2}}, \end{aligned} \tag{14}$$

to $O(\exp(-\pi^2/\gamma))$, i.e. for $0.25 < z < 1$. In the common range $0.25 < z < 0.7$, we obtain a simple, analytic expression,

$$\mu_2 p_6^2 = \frac{1}{2\pi}, \tag{15}$$

which had already been guessed (as $\mu_2 p_6^2 = 0.15$) by Le Caër and Delannay [13]. For $z < 0.25$, the Gaussian sum reduces to the first term (to $O(z^4)$) and one has the trivial virial equation ($P/kT = \rho$). In summary, here is the universal Lemaitre's law

$$\begin{aligned} \mu_2 &= 1 - p_6 & & 0.66 < p_6 \leq 1 \\ \mu_2 p_6^2 &= 1/(2\pi) & & 0.34 < p_6 < 0.66. \end{aligned} \tag{16}$$

Note the universality ($1/2\pi$), the range and the non-triviality. Known foams departing from the universal curve (Feynman diagrams, random fragmentation, Poisson, random $T1$, Sierpinski foams [14–17,1], all artificially generated) have $p_6 < 0.22$, large μ_2 and dominant contribution of cells with very large n (≥ 10) which are unlikely to be equiprobable *a priori*. Lemaitre's law does indeed control disorder, in a universal fashion.

A final remark. The distribution p_n is not yet fully determined, and not a Gaussian in general. By concentrating on the even moments, we have obtained $p_5 + p_7$, $p_4 + p_8$ and $p_3 + p_9$ binned together. Differences between p_5 and p_7, etc., are free, restricted only by the odd moment constraint $\langle n \rangle = 6$. The distribution has two adjustable parameters, one of which can be its third moment [2]. These two adjustable parameters are not universal. They vary from foam to foam.

References

1. B. Dubertret, N. Rivier, and M. A. Peshkin, "Long-range geometrical correlations in two-dimensional foams," *J. Phys. A: Math. Gen*, **31**, pp. 879–900, 1998. See especially Appendix and §2.
2. B. Dubertret, T. Aste, O. Ohlenbusch, and N. Rivier, "Two-dimensional froths and the dynamics of biological tissues," *Phys. Rev. E*, **58**, 1998. to appear in Nov.
3. J. Lemaitre, A. Gervois, J.-P. Troadec, N. Rivier, M. Ammi, L. Oger, and D. Bideau, "Arrangements of cells in Voronoï tesselations of monosize packings of discs," *Phil. Mag. B*, **67**, pp. 347–362, 1993.
4. N. Rivier, "Maximum entropy for random cellular structures," in *From Statistical Physics to Statistical Inference and Back*, P. Grassberger and J.-P. Nadal, eds., pp. 77–93, Kluwer, 1994.
5. D. Weaire and N. Rivier, "Soap cells and statistics -Random patterns in two dimensions," *Contemp. Phys.*, **25**, pp. 59–99, 1984.
6. N. Rivier, "Order and disorder in packings and froths," in *Disorder and Granular Media*, D. Bideau and A. Hansen, eds., pp. 55–102, North Holland, 1993.
7. N. Rivier, B. Dubertret, and G. Schliecker, "The stationnary state of eptihelial tissues," in *Dynamics of Cell and Tissue Motion*, W. Alt, G. Dunn, and A. Deutsch, eds., pp. 275–282, Birkhaeuser, Basel, 1997.
8. N. Rivier and T. Aste, "Curvature and frustration in cellular systems," *Phil. Trans. R. Soc. Lond. A*, **354**, pp. 2055–2069, 1996.
9. R. Delannay, "Quelques utilisations de la statistique et des probabilités en mécanique (et ailleurs...)," 1996. Mémoire d'habilitation, Nancy, INPL.
10. G. Le Caër and R. Delannay, "Topological models of 2d fractal cellular structures," *J. Phys. I*, **5**, pp. 1417–1429, 1995.
11. T. Aste, D. Boosé, and N. Rivier, "From one bubble to the whole froth: A dynamical map," *Phys. Rev. E*, **53**, pp. 6181–6191, 1996.
12. N. Rivier, "Geometry and evolution of biological tissues," in *Continuum models of discrete systems*, K.-H. Anthony and H.-J. Wagner, eds., pp. 383–392, Trans Tech Publication, 1993.
13. G. Le Caër and R. Delannay, "Correlations in topological models of 2d random cellular structures," *J. Phys. A*, **26**, pp. 3931–3954, 1993.
14. C. Godrèche, I. Kostov, and I. Yekutieli, "Topological correlations in cellular structures and planar graph theory," *Phys. Rev. Lett.*, **69**, pp. 2674–2677, 1992.
15. R. Delannay and G. Le Caër, "Topological characteristics of 2d cellular structures generated by fragmentation," *Phys. Rev. Lett.*, **73**, pp. 1553–1556, 1994.
16. M. Fortes, "Applicability of the Lewis and Aboav-Weaire law to 2d and 3d cellular structures based on poisson partitions," *J. Phys. A: Math Gen.*, **28**, pp. 1055–1068, 1995.
17. M. Peshkin, K. Strandburg, and N. Rivier, "Entropic prediction for cellular networks," *Phys. Rev. Lett.*, **67**, pp. 1803–1806, 1991.

TEXTURE RECONSTRUCTION IN NOISY IMAGES

M. WALESSA AND M. DATCU
Deutsches Zentrum für Luft und Raumfahrt (DLR) e.V.
Deutsches Fernerkundungsdatenzentrum (DFD), Oberpfaffenhofen
D-82234 Weßling †

Abstract. An approach for texture reconstruction in noisy images is presented. The reconstruction is achieved by a denoising algorithm using Gauss Markov random fields as texture models and taking advantage of both the first and the second level of Bayesian inference to obtain a MAP estimate of the noise-free image and an estimate of the model parameters, respectively. For noise-free images exist several techniques for parameter estimation. In the noisy case, however, parameter estimation becomes much more difficult. In order to obtain the best possible texture reconstruction we propose an iterative algorithm to estimate the parameters that provide the highest evidence. This results in an optimization problem where we use the current MAP estimate of the noise-free image to calculate an approximated model evidence for a given set of parameters and adjust these parameters accordingly.

Key words: Speckle noise, synthetic aperture radar (SAR), Gauss Markov random fields, Bayesian inference, evidence, texture, model parameter estimation.

1. Introduction

Synthetic aperture radar (SAR) images are affected by strong speckle noise. Especially in high resolution imagery it is important to preserve textural information while removing speckle noise in order to simplify subsequent image interpretation. To achieve this goal we use a Bayesian approach for image reconstruction relying on texture models to be able to preserve textural features [2]. We calculate the MAP estimate of the noise-free image using adaptive model parameters. These parameters must be estimated taking into account the characteristics of the noise making this a difficult and usually unfeasable task. In this article we present how this problem can be overcome for a particular class of models by introducing several reasonable approximations. We give a short overview of the problem by introducing the likelihood function, the Gauss Markov random field model and how a MAP estimate can be obtained. Then, in the main part of this paper, we

†Email: Marc.Walessa@dlr.de

W. von der Linden et al. (eds.), Maximum Entropy and Bayesian Methods, 65–72.

address the problem of model parameter estimation, outline the used algorithm and provide some images for illustration before concluding.

2. Characteristics of Speckle Noise

The noise-free signal x is affected by speckle noise s which acts in a multiplicative manner. This multiplicative noise is characteristic for all coherent imaging systems [6] and generally is of very high variance compared to the mean image amplitude. As a result, the probability density function for the likelihood of the observed amplitude $y = x \cdot s$ is described by

$$p(y|x) = 2\left(\frac{y}{x}\right)^{2L-1} \frac{L^L}{x\Gamma(L)} \exp\left(-L\left(\frac{y}{x}\right)^2\right), \tag{1}$$

where L denotes the degree of noise given by the so-called equivalent number of looks (ENL), a term used in the SAR community. The highest noise level is observed for $L = 1$ while the noise variance decreases with L growing.

3. Gauss Markov Random Fields as Texture Models

To preserve textures we use a texture model as prior information to despeckle noisy images using a Bayesian approach. Due to its analytical tractability we have chosen the Gauss Markov random field model (GMRF) [11,15] which represents an autoregressive process. Transforming this process into its corresponding probability density function yields the following equation for the stochastic Gauss Markov random field model:

$$p(x|x_\eta, \theta) = \frac{1}{\sqrt{2\pi\sigma^2}} \exp\left(-\frac{(x - \sum_{i\in\eta} \theta_i x_i)^2}{2\sigma^2}\right), \tag{2}$$

where θ_i and σ are model parameters describing textural information and the prediction uncertainty of the model, respectively. By η we denote a neighbourhood system around a central pixel x. The complexity of the model is determined by the size of this neighbourhood system. In the following, we also use the vector notation $\boldsymbol{\theta}$ with its dimension depending on the size of the considered neighbourhood to denote the set of texture parameters θ_i and the scale parameter σ.

4. MAP Estimate

In order to reduce speckle we use the rules of Bayesian inference [7,16], i.e. we try to estimate the noise-free image which best explains the noisy observation assuming some prior information. Here the prior comes in the form of the assumed texture model. Using the Bayes equation at the first level of inference for single pixels, i.e. for local characteristics due to the Gibbs-Markov equivalence [4,17]

$$p(x|y, x_\eta, \boldsymbol{\theta}) = \frac{p(y|x)p(x|x_\eta, \boldsymbol{\theta})}{p(y|x_\eta, \boldsymbol{\theta})} \tag{3}$$

we find the following proportionality (the denominator only serves as a normalization constant in this context but becomes important later)

$$p(x|y, x_\eta, \boldsymbol{\theta}) \propto p(y|x)p(x|x_\eta, \boldsymbol{\theta}) \tag{4}$$

which has to be maximized as a function of x. This can be achieved by setting the first derivative to zero. The logarithm may be used here for simplification since it is a monotonic function not changing the location of any maxima:

$$\frac{\partial}{\partial x} \log p(x|y, x_\eta, \boldsymbol{\theta}) = -\frac{2L}{x} + \frac{2Ly^2}{x^3} - \frac{x - \sum_{i\in\eta} \theta_i x_i}{\sigma^2} = 0\,. \tag{5}$$

For the proposed model and the presented likelihood function we obtain a fourth order polynomial yielding four solutions for x_{MAP} depending on the model parameter vector $\boldsymbol{\theta}$ and the current neighbourhood configuration η of x:

$$x^4 - x^3 \sum_{i\in\eta} \theta_i x_i + 2L\sigma^2 x^2 - 2L\sigma^2 y^2 = 0\,. \tag{6}$$

Finally, a case study between the four solutions provides the desired result for x_{MAP}. Because of their complexity equations are not given here. However, due to inter-dependencies with neighbouring pixels x_i via $\boldsymbol{\theta}$ an iterative algorithm still has to be applied for relaxation. We obtained good results using a few iterations of a simple ICM algorithm.

5. Model Parameter Estimation under Noise

Being now able to calculate a MAP estimate of the noise-free signal for given parameters, a crucial problem still remains: How to choose the model parameter vector $\boldsymbol{\theta}$? The noise level L is generally known. Here, we take advantage of the second level of Bayesian inference to estimate the set of parameters that best explain the noisy image y. Using the vector notation to consider a set of data within a window, we obtain the global Bayes equation compared to the local one of (3)

$$p(\boldsymbol{\theta}|\mathbf{y}) = \frac{p(\mathbf{y}|\boldsymbol{\theta})p(\boldsymbol{\theta})}{p(\mathbf{y})}, \tag{7}$$

which again, assuming a uniform prior $p(\boldsymbol{\theta})$, may be reduced to the following proportionality where the integral has to be performed over the whole space of $\mathbf{x}$:

$$p(\boldsymbol{\theta}|\mathbf{y}) \propto p(\mathbf{y}|\boldsymbol{\theta}) = \int p(\mathbf{y}|\mathbf{x})p(\mathbf{x}|\boldsymbol{\theta})d\mathbf{x} \tag{8}$$

We note that the right-hand side of this equation is identical to the normalization constant neglected in (4). This term is called evidence and reflects the probability of the data y given an assumed model characterized by $\boldsymbol{\theta}$. The task now consists in maximizing this evidence as a function of $\boldsymbol{\theta}$, i.e. find the most plausible parameter vector. Unfortunately, this maximization is generally a quite difficult task, especially for non-gaussian multi-dimensional functions, as they normally occur in image processing. A closed analytical expression for the integral over the posterior $p(\mathbf{y}|\mathbf{x},\boldsymbol{\theta})p(\mathbf{x}|\boldsymbol{\theta})$ is often impossible to be obtained, so several approximations are needed to make the problem tractable again, unless a MCMC means [9,14] is used:

- As a first approximation, we consider the integrand of (8) to consist of mutually independent processes, breaking the joint probability density functions into the products of their components. Of course, this statistical independence is not given but has been shown to be a good approximation for large N, i.e. number of pixels within a window, as it is similar to the maximum pseudo-likelihood approach presented in [8].
- Furthermore, we approximate the resulting multidimensional probability density function by a multivariate Gaussian distribution centred around the MAP estimate of $\mathbf{x}$, i.e. around the maximum of the posterior distribution [10,16]. This can be shown to be a good approximation for the used likelihood function (1) together with the Gauss Markov random field model (2).

Hence, we find

$$\begin{aligned} p(\mathbf{y}|\mathbf{x})p(\mathbf{x}|\boldsymbol{\theta}) &\approx \prod_{i=1}^{N} p(y_i|x_i)p(x_i|x_{i_\eta},\boldsymbol{\theta}) \qquad (9)\\ &\approx \prod_{i=1}^{N} p(y_i|x_{i_{MAP}})p(x_{i_{MAP}}|x_{i_\eta},\boldsymbol{\theta}) \exp\left(-\frac{1}{2}\boldsymbol{\Delta}\mathbf{x}^T\mathbf{H}\boldsymbol{\Delta}\mathbf{x}\right), \end{aligned}$$

where $x_{i_{MAP}}$ is the MAP estimate of x_i obtained using a fixed parameter vector $\boldsymbol{\theta}$, $\boldsymbol{\Delta}\mathbf{x} = \mathbf{x} - \mathbf{x}_{\mathbf{MAP}}$ and $\mathbf{H}$ is the Hessian matrix given by

$$\begin{aligned} \mathbf{H} &= -\boldsymbol{\nabla}\boldsymbol{\nabla} \log \prod_{i=1}^{N} p(y_i|x_i)p(x_i|x_{i_\eta},\boldsymbol{\theta}) \qquad (10)\\ &= -\boldsymbol{\nabla}\boldsymbol{\nabla} \sum_{i=1}^{N} \log(p(y_i|x_i)p(x_i|x_{i_\eta},\boldsymbol{\theta})). \end{aligned}$$

Applying these simplifications, we are now able to perform an integration of the approximating resulting function being described in the following way:

$$\begin{aligned} p(\mathbf{y}|\boldsymbol{\theta}) &= \int p(\mathbf{y}|\mathbf{x})p(\mathbf{x}|\boldsymbol{\theta})d\mathbf{x} \qquad (11)\\ &\approx \int \prod_{i=1}^{N} p(y_i|x_{i_{MAP}})p(x_{i_{MAP}}|x_{i_\eta},\boldsymbol{\theta}) \exp\left(-\frac{1}{2}\boldsymbol{\Delta}\mathbf{x}^T\mathbf{H}\boldsymbol{\Delta}\mathbf{x}\right) d\mathbf{x}\\ &\approx \frac{(2\pi)^{\frac{N}{2}}}{\sqrt{|\mathbf{H}|}} \prod_{i=1}^{N} p(y_i|x_{i_{MAP}})p(x_{i_{MAP}}|x_{i_\eta},\boldsymbol{\theta}). \end{aligned}$$

Finally, since we prefer to use the logarithmic form of $p(\mathbf{y}|\boldsymbol{\theta})$ for numerical reasons, this can be expressed as:

$$\log p(\mathbf{y}|\boldsymbol{\theta}) \approx \frac{1}{2}(N \log 2\pi - \log|\mathbf{H}|) + \qquad (12)$$

$$\sum_{i=1}^{N}(\log p(y_i|x_{i_{MAP}}) + \log p(x_{i_{MAP}}|x_{i_\eta}, \boldsymbol{\theta}))$$

$$\approx \sum_{i=1}^{N}\left(\frac{1}{2}(\log 2\pi - \log h_{ii}) + \log p(y_i|x_{i_{MAP}}) + \log p(x_{i_{MAP}}|x_{i_\eta}, \boldsymbol{\theta})\right),$$

where another approximation was made in the final step:

$$|\mathbf{H}| \approx \prod_{i=1}^{N} h_{ii}\,. \tag{13}$$

This approximation is assumed to be valid since it again implies that all covariances resulting from this sparsely set matrix are zero (off-main-diagonal values are neglected) being in accordance with the already made inherent assumption of statistical independence in (9). Moreover, this simplifies practical calculations preventing us from the computation of determinants of dimension $N \times N$ which are in the order of 1024×1024 in our case. In this way, only the components on the main-diagonal of the matrix $\mathbf{H}$ which are found to be given by

$$\begin{aligned} h_{ii} &= -\frac{\partial^2}{\partial x_i^2}\sum_{j=1}^{N}\left(-2L\log x_j - \frac{Ly_j}{x_j} - \frac{(x_j - \sum_{k\in\eta_j}\theta_k x_k)^2}{2\sigma^2}\right) \\ &= \frac{6Ly_i^2}{x_i^4} - \frac{2L}{x_i^2} + \frac{1}{\sigma^2}\left(1 + \sum_{j\in\eta_i}\theta_j^2\right) \end{aligned} \tag{14}$$

are needed.

6. Implementation and Examples

We now briefly describe the texture reconstruction and parameter estimation technique and present some obtained results: In our implementation the image to be reconstructed is scanned using windows, either overlapping or non-overlapping, of fixed size. In each window the most evident model parameters are determined by maximizing $\log p(\mathbf{y}|\boldsymbol{\theta})$ by means of a gradient search as a function of $\boldsymbol{\theta}$. This parameter vector is used to calculate the corresponding MAP estimate of the noise-free data within a window as described above.

For test purposes we present two examples showing a pure texture image from the Brodatz album and a textured SAR image also comprising non-textured areas. The complexity of the employed model was reduced to the second order neighbourhood, i.e. to the eight adjacent pixels of a centre pixel but can easily be extended. As quality measures we used the mean squared error (MSE) when the noise-free image was available and the cross-entropy [1,2]

$$H_c = \int p(s)\log\frac{p(s)}{q(s)}ds \tag{15}$$

between the true speckle statistics $p(s)$ and the statistics $q(s)$ of the speckle obtained from the MAP estimate via $s = y/x_{MAP}$. The cross-entropy describes the divergence between these two distributions with $H_c = 0$ for $q(s) = p(s)$. We considered only one-dimensional distributions neglecting spatial correlations between neighbouring pixels of the extracted speckle noise.

EXAMPLE 1

In the first example shown in Fig. 1 we have applied the algorithm to a homogeneously textured image to demonstrate its performance for one-look speckle noise ($L = 1$). The reconstruction highly preserves textural information and is able to generate an image visually difficult to distinguish from the original with a mean squared error of $MSE = 214$. The experiment was repeated for a commonly used speckle filter giving visually much less attractive results and a mean squared error of $MSE = 509$. Concerning the cross-entropies we obtained values of $H_c = 0.048$ for the texture reconstruction and $H_c = 0.081$ for the normal filter, respectively.

Figure 1. Left: Original texture from the Brodatz album. Centre: Original texture affected by speckle noise. Right: Reconstructed texture using the presented algorithm (MSE=214 and $H_c = 0.048$ compared to MSE=509 and $H_c = 0.081$ for a normal speckle filter).

EXAMPLE 2

The second example of Fig. 2 shows a high-resolution SAR image from an airborne sensor containing both highly textured areas and areas of uniform backscatter. For the latter another more suitable δ-model $p(x|\theta_m) = \delta(x - \theta_m)$ with one free parameter θ_m was added. The algorithm was extended to choose among the two models the one with the highest evidence. As quality measures we obtained $H_{c_t} = 0.045$ and $H_{c_u} = 0.037$ for the model based reconstruction and $H_{c_t} = 0.126$ and $H_{c_u} = 0.070$ for conventional filtering. Here we have distinguished between textured and uniform areas denoted by the indices t and u, respectively. It results that we achieve very high smoothing for uniform areas where the δ-model is valid while the positive properties of the texture model are exploited in areas of higher variance preserving very well textural information.

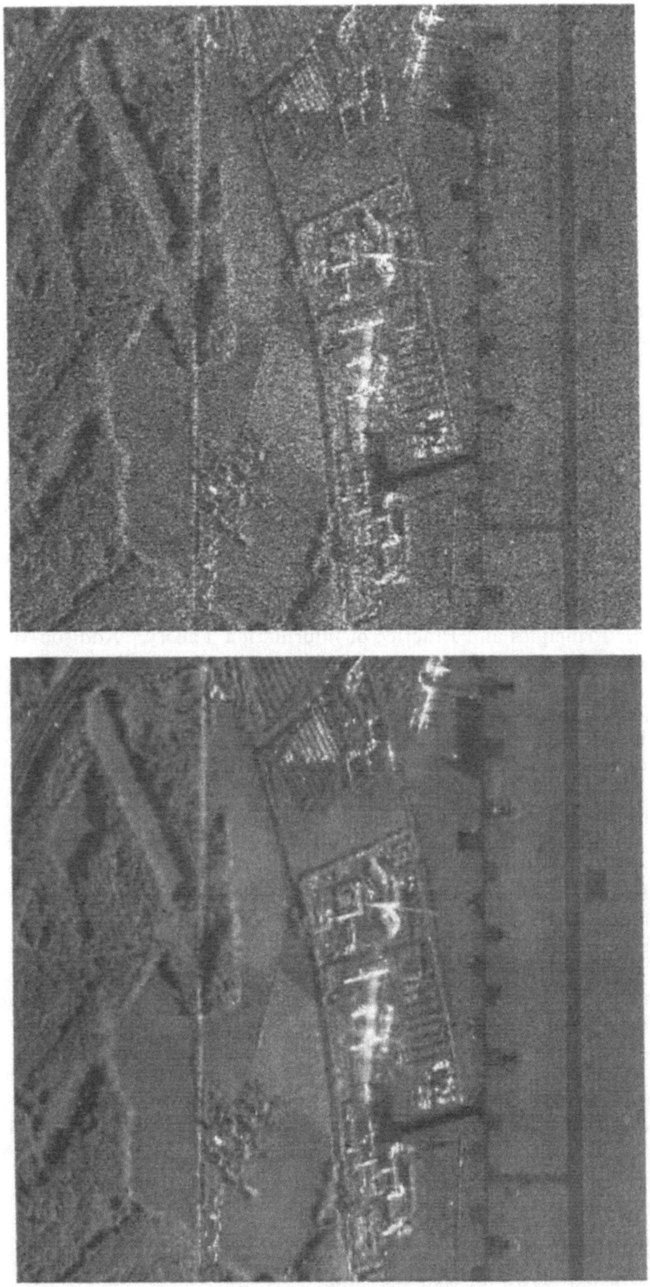

Figure 2. Top: Original SAR image with textured areas such as forest, buildings and parking lots and uniform areas such as agricultural areas and airfield. Bottom: Despeckled image using the described algorithm extended by a δ-model for uniform areas ($H_{c_t} = 0.045$, $H_{c_u} = 0.037$ compared to $H_{c_t} = 0.126$ and $H_{c_u} = 0.070$ for a common speckle filter).

7. Conclusions

We have presented a way to estimate model parameters in noisy conditions. To achieve this goal we have seen that several approximations are needed to make this optimization problem computationally feasible. Nevertheless, the resulting texture-preserving despeckling approach is still computationally demanding, especially compared to commonly used algorithms which are rather fast. On the other hand, the obtained results are of much higher quality for highly textured images, i.e. high-resolution imagery. An extension of this concept by using several models and choosing the one with the highest evidence was demonstrated and gave promising results. Further work will concentrate on the use of a library of models including more complex, non-linear ones.

Acknowledgments

We like to thank Aerosensing GmbH for providing the high-resolution SAR image used in our experiments. The texture image was taken from the Brodatz album.

References

1. R.E. Blahut, "Principles and Practice of Information Theory," Addison-Wesley.
2. M. Datcu,K. Seidel, M. Walessa, "Spatial Information Retrieval From Remote Sensing Images: Part A. Information Theoretical Perspective," IEEE Trans. Geosci. Remote Sensing, Sept. 1998, to be published.
3. H. Derin, H. Elliott, "Modeling and Segmentation of Noisy and Textured Images using Gibbs Random Fields," IEEE Trans. PAMI, vol. PAMI-9, pp. 39-55, Jan. 1987.
4. S. Geman, D. Geman, "Stochastic Relaxation, Gibbs Distributions and the Bayesian Restoration of Images," IEEE Trans. PAMI, vol. 6, No.6, pp. 721-741, Nov. 1984.
5. D. Geman, "Random Fields and Inverse Problems in Imaging," Lecture Notes in Mathematics, Springer-Verlag, 1988, pp. 117-193.
6. J.W. Goodman, "Statistical Properties of Laser Speckle Patterns," in Laser Speckle and related Phenomena, Edit. J.C. Dainty, Springer, 1975.
7. E.T. Jaynes, "Probability Theory: The Logic of Science," in preparation, ftp://bayes/wustl.edu/Jaynes.book.
8. S. Lakshmanan, H. Derin, "Simultaneous Parameter Estimation and Segmentation of Gibbs Random Fields Using Simulated Annealing," IEEE Trans. PAMI, vol. 11, No. 8, pp. 799-813, Aug. 1989.
9. W. von der Linden, R. Preuss, V. Dose, "The Prior-Predictive Value: A Paradigm of nasty multidimensional Integrals," in Proceedings of the MaxEnt Workshop 1998.
10. D.J.C. MacKay, "Bayesian Interpolation," Neural Computation, Vol. 43, 1992.
11. B.S. Manjunath, R. Chellappa, "Unsupervised Texture Segmentation using Markov Random Field Models," IEEE Trans. PAMI, vol. 13, No. 5, pp. 478-482, May 1991.
12. C.J. Oliver, "Information from SAR Images," J. Phys. D: Appl. Phys. 24, pp. 1493-1514, 1991.
13. C.J. Oliver, A. Blake, R.G. White, "Optimum Texture Analysis of Synthetic Aperture Radar Images," SPIE vol. 2230, pp. 389-398, 1994.
14. J. Skilling, "Annealing and Ensembles," in Proceedings of the MaxEnt Workshop 1998.
15. A.H. Schistad Solberg, A.K. Jain, "Texture Fusion and Feature Selection applied to SAR Imagery," IEEE Trans. Geosci. Remote Sensing, vol. 35, No. 2, pp. 475-479, March 1997.
16. D.S. Sivia, "Data Analysis: A Bayesian Tutorial," Clarendon Press, Oxford, 1996.
17. G. Winkler, "Image Analysis, Random Fields and Dynamic Monte Carlo Methods: A mathematical Introduction,", Springer, 1995.

BAYESIAN 3D X-RAY RECONSTRUCTION FROM INCOMPLETE NOISY DATA

V. VENGRINOVICH AND YU. DENKEVICH
Institute of Applied Physics
Akademicheskaya 16, Minsk 220072, Belarus [‡]

AND

G.-R. TILLACK
Federal Institute for Materials Research and Testing (BAM)
Unter den Eichen 87, 12205 Berlin, Germany [§]

Abstract. Here the 3D X-Ray Bayesian reconstruction (BR) with Gibbs priors (GP) is considered in the form of quadratic functional (QF) representation of two mechanical and one statistical model being applied to the problem of image restoration from strongly incomplete noisy data. The BR with GP in the form of mechanical models as priors can result in an image restoration using 10-100 times less number of projections compared to cone-beam Computer Tomography. The quality of reconstruction as well as the computing time are shown to be strongly dependent upon the form of the a priory model and the noise level. Additionally a new prior is introduced in terms of a statistical model. Results are presented investigating the influence of the applied GP form and the noise level on quality of the reconstruction procedure for which quantitative estimates are provided. Restoration capabilities are compared for three types of models: cluster, plane and phase support algorithm. The influence of the following variables involved in the reconstruction procedure of binary and three level object are investigated: (*i*) the number of cone-beam projections up to 36, (*ii*) the level of Gaussian distributed white noise imposed on the 2D projections having the variance up to the value four times larger than the grey level resulted from projecting of one defect voxel, (*iii*) the form of the prior constraint, (*iv*) the value of the regularization parameter, and (*v*) the zero-level approximation before starting the minimization procedure. The error of the reconstruction and the computing time are choosen to be the final estimators of the capability of the method in all cases. Finally, conclusions are drawn about the potential of the method for multi-step BR with GP.

Key words: Bayesian reconstruction, Gibbs prior

[‡]Email: veng@iaph.bas-net.by
[§]Email: Gerd-Ruediger.Tillack@bam.de

W. von der Linden et al. (eds.), Maximum Entropy and Bayesian Methods, 73–82.

1. Introduction

In the problem of image and data restoration, the Maximum-Entropy Method (MEM) is directed to the solution of operator-type equation $P = \mathbf{A}f$ with regard to the function f. It stands apart from classical regularization methods in the sense of its self-sufficiency for introducing the *a priori* knowledge originating from Boltzman statistics [1]. The price which should be paid for this self-sufficiency is the non-linearity of the variational procedures and excessive commonality of the entropy expression in the form of $\mu \ln \mu$, which does not take into consideration the structural peculiarities of the object under restoration. This specifically leads to the known difficulties in computerized restoration of e.g. 3D images from 2D data set making the final solution actually unattainable.

As mentioned, in MEM the "prior" is based on Boltzman statistics, which assumes that each luminescence quantum has an equal probability of being in any cell of the space under consideration. Together with the condition of the constant number of all quanta participating in the event, this hypothesis is sufficient to justify the expression for the entropy of the system splitted into equal cells. While maximizing the entropy, having given experimental data, one supports the smoothness and the posivity of the image, without the necessity of additional deterministic constraints. Hence, the nonlinearity and neglect of the correlation between neighboring cells are main objectives of the prior introduced by MEM.

Gibbs statistics and Gibbs measures have been successfully applied to the Ising problem and introduced in inverse theory by Besag [2], particularly for the 2D and 3D image restoration, which made it possible to overcome the difficulties caused by Boltzman statistics. This theory assigns to the considered 2D or 3D lattices some Gibbs "energies" or Gibbs measures which expressions are completely defined by the assumed interactions between neighboring cells in the lattices. Some applications of this approach to the solution of inverse problems in non-destructive evaluation are considered in [3]. In the series of articles [4,5] we have analysed more deeply this approach being applied to the problem of 3D image reconstruction from incomplete noisy 2D data. The present report summarizes these results and gives new evidence for the efficiency of this method.

2. General Considerations

Let's assume cubic voxels, assigning to their geometrical centers the values of the attenuation coefficient μ_j. This gives a 3D lattice with different values in M nodes. For such a lattice a potential V_k can be assigned as the nearest neighbour Gibbs potential, similar to that for Markov fields [6]. For this field, points and adjacent points contribute to the value of the potential V_k and therefore to the energy, and these are the only cliques other than the empty set which are accounted. The energy $U(\boldsymbol{\mu})$ of the configuration $\boldsymbol{\mu}$ is defined by the sum of potentials:

$$U(\boldsymbol{\mu}) = \sum_k V_k(\boldsymbol{\mu}), \tag{1}$$

where the sum is taken over all cliques in the lattice for fixed $\boldsymbol{\mu}$.

The Gibbs measure, which is the probability measure for the particular configuration, is thus given by

$$P(\boldsymbol{\mu}) = \frac{e^{-U(\boldsymbol{\mu})}}{Z} = \frac{1}{Z}\prod_k e^{-V_k(\boldsymbol{\mu})}, \tag{2}$$

where the normalization constant Z is a Gibbs measure for all configurations:

$$Z = \sum_{\boldsymbol{\mu}} e^{-U(\boldsymbol{\mu})}. \tag{3}$$

The Gibbs measure 2 of the potential lattice can be considered as a nontrivial prior in the Bayes formula for the conditional probability, applied to the problem of image restoration:

$$P(\boldsymbol{\mu}|\boldsymbol{p}^m, T) = P(\boldsymbol{p}^m|\boldsymbol{\mu}, T)\,\frac{P(\boldsymbol{\mu}|T)}{P(\boldsymbol{p}|T)}, \tag{4}$$

where $P(\boldsymbol{p}^m|\boldsymbol{\mu}, T)$ is the probability of the data $\boldsymbol{p}^m$, given the hypothesis $\boldsymbol{\mu}$ and the distortion T (this term will be omitted below but implied). This probability proceeds exactly as in the maximum likelihood method for Gaussian errors

$$P(\boldsymbol{p}^m|\boldsymbol{\mu}) = \chi^2 = \sum_{n=1}^{N}\sum_{i=1}^{I_n} \frac{\left(p_{i,n}^c - p_{i,n}^m\right)^2}{\sigma_i^2}, \tag{5}$$

where χ^2 measures how close, in a least-square sense, are the measured and calculated data $\boldsymbol{p}^m$ and $\boldsymbol{p}^c$, correspondingly. N is the number of projections and I_n is the number of pixels in projection n.

Hence, the final formula which controls the variational procedure is:

$$\text{minimize}: \quad -\ln\left[\mathrm{P}(\boldsymbol{\mu}|\mathrm{p}^{\mathrm{m}})\right] = \frac{1}{2}\chi^2 + \alpha\sum_{\mathrm{k}} \mathrm{V_k}(\boldsymbol{\mu}) = \frac{1}{2}\chi^2 + \alpha \mathrm{U}(\boldsymbol{\mu}), \tag{6}$$

where α incorporates all normalizing constraints we omitted and some ability of compromise between entropy solution (default model) and data constraints χ^2.

It is seen that superseding the functional $B(\boldsymbol{\mu})$ in the form of Gibbs measure 2 ensures the linearity of the minimization procedure (if $V(\boldsymbol{\mu})$ is presented in the square form), simplifies the iteration procedure, and naturally provides the support of any expected feature in the image. The price for this is that the *a priori* information is introduced in a more biased, but quite natural form.

3. Multi Step Reconstruction (MSR) Technique and the Forms of Potentials $V_k(\boldsymbol{\mu})$ for 3D Lattices

MSR concept has its aim to provide fast convergence of the solution given by Maximum Entropy Method (MEM) to the most probable one. This is reached by introducing in the first step of reconstruction a zero level approximation of attenuation coefficients spacial distribution and by selecting prior information in the form of Gibbs energy. The first step was considered as a way to get a connected set of solutions D, admissible in the Virtual Defects Space (VDS) which is defined by

$$\mathrm{VDS} = \left\{ j : \sum_{n=1}^{N} \eta\left(x\right) \frac{p_n^j}{\epsilon_n^j} \geq N_\epsilon \right\}, \tag{7}$$

where p_n^j is the gray level in the pixel, which corresponds to voxel j on projection n assuming that the corresponding beam passes through the center of voxel j, ϵ_n^j is the available estimation of the error in determining p_n^j, $\eta\left(x\right)$ is the threshold function

$$\eta\left(x\right) = \begin{cases} 1 & \text{for} \quad x > 1 \\ 0 & \text{for} \quad x \leq 1 \end{cases}, \tag{8}$$

and N_ϵ is a number which satisfies the condition: $1 \leq N_\epsilon \leq N$.

According to eq. 7 a voxel is assigned to be part of VDS if the gray levels in the corresponding pixels exceed the error in N_ϵ projections. Unlike the procedure, eq. 8 places among VDS those voxels which remain meaningful at least at one from N projections. The nature of the errors ϵ_n^j can be diverse. It includes random noise, accuracy of readings, as well as systematic error caused by uncompensated shadows, etc.

For the case when incomplete number of projections and limited angle of observation are intensified by the incomplete data set on the projections (or the 2D image is overlapped by strong noise), it can be shown that N_ϵ can be selected to be equal 1.

For the solution of eq. 3 we will exploit the iteration procedure, for which the zero level approximation $\mu^{(0)}$ is needed. We consider this step as the first step reconstruction which can be provided by searching VDS.

At the second step of the image restoration the BR with GP in the forms, which will be considered below, is implemented. It provides the minimization of the functional M^α which is a sum of two functionals weighted by α: A - *a posteriori* and B - *a priori* (see [6]).

Like the forms of the potentials mechanical models are used. Two of them, namely cluster support algorithm (CSA) and plane support algorithm (PSA), were described in details in [3,4]. Investigations with simulated and experimental data have shown that the iteration procedure yields the sweeping of the structures which are not volumetric-like or thin surface like, correspondingly. While the number of required projections for the reconstruction is reduced by 10 - 100 times, the quality of reconstruction, estimated quantitatively, remained quite comparative (sometimes even with less artefacts) with results obtained by classic Computer Tomography (CT).

To illustrate the efficiency of CSA included in the MSR procedure developed in [4,5], results are shown of its application to restore the 3D image of a structure made from a set of different metallic articles like wire skein, rings and stars with inner and outer teeth, inserted into a plastic matrix (fig. 1). Four X-ray projections

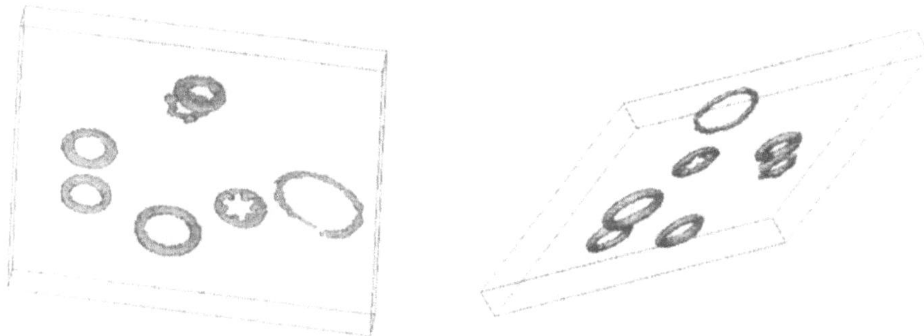

Figure 1. Two spatial views of the restored image of the object using 4 projections and CSA

were produced using digital laminography equipment[1]. It is seen that even the image of the wire having diameter of 0.5 mm is restored. The number of used projections was reduced more then hundred times in comparison with CT.

Often the *a priori* knowledge about the structures under restoration consists of the knowledge that it contains two or more different materials or structures of one material which we will define as phases. Then, the problem of phase segmentation having measured data is quite actual. To explain the mathematical formulation of this prior let's consider the following Gibbs energy:

$$U(\mu) = \sum_{r=1}^{R} (m_r - m_{r-1})^2 \sum_{j} \sum_{a=1}^{26} \left(\mu'_{j,r,a} - \mu'_{j,r}\right) + \beta \mu'^{2}_{j,r} \left(\mu'_{j,r} - 1\right)^2 \quad (9)$$

$$\text{with } \mu'_{j,r} = \frac{\mu_j - m_{r-1}}{m_r - m_{r-1}},$$

where m_r is the r-th level of attenuation coefficients in the structure with $m_0 = 0$. The subscript a is assigned to the voxel adjacent to voxel j, and β is constant. In eq. 9, the smoothing constraint, given by the first term in square brackets, is formulated in the simplest form of neighborhood potential, related to the square of the discrete first space derivative of the attenuation coefficient in voxel j relatively to all surrounding 26 voxels in the 3D lattice. The second term in brackets obviously has two minima at which the values of attenuation coefficients in the voxels equal to that inherent to the phases. At the same time the boundary transition is pushed from one that is smooth to one that the jumps from voxel to voxel are uniformly distributed.

While minimizing $B(\mu)$ this second term pushes the solution in the direction of subdividing the phases or structures by smooth boundaries. Affecting simultaneously these two parts support in the weighted form both, the subdivision of the phases and the constant value in each one. It is quite understandable that during the iteration procedure this process is in competition with the experimental data

[1]The experiments were performed by U. Ewert and B. Redmer in the laboratory VIII.31 Radiation Techniques, Federal Institute of Material Research and Testing (BAM), Berlin, Germany

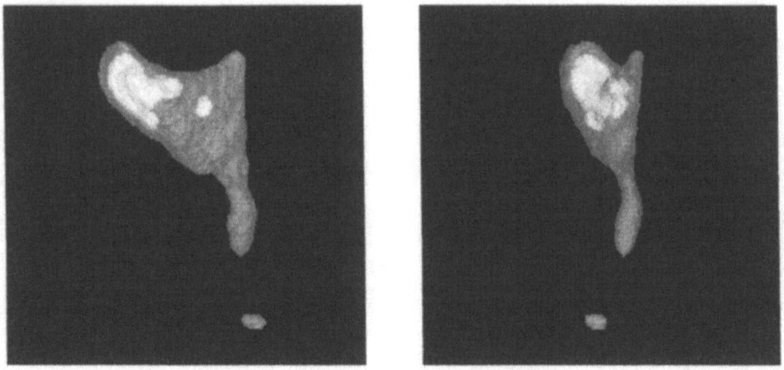

Figure 2. Two views of the restored image of complex structure (porosity and slag inclusion in ferritic weld) applying the phase support algorithm 9 (PhSA)

at any iteration. Fig. 2 a and b show the results of reconstruction of three phases structure (porosity and slag inclusion in the ferritic weld) using 10 projections[2] processed by MEM with *a priori* functional in the form 9. The quality of reconstruction is comparable to that obtained by CT from 720 projections data.

4. Simulated Object Superimposed by Noise and Noise Influence

To investigate the noise influence, we simulated the phantom shown in fig. 3, consisted of two artificial crack-like and three volumetric flaws, two of them having one voxel gap between to check the resolution of the reconstruction procedure. Simulated X-ray projections were computed using the technique described in [4]. The point source was placed at up to 36 (namely 4, 8, 16 and 36) equidistant positions on a circle parallel to the detector plane. The detector was split into 400×400 pixels, with size 45×45 μm.

The Gaussian white noise with zero mean and rms variation n which varied from zero to 4 (every unit of the noise variance value corresponds to the intensity in the detector caused by X-ray passing through one voxel belonging to the detector) was superimposed on the image.

The threshold level in the first step was selected with respect to the number of projections, particularly 4, 7, 13, and 29 for 4, 8, 16 and 36 projections, respectively. The noise in the projections resulted in appearance of a variety of separated agregates imposed on the restored VDS, but the last weakly influenced the quality of final reconstruction as they were smoothed by the prior support. A simple $3 \times 3 \times 3$ median filter after first step facilitated this process.

It is seen that the increasing noise subsequently "contaminates" the VDS but the second step provides the reconstruction with the accuracy improved by the prior support given by PhSA.

[2]The experiments were performed by J. Goebbels and H. Riesermeier in the laboratory of Computer Tomography, Federal Institute of Material Research and Testing (BAM), Berlin, Germany

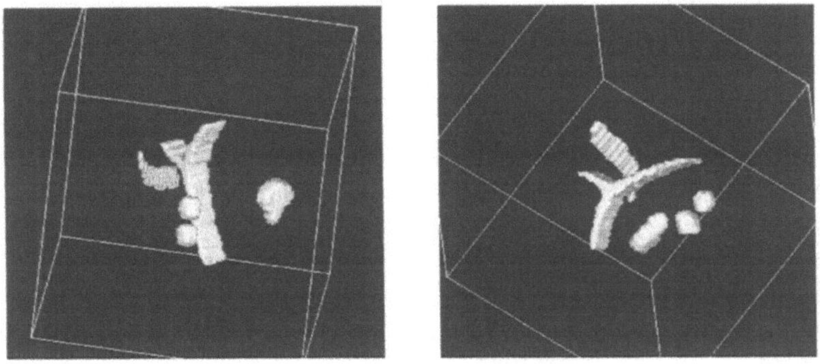

Figure 3. Two views of a simulated object

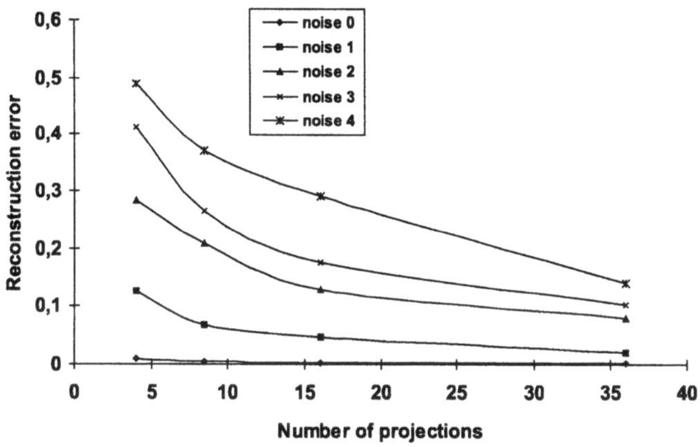

Figure 4. Reconstruction error via the number of projections for different noise level using PhSA

The influence of number of projections and the noise level on the error of reconstruction using PhSA can be observed in fig. 4 where noise was imposed on the projections after first step.

It is seen that progressive increasing of the noise level requires an increasing number of projections to preserve the same quality of reconstruction, while for comparatively moderate noise, 16 and even 8 projections seem to be sufficient to stabilize the error.

5. Influence of the Parameter α on the Error of Reconstruction from Noisy Data

The parameter α, as well known in inverse theory, defines the weight of default model in the restored image. The error of reconstruction as function of the parameter α for three supports, namely CSA, PSA and PhSA, is shown in fig. 5 a-c,

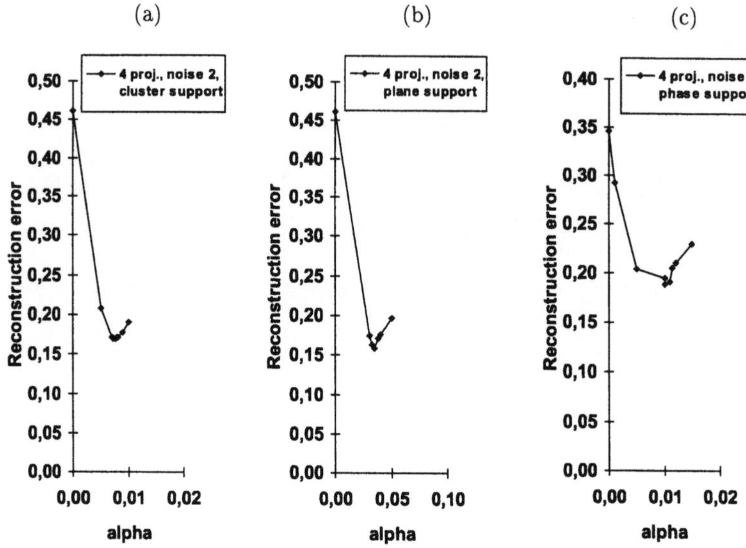

Figure 5. Reconstruction error via the parameter alpha for (a) CSA, (b) PSA, and (c) PhSA

respectively. For all algorithms the distinct minima in the error is observed, provided α_{min} values of different levels. The values for the errors provided by all three algorithms are quite close. All previous and successive results of reconstruction were accessed by using optimal α-values.

The number of iterations influences the computing time. It was shown that the PhSA has obvious advantages over both other algorithms considering also computer time advantage more then 3 times.

6. Conclusions

Application of Bayesian reconstruction and multi-step reconstruction is advantageous for the image restoration from strongly incomplete noisy data.

Among three examined algorithms, particularly cluster, plane or phase support, the last one has distinct advantages in providing substantial decreasing of the time required for computations and number of necessary iterations, while their noise-suppressing capabilities are very closed to one another. The optimal α-value for the PhSA is 4-7 times smaller then for CSA and PhSA which is the additional evidence of its efficiency.

Increasing the number of projections to stabilize the quality of reconstruction is appropriate to some limit which also grows with increasing the noise level.

The procedure for optimizing of parameter α-value is a necessary part of the MSR technique to provide the minimization of the reconstruction error.

Acknowledgement

The authors deeply appreciate the Ministry of Economics of Germany and the Belarus National Academy of Sciences, which supported this work.

References

1. W. von der Linden, "Maximum-entropy data analysis," *J. Appl. Phys.*, **A60**, pp. 155–165, 1995.
2. J. Besag, "Spacial interaction and the statistical analysis of lattice systems," *J. R. Statist. Soc.*, **B36**, pp. 192–236, 1974.
3. V. Vengrinovich, Y. Denkevich, and G.-R. Tillack, "Limited projections and views bayesian 3d reconstruction using gibbs prior," in *Proceedings of the 7th European Conference on Nondestructive Testing*, in print, 1998.
4. V. Vengrinovich, Y. Denkevich, G.-R. Tillack, and C. Nockemann, "Multi step 3d x-ray tomography for a limited number of projections and views," in *Review of Progress in Quantitative Nondestructive Evaluation*, D. O. Thompson and D. E. Chimenti, eds., vol. 16, (N. Y.), pp. 317–323, Plenum Press, 1997.
5. V. Vengrinovich, Y. Denkevich, and G.-R. Tillack, "Limited projection 3d x-ray tomography using the maximum entropy method," in *Review of Progress in Quantitative Nondestructive Evaluation*, D. O. Thompson and D. E. Chimenti, eds., vol. 17, (N. Y.), pp. 403–410, Plenum Press, 1998.
6. R. Kindermann and J. L. Snell, *Markov Fields and Their Applications*, vol. 1 of *Contemporary Mathematics*, American Mathematical Society, 1980.

Arnold Zellner and Richard Silver.

A BAYESIAN REFLECTION ON SURFACES

The multiresolution inference of continuous-basis fields

DAVID R. WOLF
NASA Ames Research Center
MS 269-2 Moffet Field
CA 94035-1000†

Abstract. The topic of this paper is a novel continuous-basis field representation and inference framework applied to the inference of continuous surfaces from measurements (for example camera image data). Traditional approaches to surface representation and inference are briefly reviewed. The new field representation and inference paradigm is then introduced within a maximally informative (MI) (see [1]) inference framework. The knowledge representation is introduced and discussed in the context of MI inference. Then, using the MI inference approach, the here-named Generalized Kalman Filter (GKF) equations are derived. The GKF equations allow the update of field knowledge from previous knowledge *at any scale*, and new data, to new knowledge *at any other scale*. The GKF equations motivate a location-dependent scale or multigrid approach to the MI inference of continuous-basis fields.

Several problems are uniquely solved: The MI inference of fields, where the basis for the field is itself a continuous object and generally is not representable in a finite manner; the tradeoff between accuracy of representation in terms of information learned, and memory or storage capacity in bits; the approximation of probability distributions so that a maximal amount of information about the object being inferred is preserved by the approximation.

Key words: Field, inference, manifold, surface, Bayesian inference, Kalman Filter, multigrid, maximally informative inference, stochastic process, adaptive scale inference, knowledge representation, minimum description length

1. Introduction

1.1. TRADITIONAL SURFACE REPRESENTATION METHODS

Many methods for representing surfaces have been utilized previously, however all of these methods involve representing the surface by a *discrete* field, perhaps with a deterministic interpolation defined (bi-linear, tensor B-splines, etc.) to provide

†Email: greywolf@ptolemy.arc.nasa.gov, Web: http://dino.ph.utexas.edu/ ˜ wolf

W. von der Linden et al. (eds.), Maximum Entropy and Bayesian Methods, 83–90.

a definition for the surface at points intermediate to the discrete field. Probability distributions or densities of these discrete fields then often take the form of normalized exponentials of sums of clique energy functions, and produce a construct commonly known as a Markov Random Field. (See Geman [2], for the most often cited example.) There are several immediate observations on these approaches:

- The surface remains unspecified at points intermediate to the discrete field, except by the often undefined notion of interpolation.
- When interpolation is *not* defined, the discrete field probability distribution says nothing about the probability distribution of surface at points intermediate to the discrete field points.
- When interpolation is defined then, given a value of the discrete field, there is *no uncertainty* in the surface intermediate to the discrete field points. This is generally not a good representation of knowledge of physical fields.
- The surface distribution is not an intrinsic property of any physical object, rather a post-hoc imposition of the analyst attempting a useful regularization. The necessary scaling properties are ignored. See next section.

1.2. SCALING CONSISTENCY

The consistency condition mentioned in the last section, which must be imposed on probability distributions for continuous fields is:

Scaling of sample points consistency: For $S \subset A$ indices of discrete field variables,

$$P(X_S) = \int P(X_A)\, dX_{A\setminus S}. \tag{1}$$

1.3. ELEMENTS OF THE NEW PARADIGM

In the rest of this paper we discuss a new approach to continuous field inference which corrects the deficiencies, including the intermediate value and scaling problems, of traditional discrete-basis approaches to discrete height fields, for example. There are four objects of central importance within the inference approach described in this paper, one of which is a new object to Bayesian inference:

- The prior distribution for field. The prior holds all information about the field before any data is observed.
- The likelihood distribution. The likelihood is predictive for data, given the field. It incorporates all of the physics of the measurement process.
- The knowledge-representation (KR) distribution. Within the usual Bayesian point of view, the KR distribution is a new object. In the paradigm described in this paper the KR distribution is the object most directly updated when new data arrives.
- The posterior distribution. The posterior distribution summarizes everything known about the field given the prior knowledge and all data. The prior distribution and the KR distribution determine an approximation (possibly exact) to the surface posterior distribution.

In section 2 we make explicit the continuous field representation and inference paradigm via an example on surface inference.

2. Surface representation and inference

In this section the main ideas of the field representation and inference paradigm presented in this paper are given via the example of continuous-basis height field inference. The technique is general, though: it will become obvious that it extends to an arbitrary-basis, arbitrary-dimension field.

2.1. SURFACE DISTRIBUTIONS

The surface and height field distributions (the prior, likelihood, and posterior surface and height field distributions) are discussed in this section. As to notation, $\boldsymbol{s}$ represents a continuous-basis height field (surface), $\boldsymbol{h}$ represents a vector of heights (necessarily discrete), and $\boldsymbol{v}$ is reserved for the position basis-vector for those heights. The symbol $\boldsymbol{x}$ is reserved for measurements, θ represents parameters for the prior, $\Theta = (\boldsymbol{\mu}, \Sigma)$ represents parameters for the KR distribution. and ϕ represents parameters which determine the measurement operator, M. The symbol P is reserved for probability distributions or densities, as appropriate, and the hatted-P is associated with the KR distribution. The vector delta-function is defined as the product of the delta-functions for the components as in $\boldsymbol{\delta}(\boldsymbol{h}_v - \boldsymbol{h}(\boldsymbol{s},\boldsymbol{v})) = \Pi_{i=1}^{n} \delta(h_{v,i} - h_i(\boldsymbol{s},\boldsymbol{v}))$.

2.1.1. *Surface and height field prior distributions*

Consider a set S of surfaces where each element $\boldsymbol{s} \in S$ is a height field, i.e. such that $\boldsymbol{s} = s(x,y)$ is scalar function of two variables. Write the prior probability distribution for surfaces in S given the parameters θ which determine the prior distribution as (justification for the treatment of probability distributions over continuous-basis $\boldsymbol{s}$ appears in [3].

$$P(\boldsymbol{s} \mid \theta). \tag{2}$$

Consider a vector $\boldsymbol{v} = (v_1, \ldots, v_n)$ of discrete (x,y) points, $v_i = (x_i, y_i)$, and for any given surface $\boldsymbol{s}$ denote the associated vector of heights by $\boldsymbol{h}(\boldsymbol{s},\boldsymbol{v}) = (h_1(\boldsymbol{s},\boldsymbol{v}), \ldots, h_n(\boldsymbol{s},\boldsymbol{v}))$. Write the prior distribution of the surface heights at the chosen points v as $P(\boldsymbol{h}_v \mid \theta)$. This discrete height distribution may be found as follows:

$$P(\boldsymbol{h}_v \mid \theta) = \int P(\boldsymbol{h}_v \mid \boldsymbol{s}, \theta)\, P(\boldsymbol{s} \mid \theta)\, d\boldsymbol{s} \tag{3}$$

$$= \int P(\boldsymbol{h}_v \mid \boldsymbol{s})\, P(\boldsymbol{s} \mid \theta)\, d\boldsymbol{s} \tag{4}$$

$$= \int \boldsymbol{\delta}(\boldsymbol{h}_v - \boldsymbol{h}(\boldsymbol{s},\boldsymbol{v}))\, P(\boldsymbol{s} \mid \theta)\, d\boldsymbol{s} \tag{5}$$

Now, given that what is known is the surface heights $\boldsymbol{h}_v$ at a vector $\boldsymbol{v}$ of discrete (x,y) points, the posterior distribution of surfaces is found from Bayes' theorem as

$$P(\boldsymbol{s} \mid \boldsymbol{h}_v, \theta) = \frac{P(\boldsymbol{h}_v \mid \boldsymbol{s}, \theta)\, P(\boldsymbol{s} \mid \theta)}{P(\boldsymbol{h}_v \mid \theta)} \tag{6}$$

$$= \frac{P(\boldsymbol{h}_v \mid \boldsymbol{s})\, P(\boldsymbol{s} \mid \theta)}{P(\boldsymbol{h}_v \mid \theta)} \tag{7}$$

$$= \frac{\delta(\boldsymbol{h}_v - \boldsymbol{h}(\boldsymbol{s}, \boldsymbol{v}))\, P(\boldsymbol{s} \mid \theta)}{\int \delta(\boldsymbol{h}_v - \boldsymbol{h}(\boldsymbol{s}, \boldsymbol{v}))\, P(\boldsymbol{s} \mid \theta)\, d\boldsymbol{s}} \tag{8}$$

where the denominator distribution was found in equation 5.

2.1.2. *Measurements: The Likelihood*

In general, a surface $\boldsymbol{s}$ and some other parameters ϕ not dependent upon $\boldsymbol{s}$ (i.e. camera point spread function, camera position and direction, lighting position and direction, etc.) specify the probability distribution for data (likelihood)

$$P(\boldsymbol{x} \mid \boldsymbol{s}, \phi, \theta) = P(\boldsymbol{x} \mid \boldsymbol{s}, \phi) \tag{9}$$

where the data distribution is independent of θ once $\boldsymbol{s}$ is known. In practice an approximate model for the surface is utilized to make the computation of the likelihood tractable, as in a rendering step. Much confusion will result if the reader fails to distinguish the surface model for approximating the likelihood with the actual surface on which the likelihood depends.

2.1.3. *Conditioning on data*

Given data, we may utilize the likelihood of the previous section and steps similar to those of equations 3–5 and equations 6–8 to find $P(\boldsymbol{s} \mid \boldsymbol{x}, \phi, \theta)$, $P(\boldsymbol{h}_v \mid \boldsymbol{x}, \phi, \theta)$, and $P(\boldsymbol{s} \mid \boldsymbol{h}_v, \boldsymbol{x}, \phi, \theta)$. We refer the reader to [3] for the details.

2.2. KNOWLEDGE REPRESENTATION

The solution to the surface representation problem presented here addresses the competition for representational resources (computer memory) issue in a unique manner. The full posterior may be written in the form

$$P(\boldsymbol{s} \mid \boldsymbol{x}, \phi, \theta) = \int P(\boldsymbol{s} \mid \boldsymbol{h}_v, \boldsymbol{x}, \phi, \theta)\, P(\boldsymbol{h}_v \mid \boldsymbol{x}, \phi, \theta)\, d\boldsymbol{h}_v \tag{10}$$

where the distributions inside the integral appear in section 2.1.3. The following approximation defines the KR distribution. The prior conditioned on a set of heights, along with a new distribution, the *knowledge representation* distribution $\hat{P}(\boldsymbol{h}_v \mid \boldsymbol{x}, \phi, \theta)$, are substituted for the distributions inside the integral of equation 10 to approximate the posterior as

$$\hat{P}(\boldsymbol{s} \mid \hat{P}(\boldsymbol{h}_v \mid \boldsymbol{x}, \phi, \theta)) = \int P(\boldsymbol{s} \mid \boldsymbol{h}_v, \theta)\, \hat{P}(\boldsymbol{h}_v \mid \boldsymbol{x}, \phi, \theta)\, d\boldsymbol{h}_v. \tag{11}$$

Commentary on conditioning on the KR distribution, and a proof of the ability of the KR distribution to achieve a good approximation to the posterior appears in

[3]. In practice, it is useful to take a multinormal distribution over a discrete-point height field as the KR distribution; then the parameters for the KR distribution may taken as $\Theta_v(\boldsymbol{x}) = (\boldsymbol{\mu}_v(\boldsymbol{x}), \Sigma_v(\boldsymbol{x}))$ so that the posterior for surfaces is then approximated by

$$\hat{P}(\boldsymbol{s} \mid \Theta_v, \theta) = \int P(\boldsymbol{s} \mid \boldsymbol{h}_v, \theta)\, \hat{P}(\boldsymbol{h}_v \mid \Theta_v)\, d\boldsymbol{h}_v. \tag{12}$$

2.3. UPDATING THE KNOWLEDGE REPRESENTATION

Now we discuss updating the KR parameters Θ_v when new data are acquired and write GKF equations which are the appropriate parameter update. Temporarily restrict attention to the fixed v case.

2.3.1. *Bayes' theorem*

Having acquired $\Theta_v^n = \Theta_v(x^n)$, from previously seen data $\boldsymbol{x}^n = (\boldsymbol{x}_1, \ldots, \boldsymbol{x}_n)$ and upon seeing new data $\boldsymbol{x}_{n+1}$, the goal is to find Θ_v^{n+1} such that the surface distribution given Θ_v^{n+1} approximates the surface distribution given $\boldsymbol{x}_{n+1}$ and Θ_v^n. Given new data $\boldsymbol{x}_{n+1}$ in the context of the previously seen data $\boldsymbol{x}^n$ summarized by Θ_v^n, our updated surface distribution is found via Bayes' theorem

$$\begin{aligned}
\hat{P}(\boldsymbol{s} \mid \boldsymbol{x}_{n+1}, \Theta_v^n, \phi, \theta) &= \frac{P(\boldsymbol{x}_{n+1} \mid \boldsymbol{s}, \Theta_v^n, \phi, \theta)\hat{P}(\boldsymbol{s} \mid \Theta_v^n, \phi, \theta)}{\hat{P}(\boldsymbol{x}_{n+1} \mid \Theta_v^n, \phi, \theta)} \\
&= \frac{P(\boldsymbol{x}_{n+1} \mid \boldsymbol{s}, \phi)\hat{P}(\boldsymbol{s} \mid \Theta_v^n, \theta)}{\hat{P}(\boldsymbol{x}_{n+1} \mid \Theta_v^n, \phi, \theta)} \\
&= \frac{P(\boldsymbol{x}_{n+1} \mid \boldsymbol{s}, \phi)\hat{P}(\boldsymbol{s} \mid \Theta_v^n, \theta)}{\int P(\boldsymbol{x}_{n+1} \mid \boldsymbol{s}, \phi)\hat{P}(\boldsymbol{s} \mid \Theta_v^n, \theta)\, d\boldsymbol{s}}
\end{aligned} \tag{13}$$

where we defined

$$\hat{P}(\boldsymbol{x}_{n+1} \mid \Theta_v^n, \phi, \theta) = \int P(\boldsymbol{x}_{n+1} \mid \boldsymbol{s}, \phi)\hat{P}(\boldsymbol{s} \mid \Theta_v^n, \theta)\, d\boldsymbol{s}. \tag{14}$$

The updated posterior $\hat{P}(\boldsymbol{s} \mid \Theta_v^n, \boldsymbol{x}_{n+1}, \phi, \theta)$ will be approximated by the Θ_v^{n+1} parameterized KR distribution of equation 12 as

$$\hat{P}(\boldsymbol{s} \mid \Theta_v^{n+1}, \theta) \;=\; \int P(\boldsymbol{s} \mid \boldsymbol{h}_v, \theta)\, \hat{P}(\boldsymbol{h}_v \mid \Theta_v^{n+1})\, d\boldsymbol{h}_v. \tag{15}$$

The approximation condition for determining Θ_v^{n+1} is then written

$$\hat{P}(\boldsymbol{s} \mid \Theta_v^{n+1}, \theta) \;\approx\; \hat{P}(\boldsymbol{s} \mid \boldsymbol{x}_{n+1}, \Theta_v^n, \phi, \theta) \tag{16}$$

Equation 16 suggests we minimize various measures of the closeness of the two distributions. For example, one measure is the average square difference of the two distributions,

$$\int |P_1(\boldsymbol{s}) - P_2(\boldsymbol{s})|^2\, d\boldsymbol{s} \tag{17}$$

but there is no good first-principles reason to use this form. In the next section we discuss the measure of distance which leads to the *MI* choice of Θ_v^{n+1}.

2.3.2. *Maximally Informative inference*
The measure of distance which leads to the Θ^{n+1} providing the most information about the surface distribution is the *MI* choice for the statistic Θ^{n+1}. The condition for being MI, see [1], is

Find the Θ^{n+1} such that

$$\partial_{\Theta_v^{n+1}} \int \hat{P}(s \mid \Theta_v^n, x_{n+1}, \phi, \theta) \, log \left(\frac{\hat{P}(s \mid \Theta_v^n, x_{n+1}, \phi, \theta)}{\hat{P}(s \mid \Theta_v^{n+1}, \theta)} \right) ds = 0 \tag{18}$$

while at the Θ_v^{n+1} satisfying the derivative condition above the corresponding Hessian is negative definite and the extremum is a local maximum. If possible, choose the global maximum.

Note that the Kullback-Leibler distance above is asymmetric and that it is highly relevant which distribution contains the prior information and which distribution is being updated, a fact which many authors appear not to note. Maximum entropy techniques reverse the roles of the distributions which appear here. For a detailed explanation see [1]. In the following section the GKF update equations which follow from the MI approach are presented.

3. The Generalized Kalman Filter equations.

In this section the MI update equations for the KR, the GKF equations, are written. Details of the calculation and numerous appendices utilized in the derivation are found in [3]. Note the parameters of the updated KR (here $\overline{\boldsymbol{v}}$) need not have the same dimension nor position as those of the previous KR (i.e. $\boldsymbol{v}$), solving the problem of how to allow updates from one representation to the next, (finer, same, or coarser sampling), representation, i.e. it is not necessarily the case that $v_i \in \{\overline{v}_j\}$ or that $\overline{v}_i \in \{v_j\}$. Note that $\theta = (\boldsymbol{\mu}_s, \Sigma_s)$.

Denote the union of the components of $\boldsymbol{v}$ and $\overline{\boldsymbol{v}}$ by $\boldsymbol{v} \cup \overline{\boldsymbol{v}}$. Let A_{v_1,v_2} denote the projection from $\boldsymbol{v}_2$ to $\boldsymbol{v}_1$, for arbitrary position vectors $\boldsymbol{v}_1$ and $\boldsymbol{v}_2$. Skipping the mathematical details found in [3] the GKF equations are

$$\begin{aligned} \Theta_{\overline{v}}^{n+1} &= (\boldsymbol{\mu}_{\overline{v}}^{n+1}, \Sigma_{\overline{v}}^{n+1}) \\ \boldsymbol{\mu}_{\overline{v}}^{n+1} &= \boldsymbol{\mu}_{\overline{v}}^{R} \\ \Sigma_{\overline{v}}^{n+1} &= \Sigma_{\overline{v}}^{R} \end{aligned} \tag{19}$$

where

$$\begin{aligned} \boldsymbol{\mu}_{\overline{v}}^{R} &= \Sigma_R(\Sigma_Q^{-1} \boldsymbol{\mu}_{\overline{\boldsymbol{v}}}{}^Q + (\Sigma_{\overline{v}}^{n})^{-1} \boldsymbol{\mu}_{\overline{v}}^{\boldsymbol{n}} - \Sigma_{\overline{v}}^{-1} \boldsymbol{\mu}_{\overline{\boldsymbol{v}}}) \\ \Sigma_R^{-1} &= \Sigma_Q^{-1} + (\Sigma_{\overline{v}}^{n})^{-1} - \Sigma_{\overline{v}}^{-1} \end{aligned} \tag{20}$$

and where

$$\begin{aligned} \boldsymbol{\mu}_{\overline{v}}^{Q} &= A_{\overline{v},v\cup\overline{v}} A_{v\cup\overline{v},s} \boldsymbol{\mu}_{s}^{P} \\ \Sigma_{Q}^{-1} &= A_{\overline{v},v\cup\overline{v}} A_{v\cup\overline{v},s} \Sigma_{P}^{-1} A_{v\cup\overline{v},s}^{T} A_{\overline{v},v\cup\overline{v}}^{T} \end{aligned} \tag{21}$$

$$\begin{aligned} \boldsymbol{\mu}_{s}^{P} &= \Sigma_{P}(\Sigma_{s}^{-1}\boldsymbol{\mu}_{s} + M^{T}\Sigma_{\epsilon}^{-1}\boldsymbol{x}_{n+1}) \\ \Sigma_{P}^{-1} &= \Sigma_{s}^{-1} + M^{T}\Sigma_{\epsilon}^{-1}M \end{aligned} \tag{22}$$

$$\begin{aligned} \boldsymbol{\mu}_{\overline{v}}^{n} &= A_{\overline{v},v}\boldsymbol{\mu}_{v}^{n} \\ (\Sigma_{\overline{v}}^{n})^{-1} &= A_{\overline{v},v}(\Sigma_{v}^{n})^{-1}A_{\overline{v},v}^{T} \end{aligned} \tag{23}$$

$$\begin{aligned} \boldsymbol{\mu}_{\overline{v}} &= A_{\overline{v},v}\boldsymbol{\mu}_{v} \\ \Sigma_{\overline{v}}^{-1} &= A_{\overline{v},v}\Sigma_{v}^{-1}A_{\overline{v},v}^{T} \end{aligned} \tag{24}$$

$$\begin{aligned} \boldsymbol{\mu}_{v} &= A_{v,s}\boldsymbol{\mu}_{s} \\ \Sigma_{v}^{-1} &= A_{v,s}\Sigma_{s}^{-1}A_{v,s}^{T} \end{aligned} \tag{25}$$

Equations 19–25 are the GKF update equations. Standard KF equations are discussed in many places, see for example [4]. Another helpful paper is [5]. Having these update equations allows one to consider updating a representation of any dimension and position relative to the original representation. Finally, the GKF equations would not have been discovered without the MI inference approach, which led directly to the the correct interpretation of the Kullback Leibler distance, with application here. Evaluating the GKF update across new representations, a multigrid-like search algorithm for representation optimization, and the characterization of the measurement operator M for nonlinear systems are discussed in information-theoretic terms in [3]. This appears to provide the first information-theoretic justification for multi-grid approaches to surface or image inference.

4. Conclusion

Field inference has been generalized from the typical discrete fixed-basis setting to a continuous-basis setting. The problem of surface inference was solved in the context of continuous field inference. Using the approach of acquiring the MI KR distribution, the GKF equations were found. The GKF allows the updated KR parameters to be found at any scale and any position. The approach allows the learning of information at the relevant scales desired. It provides an information-theoretic justification for location-dependent adaptive multi-grid inference. To the knowledge of the author, this is the first time that the inference of continuous objects has been rigorously justified.

5. Acknowledgements

This work was supported by the NASA Ames Center for Excellence in Information Technology. I thank the members of the Ames Data Understanding group for their interest and comments, especially the invaluable contributions of Dr. Robin D. Morris, who checked the results, and Dr. Peter Cheeseman, whose comments were thoughtful.

References

1. D. R. Wolf and E. I. George, "Maximally informative statistical inference," in *Bayesian Statistics, 6th Valencia Meeting*, vol. 6, 1998.
2. S. Geman and D. Geman, "Stochastic relaxation, Gibbs distribution, and the Bayesian restoration of images," *IEEE Trans. Patt. Anal. Mach. Intell.*, **PAMI-6**, (6), pp. 721–741, 1984.
3. D. R. Wolf, "A Bayesian reflection on surfaces," *http://dino.ph.utexas.edu/~ wolf*, 1998.
4. R. G. Brown, *Introduction to Random Signal Analysis and Kalman Filtering*, Wiley, NY, 1983.
5. H. Zhu, C. K. Williams, R. Rohwer, and M. Morciniec, "Gaussian regressian and optimal finite dimensional linear models," *http://www.ncrg.aston.ac.uk/*, 1997.

QUANTUM ENTROPY REGULARIZATION

R. N. SILVER
MS B262, Los Alamos National Laboratory
Los Alamos, New Mexico 87545†

Abstract. By generalizing from classical entropy to quantum entropy, the maximum entropy method may be applied to non-parametric function estimation problems requiring smoothing, such as density estimation. I describe the use of quantum entropy as a regularization functional. Then I introduce a maximum Fourier entropy model whose behavior is similar to maximum quantum entropy, but which is easier to calculate and interpret.

Key words: quantum entropy, maximum entropy, density estimation, inverse problems, smoothing, Fourier methods, non-parametric function estimation

1. Introduction

Boltzmann introduced the concept of entropy in classical statistical physics as a measure of disorder. Von Neumann extended entropy to quantum statistical physics, introducing the density matrix. Shannon reinterpreted entropy as the measure of optimal code length in communication channels, initiating the development of information theory. Jaynes applied entropy to statistical inference, proposing the maximum entropy principle (MEP) to estimate probability distributions from incomplete information.

Most applications of the MEP to statistical inference use classical entropy, $S_C \equiv -\sum_i p_i \ln(p_i)$, where p is a probability distribution. Unfortunately, classical entropy is inadequate for non-parametric function estimation problems, such as density estimation, which require control over the smoothness of the function being estimated. A few years ago I proposed how to adapt quantum entropy for function estimation tasks using the MEP [1–3]. The additional feature added by quantum entropy is control of the smoothness of the estimate, tuning a variable analogous to $\hbar$. Quantum entropy is represented by $S_Q = -Tr\{\rho \ln(\rho)\}$, where ρ is a density matrix. A probability density f is taken to be the diagonal elements of ρ. Quantum entropy reduces to classical entropy in the limit $\hbar \to 0$. Using some mathematical formalism originating in quantum statistical physics, I applied maximum quantum

†Email: rns@loke.lanl.gov

W. von der Linden et al. (eds.), Maximum Entropy and Bayesian Methods, 91–98.

entropy models to a variety of common statistical inference tasks including density estimation, ill-posed inverse problems, spectrum estimation, interpolation, etc.

This paper reviews and develops further the applications of quantum entropy to non-parametric function estimation. Section 2 describes maximum quantum entropy models. Section 3 discusses a maximum Fourier entropy model, which behaves similarly to maximum quantum entropy, but which is easier to calculate and interpret. Section 4 briefly discusses hyperparameter selection. Section 5 presents a density estimation example. Section 6 concludes.

But first, let me emphasize what this paper is not: *I do not use intrinsically quantum phenomena such as entangled states or interference; I do not consider EPR paradoxes, Bells inequalities, quantum logic, qubits, etc.; I am not addressing quantum computers, teleportation, cryptography, quantum state determination, etc.; I am not suggesting yet another interpretation of quantum mechanics; I am not, yet, suggesting statistical inference applications for quantum phenomena which violate the axioms of probability theory.*

2. Maximum Quantum Entropy

Consider the non-parametric estimation of a one-dimensional density, $f(x) \geq 0$ defined on $0 \leq x \leq 1$ and normalized to $\int f(x)dx = 1$. To apply quantum entropy, identify $f(x) = \rho(x,x)$, where $\rho(x,x')$ is a positive definite density matrix. I consider the canonical form

$$\widehat{\rho} = \frac{e^{-\widehat{H}}}{Z} \quad Z \equiv Tr\{e^{-\widehat{H}}\} \quad , \tag{1}$$

where

$$\widehat{H} = \widehat{T} + \widehat{U} + \widehat{U}_o \quad ; \quad \widehat{T}\widehat{\rho} \equiv -\gamma^{2n}\left(-\frac{\partial^2}{\partial x^2}\right)^n \rho(x,x') \quad ; \quad \widehat{U} \equiv u(x)\delta(x-x') \tag{2}$$

$\widehat{H}$ is analogous to a Hamiltonian, $\widehat{T}$ to a generalized kinetic energy operator, and $\widehat{U} + \widehat{U}_o$ to a potential energy operator. If $\widehat{\rho}_o$ is the density matrix corresponding to a default model $f_o(x)$, then $\widehat{U}_o$ is the corresponding default 'potential'.

Eqs. (1,2) may be derived by maximizing the relative quantum entropy,

$$S[\rho||\rho_o] \equiv -Tr\{\widehat{\rho}(\ln\widehat{\rho} - \ln\widehat{\rho_o})\} \quad , \tag{3}$$

with respect to $\widehat{\rho}$ subject to constraints of the form

$$1 = Tr\{\widehat{\rho}\} = \int f(x)dx \quad ; \quad Tr\{\widehat{U}\widehat{\rho}\} = \int u(x)f(x)dx \quad ; \quad Tr\{\widehat{T}\widehat{\rho}\} \tag{4}$$

The first constraint is normalization. The second constraint is moments of f. The third constraint is the unique feature of quantum entropy, implicitly controlling the smoothness of f.

Because quantum entropy is a convex function, it satisfies Legendre transform duality, just like classical entropy. Define $I \equiv -S$ as the relative information. It obeys $I \geq 0$ with equality *iff* $\widehat{\rho} = \widehat{\rho}_o$. Define a Legendre transform F, analogous to free energy, by

$$F \equiv I + Tr\{\widehat{U}\widehat{\rho}\} \tag{5}$$

Then, the duality relations are

$$\frac{\partial I}{\partial f(x)} = -u(x) \quad ; \quad \frac{\partial F}{\partial u(x)} = f(x) \quad . \tag{6}$$

From the MEP viewpoint, we seek the f which maximizes S subject to the constraints of the data. For a log-Likelihood function $L[f|D]$, where D represents the data, we maximize $S + L/\alpha$ where $1/\alpha$ is a Lagrange multiplier. From the maximum likelihood viewpoint, S may be regarded as a *statistical regularization functional* and α as a statistical regularization parameter. Using Eq. (6),

$$u(x) = -\frac{1}{\alpha}\frac{\partial L}{\partial f(x)} \tag{7}$$

For example, for density estimation one has

$$L[f|D] = \sum_i \ln f(x_i) \quad ; \quad u(x) = -\sum_i \frac{\delta(x - x_i)}{\alpha f(x_i)} \quad , \tag{8}$$

where the $\{x_i\}$ are *iid* samples drawn from f. Eq. (7) is the same using either classical or quantum entropy. For familiar tasks such as density estimation and linear inverse problems, it requires the solution of a convex non-linear optimization problem. Methods for solving such problems are well-developed. It is best to use a dual-space algorithm with Lagrange multipliers rather than f as the parameters.

The choice of n in Eq. (2) dictates the degree of smoothness of the corresponding f. Consider the u in Eq. (8). In the classical entropy limit, the smoothness of f is the same as the smoothness of u. Classical maximum entropy would yield an f as a sum of δ-functions - a nonsense result. However, for quantum entropy with $n = 1$, Eq. 2 is the standard Hamiltonian for the Schroedinger equation, with $\gamma \propto \hbar$, and f must be continuous. If $n = 2$, the second derivatives of f must be continuous. Such smoothness properties are essential to success with density estimation. Other possible choices for $\widehat{T}$ correspond to varying smoothness constraints on f.

An analogy can be developed with the kernel methods. The second derivatives of F are

$$\widehat{K} \equiv K(x, x') \equiv -\frac{\partial^2 F}{\partial u(x)\partial u(x')} = -\frac{\partial f(x)}{\partial u(x')} \quad . \tag{9}$$

By the second equality $\widehat{K}$ is the impulse response of f to a change in u, so it may be regarded as a *kernel function*. For $\gamma \to 0$, the classical limit, $\widehat{K} \to f(x)\delta(x - x') - f(x)f(x')$. For $\gamma \neq 0$, $\widehat{K}$ has a width and a shape that depends on γ and n. Maximum quantum entropy f are broadened compared to the maximum classical entropy f. The kernel function is positive definite due to the convexity of S. A measure of the number of degrees of freedom is

$$N_f \equiv Tr\{\widehat{G}\} \quad ; \quad \widehat{G} \equiv \frac{K(x, x')}{\sqrt{f(x)}\sqrt{f(x')}} \tag{10}$$

N_f is proportional to $1/\gamma$, diverging in the classical entropy limit. The eigenvalue spectrum of $\widehat{G}$ has approximately $1/\gamma$ eigenvalues near one, with all other eigenvalues decreasing rapidly with increasing rank. The kernel determines the covariance, C,

$$\widehat{C}^{-1} \equiv -\frac{\partial^2(L-\alpha I)}{\partial\sqrt{f(x)}\partial\sqrt{f(x')}} \tag{11}$$

For density estimation the covariance is

$$\widehat{C} \equiv \frac{1}{4}\mathbf{E}\{\delta\sqrt{f(x)}\delta\sqrt{f(x')}\}$$
$$= G(x,x') - \sum_{k,l} G(x,x_k)\left(1+\widehat{G}\right)^{-1}_{k,l} G(x_l,x')$$

This is a positive definite matrix. The second term is the reduction in covariance due to the measurements.

3. Maximum Fourier Entropy

The new computational task which quantum entropy presents is to calculate f as a function of u. Fortunately, a wide variety of efficient numerical methods have already been developed in the quantum mechanics literature. The computational resources scale linearly in the number of pixels used, which is much better than the cubic scaling of Hamiltonian diagonalization [4]. Nevertheless, the implicit relation between f and u obscures understanding of how maximum quantum entropy works. In this section I present a simpler model which captures the essential behavior of maximum quantum entropy models while making explicit the relation between f and u.

Consider a cosine expansion of f

$$f(x) = 1 + 2\sum_{m=1}^{\infty} \mu_m \cos(\pi m x) \qquad \mu_m = \int_0^1 \cos(\pi m x) f(x) dx \quad . \tag{12}$$

A 'smoothed' approximation to f could be described as 'band limited', having only a finite number M of non-zero moments μ_m. To avoid the Gibbs phenomena in truncating a Fourier series, use Gibbs damping factors g_m. The *band limited approximation* to f is

$$f_M(x) = 1 + 2\sum_{m=1}^{M} g_m \mu_m \cos(\pi m x) = \int_0^1 \delta_M(x-x') f(x') dx' \quad . \tag{13}$$

The resolution function is

$$\delta_M(x) = 1 + 2\sum_{m=1}^{M} g_m \cos(\pi m x) \quad . \tag{14}$$

The g_m can be chosen [5] so that δ_M is positive, normalized, peaks at $x = 0$, has minimal variance, and has width $\Delta x \propto 1/M$. Then introduce parameters $p_i \geq 0$ by

$$f_M(x) = \sum_{i=0}^{M} p_i \delta_M(x - z_i) \qquad z_i \equiv \frac{i + 0.5}{M + 1} \quad . \tag{15}$$

The p_i are related to the μ_m by discrete Fourier transform. The resolution function satisfies

$$\int_0^1 \delta_M(x - z_i)dx = 1 \quad ; \quad \sum_{i=0}^{M} \delta_M(x - z_i) = 1 + M \quad . \tag{16}$$

So the normalization is

$$\int_0^1 f_M(x)dx = \sum_{i=0}^{M} p_i = 1 \quad . \tag{17}$$

Hence, the p_i are probabilities.

One can now construct a Fourier relative entropy

$$-S = I = \sum_{i=0}^{M} p_i \ln\left(\frac{p_i}{p_i^o}\right) \quad . \tag{18}$$

One can also define a dual variable u_i by

$$p_i = \frac{p_i^o e^{-u_i}}{Z} \quad ; \quad Z = \sum_i p_i^o e^{-u_i} = -\ln F \quad . \tag{19}$$

F is the Legendre transform to I, satisfying duality relations

$$\frac{\partial I}{\partial p_i} = -u_i \quad ; \quad \frac{\partial F}{\partial u_i} = p_i \quad . \tag{20}$$

The description is completed by

$$u_i = \int_0^1 \delta_M(x - z_i)u(x)dx \quad . \tag{21}$$

This parameterization also satisfies the same Legendre relations for continuous f and u as quantum entropy. The kernel is

$$K(x, x') = \sum_{i=0}^{M} \delta_M(x - z_i)p_i\delta_M(x' - z_i) - f(x)f(x') \quad . \tag{22}$$

Note that $N_f = Tr\{\widehat{G}\} \propto M$, so the number of parameters M is analogous to $1/gamma$.

In this maximum Fourier entropy model, a 'band limited' f can be described by a finite number M of parameters, as expected by the Shannon sampling theorem. The standard Shannon parameters equal the value of f at equally spaced

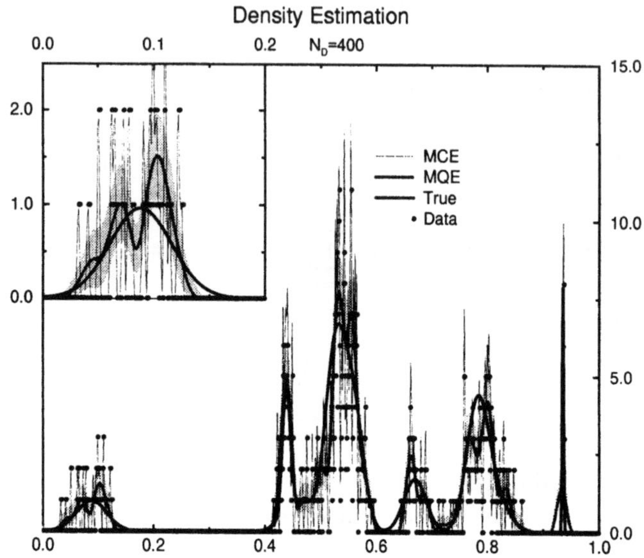

Figure 1. Density Estimation Example.

sampling points x_i. However, this parameterization is not suitable for us because their influence on neighboring x via the Shannon interpolation formula is not local; it decays approximately as $1/|x - x_i|$. Instead, we choose the M parameters p_i which, while not equal to f at any sampling point, have only local influence on neighboring x, decaying faster than any power law. The more moments, the more local.

The maximum Fourier entropy model is not a special case of a quantum entropy model. However, it behaves similarly. Both have a finite number of degrees of freedom, obey the same Legendre relations, can be applied to the same nonparametric function estimation problems requiring smoothing, and produce similar looking results. An advantage of the maximum Fourier entropy model is the ease of computation by fast Fourier transform. An advantage of the quantum entropy model is the much greater choice of smoothness properties; however, in applications so far this added flexibility has not been important.

4. Hyperparameter Selection

To test these models, we choose hyperparameters by maximizing

$$Q = L[f|D] - \alpha I - O \quad . \tag{23}$$

The 'Occam' or 'complexity' term,

$$O \equiv \frac{1}{2} \ln \left(\frac{\det[\widehat{C}^{-1}]}{\det[\alpha \widehat{K}^{-1}]} \right) \quad . \tag{24}$$

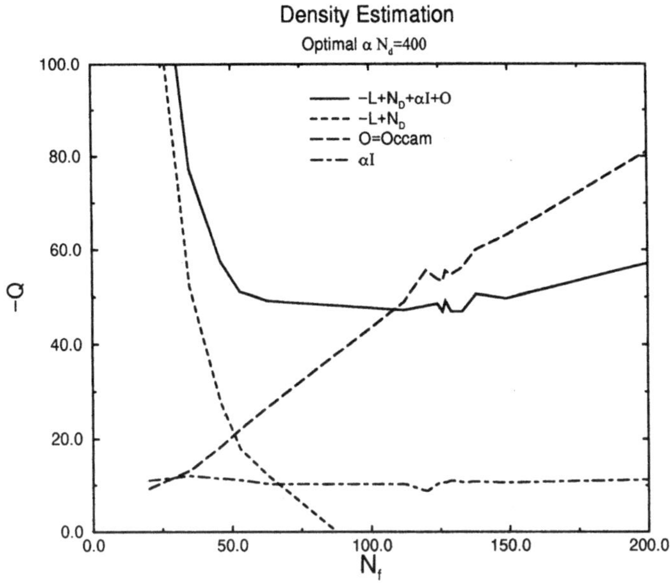

Figure 2. Optimization of Hyperparamaeters

The Occam term is the log of the square root of the ratio of determinants of the Fisher information matrix before and after the measurement. It is strictly positive, and it penalizes the number of measured parameters. For density estimation,

$$O \Rightarrow \frac{1}{2}\ln\left(\det\left[1+\frac{\widehat{\mathcal{M}}}{\alpha}\right]\right) \quad ; \quad \mathcal{M}_{k,l} \equiv \frac{K(x_k, x_l)}{f(x_k)f(x_l)} \tag{25}$$

A Bayesian interpretation of Eqs. (23,24) is that Q is the log of a Gaussian approximation to the marginal-likelihood, calculated from a prior probability proportional to $\exp[-\alpha I]$. A difficulty is that the Gaussian approximation to the marginalization integral is known to be terrible in the classical entropy limit where $N_f \to \infty$. I believe these approximations are better controlled for quantum entropy where N_f is finite [3]. An alternative justification for similar formulae comes from the minimum message length (or minimum description length) approaches motivated by information theory. There $-Q$ is proportional to the number of bits required to encode a probabilistic model, along with the data using the probabilities specified by the model. The optimal model should require the fewest number of bits. The Occam factor is the penalty for the precision with which parameter values should be quoted. In information theory the Occam factor is not regarded as an approximation, but rather as 'the answer'.

5. Example

The behavior of the maximum Fourier entropy and the maximum quantum entropy models are quite similar, so much so that one example is sufficient to illustrate both. Fig. 1 shows a density estimation example, in which 400 samples are drawn *iid* from a true f (True). The data are represented as a histogram with narrow bins. Maximum classical entropy (MCE), using an α optimized per Section 4, yields the nonsense result of fitting only the sampled points. Maximum quantum entropy (MQE) using $N_f \approx 100$ provides a reasonable smoothed estimate of f. The shaded area indicates the one standard deviation variance of the MQE estimate.

The behavior of the hyperparameter selection criteria as a function of N_f is shown if Fig. 2. Here L is the log-Likelihood, I is the relative entropy, and O is the Occam factor. The parameter α has already been chosen to maximize Q, given N_f. The regularization term is small and slowly varying with N_f. Hyperparameter selection is dominated by the competition between the Likelihood function and the Occam factor (or model complexity). Notice that the minimum in N_f is rather flat, which is often the case for true f containing a mixture of sharp and smooth structures. If f contains structures with comparable smoothness, then the minimum in $-Q$ is much sharper.

I have tried this method as well on a variety of inverse problems with resolution functions of varying degrees of smoothness, with reasonable success. But space does not permit me to present those results here.

6. Conclusion

I have presented two different methods for incorporating smoothness into the maximum entropy modeling of data, one based on quantum entropy and the other based on a Fourier entropy. They extend the maximum entropy principle to non-parametric function estimation with smoothing. Applications include density estimation, spectrum estimation, inverse problems, interpolation, etc. They preserve attractive maxent features such as convexity along with Legendre transform duality, default models, positivity, the ability to fuse data from different sources using likelihood functions, etc. The only feature of classical maximum entropy abandoned is strict locality. The maximum Fourier entropy model may be calculated efficiently using FFT methods. The next step is to make smoothing adaptive.

References

1. R. N. Silver, "Quantum statistical inference," in *Physics & Probability: Essays in honor of Edwin T. Jaynes*, J. W. T. Grandy and P. W. Milonni, eds., p. 223, Cambridge University Press, 1993.
2. R. N. Silver, "Density estimation by maximum quantum entropy," in *Physics & Probability: Essays in honor of Edwin T. Jaynes*, J. W. T. Grandy and P. W. Milonni, eds., p. 223, Cambridge University Press, 1993.
3. R. N. Silver and H. F. Martz, "Applications of quantum entropy to statistics," in *Proceedings of the Statistical Computing Section 1994*, p. 61, American Statistical Association, 1994.
4. R. N. Silver and H. Roeder, "Density of states of megadimensional hamiltonian matrices," *Int. J. Mod. Phys. C*, **5**, p. 735, 1994.
5. R. N. Silver, H. Roeder, A. F. Voter, and J. D. Kress, "Kernel polynomial approximations for densities of states and spectral functions," *J. Comp. Phys.*, **124**, p. 115, 1996.

BAYESIAN RECONSTRUCTION OF ELECTRON ENERGY DISTRIBUTIONS FROM EMISSION LINE INTENSITIES

R. FISCHER, W. JACOB, W. VON DER LINDEN AND V. DOSE
Max-Planck-Institut für Plasmaphysik, EURATOM Association
POB 1533, D-85740 Garching, Germany[†]

Abstract. Low-pressure plasmas are nowadays widely used for technical applications of plasma-surface interactions, such as plasma etching, material deposition, sputtering, etc. For a thorough understanding of individual processes in plasma processing the electron energy distribution (EED) function in the bulk plasma is of great importance. The EED determines the rates of all electron induced reactions as ionization, excitation or dissociation of molecules. The ubiquitous assumption of a Maxwellian EED becomes progressively worse for hot and low-density plasmas. Measurements of the EED with probes penetrating the plasma result in deteriorating effects on the plasma and the probe, thus measurements without plasma contact are of great interest. A non-destructive measurement is the detection of radiation emitted by the plasma.

The form-free reconstruction of the EED from a small number of measured emission intensities results in an ill-posed inversion problem. In order to avoid spurious features due to overfitting of the data (*ringing*) we apply Bayesian probability theory along with the adaptive-kernel method. The Bayesian approach will be applied to emission lines of helium, since in this case the relevant atomic input quantities are best known.

Key words: Electron Energy Distribution, Low-Pressure Plasma, Inverse Problem, Over-Fitting, Adaptive Kernels, Occam's Razor

[†]Email: Rainer.Fischer@ipp.mpg.de

W. von der Linden et al. (eds.), Maximum Entropy and Bayesian Methods, 99–106.

1. Introduction

The knowledge of the electron energy distribution (EED) in plasmas is essential for modeling the chemical reactions in plasmas, growth processes in thin film deposition, surface treatment with plasmas and wall erosion and material migration in fusion experiments. The relevant quantity is the rate of electron-impact induced processes. This rate is determined by the integral

$$< \sigma v >= \int_0^\infty dv \; f(v) \, v \, \sigma(v) \tag{1}$$

with the EED $f(v)$, the electron velocity v and the cross section $\sigma(v)$. Fig. 1

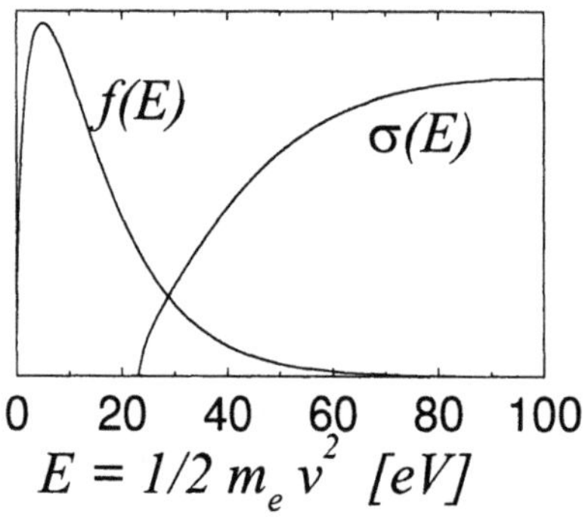

Figure 1. Sketch of a typical electron energy distribution and a typical electron-impact cross-section. The overlap of the two curves overlap only in a small energy range.

shows a sketch of a Maxwellian EED f and a cross-section σ for electron-impact excitation of helium from the ground state. The two curves overlap only in the high-energy tail of the EED. Below the threshold energy of this cross section the rate contains no information about the EED. Excitations between excited states with lower threshold energies couple the EED at lower energies to the rate but this is expected to be rather weak since excitations between excited states are less frequent.

It is common practice to assume thermodynamic equilibrium and hence a Maxwell distribution for the EED. In this case there is only one free parameter, the electron temperature T. The information content in the emission experiment is sufficient to determine T unambiguously. For non-equilibrium situations the generalized Druyvesteyn distribution characterized by the electron temperature T and an exponent ν, $f(E) \propto \sqrt{E} \exp(-(E/kT)^\nu)$, is used [1]. The exponent allows for a steep decrease in an energy region where energy losses due to excitation of atoms or molecules provide a loss channel for the EED. In order to describe multiple energy loss channels a superposition of two or more such functions may be required. In this work the rules of probability theory are used for a form-free reconstruction of the EED.

2. Experiment

The low-pressure plasma is generated by electron cyclotron resonance heating with a microwave. The microwave accelerates those electrons which gyrate in a magnetic field with the frequency of the electric field. The heated electrons leave the resonance zone by drift or impact with other particles. Details can be found in [2].

Two established methods for measuring the EED are given by Langmuir probes and by the detection of radiation emitted by the plasma. A Langmuir probe is a thin wire which penetrates the plasma. The values of the electron temperature and electron density is fitted to the current-voltage characteristic of the wire. An advantage is that the measurement can be done for different locations in the plasma. Disadvantages arise due to the direct contact of the wire with the plasma. Local disturbances of the plasma by the wire can occur and film growth on the probe surface can distort the current-voltage characteristic. The thermal load on the probe restricts the parameter range and interpretation problems in strong magnetic fields can arise.

A measurement method, which is non-destructive for both the plasma and the probe, is the detection of radiation emitted by the plasma. The intensity of an emission line depends on the density of the excited state n_i which emits the radiation integrated over the line of sight,

$$I_{ij} = \frac{1}{4\pi}\frac{hc}{\lambda_{ij}} A_{ij} \int_0^L dx\, n_i \quad , \tag{2}$$

where λ_{ij} is the wavelength and A_{ij} is the transition probability from level i to level j. In this work we have only one line of sight. Thus we have no spatial resolution.

We analyzed the intensity of 8 helium emission lines from principal quantum numbers 3 and 4 to principal quantum number 2 in the photon energy range 2.1-3.2 eV. There are four measured optical transitions in the singlet and triplet term systems of helium, each. The majority of transitions are not recorded due to experimental restriction in the photon energy range.

3. Collisional-radiative model

For determining the intensities of spectral lines, the densities of excited levels n_i are calculated with the collisional-radiative model. $\boldsymbol{n}$ in the stationary state is given by a balance of population and de-population processes:

- e^--impact excitation and de-excitation $i \to j$: $\pm n_i n_e < \sigma v >_{ij}$
- e^--impact ionization $i \to \infty$: $n_i n_e < \sigma v >_{i\infty}$
- population and de-population by radiative decay: $\pm n_i A_{ij}$
- excitation by re-absorption of radiation: $A_{ij} \to \Theta_{ij} A_{ij}$
- de-population by wall collisions: $n_i \Gamma_i$
- dielectronic and radiative recombination: $\beta_{DE/Rad}\, n_{ion} n_e$

n_e denotes the electron density, $0 < \Theta_{ij} < 1$ is the escape factor of radiation from the plasma, Γ_i is the wall-collision frequency and $\beta_{DE/Rad}$ denote the dielectronic and radiative rates. The atomic input data are fairly well known for helium. For a given EED $f(E)$ we have to calculate the rate coefficients $< \sigma v >_{ij}$ of the electron-impact processes and solve the linear equation for the density of excited levels,

$$\begin{aligned} \boldsymbol{C n} &= \boldsymbol{r} && (3)\\ \mathbf{C} &= \mathbf{C}\left(< \sigma v >_{ij}, n_e, A_{ij}, \Theta_{ij}, \Gamma_i, \beta_{DE/Rad}\right) && (4)\\ \boldsymbol{r} &= \boldsymbol{r}\left(< \sigma v >_{ij}, n_e, n_{He}, \beta_{DE/Rad}\right) && (5) \end{aligned}$$

for the helium density n_{He}. With the assumption of a density profile independent of the level number we can calculate the intensity of the emission lines from eqn. 2. Details of this model can be found in [3].

Since we have only 8 measured data and a form-free reconstruction with about 100 degrees-of-freedom (DOF) we are dealing with an ill-posed inversion problem.

4. Bayesian Probability Theory

The ill-posed inversion problem is tackled by Bayesian Probability Theory (BPT).

In our case of independent measurement errors σ_i, which are estimated from N_d data d_i, a Gaussian likelihood with the χ^2 misfit applies.

$$\begin{aligned} P(\boldsymbol{d}|\boldsymbol{f},\boldsymbol{\sigma},I) &= \frac{1}{\prod_{i=1}^{N_d}\sqrt{2\pi}\sigma_i}\, exp\left(-\frac{1}{2}\chi^2\right) && (6)\\ \chi^2 &= \sum_{i=1}^{N_d}\left(\frac{d_i - I_i(f(E))}{\sigma_i}\right)^2 && (7) \end{aligned}$$

In order to apply Bayes' theorem we need a prior probability for the EED f. The appropriate prior for a positive and additive distribution is the entropic prior [4].

$$\begin{aligned} P(\boldsymbol{f}|\alpha,I) &= \frac{1}{Z} exp\,(\alpha S) && (8)\\ S &= \sum_{j=1}^{N} f_j - m_j - f_j\, ln\left(\frac{f_j}{m_j}\right) && (9)\\ Z &= \int d^N\boldsymbol{f}\, P(\boldsymbol{f}|\alpha,I) && (10)\\ P(\boldsymbol{f}|I) &= \int d\alpha\, P(\boldsymbol{f}|\alpha,I)\, P(\alpha) && (11) \end{aligned}$$

$\boldsymbol{m}$ denotes the default model. Marginalization over the hyper-parameter α completes the prior.

It has been demonstrated by the authors [5] that even the optimal treatment of the α-marginalization leaves some residual *ringing* and noise fitting. For form-free density estimation, Quantified Maximum Entropy (QME) yields disappointing results [6]. The reason in both cases is the large number of DOF of a form-free distribution which is not sufficiently restricted by the entropic prior since it contains no correlation between the image cells. We need an approach that reproduces significant structures in the data while noise-fitting by redundant DOF is avoided. In addition, in real world we always have functions containing a mixture of sharp and smooth structures, for example signal and background. We therefore need a method that adopts to the local information content. An Occam factor arising self-consistently from BPT penalizes the complexity of the model reducing the DOF to the minimum amount necessary to describe the data.

Closely related to the preblur concept of John Skilling [4] we convolve a hidden image with a smoothing kernel but with locally varying kernel widths [7].

$$f(x) = \int dy B\left(\frac{x-y}{b(y)}\right) h(y) \tag{12}$$

$$B\left(\frac{x-y}{b(y)}\right) = \frac{1}{\sqrt{2\pi}b(y)} \exp\left[-\frac{1}{2}\left(\frac{x-y}{b(y)}\right)^2\right] \tag{13}$$

We assume a Gaussian kernel. The rules of BPT are used to determine the posterior probability density. To this end we marginalize the hyper-parameters: hidden image $\boldsymbol{h}$, blurring widths $\boldsymbol{b}$ and α.

$$P(\boldsymbol{f}|\boldsymbol{d},I) = \int_0^\infty \cdots \int_0^\infty d^N\boldsymbol{h}\, d^N\boldsymbol{b}\, d\alpha\, P(\boldsymbol{f},\boldsymbol{h},\boldsymbol{b},\alpha|\boldsymbol{d},I) \tag{14}$$

$$\propto \int_0^\infty \cdots \int_0^\infty d^N\boldsymbol{h}\, d^N\boldsymbol{b}\, d\alpha\, P(\boldsymbol{d}|\boldsymbol{h},\boldsymbol{b},I)\, P(\boldsymbol{f}|\boldsymbol{h},\boldsymbol{b},I) \times \tag{15}$$

$$P(\boldsymbol{h}|\alpha,I)\, P(\boldsymbol{b}|I)\, P(\alpha|I) \tag{16}$$

This can be done by Markov-Chain Monte-Carlo integration. Alternatively, the cumbersome multi-dimensional integral can be tackled with the evidence approximation determining the most probable α^*, $\boldsymbol{b}^*$ and $\boldsymbol{h}^*$ [7].

The kernel-width distribution encodes the complexity of the model. A measure for the effective DOF (eDOF) of the model parameters, which serves as a diagnostic tool, is given by the sum of the eigenvalues of B which may be illustrated in the two limits: If the kernel widths are all very small the reconstruction is determined pointwise with eDOF=N: $f_i = h_i$. We have no correlations and obtain the conventional result with the uncorrelated entropic prior. If the kernel widths go to infinity the image is described solely by the mean of the density $\boldsymbol{h}$ with eDOF=1: $f_i = \sum_i^N h_i/N$. The large eigenvalues of B define eigenvectors which are essential for describing the data and the small eigenvalues define eigenvectors describing insignificant contributions due to noise.

The quantity that governs Occam's razor is the volume of the prior covered by the high-likelihood region picked by the data. The essential contribution is given by the integral over the hidden image.

$$\int d^N\boldsymbol{h}\, P(\boldsymbol{d}|\boldsymbol{h},\boldsymbol{b},I)\, P(\boldsymbol{h}|\alpha,I) \propto det^{-\frac{1}{2}}(B(\boldsymbol{b})^T\, H_L\, B(\boldsymbol{b}) + \frac{\alpha}{\boldsymbol{h}^*}) \tag{17}$$

The first term in the determinant is the Hessian of the logarithm of the likelihood function H_L with the kernel B and the second term arises from the Hessian of the entropy. There is a tradeoff between the likelihood function which favors small kernel widths in order to fit the data and Occam's razor which penalizes small kernel widths in order to reduce complexity.

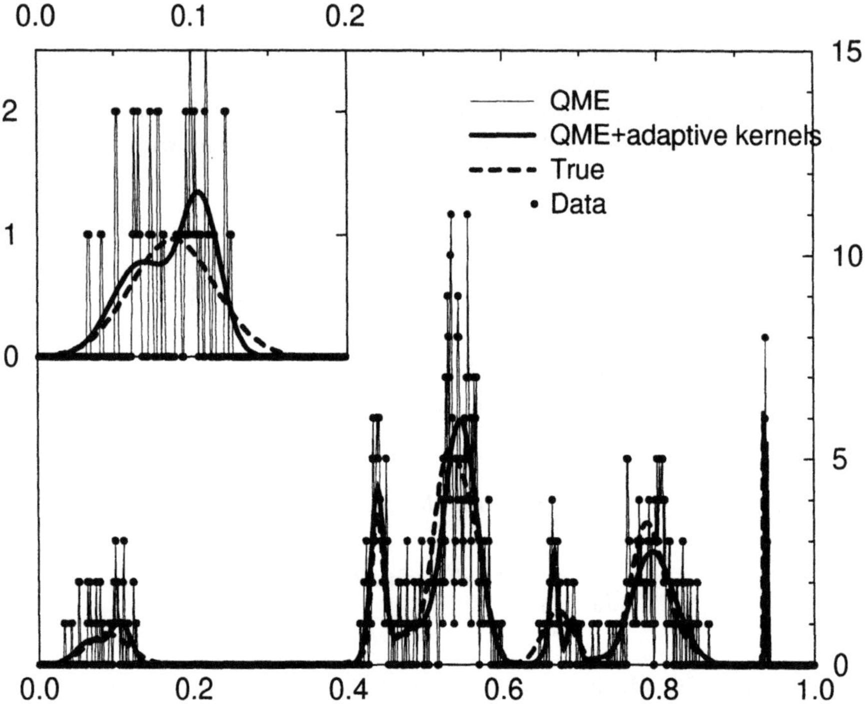

Figure 2. Density Estimation Example.

To illustrate the adaptive-kernel approach, Fig. 2 shows a mock density estimation example introduced by Silver [6], in which 800 samples are drawn *iid* from a true f (True). The data are represented as a histogram with narrow bins. Quantified Maximum Entropy (QME) fits only the sampled points. To overcome this problem Silver proposed the Quantum Entropy concept [6]. We compare his result with that of our approach. The adaptive-kernel result shows no fine structure in the left-most peak. There is only a small shoulder which reflects the double structure in the data. The peak on the right is fully resolved which is hardly obtainable with one smoothing parameter for the whole density range. The most important ingredient in avoiding ringing or noise fitting is the adaptive concept.

5. Results

Next we turn to the form-free reconstruction of the EED in a plasma. The results for a real data set consisting of 8 emission line intensities is shown in Fig. 3. The dashed line depicts a Maxwell distribution of the temperature measured with the

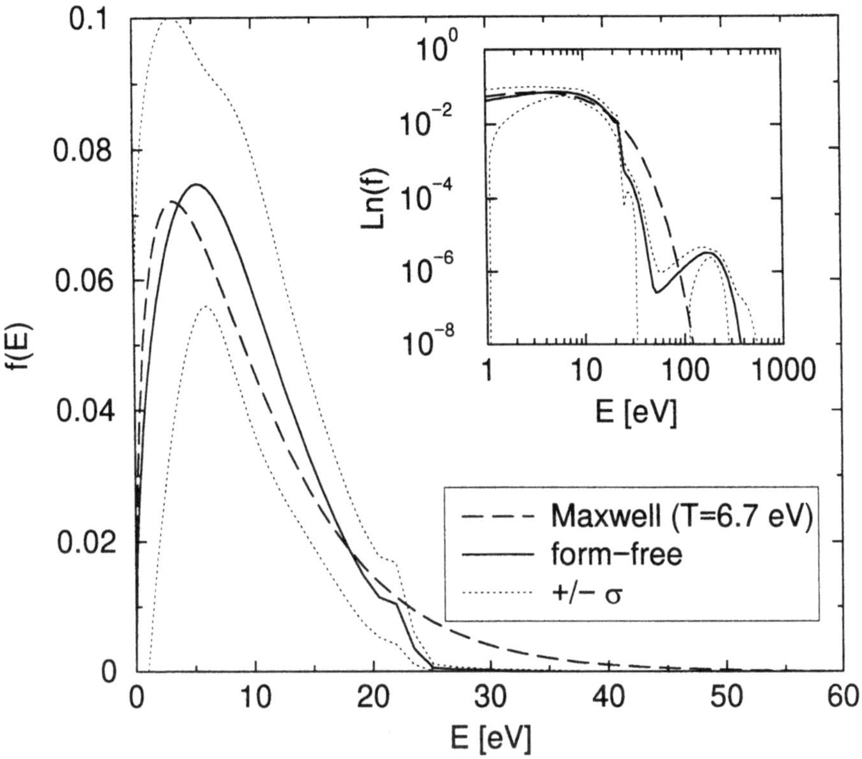

Figure 3. The dashed line depicts a Maxwell distribution of the temperature measured with the Langmuir probe. The solid line shows the form-free estimation of the EED with the adaptive-kernel approach. The steep decrease at an energy of about 20 eV arises due to a loss channel by excitation of helium. The peak at about 100-200 eV is interpreted as an accumulation of high energy electrons due to the heating mechanism.

Langmuir probe. The solid line shows the form-free estimation of the EED with the adaptive-kernel approach. The dotted line indicates the one standard deviation variance of the form-free estimation. The significant difference in the form of the distributions is quite obvious. There are two major results: The first is the steep decrease of the distribution at an energy of about 23 eV, which is even more obvious on a log-log scale. This is not a result of insufficient information in the data which would yield a smoother reconstruction compared to the Maxwell distribution. The steep decrease is driven by significant information in the data.

The threshold energies of excitation of an electron from helium in the ground state into the excited states is between 20 eV and the ionization energy of 24.6 eV. Since most helium atoms are in the ground state, the dominance of the loss channel by inelastic collisions is not surprising. At very low degrees of ionization, here less than 10^{-3}, the electron energy gained in the microwave field is essentially transferred to helium atoms by inelastic collisions. In retrospect, instead of using a Maxwellian EED a Druyvesteyn distribution fitted to the current-voltage characteristic of the Langmuir probe would do a better job.

The second new result is a small but significant peak between 100 and 200 eV which is shown in the inset. We interpret this significant additional structure as an accumulation of high energy electrons due to the heating mechanism by electron cyclotron resonance. The energetic position and amplitude of this structure is not yet understood.

Please note, that the decrease of the EED at the energy $E = 0$ is not driven by the data, but enforced by boundary conditions imposed due to the vanishing phase space in three-dimensional electron velocity space at $v = 0$, which is exact prior knowledge.

6. Summary

In summary, we have shown that ringing and noise fitting, intrinsic to ill-posed inversion problems, can be reduced drastically by the introduction of adaptive resolution into the Bayesian analysis. Occam's razor favors the smallest effective degree-of-freedom which is necessary to describe the data. The form-free estimation of the EED shows large deviations from a Maxwell distribution. There is a strong decrease where excitation of helium provides a loss channel. An additional structure arises which is interpreted as an accumulation of high energy electron as a consequence of the electron cyclotron resonance heating.

References

1. K. Behringer and U. Fantz, "Spectroscopic diagnostics of glow discharge plasmas with non-Maxwellian electron energy distributions," *Appl. Phys.*, **27**, p. 2128, 1994.
2. P. Reinke, S. Schelz, W. Jacob, and W. Möller, "Influence of a direct current bias on the energy of ions from an electron cyclotron resonance plasma," *J. Vac. Sci. Technol. A*, **10**, p. 434, 1992.
3. D. Bates, A. Kingston, and R. McWhirter *Proc. Roy. Soc. A*, **270**, p. 155, 1962.
4. J. Skilling, "Fundamentals of maxent in data analysis," in *Maximum Entropy in Action*, B. Buck and V. Macaulay, eds., p. 19, Clarendon Press, Oxford, 1991.
5. R. Fischer, W. von der Linden, and V. Dose, "On the importance of α marginalization in maximum entropy," in *Maximum Entropy and Bayesian Methods*, R. Silver and K. Hanson, eds., p. 229, Kluwer Academic Publishers, Dordrecht, 1996.
6. R. Silver, "Quantum entropy regularization," in this proceedings.
7. R. Fischer, W. von der Linden, and V. Dose, "Adaptive kernels and occam's razor in inversion problems," in *MAXENT96 - Proceedings of the Maximum Entropy Conference 1996*, M. Sears, V. Nedeljkovic, N. E. Pendock, and S. Sibisi, eds., p. 21, NMB Printers, Port Elizabeth, South Africa, 1996.

DEPTH PROFILE RECONSTRUCTION FROM RUTHERFORD BACKSCATTERING DATA

U. V. TOUSSAINT, K. KRIEGER, R. FISCHER, V. DOSE
Max-Planck-Institut für Plasmaphysik, EURATOM Association
POB 1533, D-85740 Garching, Germany †

Abstract. An adaptive kernel method in the Bayesian framework together with a new simulation program for Rutherford backscattering spectroscopy (RBS) have been applied to the analysis of RBS data. Even in the case of strongly overlapping RBS peaks a depth profile reconstruction without noise fitting has been achieved. The adaptive kernel method leads to the simplest depth profile consistent with the data. Erosion and redeposition rates of carbon divertor plates in the fusion experiment ASDEX Upgrade could be determined by RBS-analysis of thin film probes before and after exposition to plasma discharges.

Key words: Rutherford Backscattering, Inverse Problem, Deconvolution

1. Introduction

Rutherford backscattering is one of the most important and most commonly applied techniques in surface analysis. Its main advantages are that it is fully quantitative and that precisions less than 1% can be achieved [1]. The interpretation of the data, however, is in many cases not straightforward. During the last decade several computer programs for the simulation and analysis of spectra obtained from RBS were developed, such as RUMP [2] or SIMNRA [3]. With these programs the determination of a depth profile is, however, a matter of trial and error. The user has to prescribe depth profiles of all elements and has to compare the simulated spectrum calculated from the input profiles with the data. The depth profiles are then adjusted until one obtains a reasonable agreement of simulated and measured data. Obviously this evaluation procedure has several shortcomings. It is a time-consuming cumbersome task, the accuracy of the achieved depth profile is unknown and in many cases there is an ambiguity between different depth profiles which fit the data equally well. The combination of the adaptive kernel method in the Bayesian framework [4] with an RBS-simulation program allows to overcome these disadvantages and extends the potential of Rutherford backscattering spectroscopy.

†Email: udo.v.toussaint@ipp.mpg.de

W. von der Linden et al. (eds.), Maximum Entropy and Bayesian Methods, 107–114.

2. Basic Concepts of Rutherford Backscattering

In RBS-analysis, a sample is exposed to a beam of ions with mass m_0 (e.g. He-particles) with a well defined energy E_0 in the order of MeV. Ions undergoing elastic Coulomb collisions with sample atoms are recorded in a solid state detector which views at a fixed deflection angle θ. The Rutherford cross-section for this coulombic projectile-target interaction is quantitatively known. The energy E' of the backscattered ions depends on the energy E before the collision, the mass of the ions m_0, the mass of their colliding partner M_i and the deflection angle θ :

$$E' = E \left[\frac{\sqrt{1 - \left(\frac{m_0}{M_i}\right)^2 \sin^2\theta} + \frac{m_0}{M_i}\cos\theta}{1 + \frac{m_0}{M_i}} \right]^2 . \quad (1)$$

From Eq. 1 we see that ions undergoing a collision with a heavy target atom loose less energy than ions colliding with a target atom of lower atomic mass. In addition, both primary ions and scattered ions loose energy on their way through the sample, depending on the stopping power. This is the main reason which enables RBS to be depth sensitive. The stopping power depends on the energy of the particles and the composition of the sample.
Fig. 1a depicts a typical RBS experiment. A thin overlayer (A) of atoms with a

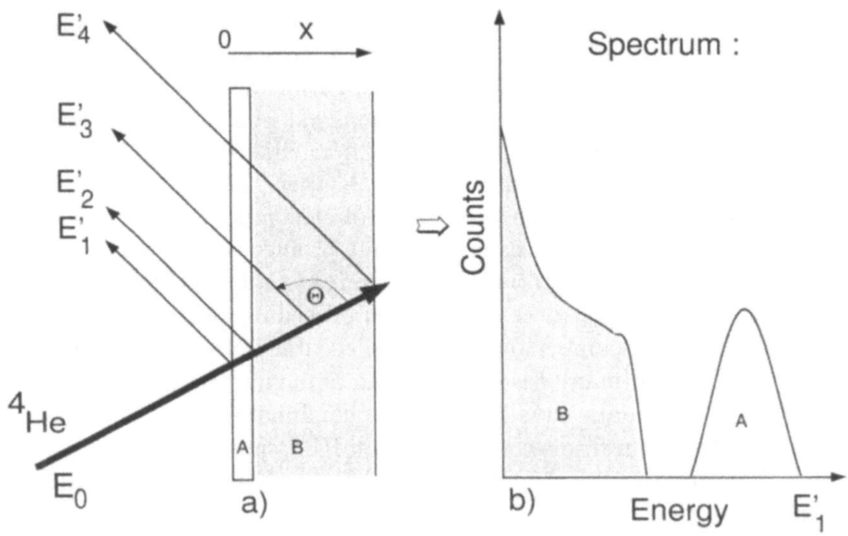

Figure 1. Schematic diagram of a RBS-experiment a) and the corresponding spectrum b).

high atomic mass M_A is on top of the bulk substrate (B) with a lower atomic mass M_B. In the energy spectrum of backscattered particles (Fig. 1b), the film A leads to a spectral peak at higher energies, broadened by the apparatus transfer function and the statistical fluctuations of the energy loss of the ions. Scattering from B produces a broadened step at lower energies. The high energy side of this step originates from scattering from the topmost B-Layer. The increase of the spectrum with decreasing energy results mainly from the $\frac{1}{E^2}$ dependence of the Rutherford cross section.

3. Simulation of RBS-Spectra

For a spectrum synthesis the sample is divided into sub-layers with thickness Δx. The spectrum is calculated from the superimposed contributions of scattering processes from all elements in all sub-layers of the sample. For each sub-layer the concentrations on the layer-boundaries must be given. Inside the sub-layer the concentration profile is assumed to interpolate linearly. In each sub-layer the energy loss of the ions inside this layer and the cross-sections are determined.

Cross-Section Data: The actual cross-section deviates from the well known Rutherford cross-section [5] at both, high and low energies. The low-energy discrepancy is caused by partial screening of the nuclear charges by the electronic shells [5]. This screening is taken into account by a correction factor $C(E, \Theta)$ [6]. At high energies the cross sections deviate from the Rutherford cross-section due to the influence of the nuclear force [7]. This is unimportant in the present case.

Stopping Power Data: The two dominant processes of energy loss of a penetrating ion are the interactions of the moving ion with bound or free electrons in the target, and the interactions of the moving ion with the screened or unscreened nuclei of the target atoms. The electronic stopping power data are taken from Ziegler, Biersack and Littmark [8]. The nuclear stopping power for helium is calculated from [8]. In compound materials, Bragg's rule is used,

$$\left(\frac{dE}{dx}\right)_{total} = \sum_i c_i \left(\frac{dE}{dx}\right)_i , \tag{2}$$

to calculate the effective stopping power $\left(\frac{dE}{dx}\right)_{total}$ from the concentrations c_i and the stopping power $\left(\frac{dE}{dx}\right)_i$ of each individual component i. The key assumption of Bragg's rule that the interaction between the ion and a target atom is independent of the environment holds in most cases. In some compounds such as oxides the deviations from Bragg's rule predictions may, however, be of the order of 10% to 20% [9].

Energy Loss Straggling: The energy loss of charged particles penetrating material is accompanied by a spread of the beam energy which is due to statistical fluctuations of the energy transfer in the loss channels. As the number of interactions is high, the energy broadening is well described by a Gaussian. The program uses Bohr's theory of energy-loss straggling [10], together with corrections by Chu [11], which include the electron binding in the target atoms. The energy dependence of the stopping power results further in a non-stochastic broadening (or squeezing) of the energy distribution of the ion beam. The energy width ΔE_f after passing the sub-layer is given by [12]:

$$\Delta E_f = \frac{S(E_f)}{S(E_i)} \Delta E_i \tag{3}$$

with E_i, E_f as the mean energies and $S(E_i)$, $S(E_f)$ as the stopping powers at the entrance and exit of the sub-layer, respectively.

4. Experiment

The interpretation of RBS data is required for the analysis of erosion measurements of plasma facing materials in fusion experiments. The solid inner walls surrounding the plasma are subjected to an intense bombardment by plasma particles because the confinement of the plasma by the confining magnetic field is not perfect. The surfaces of the inner walls are mainly modified by ion implantation, erosion and by deposition of material from other wall areas.

One major problem in fusion research is to find a wall material where wall erosion rate and wall modifications are small and tolerable [13]. The importance of this problem for planned fusion power plants is emphasized by an erosion analysis for ITER [14]. The modeled gross erosion yield of a carbon-divertor could reach a maximum of 5m/burning-year, which is reduced by redeposition down to about 0.5m/burning-year. The modeling, however, faces exceptional difficulties due to complex hydrocarbon transport phenomena and the lack of input data (e.g. for low energy sputtering). Therefore experimental determination of erosion and redeposition yields is necessary to validate the modeling and to improve the quantitative knowledge of the fundamental erosion processes.

To determine carbon erosion rates in the divertor of ASDEX Upgrade, graphite probes which were covered with a 150nm layer of ^{13}C were exposed to single plasma discharges. ^{13}C was used because chemical erosion is unaffected by isotope substitution and to allow the measurement of redeposited ^{12}C eroded at other plasma facing components. Furthermore the stopping power in ^{13}C and ^{12}C is the same and so the limited accuracy of the stopping power in the simulation cancels. The sample was introduced in the outer divertor of ASDEX Upgrade (circle in Fig. 2) covering in particular the strike point, which is the point where the outermost last closed magnetic flux line touches the plate surface with a corresponding maximum of the power load.

The samples were analyzed before and after plasma exposure with a total exposure time of 4 seconds using RBS with 2.0 MeV ^{4}He ions. The backscattered particles were detected at a scattering angle of $\Theta = 165°$. The width of the apparatus transfer function is about 19keV FWHM [15]. Fig. 3 shows typical spectra before and after plasma exposure. Before plasma exposure the signal from the ^{13}C-layer at higher energy is separated by a gap from the part of the spectrum corresponding to the underlying ^{12}C-bulk material. After plasma exposure the high energy edge of the signal from ^{13}C has shifted towards lower energies. This indicates that there is no longer ^{13}C at the surface of the sample. The peak at 430 keV is due to the ^{12}C at the sample surface and from the ^{13}C fraction below the surface. The difference

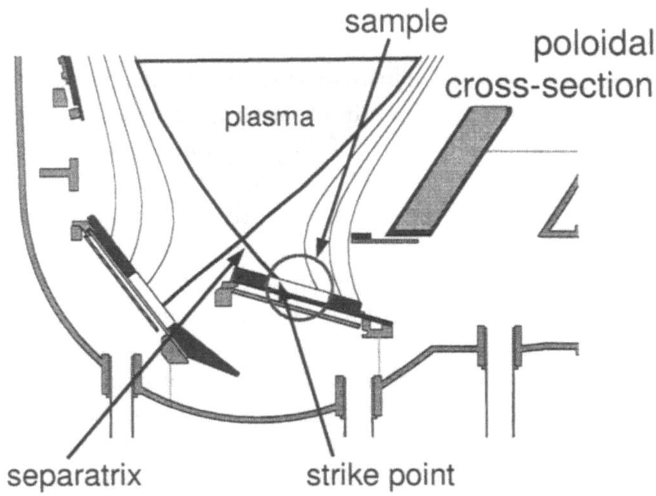

Figure 2. *Poloidal cross-section of ASDEX-Upgrade. The circle indicates the position of the sample on the outer divertor in ASDEX-Upgrade. The separatrix is the outermost closed magnetic flux line. The point the separatrix touches the divertor is called the strike point*

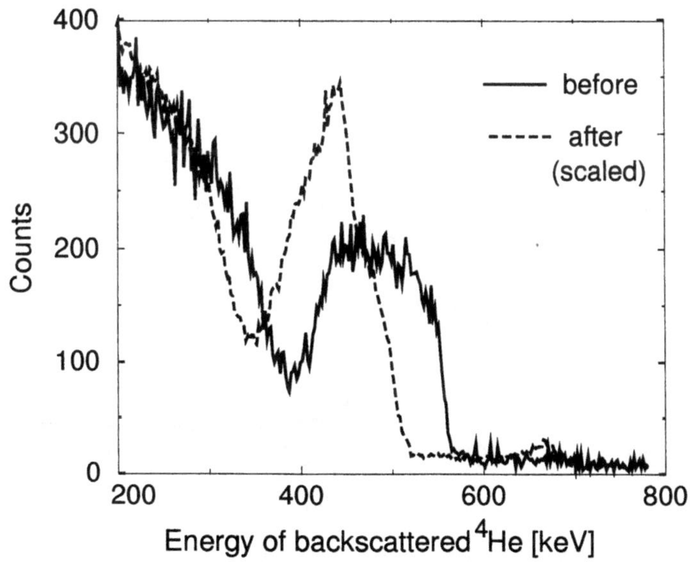

Figure 3. *RBS-spectra before and after plasma exposure. The shift of the high energy edge is clearly visible.*

of the RBS-spectra before and after exposure contains the information about the erosion and redeposition yields.

5. Results

To determine the concentration depth profiles from the measured RBS data a simple χ^2-fit is insufficient and results in useless rapidly oscillating depth profiles. This is due to the ill-conditioned nature of the inversion problem which results from the energy-straggling broadening, the finite apparatus-induced energy resolution and the counting statistics. Furthermore the optimal grid, given by the thickness of the sub-layers the sample is divided in, is unknown.
For this kind of problems the adaptive kernel method is well suited. The concept of adaptive kernels provides local smoothness which makes the result robust against noise corruption. The locality of the information content of the data is taken into consideration by the local varying kernel widths. Constraints like positivity or other prior knowledge (like bulk concentrations) are easy to include. The used adaptive kernel method is presented in detail in this proceeding [16].
Fig. 4a shows the reconstructed ^{12}C and ^{13}C−depth profiles of a sample before

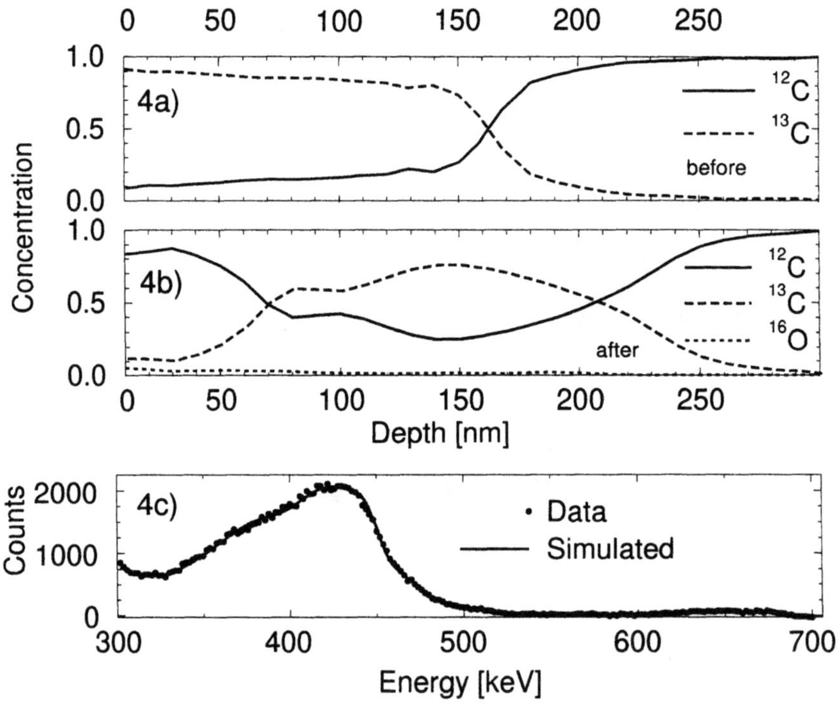

Figure 4. panels a) and b): ^{12}C and ^{13}C-distribution before and after plasma exposure. Panel c): RBS-data (black dots) and the calculated RBS-spectrum (grey line) from the depth profile in the panel b).

plasma exposure. The concentrations in each layer sum up to one. The surface concentration of ^{13}C (on the left side) is above 90% and decreases only slightly to a depth of about 150nm. The remaining 10% fraction of ^{12}C is caused by impurities in the coating process. The broad transition between the ^{13}C−layer and the ^{12}C−bulk can be explained by the interface roughness of the virgin sample.

After 4 seconds of plasma exposure the depth profiles have changed dramatically, as shown in Fig. 4b. There is a ^{12}C–layer with a thickness of about 70nm on top of the ^{13}C. The maximum concentration of ^{13}C has decreased, however, the thickness of the ^{13}C-layer is with about 170nm nearly unchanged. Furthermore, there is a continuous level of ^{12}C in the whole sample with a minimum concentration of 20%. Since diffusion due to thermal effects could be excluded, the impacting ^{12}C-atoms must have mixed the material. Fig. 4c shows the RBS-data as black dots and the calculated RBS-spectrum (solid line) based on the depth profile shown in Fig. 4b. The agreement is within the counting statistics.
With samples in different distances to the strike point we achieved a laterally resolved determination of erosion and deposition as shown in Fig. 5. The height of the ^{13}C-tracer was 153nm before exposure (dashed line in Fig. 5). The grey shaded

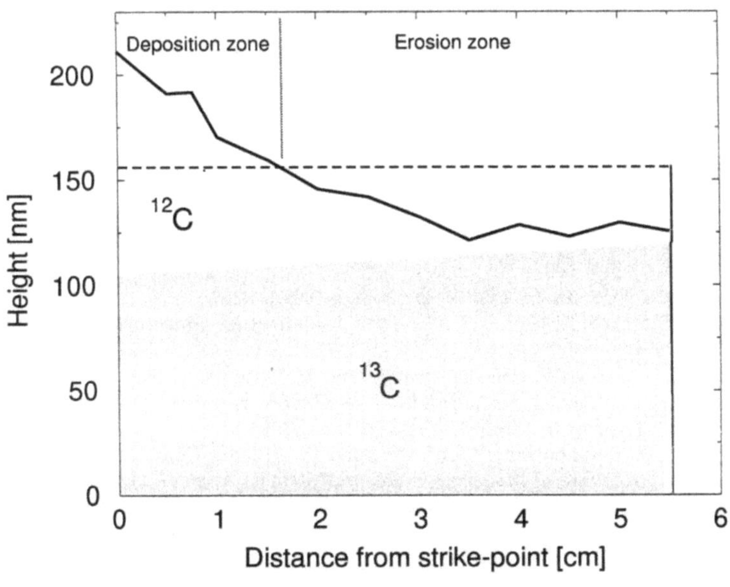

Figure 5. Schematic picture of the ^{12}C and ^{13}C distribution before and after plasma exposure. The grey dashed line gives the height of the ^{13}C-tracer before plasma exposure. The grey shaded area marks the height of ^{13}C after plasma exposure and the difference between the upper black line and the grey shaded area gives the height of deposited ^{12}C.

area marks the thickness of the ^{13}C-layer after plasma exposure. The highest erosion of 40nm was observed at the strike point. With increasing distance the erosion reduces slightly to $\simeq$ 30nm in 5cm distance. The solid line represents the joint height of the ^{13}C and deposited ^{12}C under the assumption that no ^{12}C from the bulk was eroded. The difference between the solid line and the grey shaded area of ^{13}C is the height of deposited ^{12}C. The amount of ^{12}C which covers the ^{13}C is largest at the strike point with over 100nm and reduces down to 10nm in a distance of 5.5cm. Near the strike point the redeposition of carbon is larger than the erosion, which makes this location a net deposition zone. By contrast, in a distance larger than 1.5cm from the strike point there is a net erosion area.

Fig. 5 is only a schematic representation which shows the total amount of ^{12}C and ^{13}C in a simplified distribution. It can be seen from the depth profiles in Fig. 4 that after plasma exposure there are no longer clearly separated layers of the two different isotopes and pronounced mixing has occurred. The large spatial variation of erosion and deposition rates shows, that the lifetime of plasma facing components can only be evaluated for specific local conditions.

6. Conclusions

With the used combination of the RBS-simulation program and the adaptive kernel method the capabilities of RBS-data evaluation have been considerably extended. This allows to study erosion, deposition and mixing of carbon as inner wall material in fusion experiments by using different isotopes which have no influence on the chemical erosion. The experiment shows a spatially varying net erosion/deposition rate with large mixing. Further investigations are necessary to answer the question of the long-time behavior of the erosion of the inner wall materials facing different plasma conditions.

References

1. C. Jeynes, Z. H. Jafri, R. P. Webb, A. C. Kimber, and M. Ashwin *Surface and Interface Analysis*, **25**, p. 254, 1997.
2. L. R. Doolittle *Nucl. Instr. and Meth. B*, **9**, p. 291, 1985.
3. M. Mayer, "Simnra user's guide," *Max-Planck-Institut für Plasmaphysik, Technical Report*, **9**, p. 113, 1997.
4. R. Fischer, M. Mayer, W. von der Linden, and V. Dose *Phys. Rev. E*, **55**, p. 6667, 1997.
5. J. Tesmer and M. Nastasi, eds., *Handbook of Modern Ion Beam Materials Analysis*, Materials Research Society, Pittsburgh, Pennsylvania, 1995.
6. H. Anderson, F. Besenbacher, P. Loftager, and W. Moeller *Phys. Rev. A*, **21**, p. 1891, 1980.
7. M. Bozoin in *Handbook of Modern Ion Beam Materials Analysis*, J. Tesmer and M. Nastasi, eds., Materials Research Society, Pittsburgh, Pennsylvania, 1995.
8. J. Ziegler, J. Biersack, and U. Littmark, *The Stopping and Ranges of Ions in Matter*, vol. 1, Pergamon Press, New York, 1985.
9. J. Ziegler and J. Manoyan *Nucl. Instr. Meth. B*, **35**, p. 215, 1988.
10. N. Bohr *Mat. Fys. Medd. Dan. Vid. Selsk.*, **18**, 1948.
11. W. Chu *Phys. Rev.*, **13**, p. 2057, 1976.
12. E. Szilàgy, F. Pàszti, and G. Amsel *Nucl. Instr. Meth. B*, **100**, p. 103, 1995.
13. R. Behrisch *Phys. Res.*, **8**, p. 569, 1988.
14. J. N. Brooks, D. Alman, G. Federici, D. N. Ruzic, and D. G. White, "Erosion/redeposition analysis: status of modeling and code validation for semi-detached edge plasmas," in *13th International Conference on Plasma Surface Interactions in Controlled Fusion Devices*, San Diego, California, 1998.
15. V. Dose, R. Fischer, and W. von der Linden, "Deconvolution based on experimentally determined apparatus functions," in *Maximum Entropy and Bayesian Methods*, J. Rychert, G. Erickson, and R. Smith, eds., Kluwer Academic Publishers, Dordrecht, 1998.
16. R. Fischer, W. Jacob, W. von der Linden, and V. Dose, "Bayesian reconstruction of electron energy distributions in helium plasmas from emission line intensities," in *Maximum Entropy and Bayesian Methods*, V. Dose, W. von der Linden, R. Fischer, and R. Preuss, eds., Kluwer Academic Publishers, Dordrecht, 1999.

BAYESIAN IMAGE SEGMENTATION USING A DYNAMIC PYRAMIDA STRUCTURE

H. REHRAUER AND K. SEIDEL

Swiss Federal Institute of Technology ETH Zürich,
Communication Technology Lab, Image Science Group
Gloriastr. 35, CH-8092 Zürich[‡]

AND

M. DATCU

German Aerospace Center DLR,
German Remote Sensing Data Center DFD, Oberpfaffenhofen
D-82234 Weßling[§]

Abstract. We present a Bayesian segmentation algorithm which is part of a fully Bayesian approach for automatic information extraction from satellite images. It was shown that pyramidal image models based on multi-scale Markov random fields in combination with a texture model yield good classification and segmentation results. The texture model is used for an initial characterization and then an optimal segmentation is inferred using the multi-scale random field defined on a pyramid structure. Segment probabilities are calculated in a fine-to-rough analysis and segmentation is performed by a rough-to-fine decision algorithm that maximizes the a posteriori probability for the pyramid. The procedure is iterated until it converges to a stable solution.

We improve the maximization procedure by optimizing the underlying pyramidal structure of the multi-scale Markov random field. Neighborhood dependencies are switched on and off according to the image data.

Key words: Image classification, segmentation, multi-scale Markov random fields, hierarchical modeling, Bayesian inference, image pyramid The hierarchical organization allows a fast computation and the segmentations obtained are smooth, even at coarse scales. Additionally it has the advantage to be exactly tractable.

[‡]Email: {rehrauer,seidel}@vision.ee.ethz.ch
[§]Email: mihai.datcu@dlr.de

W. von der Linden et al. (eds.), Maximum Entropy and Bayesian Methods, 115–122.

1. Introduction

Image segmentation algorithms are an essential component of any image understanding, image analysis or image description system. These systems are in use for a wide variety of purposes, e.g. object detection, motion analysis, scene interpretation, and under current development for purposes like content-based query in image archives [1]. The algorithm presented is designed to work on large images in the frame of an image information extraction system for remote-sensing image archives [2].

Most of the approaches presented in the past use the two-stage scheme [3] [4]:

1. Features are extracted from the pixels and associated with image content.
2. The image is segmented into regions of homogeneous content.

In this paper we assume that there exists a stochastic *emission* model that defines probabilities for pixel values if the local image content is given. The focus of our contribution lies on the second stage: The determination of segments of *homogeneous* content. ('Homogeneous' is to understand as 'homogeneous with respect to the emission model'.) Existing methods model the shape of regions using Markov random fields with smooth spatial behavior [3] . They are usually computationally expensive requiring simulated annealing or similar iterative schemes. The extension to multi-scale Markov random fields (MSRF) which are hierarchically ordered, reduces the computing time significantly [5] [6] [7]. As a disadvantage some MSRF models are either not exactly tractable or produce rather blocky regions.

Here we model within a Bayesian framework the pixel labels as a MSRF. As an essential extension of the approach of Bouman and Shapiro [5], we optimize dynamically the structure of the dependencies of the MSRF according to the data. This results in smoother and more realistic segmentations.

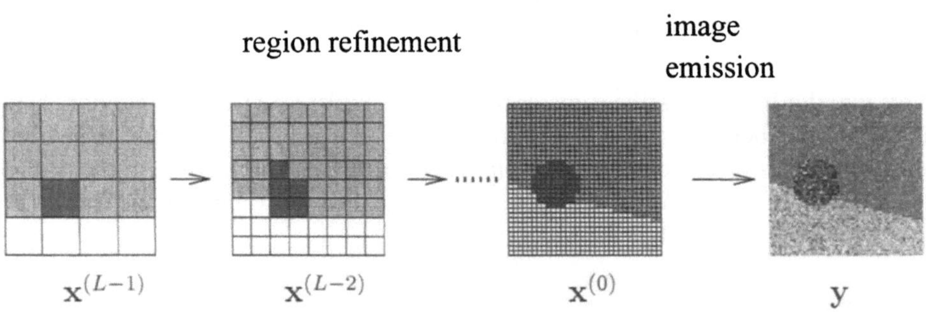

Figure 1. Image model. The process starts on the left at scale $L-1$ with a coarse field of labels $\mathbf{x}^{(L-1)}$ that give the region type. Through Markov transitions new label fields are created. With each transition the number of labels is quadrupled and the regions are refined. At the final resolution an emission process creates the actual image $\mathbf{y}$. In the example the image consists of three regions with different type of Gaussian noise.

2. Pyramidal Image Model

For the image analysis we use a doubly stochastic image model composed of two different random processes. As shown in Fig. 1, there is first a region process that establishes the regions of homogeneous image content. The second process produces the actual pixel values given the image content; in our context this is called the *emission* process.

The region process is modeled by a multi-scale Markov random field. The random variables (RVs) are called ***labels.*** They indicate the image content of a certain image region. The process starts with a 2D coarse-scale random field giving the region labels at coarse resolution. The field is successively refined until the final resolution of the image is reached. Labels at the finest scale give the image content of individual pixels.

The RVs of the MSRF are described by a set of 2D random fields $\{\mathbf{x}^{(n)} \mid n = 0, \ldots, L-1\}$. Each field is defined on a 2D lattice $S^{(n)}$, $n = 0, \ldots, L-1$. An individual RV is denoted by $X_s^{(n)}$ where $s \in S^{(n)}$. Random quantities are always denoted by upper case letters while realizations are lower case. The number of points in the zeroth lattice $|S^{(0)}|$ is equal to the number of image pixels. When moving from scale n to scale $n+1$ the number of lattice points shrinks by a factor of 4. The possible values of the labels are $x_s^{(n)} = 0, 1, \ldots, M-1$ where M is the maximum number of image content types that might be in the image. The Markovianity condition is

$$\mathrm{P}[\,X_s^{(n)} = x_s^{(n)} \mid \mathbf{X}^{(m)} = \mathbf{x}^{(m)},\, m \geq n\,] = \mathrm{P}[\,X_s^{(n)} = x_s^{(n)} \mid X_{\mathcal{P}(s)}^{(n+1)} = x_{\mathcal{P}(s)}^{(n+1)}\,] \quad (1)$$

or in the shorthand notation used throughout the paper

$$\mathrm{P}[\,x_s^{(n)} \mid \mathbf{x}^{(m)},\, m \geq n\,] = \mathrm{P}[\,x_s^{(n)} \mid x_{\mathcal{P}(s)}^{(n+1)}\,] \quad (2)$$

where $\mathcal{P}(\cdot)$ is the parent operator. It returns the index of the parent of a RV $X_s^{(n)}$ and defines the dependencies (parent-child links) of the RVs. The parent of an RV at scale n is always located at scale $n+1$. Eq. (1) means that if the direct parent of $x_s^{(n)}$ is given, then $X_s^{(n)}$ is independent of all other RVs in the past (at coarser scales) and the present (current scale).

Fig. 2 shows the MSRF stacked as layers of a pyramid. The indicated parent-child links form the classical quadtree structure. Note that we didn't specify further the parent operator. It can also produce rather irregularly connected pyramids.

The emission process is not specified here. Its underlying model may be a stochastic multispectral or texture model. The only requirements are: It must be able to provide probabilities for the pixel values if the content is known and these probabilities must factorize

$$\mathrm{P}[\,\mathbf{y} \mid \mathbf{x}^{(0)}\,] = \prod_{s \in S^{(0)}} \mathrm{P}[\,y_s \mid x_s^{(0)}\,]. \quad (3)$$

This assumes that pixel values are independent if their region type is known. While this assumption would produce image discontinuities if the model was used

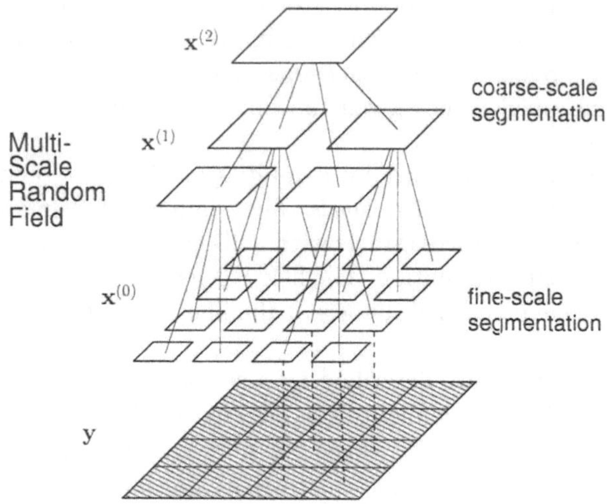

Figure 2. Representation of the MSRF as a pyramid. The connections between the layers show the dependencies of the RVs. Here a uniform quadtree structure is shown. Below the pyramid we show the actual image $\mathbf{y}$.

to synthesize images, we accept it for analysis purposes. For texture models the term 'pixel' refers to the set of pixels in an estimation window.

The probability for an entire configuration—MSRF and image—is

$$\mathrm{P}[\mathbf{y}, \{\mathbf{x}^{(n)}\}] = \mathrm{P}[\mathbf{y} \,|\, \mathbf{x}^{(0)}]\, \mathrm{P}[\mathbf{x}^{(0)} \,|\, \mathbf{x}^{(1)}] \cdots \mathrm{P}[\mathbf{x}^{(L-1)}]. \tag{4}$$

An important property of the transition probabilities is that they factorize

$$\mathrm{P}[\mathbf{x}^{(n)} \,|\, \mathbf{x}^{(n+1)}] = \prod_{s \in S^{(n)}} \mathrm{P}[x_s^{(n)} \,|\, x_{\mathcal{P}(s)}^{(n+1)}] \tag{5}$$

which means that it is an effective 1D Markov model with a clearly defined past and future.

3. Bayesian Segmentation Algorithm

Within our model we define the segmentation problem as:

> *Find the region labels and parent-child connections that are most likely given the image.*

We perform this maximization in the following way:

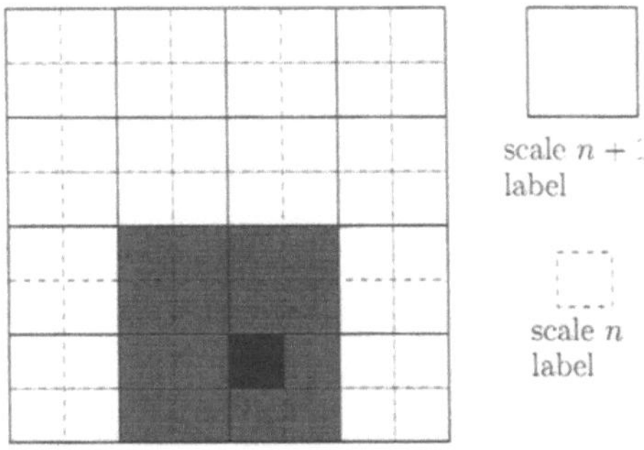

Figure 3. Potential parents of a label. The fine-scale labels are delineated by dashed lines, the coarse-scale labels by solid lines. The darkly shaded fine-scale label can select its parent only among the shaded coarse-scale labels. This keeps regions localized and reduces computing time.

1. Start with given parent-child links.
2. Calculate the conditioned image likelihoods

$$\mathrm{P}[\,\mathbf{y}\,|\,\mathbf{x}^{(n)}\,] \qquad \text{, for } n = 0, 1, \ldots, L-1$$

3. Find the labels $\hat{\mathbf{x}}^{(L-1)}$ that maximize the posterior probability:

$$\max_{\mathbf{x}^{(L-1)}} \mathrm{P}[\,\mathbf{x}^{(L-1)}\,|\,\mathbf{y}\,].$$

   This gives the segmentation at the coarsest scale.
4. For $n = L-2, L-3, \ldots, 0$: Find the segmentation at scale n and the parent operator that maximize:

$$\max_{\mathbf{x}^{(L-1)},\,\mathcal{P}} \mathrm{P}[\,\mathbf{x}^{(n)}\,|\,\mathbf{y},\,\mathbf{x}^{(n+1)}\,].$$

5. Repeat steps 2–4 until no more parent-child links change.

In principle the initial parent-child links can be chosen arbitrarily. However we always start with the uniform quadtree configuration depicted in Fig. 2. During the optimization in step 3 we have to find the optimal parent of a label $x_s^{(n)}$. As potential parents we consider only those RVs at the scale $n+1$ that are located in the spatial neighborhood of the lattice point s. The exact size of the neighborhood is shown in Fig. 3. The restriction to the neighborhood keeps regions localized and makes the search computationally feasible.

Calculation of image likelihoods is done recursively. Starting with the emission probability

$$\mathrm{P}[\,\mathbf{y}\,|\,\mathbf{x}^{(0)}\,] = \prod_s \mathrm{P}[\,y_s\,|\,x_s^{(0)}\,]. \tag{6}$$

The recursion can be stated as

$$\mathrm{P}[\mathbf{y}\,|\,\mathbf{x}^{(n)}\,] = \sum_{\mathbf{x}^{(n-1)}} \mathrm{P}[\mathbf{y}\,|\,\mathbf{x}^{(n-1)}\,]\,\mathrm{P}[\mathbf{x}^{(n-1)}\,|\,\mathbf{x}^{(n)}\,]. \tag{7}$$

It can be performed independently for each label

$$\mathrm{P}[\,y_s^{(n)}\,|\,x_s^{(n)}\,] = \prod_{r\in\mathcal{C}(s)} \sum_{x_r^{(n-1)}} \mathrm{P}[\,y_r^{(n-1)}\,|\,x_r^{(n-1)}\,]\,\mathrm{P}[\,x_r^{(n-1)}\,|\,x_s^{(n)}\,] \tag{8}$$

where $y_s^{(n)}$ is the set of pixels dependent on $x_s^{(n)}$ and $\mathcal{C}(s)$ is set of direct children of $x_s^{(n)}$. Since all likelihoods factorize we are in the position to carry out the maximization for each label of a scale independently. At the coarsest scale we invert the probabilities using Bayes rule

$$\mathrm{P}[\,x_s^{(L-1)}\,|\,y_s^{(L-1)}\,] = \frac{\mathrm{P}[\,y_s^{(L-1)}\,|\,x_s^{(L-1)}\,]\,\mathrm{P}[\,x_s^{(L-1)}\,]}{\mathrm{P}[\,y_s^{(L-1)}\,]}. \tag{9}$$

This posterior probability is easily maximized over the set of possible values $x_s^{(L-1)} \in [\,0, 1, \ldots, M-1\,]$. At intermediate scales we maximize

$$\mathrm{P}[\mathbf{x}^{(n)}\,|\,\mathbf{y},\,\hat{\mathbf{x}}^{(n+1)}\,] \;=\; \frac{\mathrm{P}[\mathbf{y}\,|\,\mathbf{x}^{(n)}\,]\,\mathrm{P}[\mathbf{x}^{(n)}\,|\,\hat{\mathbf{x}}^{(n+1)}\,]}{\mathrm{P}[\mathbf{y}\,|\,\hat{\mathbf{x}}^{(n+1)}\,]}$$

which again is done for all lattice points $s \in S^{(n)}$ independently. We have to maximize with respect to the RV $x_s^{(n)}$ and the parent operator $\mathcal{P}$:

$$\max_{x_s{}^{(n)}\,\mathcal{P}(s)} \;\mathrm{P}[\,y_s^{(n)}\,|\,x_s{}^{(n)}\,]\;\mathrm{P}[\,x_s^{(n)}\,|\,\hat{x}_{\mathcal{P}(s)}^{(n+1)}\,] \tag{10}$$

An important point is the choice of the transition probabilities $\mathrm{P}[\,x_s^{(n)}\,|\,x_{\mathcal{P}(s)}^{(n+1)}\,]$. It can introduce additional prior knowledge of the relationships between region types. We use the functional form:

$$\mathrm{P}[\,x_s^{(n)}\,|\,x_{\mathcal{P}(s)}^{(n+1)}\,] = \begin{cases} \theta & \text{if } x_s^{(n)} = x_{\mathcal{P}(s)}^{(n+1)} \\ \frac{1-\theta}{M-1} & \text{else} \end{cases} \tag{11}$$

This choice says that if there is a certain region type at coarse scale then with probability θ the same region type is present at the fine scale. Different region types may occur at the fine scale, but there is no special preference among them.

4. Applications

The scheme has been applied to synthetic apperture radar (SAR) images from the XSAR sensor. This is a German sensor which was on board of several space shuttle missions. As image emission model we used a texture model based on Gibbs Markov random fields [8]. This model is combined with an unsupervised soft clustering algorithm that estimates the number of different textures, i.e. contents, in the image and returns the probabilities that are needed as input for the segmentation. For segmentation the parameter of Eq. (11) was set to $\theta = 0.7$. We compare our model to the model proposed by Bouman and Shapiro [5] that works with a quadtree with fixed dependencies.

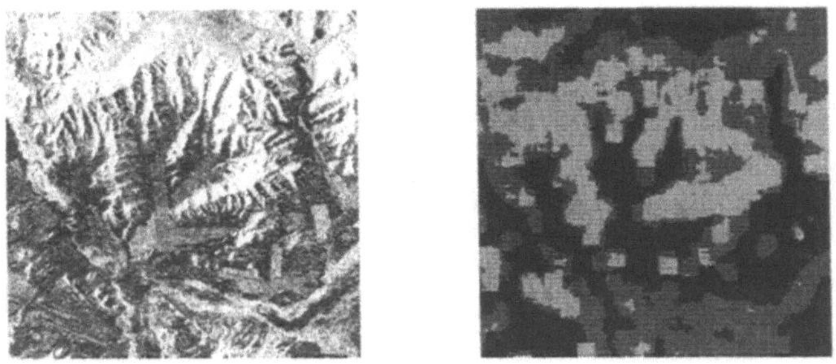

Figure 4. XSAR image and a segmentation obtained by clustering the texture features in the feature space.

A typical example image and its maximum likelihood segmentation derived from a hard clustering of the texture features are shown in Fig. 4. The multi-scale segmentations shown in Fig. 5 contain significantly fewer one-pixel regions than the maximum likelihood segmentation. This is a desired feature since these regions represent very often noise and are irrelevant for a compact and condensed image description. The regions at scale 0 in Fig. 5 are also more compact and less frayed out than those of Fig. 4. Further the example shows that our model produces coarse-scale segmentations that have a high fidelity to the true region boundaries and are smoother than in the Bouman/Shapiro model.

5. Summary

We have presented a fast hierarchical image segmentation algorithm based on a multi-scale Markov random field model. This model implicitly takes into account that neighboring image pixels are correlated and have a high probability to belong to the same region type. This correlation is not *imposed by the model* but *learned from the data* during the segmentation. The algorithm combines properties of region growing schemes and feature space clustering.

Experiments show that it produces segments with properties suitable for a compact image description. The segments have rather smooth boundaries and are well localized. Additionally it removes undesired single-pixel regions. The returned coarse-scale segmentations are of importance since they are reduced versions of the original segmentation which can be fast browsed – a favorable aspect for large image databases providing content-based services.

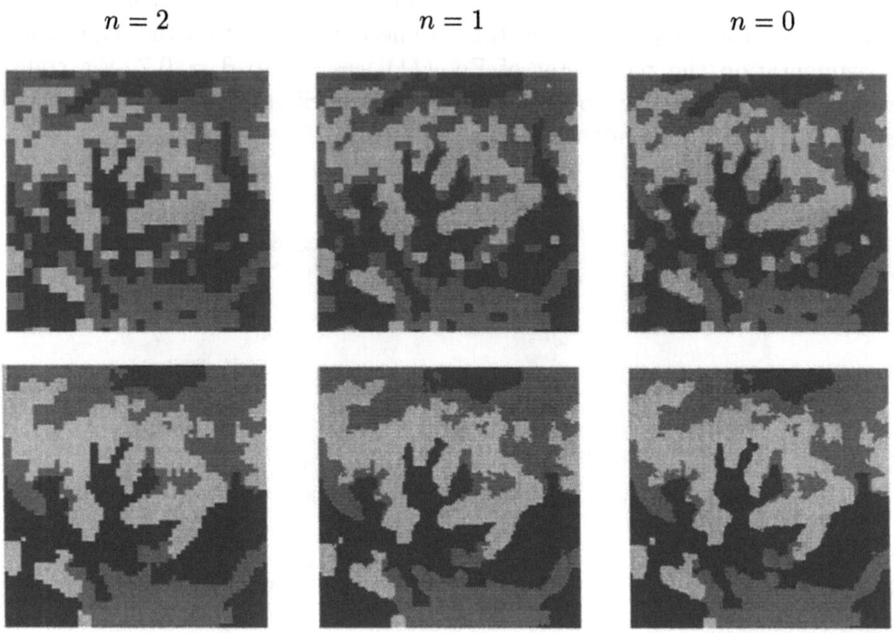

Figure 5. Segmentations at multiple scales. The upper row shows the result of the fixed quadtree model, the lower row shows the segmentations obtained with the dynamic pyramid. The noisy output of the texture model results in an also noisy full scale segmentation. However the dynamic structure MSRF produces smooth coarse scale segmentations with high fidelity to the real region boundaries.

References

1. M. Flickner, H. Sawhney, W. Niblack, J. Ashley, B. D. Q. Huang, M. Gorkani, J. Hafner, D. Lee, D. Petkovic, D. Steele, and P. Yanker, "Query by image and video content: the QBIC system," *IEEE Computer*, **28**, (9), pp. 23–32, 1995.
2. K. Seidel, R. Mastropietro, and M. Datcu, "New architectures for remote sensing image archives," in *A Scientific Vision for Sustainable Development, IGARSS'97*, T. I. Stein, ed., vol. 1, pp. 616–618, 1997.
3. S. Geman and D. Geman, "Stochastic relaxation, Gibbs distributions, and the Bayesian restoration of images," *IEEE Trans. on PAMI*, **6**, pp. 721-741, Nov. 1984.
4. J. Mao and A. K. Jain, "Texture classification and segmentation using multiresolution simultaneous autoregressive models," *Pattern Recognition*, **25**, (2), pp. 173–188, 1992.
5. C. A. Bouman and M. Shapiro, "A multiscale random field model for Bayesian image segme ntation," *IEEE Trans. on Image Processing*, **3**, pp. 162–177, Dec. 1994.
6. J. Puzicha and J. Buhmann, "Multiscale annealing for real-time unsupervised texture segmentation," in *Proceedings of the Int. Conf. Comp. Vision*, IEEE, 1998.
7. H. Rehrauer, K. Seidel, and M. Datcu, "Multiscale markov random fields for large image datasets representation," in *A Scientific Vision for Sustainable Development, IGARSS'97*, T. I. Stein, ed., vol. 1, pp. 255–257, 1997.
8. M. Schröder, H. Rehrauer, K. Seidel, and M. Datcu, "Spatial information retrieval from remote sensing images: B. Gibbs Markov random fields," *IEEE Tr. on Geoscience and Remote Sensing*, 1998. submitted 11/97.

ADAPTIVE IMAGE RECONSTRUCTION WITH PREDICTIVE MODEL

V. M. ARTEMIEV AND A. O. NAUMOV
Institute of Applied Physics
Akademicheskaya 16, Minsk 220072, Belarus ‡

AND

G.-R. TILLACK
Federal Institute for Materials Research and Testing (BAM)
Unter den Eichen 87, 12205 Berlin, Germany §

Abstract. Prior knowledge concerning information about the image and noise properties strongly influence the performance of image reconstruction from projections in computerized tomography. The authors propose an adaptive recursive 2D image reconstruction under uncertain conditions for the image and statistical noise properties. The Reconstruction is considered as adaptive estimation problem on the basis of empirical data generated by a predictive image model. The projection model is introduced by the vector $\boldsymbol{y}(n)$ given by the components $y_m(n) = \sum_{r_1=1}^{R} \sum_{r_2=1}^{R} x_{mr}(n)\, a_r + \xi_m(n)$, $r = r_1, r_2$ with the random image values a_r on a rectangular grid of size $R \times R$, the number of current projection n ($n = 1, 2, 3, \cdots$), the detector number $m = 1, 2, \cdots, M$, the elements of the $M \times R^2$ projection matrix $x_{mr}(n)$, and the noise component $\xi_m(n)$. For the reconstruction step n only the new information provided by the projection $y_m(n)$ ($m = 1, 2, \cdots, M$) and the previous estimations $\widehat{a}_r(k)$ ($k = n-1, n-2, \cdots, n-1-n_0$) are used with predictive image model algorithm $\boldsymbol{a}^*(n) = \boldsymbol{\Phi}[\widehat{\boldsymbol{a}}(k), k = n-1, \cdots, n-1-n_0] = \boldsymbol{\Phi}(\widehat{\boldsymbol{a}}, n_0)$, where n_0 gives the order of the prediction model. For the solution of this problem the cost function $F(\widehat{\boldsymbol{a}}, \boldsymbol{a}, n)$ with the constraints $Q_m(\widehat{\boldsymbol{a}}, \boldsymbol{y}, n) = 0$ is employed. Because the prior PDD function for the image parameters and the noise properties are unknown, the empirical PDD is used which is determined from predicted image data. Therefore the adaptive, empirical data-based estimation criteria has the form $\min_{\widehat{a}_r(n)} F(\widehat{\boldsymbol{a}}, \boldsymbol{a}^*, n)$, $Q_m(\widehat{\boldsymbol{a}}, \boldsymbol{y}, n) = 0$, $\boldsymbol{a}^*(n) = \boldsymbol{\Phi}(\widehat{\boldsymbol{a}}, n_0)$. The proposed reconstruction algorithm is applied to simulated and experimental projection data, and compared with standard CT algorithms.

Key words: reconstruction algorithm, empirical data, adaptive procedure

‡Email: artemiev@iaph.bas-net.by
§Email: Gerd-Ruediger.Tillack@bam.de

W. von der Linden et al. (eds.), Maximum Entropy and Bayesian Methods, 123–130.

1. Introduction

Image reconstruction from projections is the main procedure in the X-Ray computer tomography [1]. Often it is required to perform the reconstruction task parallel to the sequence of projection measurements and to be applicable for objects with very different geometry and material properties. For this case, the use of recursive reconstruction algorithms with small calculation expenses is proposed which include a certain flexibility to handle the variations of the object properties. Because the reconstruction can be interpreted as a kind of parameter estimation problem the solution can be found among the class of recursive adaptive procedures. The singularity of the reconstruction is given by the ill-posed properties of the projection matrix which requires the application of a regularization procedure. In present work a statistical approach is used due to the probabilistic nature of the image forming process including noise.

Statistical estimation and regularization techniques need prior knowledge about the image and noise probability density distribution (PDD) functions. But for most of the practical cases this kind of information is not available. One way to avoid these difficulties is to employ principles of adaptive estimation as developed in automatic control theory [2]. The general idea used from adaptive estimation principles is the application of recursive optimization on the basis of empirical data composed from the current projection measurement and previously obtained image parameter estimates. To include previously treated empirical data so called predictive image models are introduced. This proposal is an extension of automatic control theory [2] which supports the restoration of the posterior image PDD function from empirical data. Another approach to the formulated problem based on pseudo-inverse linear filtering is described in [3].

2. Mathematical Model

The 2-D image model is represented as a discrete random field with components $a_{r_1 r_2}$ on a rectangular grid with space coordinates $r_1, r_2 = 1, 2, \cdots, R$. In X-ray tomography the components $a_{r_1 r_2}$ correspond to the linear attenuation coefficients of the object material. The reconstruction procedure is based on the treatment of projection measurements. The geometry for projection forming process is shown in fig. 1 with the following notations: current position number n of the X-ray point source, current number m of the photon detector, and the total number M of detectors.

The responce $y_m(n)$ of detector m for source position n is called ray sum and represented by

$$y_m(n) = \sum_r x_{mr}(n)\, a_r + \xi_m(n), \tag{1}$$

with the notations $r = (r_1, r_2)$, $m = 1, 2, \cdots, M$, $n = 1, 2, \cdots$, $a_r = a_{r_1 r_2}$, $x_{mr}(n) = x_{mr_1 r_2}(n)$, and $\sum_r = \sum_{r_1=1}^{R} \sum_{r_2=1}^{R}$. The elements of the projection

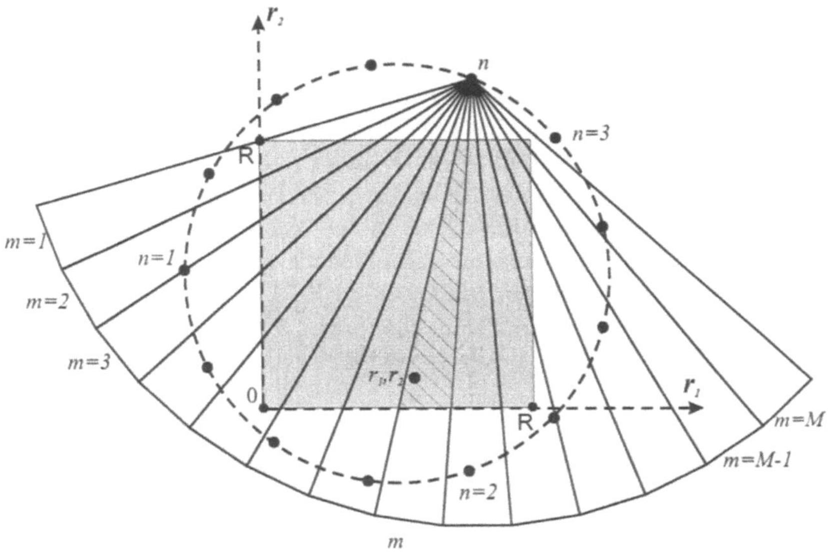

Figure 1. Geometrical setup

matrix $x_{mr}(n)$ are given by

$$x_{mr}(n) = \begin{cases} 1 & \text{for } a_r \text{ inside or on the right border of ray } m \\ 0 & \text{else} \end{cases} \tag{2}$$

As a result the sum $\sum_r x_{mr}(n) = C_m(n)$ gives the number of image elements in ray m and $\sum_{m=1}^{M} C_m(n) = R^2$ is equal to the total number of image elements. The contribution $\xi_m(n)$ on the right side of eq. 1 describes the noise component of the detector responce and is supposed to be random.

For any fixed source position n a set of ray sums $y_m(n)$ defines projection n by $\boldsymbol{y}(n) = (y_1(n), \cdots, y_M(n))$. Each projection contains information about the image parameters a_r. Because of the properties of the projection matrix given by 2 any particular image parameter belongs only to one ray sums 1.

Due to the ill-posed properties of the stated problem two approaches can be applied to solve the reconstruction problem: (*i*) variational approach, and (*ii*) statistical approach in general discussed in [4,5]. For both cases prior knowledge is required given in different form. The variational methods need knowledge about the smoothness properties of the image represented in a special form of the stabilizing functional as a part of the optimization criteria. The stochastic methods apply prior PDD functions to determine posterior distributions for the Bayesian estimation procedure for the minimization of the average value of the cost functional [6]. In the present work a kind of combination of this two approaches is used. The structure of the criteria derived from the variational methods but the cost functional is introduced instead of stabilizing functional. Since the adaptive reconstruction is examined the averaging of cost functional has to be carried out with the empirical PDD function.

3. Optimization Methodology

The optimization methodology has to include the above stated two properties of the reconstruction, namely recursivnes and adaptation. Due to this the following methodology is proposed employing the constraint-based criteria presented in [7].

The multidimentional cost function $F(\boldsymbol{a}, \widehat{\boldsymbol{a}}(n)) = F(\boldsymbol{a}, \widehat{\boldsymbol{a}}, n)$ is introduced with the set of image parameters $\boldsymbol{a} = (a_{r_1 r_2}; r_1, r_2 = 1, \cdots, R)$ and the corresponding set of estimates $\widehat{\boldsymbol{a}} = (\widehat{a}_{r_1 r_2}(n))$ in the reconstruction step n. This cost function allows to take into account the influence of the prior information and make available the regularization.

The optimization criteria applies in the form

$$\min_{\widehat{\boldsymbol{a}} \in A_0} \langle F(\boldsymbol{a}, \widehat{\boldsymbol{a}}, n) \rangle_p \tag{3}$$

where A_0 is the existing space for the estimation and the triangular brackets denote the average operator with respect to the image posterior distribution given by

$$\langle F(\boldsymbol{a}, \widehat{\boldsymbol{a}}, n) \rangle_p = \int_{\boldsymbol{a}} \mathrm{d}a^{R^2} \; F(\boldsymbol{a}, \widehat{\boldsymbol{a}}, n)\, w_p(\boldsymbol{a}, n) \tag{4}$$

with integration over the R^2 variables a_r. The posterior PDD function $w_p(\boldsymbol{a}, n) = w_p(\boldsymbol{a}, n | \boldsymbol{y}(1), \cdots, \boldsymbol{y}(n), w_0(n))$ can be determined from the Bayesian rule including the prior PDD function $w_0(n)$.

It is assumed that the prior PDD function is unknown. Therefore the criteria 3 has to be modified due to this uncertainty. One way is to introduce an empirical PDD function instead of the posterior function. During the recursive estimation the empirical PDD describes more and more precisely the particular image and provides adaptive properties for the reconstruction process.

The reconstruction on the basis of empirical data adapts the estimation algorithm for individual properties of a particular image and makes it flexible for the application to the different objects. There exist in principle two methods to introduce empirical PDD (EPDD) functions. The direct method employ the measured projection data $\boldsymbol{y}(n)$ and, consequently, the EPDD becomes a conditional distribution depending on the projection data $w_e(\boldsymbol{a}, n) = w_e(\boldsymbol{a}, n | \boldsymbol{y}(1), \cdots, \boldsymbol{y}(n))$. For the indirect method the estimations $\widehat{\boldsymbol{a}}(n)$ are used to determine the EPDD which is a conditional function of the estimates $w_e(\boldsymbol{a}, n) = w_e(\boldsymbol{a}, n | \widehat{\boldsymbol{a}}(1), \cdots, \widehat{\boldsymbol{a}}(n))$. In the present work the indirect method is applied. For this purpose the predictive image model conception can be applied [2].

Suppose that on the recursive step $n-1$ the n_0 estimations $\widehat{\boldsymbol{a}}(k)$ $(k = n - 1 - n_0, \cdots, n-1)$ are available. The predictive image parameters $\boldsymbol{a}^*(n)$ for the n-th step can be obtained with a finite memory prediction filter in the form

$$\boldsymbol{a}^*(n) = \boldsymbol{\Phi}[\widehat{\boldsymbol{a}}(k), k = n-1-n_0, \cdots, n-1] = \boldsymbol{\Phi}(\widehat{\boldsymbol{a}}, n_0), \tag{5}$$

with $\boldsymbol{a}^*(n) = (a^*_{r_1 r_2}(n))$, $\boldsymbol{\Phi} = (\Phi_{r_1 r_2})$ for $r_1, r_2 = 1, \cdots, R$.

Expression 5 gives the n_0 order prediction function. A number of methods are known for obtaining this function, e.g. an adaptive prediction procedure. In present work it is supposed that prediction function is given. The predicted image with parameters $\boldsymbol{a}^*(n)$ can be used for constructing the empirical PDD function.

Because of the function $\boldsymbol{a}^*(n)$ is defined the EPDD $w_e(\boldsymbol{a}, n)$ can be represented as a product of the δ-functions

$$w_e(\boldsymbol{a}, n) = \prod_{r_1=1}^{R} \prod_{r_2=1}^{R} \delta\left(a_{r_1 r_2} - a^*_{r_1 r_2}(n)\right). \tag{6}$$

It is important to note that the EPDD function 6 do not depend on the measurements $\boldsymbol{y}(n)$. For this case the optimization criteria 3 has to be modified, e.g. by introducing additional constraints which depend on the data. In the present paper the following constraints are used

$$Q\left[\boldsymbol{y}(n), \widehat{\boldsymbol{a}}(n)\right] = Q(\boldsymbol{y}, \widehat{\boldsymbol{a}}, n) = 0, \tag{7}$$

where Q can be a set of constraint functions. The EPDD function 6 has to be put in eq. 4 and yielding the predictive cost function after integration

$$\langle F(\boldsymbol{a}, \widehat{\boldsymbol{a}}, n)\rangle_e = F\left[\boldsymbol{a} = \boldsymbol{a}^*, \widehat{\boldsymbol{a}}, n\right] = F(\boldsymbol{a}^*, \widehat{\boldsymbol{a}}, n). \tag{8}$$

Then the modified criteria 3 has the following form

$$\min_{\widehat{\boldsymbol{a}} \in A_0} F(\boldsymbol{a}^*, \widehat{\boldsymbol{a}}, n), \quad Q(\boldsymbol{y}, \widehat{\boldsymbol{a}}, n) = 0, \quad \boldsymbol{a}^*(n) = \boldsymbol{\Phi}(\widehat{\boldsymbol{a}}, n_0). \tag{9}$$

Different types of reconstruction algorithms can be found according to the choise of the cost function, constraints and predictive function.

The same methodology is applicable for criteria with general structure or for other type of constraints.

4. Linear Reconstruction Algorithm

For one particular case the solution can be obtained in explicit from. Suppose the cost function in quadratic form

$$F(\boldsymbol{a}, \widehat{\boldsymbol{a}}, n) = \sum_r \left(\widehat{a}_r(n) - a_r\right)^2, \tag{10}$$

and a set of linear constraints

$$Q_m(\boldsymbol{y}, \widehat{\boldsymbol{a}}, n) = y_m(n) - \sum_r x_{mr}(n)\, \widehat{a}_r(n) = 0, \quad m = 1, \cdots, M. \tag{11}$$

This type of constraints can be understood as follows: It is assumed that the noise components $\xi_m(n)$ are statistical independent random values with zero mean and their PDD functions are of even type with unknown shapes. In this case $y_m(n)$ is given by maximum likelihood estimation for the ray sum $\sum_r x_{mr} a_r(n)$. For this the constraints 11 imply that the ray sums considering the estimations $\sum_r x_{mr} \widehat{a}_r(n)$

has to be equal to the maximum likelihood estimation of $y_m(n)$ for the real ray sums.

Suppose the prediction function 5 in linear form

$$a_r^*(n) = \Phi_r(\widehat{\boldsymbol{a}}, n_0) = \sum_{i=0}^{n_0} \sum_j \alpha_{ij} \cdot \widehat{a}_{r-j}(n-1-i), \tag{12}$$

with $\sum_j = \sum_{j_1=-r_0}^{r_0} \sum_{j_2=-r_0}^{r_0}$. The value $2r_0$ is the number of image elments in the neighborhood of $\widehat{a}_r$ taken into account for the prediction. The coefficients α_{ij} are assumed to be constant and can be chosen due to the convergence properties.

The optimal reconstruction algorithm without restrictions to the estimated values $\widehat{\boldsymbol{a}}$ can be found by employing the Lagrange optimization method. The optimization criteria 9 is transformed to

$$\min_{\widehat{\boldsymbol{a}}} \left[\sum_r \left(\widehat{a}_r(n) - a_r^*\right)^2 + \sum_{m=1}^{M} \lambda_m(n) Q_m(\boldsymbol{y}, \widehat{\boldsymbol{a}}, n) \right], \quad \boldsymbol{a}^*(n) = \boldsymbol{\Phi}(\widehat{\boldsymbol{a}}, n), \tag{13}$$

with Q_m and $\boldsymbol{\Phi}$ from eqs. 11 and 12, where $\lambda_m(n)$ are the Lagrange multipliers. In the transformed criteria 13 the Lagrange multipliers operate like regularization coefficients. According to the transformed criteria the following optimization equations are obtained

$$\begin{array}{rcl} 2\left(\widehat{a}_r(n) - a_r^*(n)\right) - \lambda_m(n) \cdot x_{mr}(n) & = & 0 \\ x_m(n) - \sum_r x_{mr}(n)\, \widehat{a}_r(n) & = & 0 \\ m = 1, \cdots, M & & \end{array} \tag{14}$$

The explicite form of the solution for this set of equations is given by

$$\widehat{a}_r(n) = a_r^*(n) + x_{mr}(n)\, C_m^{-1}(n) \left[y_m(n) - \sum_{r'} x_{mr'}(n)\, a_{r'}^*(n) \right], \tag{15}$$

implicitely depending on the regularization coefficients λ_m. The expressions 12 and 15 produce a linear n_0 order reconstruction algorithm.

For this algorithm the necessary convergence conditions have the following form

$$\sum_{i=0}^{n_0} \alpha_{i0} = 1, \quad \sum_{i=0}^{n_0} \alpha_{ij} = 0 \tag{16}$$

for $j = j_1, j_2 = \pm 1, \cdots, \pm r_0$.

In the particular case $n_0 = 0$ and $a_r^*(n) = \widehat{a}_r(n-1)$ eq. 15 is equal to the well known Kaczmarz algorithm employed for reconstruction [1].

In X-Ray tomography all absorption coefficients are nonnegative, i.e. $a_r \geq 0$, which requires to use additional constraints of nonequal type. This means nonlinear programming methods has to be used in this case.

5. Simulation Results

Reconstruction results are presented for a glas fibre cable with diameter of about 1.5 cm. The projection data are obtained using a microfocus X-ray CT setup with

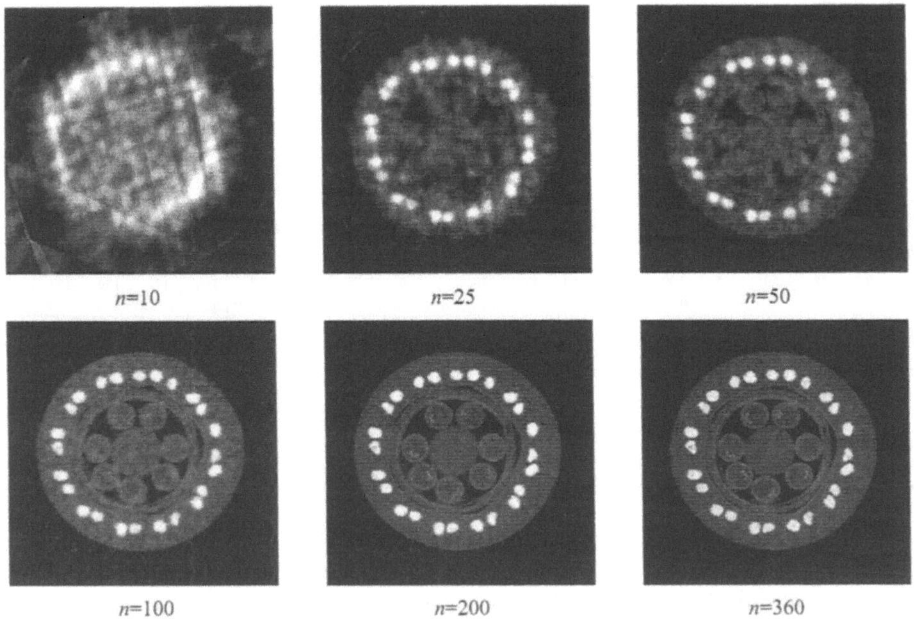

Figure 2. Reconstruction of glass fibre cable (voxel size 0.05 mm) using (a) 10, (b) 25, (c) 50 , (d) 100, (e) 200, (f) 360 randomly choosen CT projections from a set of 360 projections.

CCD camera contain 1024 × 1024 detector elements [8]. The following parameters have been used for the measurements [9]: 90 kV with 0.5 mm Cu filtering, 0.2 mA, and magnification factor 6.8 for collecting 360 projections each degree.

Figure 2 shows the reconstruction results for various number of projections using the linear algorithm introduced in eq. 15 for a simplified prediction model

$$a_r^* (n) = \sum_{i=0}^{n_0} \alpha_i \widehat{a}_r (n - 1 - i) \tag{17}$$

of second order $n_0 = 2$. The following initial conditions have been choosen

$$\widehat{a}_2 (1) = \frac{1}{M} \sum_{m=1}^{M} C_m^{-1} (1) \cdot y_m (1) = const. \tag{18}$$

The adaptive reconstruction procedure reproduces the main image properties taking into accout about 100 projections. By increasing the number of projections the quality of the reconstructed image is enhanced and comparable with the CT result using the Feldkamp algorithm.

Fig. 3 shows the convergence behaviour of the proposed linear reconstruction algorithm to show the influence of noise to compare zero-order and second-order prediction model for three different areas in the reconstructed image. The second order algorithm reacts more stable to noise influences.

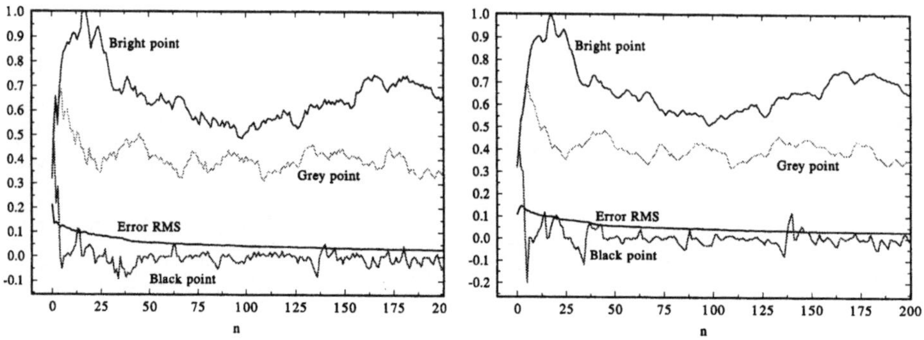

Figure 3. Convergence behavior considering noise for three different areas in the reconstructed image: left zero-order and right second order prediction algorithm. Error root mean square (RMS) calculated for the total image.

6. Conclusions

A 2D image reconstruction algorithm was proposed which is formulated as an recursive adaptive estimation problem due to the considered uncertainty of prior knowledge about the image and noise properties. It is shown that the proposed adaptive reconstruction algorithm is valuable for image reconstruction from projections without prior information even in the case of noisy data. The number of required projections can be determined by investigating the convergence properties of the reconstruction algorithm. This opens up the possibility to construct tomographic systems with variable projection scheme, which may be regular or non-regular, depending upon the actual convergence of the reconstruction algorithm and the expected gain of information from the following projections.

References

1. G. T. Herman, *Image Reconstruction from Projections*, Academic Press, New York, 1980.
2. A. A. Krasovsky, ed., *Reference Book for Automatic Control Theory (in Russian)*, Nauka, Moscow, 1987.
3. V. M. Artemiev, A. O. Naumov, and G.-R. Tillack, "Adaptive image reconstruction in x-ray tomography," in *Proceedings of the 7th European Conference on Nondestructive Testing*, in print, 1998.
4. T. Poggio, V. Torre, and C. Koch, "Computational vision and regularization theory," *Nature*, **317**, pp. 314–319, 1989.
5. D. M. Titterington, "General structure of regularization procedure in image reconstruction," *Astronomy and Astrophysics*, **144**, pp. 381–387, 1989.
6. G. Demoment, "Image reconstruction and restoration: Overview of common estimation structures and problems," *IEEE Trans. on Acoust. Speech and Sign. Process.*, **37**, (12), pp. 2024–2036, 1989.
7. R. M. Leahy and C. E. Goutis, "An optimal technique for constraint-based image restoration and reconstruction," *IEEE Trans. on Acoust. Speech and Sign. Process.*, **34**, (6), pp. 1629–1642, 1986.
8. H. Riesemeier, J. Goebbels, and B. Illerhaus, "Development and application of cone beam tomography for materials research," in *DGZfP Berichtsband*, vol. 44, (Berlin), pp. 112–119, DGZfP, 1994.
9. J. Goebbels, *private communications*, 1998.

ARE WE CRUISING A HYPOTHESIS SPACE?

C. C. RODRIGUEZ
Department of Mathematics and Statistics
University at Albany, SUNY
Albany NY 12222, USA
carlos@math.albany.edu

Abstract. This paper is about Information Geometry, a relatively new subject within mathematical statistics that attempts to study the problem of inference by using tools from modern differential geometry. This paper provides an overview of some of the achievements and possible future applications of this subject to physics.

Key words: Information Geometry, Entropic Priors, Riemannian Geometry, Entropic Connections, Information Metric

1. Introduction

It is not surprising that geometry should play a fundamental role in the theory of inference. The very idea of what constitutes a good model cannot be stated clearly without reference to geometric concepts such as size and form of the model as well as distance between probability distributions. Recall that a statistical model (hypothesis space) is a collection of probability distributions for the data. Therefore, a good model should be *big* enough to include a *close* approximation to the true distribution of the data, but small enough to facilitate the task of identifying this approximation. As Occam said: *as simple as possible but not too simple.* Moreover, regular statistical models have a natural Riemannian structure. Parameterizations correspond to choices of coordinate systems and the Fisher information matrix in a given parameterization provides the metric in the corresponding coordinate system [1,2]. By thinking of statistical models as manifolds, hypothesis spaces become *places* and it doesn't take much to imagine some of these *places* as models for the only physical place there is out there, namely: **spacetime**. In section 2 of the paper we apply the techniques of information geometry to show that the space of radially symmetric distributions admits a foliation into pseudo-spheres of increasing radius. If we think of a radially symmetric distribution as describing an uncertain physical position we discover a hypothesis space that, in many ways resembles an expanding spacetime. An isotropic, homogeneous space with pseudo-

W. von der Linden et al. (eds.), Maximum Entropy and Bayesian Methods, 131–140.

spherical symmetries and with *time* increasing with decreasing curvature radius. This admittedly simple toy model of reality already suggests a number of truly remarkable consequences for the nature of spacetime. Here are three of them:

1. The appearance of time is a consequence of uncertainty.
2. Space is infinite dimensional and only *on the average* appears as four dimensional.
3. Spin is a property of space and not of a particle so that all truly fundamental particles must have spin.

I must emphasize that, at the time of writing, there is no direct experimental evidence in favor of any of the above statements. Nevertheless there is indirect evidence that they should not be too quickly dismissed as nonsense.

With respect to the first statement. Recall that the appearance of the axis of time, in standard general relativity, is a consequence of specifying an initial and a final 3-geometry on two spacelike hypersurfaces plus evolution according to the field equation [3]. Time is therefore a consequence of 3-space geometry and the field equation of general relativity, which in turn seems to be of an statistical nature (see [4] and section 4 below). These, I believe, are facts that support, at least in spirit, the first claim above.

There is absolutely no evidence that space has infinitely many dimensions, but had this be true, it would explain why we observe only four of them. It also seems a priori desirable to have a model that produces *observed space* as a macroscopic object not unlike pressure or temperature.

With respect to the third statement. Hestenes[5] shows that many of the rules for manipulating spin have nothing to do with quantum mechanics but are just general expressions for the geometry of space. It is also worth noticing that the standard model allows the existence of elementary particles without spin but these have not yet been observed.

But there is still more. Think of the different roles that the concepts of, entropy, curvature, and, local hyperbolicity, play in statistics and in physics and you will realize that the link is a useful bridge for transporting ideas from physics to statistics and vice-versa. The following sections (3, 4, 5) of this paper do exactly that. That is, they examine the meaning of each of these concepts (entropy, curvature and local hyperbolicity) keeping the proposed link between inference and physics in mind.

The link between information geometry and general relativity promises big rewards for both, statistical inference and physics. For example, statisticians may look at the field equations of general relativity as a procedure for generating statistical models from prior information encoded in the distribution of energy and matter fields. On the other hand, physicists may see information geometry as a possible language for the elusive theory of quantum gravity since it is a language already made out of the right ingredients: *uncertainty* and *differential geometry.*

2. The Hypothesis Space of Radially Symmetric Distributions

Let $\mathcal{R}_3$ be the collection of all radially symmetric distributions of three dimensional euclidean space. The probability of an infinitesimal region around the point $x \in \mathbb{R}^3$, of volume d^3x, that is assigned by a general element of $\mathcal{R}_3$ is given by,

$$P(d^3x|\psi,\theta,\sigma) = \frac{1}{\sigma^3}\left|\psi\left(\left|\frac{x-\theta}{\sigma}\right|^2\right)\right|^2 d^3x \tag{1}$$

where, $\theta \in \mathbb{R}^3$ is a location parameter, $\sigma > 0$ is a scale parameter and ψ is an arbitrary differentiable function of $r^2 > 0$ satisfying the normalization condition:

$$\int_0^\infty r^2|\psi(r^2)|^2 dr = \frac{1}{4\pi} \tag{2}$$

Equation (2) assures that the probability assigned to the whole space by (1) is in fact 1. The derivative, ψ', must also decrease to 0 sufficiently fast so that the integrals (3) exist. Since ψ is an infinite dimensional parameter, $\mathcal{R}_3$ is also an infinite dimensional manifold but the space , $\mathcal{R}_3$ (ψ), of radially symmetric distributions for a given function ψ is a four dimensional submanifold of $\mathcal{R}_3$ parameterized by $(\theta_0,\theta_1,\theta_2,\theta_3) = (\sigma,\theta)$. The metric in $\mathcal{R}_3$ (ψ) is given by the 4×4 Fisher information matrix (see [6] p. 63) with entries:

$$g_{\mu\nu} = 4\int(\partial_\mu f)(\partial_\nu f)d^3x \tag{3}$$

where $\mu,\nu = 0,\dots,3$, the function f is the square root of the density given in (1) i.e.,

$$f(x|\theta,\sigma) = \sigma^{-3/2}\psi\left(\left|\frac{x-\theta}{\sigma}\right|^2\right) \tag{4}$$

and ∂_μ denotes the derivative with respect to θ_μ. Let us separate the computation of the metric tensor terms into three parts. The entries g_{ij}, the entries g_{0i} for $i,j = 1,2,3$ and the element g_{00}. Replacing (4) into (3), doing the change of variables $x = \theta + \sigma y$ and using the fact that $\partial_i\psi(y^2) = -2y_i\psi'(y^2)/\sigma$ we get,

$$g_{ij} = \frac{16}{\sigma^2}\int y_i y_j\left(\psi'(y^2)\right)^2 d^3y \tag{5}$$

where $y^2 = |y|^2$ is the Clifford product of the vector y by itself. Carrying out the integration in spherical coordinates we obtain, $g_{ij} = 0$ for $i \neq j$ and,

$$g_{ii} = \frac{64\pi}{3\sigma^2}\int r^4|\psi'(r^2)|^2 dr \tag{6}$$

The derivative with respect to σ of the function given in (4) is,

$$\partial_0 f = -\frac{\sigma^{-5/2}}{2}\left[3\psi + 4y^2\psi'\right] \tag{7}$$

and therefore computing as in (5) we have,

$$g_{0i} = 4\int (\partial_0 f)(\partial_i f) d^3x \propto \int [3\psi + 4y^2\psi^{'}] y_i \psi^{'} d^3y = 0 \tag{8}$$

where the value of 0 for the last integral follows by performing the integration in spherical coordinates or simply by symmetry, after noticing that the integrand is odd. Finally from (7) we get,

$$g_{00} = \frac{4\pi}{\sigma^2}\int [3\psi(r^2) + 4r^2\psi^{'}(r^2)]^2 r^2 dr \tag{9}$$

Expanding the square and integrating the cross term by parts to show that,

$$\int r^4\psi(r^2)\psi^{'}(r^2)dr = -\frac{3}{4}(\frac{1}{4\pi}) \tag{10}$$

where we took $u = \psi r^3/2$ and $v^{'} = 2r\psi^{'}$ for the integration by parts and we have used (2). We obtain,

$$g_{00} = \frac{4\pi}{\sigma^2}\left[\frac{-9}{4\pi} + 16\int r^6|\psi^{'}(r^2)|^2 dr\right] \tag{11}$$

The full matrix tensor is then, $(g) = \frac{1}{\sigma^2}\text{diag}(J(\psi), K(\psi), K(\psi), K(\psi))$, where $J(\psi)$ and $K(\psi)$ are just short hand notations for the factors of $\frac{1}{\sigma^2}$ in (11) and (6). These functions are always positive and they depend only on ψ. Straight forward calculations, best done with a symbolic manipulator like MAPLE, show that a space with this metric has constant negative scalar curvature given by $-1/J(\psi)$. It follows that for a fix value of the function ψ the hypothesis space of radially symmetric distributions $\mathcal{R}_3\ (\psi)$ is the pseudo-sphere of radius $J^{1/2}(\psi)$. We have therefore shown that the space of radially symmetric distributions has a foliation (i.e. a partition of submanifolds) of pseudo-spheres of increasing radius.

This is a mathematical theorem. There can be nothing controversial about it. What it may be disputed, however, is my belief that the hypothesis space of radially symmetric distributions may be telling us something new about the nature of real physical spacetime. What I find interesting is the fact that if we think of position subject to radially symmetric uncertainty then the mathematical object describing the positions (i.e. the space of its distributions) has all the symmetries of space plus time. It seems that time, or something like time, pops out automatically when we have uncertain positions. I like to state this hypothesis with the phrase:

there is no time, only uncertainty

2.1. UNCERTAIN SPINNING SPACE?

The hypothesis space of radially symmetric distributions is the space of distributions for a random vector $y \in \mathbb{R}^3$ of the form,

$$y = x + \epsilon \tag{12}$$

where $x \in \mathbb{R}^3$ is a non random location vector, and $\epsilon \in \mathbb{R}^3$ is a random vector with a distribution radially symmetric about the origin and with standard deviation $\sigma > 0$ in all directions. It turns out that exactly the same hypothesis space is obtained if instead of (12) we use,

$$y = x + \mathbf{i}\,\epsilon \tag{13}$$

where $\mathbf{i}$ is the constant unit pseudo scalar of the Clifford algebra of $\mathbb{R}^3$. The pseudo scalar $\mathbf{i}$ has unit magnitude, commutes with all the elements of the algebra, squares to -1 and it represents the oriented unit volume of $\mathbb{R}^3$ [7]. By taking expectations with the probability measure indexed by (x, σ, ψ) we obtain that, $E(y|x, \sigma, \psi) = x$ and,

$$E(y^2|x, \sigma, \psi) = x^2 - \sigma^2 \tag{14}$$

Equation (14) shows that, even though the space of radially symmetric distributions is infinite dimensional, on the average the intervals look like the usual spacetime intervals.

We may think of y in (13) as encoding a position in 3-space together with an uncertain degree of orientation given by the bivector part of y, i.e. $\mathbf{i}\epsilon$. In other words we assign to the point x and intrinsic orientation of direction $\hat{\epsilon}$ and magnitude $|\epsilon|$. In this model the uncertainty is not directly about the location x (as in (12)) but about its postulated degree of orientation (or spinning).

3. Entropy and Ignorance

The notion of statistical entropy is not only related to the corresponding notion in physics it is exactly the same thing as demonstrated long ago by Jaynes [8]. Entropy appears indisputable as the central quantity of information geometry. In particular, from the Kullback number (relative entropy) between two distributions in the model we obtain the metric, the volume element, a large class of connections [2], and a notion of ignorance within the model given by the so called entropic priors [9]. In this section I present a simple argument, inspired by the work of Zellner on MDIP priors [10], showing that entropic priors are the statistical representation of the vacuum of information in a given hypothesis space.

Let $\mathcal{H} = \{f(x|\theta) : \theta \in \Theta\}$ be a general regular hypothesis space of probability density functions $f(x|\theta)$ for a vector of observations x conditional on a vector of parameters $\theta = (\theta^\mu)$. Let us denote by $f(x, \theta)$, the joint density of x and θ and by $f(x)$ and $\pi(\theta)$ the marginal density of x and the prior on θ respectively. We have, $f(x, \theta) = f(x|\theta)\pi(\theta)$. Since $\mathcal{H}$ is regular, the Fisher information matrix,

$$g_{\mu\nu}(\theta) = 4 \int (\partial_\mu f^{1/2})(\partial_\nu f^{1/2}) dx \tag{15}$$

exists and it is continuous and positive definite (thus non singular) at every θ. As in (3), ∂_μ denotes the partial derivative with respect to θ^μ. The space $\mathcal{H}$ with the metric $g = (g_{\mu\nu})$ given in (15) forms a Riemannian manifold. Therefore, the invariant element of volume is given by, $\eta(d\theta) \propto \sqrt{\det\ g(\theta)} d\theta$. This is in fact a differential form [11, p. 166] that provides a notion of surface area for the manifold $\mathcal{H}$ and it is naturally interpreted as the uniform distribution over $\mathcal{H}$. This formula is known as Jeffreys rule, is often used as a universal method for building total ignorance priors. However, Jeffreys rule does not take into account the fact that a truly ignorant prior for θ should contain as little information as possible about the data x. The entropic prior in $\mathcal{H}$ demands that the joint distribution of x and θ, $f(x,\theta)$, be as difficult as possible to discriminate from the independent model $h(x)\sqrt{\det\ g(\theta)}$, where $h(x)$ is an initial guess for $f(x)$. That is, we are looking for the prior that minimizes the Kullback number between $f(x,\theta)$ and the independent model, or in other words, the prior that makes the joint distribution of x and θ to have maximum entropy relative to the measure $h(x)\sqrt{\det g(\theta)}dxd\theta$. Thus, the entropic prior is the density $\pi(\theta)$ that solves the variational problem,

$$\min_{\pi} \int f(x,\theta) \log \frac{f(x,\theta)}{h(x)\sqrt{\det\ g(\theta)}}\ dx\ d\theta \tag{16}$$

Replacing $f(x,\theta) = f(x|\theta)\pi(\theta)$ into (16), simplifying, and using a lagrange multiplier, λ, for the normalization constraint, that $\int \pi(\theta)d\theta = 1$, we find that π must minimize,

$$\int \pi(\theta) I(\theta : h)\ d\theta + \int \pi(\theta) \log \frac{\pi(\theta)}{\sqrt{\det g(\theta)}}\ d\theta + \lambda \int \pi(\theta)\ d\theta \tag{17}$$

where, $I(\theta : h)$ denotes the Kullback number between $f(x|\theta)$ and $h(x)$, i.e.,

$$I(\theta : h) = \int f(x|\theta) \log \frac{f(x|\theta)}{h(x)}\ dx \tag{18}$$

The Lagrangian $\mathcal{L}$ is given by the sum of the integrands in (17) and the Euler-Lagrange equation is then,

$$\frac{\partial \mathcal{L}}{\partial \pi} = I(\theta : h) + \log \frac{\pi(\theta)}{\sqrt{\det g(\theta)}} + 1 + \lambda = 0 \tag{19}$$

from where we obtain that,

$$\pi(\theta) \propto e^{-I(\theta\ :\ h)} \sqrt{\det g(\theta)} \tag{20}$$

The numerical values of the probabilities obtained with the formula (20) depend on the basis for the logarithm used in (16). However, the basis for the logarithm that appears in the definition of the Kullback number is arbitrary (entropy is defined only up to a proportionality constant). Thus, (20) is not just one density, but a family of densities,

$$\pi(\theta|\alpha, h) \propto e^{-\alpha I(\theta\ :\ h)} \sqrt{\det g(\theta)} \tag{21}$$

indexed by the parameter $\alpha > 0$ and the function h. Equation (21) is the family of entropic priors introduced in [2] and studied in more detail in [12],[9] and [13].

It was shown in [9] that the parameter α should be interpreted as the number of virtual observations supporting $h(x)$ as a guess for the distribution of x. Large values of α should go with reliable guesses for $h(x)$ but, as it was shown in [13], the inferences are less robust. This indicates that ignorant priors should be entropic priors with the smallest possible value for α, i.e., with,

$$\alpha^* = \inf\{\alpha > 0 : \int e^{-\alpha I(\theta, h)} \eta(d\theta) < \infty\} \tag{22}$$

Here is the canonical example.

3.1. EXAMPLE: THE GAUSSIANS

Consider the hypothesis space of one dimensional gaussians parameterized by the mean μ and the standard deviation σ. When h is an arbitrary gaussian with parameters μ_0 and σ_0 straight forward computations show that the entropic prior is given by,

$$\pi(\mu, \sigma|\alpha, \mu_0, \sigma_0) = \frac{1}{Z}\sigma^{\alpha-2} \exp\left[\frac{-\alpha\left((\mu-\mu_0)^2 + \sigma^2\right)}{2\sigma_0^2}\right] \tag{23}$$

where the normalization constant Z is defined for $\alpha > 1$ and is given by,

$$Z = \frac{2}{\sqrt{\pi}}\left(\frac{\alpha}{2}\right)^{\alpha/2} \Gamma(\frac{\alpha-1}{2})^{-1} {\sigma_0}^{-\alpha} \tag{24}$$

Thus, in this case $\alpha^* = 1$ and the most ignorant prior is obtained by taking the limit $\alpha \to 1$ and $\sigma_0 \to \infty$ in (23) obtaining, in the limit, an improper density proportional to $1/\sigma$, which makes every body happy, frequentists and bayesians alike.

4. Curvature and Information

Curvature seems to be well understood only in physics, specially from the point of view of gauge theories where the curvature form associated to a connection has been shown to encode field strengths for all the four fundamental forces of nature [14]. In statistics, on the other hand, the only thing we know (so far) about the role of curvature is that the higher the scalar curvature is at a given point of the model, the more difficult it is to do estimation at that point. This already agrees nicely with the idea of black holes , for if in a given model there is a curvature R_0 beyond which estimation is essentially impossible then the space is partitioned into three regions with curvatures, $R < R_0$, $R = R_0$ and $R > R_0$ that correspond to regular points, horizon points and points inside black holes. No body has found an example of a hypothesis space with this kind of inferential black holes yet, but no body has tried to look for one either. Before rushing into a hunt it seems necessary to clarify what exactly it is meant by the words: ***estimation is essentially impossible at a point.***

I believe that one of the most promising areas for research in the field of information geometry is the clarification of the role of curvature in statistical inference. If indeed physical spacetime can be best modeled as a hypothesis space then, what is to be learned from the research on statistical curvature will have direct implications for the nature of physical space. On the other hand, it also seems promising to re-evaluate what is already known in physics about curvature under the light of the proposed link with inference. Even a naive first look will show indications of what to expect for the role of curvature in inference. Here is an attempt at that first look.

From the classic statement: *Mass-energy is the source of gravity and the strength of the gravity field is measured by the curvature of spacetime*

We guess: *Information is the source of the curvature of hypothesis spaces. That is, prior information is the source of the form of the model*

From: *The dynamics of how mass-energy curves spacetime are controlled by the field equation:* $\mathbf{G} = \kappa\mathbf{T}$ *where* $\mathbf{G}$ *is the Einstein tensor,* $\mathbf{T}$ *is the stress-energy tensor and* κ *is a proportionality factor*

guess: *The field equation controls the dynamics of how prior information produces models*

From: *The field equation for empty space is the Euler-Lagrange equation that characterizes the extremum of the Hilbert action, with respect to the choice of geometry. That is it extremizes*

$$S_g = \int R d\Omega, \qquad d\Omega = \sqrt{|\det g|} d^4x, \tag{25}$$

where the integral is taken over the interior of a four-dimensional region Ω, R *is the scalar curvature and* g *is the metric*

guess1: *The form of hypothesis spaces based on no prior information must satisfy*

$$R_{ij} - \frac{1}{2} R g_{ij} = 0 \tag{26}$$

where g_{ij} *is the Fisher information matrix,* R_{ij} *is the Ricci tensor and* R *is the scalar curvature as above.*

guess2: *Given a hypothesis space with Fisher information matrix* $g(\theta)$, *the Einstein tensor,* G, *i.e. the left hand side of (26), quantifies the amount of prior information locally contained in the model at each point* θ.

5. Hyperbolicity

What it seems most intriguing with respect to the link between information geometry and general relativity is the role of hyperbolicity. We know from general relativity that physical spacetimes are Riemannian manifolds which are locally Lorentzian. That is, at each point, the space looks locally like Minkowski space. Or, in other words, the symmetries of the tangent space at each point are those of hyperbolic space. On the other hand, in information geometry, hyperbolicity appears at two very basic levels. First, hyperbolicity appears connected to the notion of regularity through the property of *local asymptotic normality* (LAN for short see [6]). This is in close agreement with what happens in physics. The LAN property says that the manifold of distributions of n independent and identically regularly distributed observations can be locally approximated by gaussians for large n, and since the gaussians are known to form hyperbolic spaces, the correspondence with physics is perfect. Second, in statistical inference hyperbolicity also appears mysteriously connected to entropy and Bayes' theorem! (see my *From Euclid to Entropy*[15]) and by following the link back to general relativity we obtain a completely new and unexpected result: *entropy and Bayes theorem are the source of the local hyperbolicity of spacetime!*. That entropy and thermodynamics are related to general relativity may have seem outrageous in the past, but not today. It does not seem outrageous at all when we consider that, Bekenstein found that the entropy of a black hole is proportional to its surface area [16], when we consider that Hawking discovered that black holes have a temperature [17] and specially when we consider that Jacobson showed that the field equation is like an equation of state in thermodynamics [4].

References

1. S.-i. Amari, *Differential-Geometrical Methods in Statistics*, vol. 28 of *Lecture Notes in Statistics*, Springer-Verlag, 1985.
2. C. Rodríguez, "The metrics induced by the kullback number," in *Maximum Entropy and Bayesian Methods*, J. Skilling, ed., Kluwer Academic Publishers, 1989.
3. D. S. R.F. Baierlein and J. Wheeler, "Three-dimensional geometry as carrier of information about time," *Phys. Rev.*, pp. 1864–1865, 1962.
4. T. Jacobson, "Thermodynamics of spacetime: The einstein equation of state," tech. rep., xxx.lanl.gov/abs/gr-qc/9504004, 1995.
5. D. Hestenes, *Space-Time Algebra*, Gordon and Breach, N.Y., 1966.
6. I. Ibragimov and R. Has'minskii, *Statistical Estimation*, vol. 16 of *Applications of Mathematics*, Springer-Verlag, 1981.
7. D. Hestenes and G. Sobczyk, *Clifford Algebra to Geometric Calculus*, D. Reidel, 1984. For links to online resources on Clifford algebra check, http://www.mrao.cam.ac.uk/~clifford/.
8. E. Jaynes, "Information theory and statistical mechanics," *Phys. Rev.*, **106**, p. 620, 1957. Part II; ibid, vol 108,171.
9. C. Rodríguez, "Entropic priors," tech. rep., omega.albany.edu:8008/entpriors.ps, Oct. 1991.
10. A. Zellner, "Past and reacent results on maximal data information priors," tech. rep., H.G.B. Alexander Reseach Foundation. Graduate Shool of Business, University of Chicago, 1995.
11. B. Dubrovin, A. Fomenko, and S. Novikov, *Modern Geometry–Methods and Applications, Part-I*, vol. GTM 93 of *Graduate Texts in Mathematics*, Springer-Verlag, 1984.
12. C. Rodríguez, "Objective bayesianism and geometry," in *Maximum Entropy and Bayesian Methods*, P. F. Fougère, ed., Kluwer Academic Publishers, 1990.
13. C. Rodríguez, "Bayesian robustness: A new look from geometry," in *Maximum Entropy and Bayesian Methods*, G. Heidbreder, ed., pp. 87–96, Kluwer Academic Publishers, 1996. (since Nov. 1993) in omega.albany.edu:8008/robust.ps.
14. D. Bleecker, *Gauge Theory and Variational Principles*, Addison-Wesley, 1981.
15. C. Rodríguez, "From euclid to entropy," in *Maximum Entropy and Bayesian Methods*, W. T. Grandy, Jr., ed., Kluwer Academic Publishers. omega.albany.edu:8008/euclid.ps, 1991. omega.albany.edu:8008/euclid.ps.
16. J. Bekenstein *Phys. Rev. D*, **7**, p. 2333, 1973.
17. S. Hawking *Comm. Math. Phys.*, **43**, p. 199, 1975.

Myron Tribus and Robert Fry.

FIDUCIAL INFERENCE

A Neyman–Pearson Interpretation

DIEMER SALOMÉ
University of Groningen
Department of Mathematics
P.O. Box 800
9700 AV Groningen
The Netherlands †

Abstract. Fisher's fiducial argument is a tool for deriving inferences in the form of a probability distribution on the parameter space, not based on Bayes's Theorem. Lindley established that in exceptional situations fiducial inferences coincide with posterior distributions; in the other situations fiducial inference has always been critized. In his comment on Wilkinson [1], James wrote '*Fisher's postulate of ignorance for fiducial theory should be replaced by a positive requirement that the statistical analysis is appropriate to and meets the issues at stake.*' Such a positive requirement can be obtained from a decision theoretic approach: in some interesting situations procedures provided by the fiducial argument are optimal in a class of risk–unbiased procedures.

Key words: Fiducial inference, risk-unbiasedness, weak and strong unbiasedness, proper loss function

1. Introduction

Consider the following problem. After observing the outcome $x \in \mathcal{X}$ of a random variable X, where nothing is known a priori except that the distribution function F of X is a member of the family $\mathcal{F} = \{F_\theta : \theta \in \Theta\}$, inference has to be made about the true value t, defined by $F_t = F$, in the form of a probability distribution on Θ. Let X_θ denote a random variable with distribution function F_θ, and suppose that the family $\mathcal{F} = \{F_\theta : \theta \in \Theta\}$ satisfies:

(i) F_θ is a continuous and differentiable function of (x, θ) with density $f_\theta(x) = dF_\theta(x)/dx$,
(ii) $\Theta \subset \mathbf{R}$ is convex, i.e. $\Theta = (\underline{\theta}, \overline{\theta})$
(iii) F_θ is non–decreasing in x, non–increasing in θ, and $F_\theta(x) \to 0$ (1) as $\theta \to \overline{\theta}$ ($\underline{\theta}$),
(iv) $\mathcal{F}$ has monotone likelihood ratio.

†Email : salome@math.rug.nl

W. von der Linden et al. (eds.), Maximum Entropy and Bayesian Methods, 141–148.

The Bayesian solution to this inference problem is to choose a prior measure ν on Θ, and to compute the posterior distribution $Q_\nu(x)$, with density

$$g_{\nu,x}(\theta) \propto f_\theta(x)\nu(d\theta).$$

The problem is, of course, to choose ν. Fisher [2] regarded this approach as wrong. His fiducial argument provides a method to derive inferences in the form of probability measures on Θ (distributional inferences) that does not require the specification of ν. He prescribed, without specifying the underlying principles, that the fiducial distribution $Q_{\mathrm{F}}(x)$ for statistical models satisfying (i)–(iii) should be constructed by taking the fiducial density to be

$$g_{\mathrm{F},x}(\theta) = -dF_\theta(x)/d\theta.$$

Various authors wrote about what they regarded as the essence of the fiducial argument, and made extensions to a more general context; Barnard [3] emphasized pivotal methods, and Fraser [4] worked out a theory of structural inference. In this paper a different interpretation will be given. To give a clear exposition of the ideas, the attention is restricted to statistical models that satisfy (i)–(iv).

2. The fiducial argument

Our interpretation of the fiducial argument is as follows. Any distributional inference $Q(x)$ about t can be characterized by its distribution function G_x. Such a distribution function has the interpretation that $G_x(\theta)$ is the degree of belief in the hypothesis $\mathrm{H}_\theta : t \leq \theta$ if this is tested against its logical negation $\mathrm{A}_\theta : t > \theta$. Taking the P–value $\alpha_\theta(x) = 1 - F_\theta(x)$, as such a degree of belief, the essence of the fiducial argument is identifying $G_{\mathrm{F},x}(\theta) = \alpha_\theta(x)$, i.e.

$$G_{\mathrm{F},x}(\theta) = \begin{cases} 0 & \text{if } \ \theta < \underline{\theta}, \\ 1 - F_\theta(x) & \text{if } \ \underline{\theta} \leq \theta \leq \overline{\theta}, \\ 1 & \text{if } \ \overline{\theta} < \theta. \end{cases} \tag{1}$$

Conditions (i)–(iii) imply that $G_{\mathrm{F},x}$ is, indeed, a distribution function, and $g_{\mathrm{F},x}(\theta) = dG_{\mathrm{F},x}(\theta)/dx = -dF_\theta(x)/d\theta$. Condition (iv) will be used in Section 5 to establish optimality of Q_{F}. Notice that this procedure for constructing G_x resembles the construction of confidence intervals. Our interpretation of the fiducial argument implies that if $u : \mathcal{X} \mapsto \mathbf{R}$ and $\eta : \Theta \mapsto \mathbf{R}$ are monotone transformations, then applying the fiducial argument to the transformed problem provides the same distribution as the fiducial distribution induced by the transformation.

Before studying the properties of fiducial inference in detail, the difference with Bayesian inference has to be considered. The following result due to Lindley [5] specifies exactly when fiducial inference coincides with Bayesian inference.

Theorem 1 *Suppose that conditions (i)–(iii) hold. Then there exists a measure ν such that $Q_\nu(x) = Q_{\mathrm{F}}(x)$ if and only if there exist monotone transformations $u : \mathcal{X} \mapsto \mathbf{R}$ and $\eta : \Theta \mapsto \mathbf{R}$ such that $\eta(\theta)$ is a location parameter for $u(X_\theta)$, i.e. $\mathcal{L}u(X_\theta) = \mathcal{L}(\eta(\theta) + U)$ where U is a random variable with a fixed distribution.*

If such transformations exist, then Q_F coincides with Q_ν, where ν is Lebesgue measure (in the transformed problem). To show that the existence of these transformations is restrictive, Lindley proved that for 1–parameter exponential families the normal distributions (location family) and gamma distributions (scale family) are the only possibilities.

3. Proper loss functions for evaluating distributional inferences

A procedure for making distributional inferences about t is a mapping $Q : \mathcal{X} \mapsto \Theta^*$, where Θ^* denotes the space of all probability measures on Θ. Such a procedure Q can be uniquely determined by the family of corresponding distribution functions $\mathcal{G} = \{G_x : x \in \mathcal{X}\}$. For the evaluation of distributional inferences, suitable loss functions $\mathrm{L} : \Theta \times \Theta^* \mapsto [0, \infty]$ are needed. The spread of $Q(x)$ expresses the statistical uncertainty. As the scientific goal of statistical inference is to assess the statistical uncertainty as accurately as possible, the loss function should give an incentive to do so. A key concept in this respect is properness.

Definition 2 *Let T be a random variable with values in Θ. A loss function $\mathrm{L} : \Theta \times \Theta^* \mapsto [0, \infty]$ is said to be proper if $\mathcal{L}T = P$ implies that*

$$\mathbf{E}\mathrm{L}(T, P) \leq \mathbf{E}\mathrm{L}(T, Q) \quad \forall\ P, Q \in \Theta^*.$$

It is said to be strictly proper if equality holds only if $P = Q$.

The class of proper loss functions that will be considered in this paper is given by

$$\mathrm{L}_{\tau,f}(\theta, Q) = \int_\Theta 1_{(-\infty,\theta)}(z) f_0(G(z)) + 1_{[\theta,\infty)}(z) f_1(1 - G(z))\tau(dz), \tag{2}$$

where $f = (f_0, f_1)$ satisfy: $f_i : [0, 1] \mapsto [0, \infty]$, and, for all $p \in [0, 1]$, the function

$$F_p(a) = p f_1(1 - a) = (1 - p) f_0(a) \tag{3}$$

achieves its minimum ($< \infty$) at $a = p$. If this minimum is unique for each p and if τ assigns positive mass to any non–degenerate interval, then $\mathrm{L}_{\tau,f}$ will be strictly proper. An important example of a strictly proper loss function is the quadratic loss function, where $f_0(a) = f_1(a) = a^2$ provides

$$\mathrm{L}_\tau(\theta, Q) = \int_\Theta 1_{(-\infty,\theta)}(z)(G(z))^2 + 1_{[\theta,\infty)}(z)(1 - G(z))^2\tau(dz), \tag{4}$$

It can be shown that within the class of loss functions, with f of the form $f_0(a) = f_1(a) = a^q$, the quadratic loss function is the only proper loss function. Loss functions of the form (2) have two additional properties: the loss of a perfect inference is 0, i.e. $\mathrm{L}_{\tau,f}(\theta, \delta_\theta) = 0$ for all $\theta \in \Theta$ (δ_θ denotes point mass 1 at θ), and the ordering of the real–line is taken into account.

Given a loss function $\mathrm{L}_{\tau,f}$ the risk function of a procedure $Q : \mathcal{X} \mapsto \Theta^*$ is defined as the expected loss, i.e. $\mathrm{R}_{\tau,f}(\theta, Q) = \mathbf{E}\mathrm{L}_{\tau,f}(\theta, Q(X_\theta))$, and the Bayes risk w.r.t. ν as the integrated risk function, i.e. $\mathrm{r}_{\tau,f,\nu}(Q) = \int_\Theta \mathrm{R}_{\tau,f}(\theta, Q)\nu(d\theta)$. It is well known that $Q_\nu : \mathcal{X} \mapsto \Theta^*$ is optimal in the sense that it minimizes $\mathrm{r}_{\tau,f,\nu}$, provided that a proper loss function is used.

4. Unbiasedness

Similar to the theory of point estimation, procedures will not only be evaluated on the basis of their risk, but also on the basis of their bias. The usual concept of mean–unbiasedness is not very useful in the present context, because it depends too strongly on the parameterization; it is not preserved under monotone reparameterizations. Instead Lehmann's risk–unbiasedness will be used.

Definition 3 *Let $\mathrm{L}_{\tau,f}$ be a loss function of the form (2). A procedure $Q : \mathcal{X} \mapsto \Theta^*$ is said to be risk–unbiased (w.r.t. $\mathrm{L}_{\tau,f}$) if*

$$\mathbf{E}\mathrm{L}_{\tau,f}(\vartheta, Q(X_\theta)) \geq \mathbf{E}\mathrm{L}_{\tau,f}(\theta, Q(X_\theta)) \quad \forall\, \vartheta, \theta \in \Theta.$$

Using the concept of risk–unbiasedness, it is possible to define weak and strong unbiasedness by requiring risk–unbiasedness for a class of loss functions.

Theorem 4 *Let $Q : \mathcal{X} \mapsto \Theta^*$ be a procedure for making distributional inference about t. Then (i) Q is risk–unbiased w.r.t. all proper loss functions defined in (2), with f_0 and f_1 bounded and τ finite, if and only if*

$$\mathcal{L}G_{X_\theta}(z) \leq U(0,1) \quad \text{if} \quad z \leq \theta \quad \text{and} \quad \mathcal{L}G_{X_\theta}(z) \geq U(0,1) \quad \text{if} \quad z \geq \theta.$$

(ii) Q is risk–unbiased w.r.t. all quadratic loss functions defined in (4), with τ finite, if and only if

$$\mathbf{E}G_{X_\theta}(z) \leq \tfrac{1}{2} \quad \text{if} \quad z \leq \theta \quad \text{and} \quad \mathbf{E}G_{X_\theta}(z) \geq \tfrac{1}{2} \quad \text{if} \quad z \geq \theta.$$

The proof can be found in the appendix. It is clear that strong unbiasedness (defined by requiring (i) of Theorem 4) automatically implies weak unbiasedness (defined by requiring (ii) of Theorem 4). Notice that Q_{F} is strongly unbiased, because $\mathcal{L}G_{X_\theta}(\theta) = \mathcal{L}(1 - F_\theta(X_\theta)) = U(0,1)$. Applying Bayes's Theorem will typically lead to procedures that are not weakly unbiased, and hence not strongly unbiased. This can be concluded from the following theorem (see Kroese [6]).

Theorem 5 *Suppose that conditions (i)–(iii) hold, and that ν is a finite measure on Θ that has no atoms. Then the posterior distribution Q_ν cannot be weakly unbiased.*

In fact the theorem holds also holds true if ν contains atoms in the interior of Θ. The bias function of a procedure $Q : \mathcal{X} \mapsto \Theta^*$ is defined by

$$\mathrm{B}(\theta, Q) \;=\; \mathbf{E}G_{X_\theta}(\theta) - \tfrac{1}{2}, \tag{5}$$

such that all weakly, and hence strongly, unbiased procedures have a bias function that is identically equal to 0.

5. Optimality of fiducial inference

To discuss optimality, consider the following line of thought. If all distributional inferences are represented by their distribution functions, then the prophet inference Q^* is given by $G^*(z) = 1_{[t,\infty)}(z)$. As it is the objective to construct $Q(x)$ such that it is close to the prophet inference, it should be such that $G_x(z)$ is close to 0 whenever z is smaller than t and, vice versa, $G_x(z)$ should be close to 1 if z is larger than t. This is the rationale behind the following ordering of the space of all procedures.

Definition 6 *The procedure $Q : \mathcal{X} \mapsto \Theta^*$ characterized by $\mathcal{G} = \{G_x : x \in \mathcal{X}\}$ is said be better than $\tilde{Q} : \mathcal{X} \mapsto \Theta^*$ characterized by $\tilde{\mathcal{G}} = \{\tilde{G}_x : x \in \mathcal{X}\}$ (denoted by $Q \succeq \tilde{Q}$), if*

$$\mathcal{L}G_{X_\theta}(z) \leq \mathcal{L}\tilde{G}_{X_\theta}(z) \quad \text{if} \quad z \leq \theta \quad \text{and} \quad \mathcal{L}G_{X_\theta}(z) \geq \mathcal{L}\tilde{G}_{X_\theta}(z) \quad \text{if} \quad z \geq \theta.$$

It is said to be strictly better (denoted by $Q \succ \tilde{Q}$) if at least one inequality is strict.

An optimality property for Q_{F} in terms of this ordering is given by the following theorem, of which the proof can be found in Kroese *et al.* [7].

Theorem 7 *Suppose that conditions (i)–(iv) are satisfied. Then (i) there does not exist any procedure $Q \succ Q_{\mathrm{F}}$, and (ii) for every Q that is strongly unbiased $Q_{\mathrm{F}} \succeq Q$.*

Examples can be given to show that Q_{F} is not necessarily the unique best strongly unbiased procedure, and that conditions (i)–(iii) do not suffice to guarantee optimality, see Salomé [8]. The ordering of Lemma 6 is related to the ordering defined by loss functions through the following lemma.

Lemma 8 *Let $Q : \mathcal{X} \mapsto \Theta^*$ characterized by $\mathcal{G} = \{G_x : x \in \mathcal{X}\}$, and $\tilde{Q} : \mathcal{X} \mapsto \Psi^*$ characterized by $\tilde{\mathcal{G}} = \{\tilde{G}_x : x \in \mathcal{X}\}$, be two procedures for making distributional inferences about t. Then (i) $Q \succeq \tilde{Q}$ if and only if*

$$\mathrm{R}_{\tau,f}(\theta, Q) \leq \mathrm{R}_{\tau,f}(\theta, \tilde{Q}),$$

for all proper loss functions $\mathrm{L}_{\tau,f}$ of the form (2), with f_0, f_1 bounded and τ finite, and (ii) $Q \succ \tilde{Q}$ if, in addition, strict inequality holds for the strictly proper loss functions.

The proof can be found in the appendix. Notice that the fact that there does not exist a procedure that is better than the fiducial procedure does *not* imply that it is admissible w.r.t. a particular loss function. It only tells us that there cannot exist a procedure that uniformly improves the risk of the fiducial procedure w.r.t. *all* loss functions; it can happen that for different proper loss functions there exist different procedures that uniformly improve the risk of the fiducial procedure.

6. An example

To illustrate the trade–off between risk and bias consider the following example. Let f_θ be the density of the sample correlation coefficient of, say, $n = 11$ observations from a bivariate normal distribution. Notice that $\mathcal{X} = \Theta = [-1, 1]$. Take

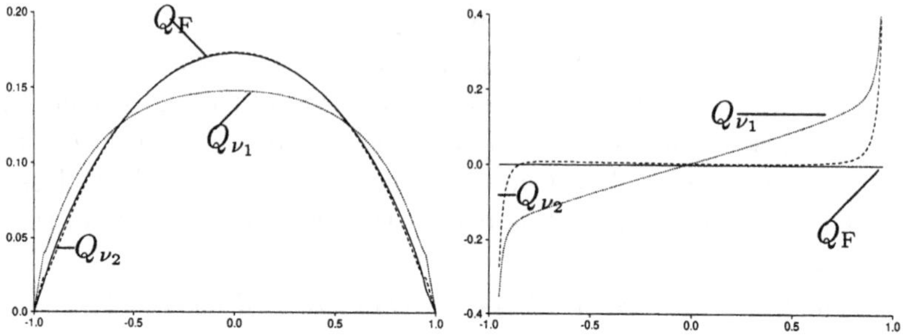

Figure 1. On the l.h.s. the risk functions of Q_F, Q_{ν_1}, and Q_{ν_2}, w.r.t. the quadratic loss function (4) with τ Lebesgue measure on $[-1, 1]$, are displayed. On the r.h.s. the bias functions are displayed. Notice that the bias function of Q_F is identically equal to 0.

ν_1 Lebesgue measure on $[-1, 1]$, and compute Q_{ν_1} as the corresponding posterior distribution. Of course, it is possible to choose ν such that Q_ν and Q_F are almost identical by using Fisher's z–transformation, Lindley's Theorem for the asymptotic problem, and the fact that monotone transformations do not essentially affect the procedures. This boils down to taking prior measure ν_2, defined by $\nu_2(d\theta) = 1/(1-\theta^2)$. In Figure 1 the bias and risk functions of Q_F, Q_{ν_1}, and Q_{ν_2} are displayed. Notice that Q_{ν_1} displays a central tendency. As a consequence, $R(\theta, Q_{\nu_1})$ close to 0 and 1 is larger than, e.g. $R(\theta, Q_F)$, but its integrated risk is smallest among all procedures.

7. Discussion

Fiducial inference is an 'all purpose' method of making inference, and hence it is natural to discuss its optimality not w.r.t. a particular loss function but w.r.t. the class of *all* reasonable loss functions. The related concept of strong unbiasedness is so restrictive that it is not to be expected that procedures satisfying this restriction are admissible w.r.t. a particular loss function. An interesting feature of strong unbiasedness, however, is that it recovers classical results like, e.g. Student's inference.

In this paper the properties of fiducial inference are discussed within the limited context of statistical models satisfying (i)–(iv). The scope of fiducial inference, however, is much wider; it can also be applied in combination with reductions by sufficiency, invariance, etc. Moreover, extensions are possible to discrete problems if strong unbiasedness is replaced by weak unbiasedness, and to problems with nuisance parameters if the P–value w.r.t. UMP tests is replaced by the P–value w.r.t. UMPU tests.

A. Proof of Theorem 4

(i) The proof is based on a theorem due to Schervish [9].

Theorem. *Let f_0 and f_1 be nonnegative extended real-valued functions on* $[0,1]$. *Assume that f_1 is right-continuous, f_0 is left-continuous, $f_i(0+) = f_i(0) = 0$, $i = 0,1$ and $f_i(1-) = f_i(1)$, $i = 0,1$. The function $F_p(a) = pf_1(1-a)+(1-p)f_0(a)$ is minimized and less than ∞ at $a = p$ if and only if there exists a measure ξ on* $[0,1)$ *such that (i) for all $a \in (0,1]$*

$$f_0(a) = \int_{[0,a)} p\ \xi(dp) \quad \text{and} \quad f_1(a) = \int_{[1-a,1)} (1-p)\ \xi(dp)$$

and (ii) f_0 and f_1 are finite on $[0,1)$. *$F_p(a)$ attains a unique minimum if ξ assigns positive mass to every non-degenerate interval.*

To show that risk-unbiasedness, for all $L_{\tau,f}$ of the form (2), with f_0 and f_1 bounded and τ finite, is equivalent to the requirement of strong unbiasedness, start by writing

$$\mathbf{E}L_{\tau,f}(\vartheta, Q(X_\theta)) = \int_\Theta 1_{(-\infty,\vartheta)}(z)\mathbf{E}f_0(G_{X_\theta}(z)) + 1_{[\vartheta,\infty)}(z)\mathbf{E}f_1(1-G_{X_\theta}(z))\tau(dz).$$

Using the characterization of f_0 and f_1, and denoting the distribution function of $G_{X_\theta}(z)$ by $H_{\theta,z}$, provides

$$\begin{aligned}
\mathbf{E}f_0(G_{X_\theta}(z)) &= \int_{[0,1]}\int_{[0,a)} p\xi(dp)H_{\theta,z}(da)\\
&= \int_{[0,1]}(1-H_{\theta,z}(p))p\xi(dp),\\
\mathbf{E}f_1(1-G_{X_\theta}(z)) &= \int_{[0,1]}\int_{[1-a,1)}(1-p)\xi(dp)H_{\theta,z}(da)\\
&= \int_{[0,1]} H_{\theta,z}(p)(1-p)\xi(dp),
\end{aligned}$$

and hence $\mathbf{E}L_{\tau,f}(\vartheta, Q(X_\theta))$ becomes

$$\int_\Theta\int_{[0,1]} 1_{(-\infty,\vartheta)}(z)(1-H_{\theta,z}(p))p + 1_{[\vartheta,\infty)}(z)H_{\theta,z}(p)(1-p)\xi(dp)\tau(dz).$$

The requirement $\mathbf{E}L_{\tau,f}(\vartheta, Q(X_\theta)) - \mathbf{E}L_{\tau,f}(\theta, Q(X_\theta)) \geq 0$, for all $\theta, \vartheta \in \Theta$, is equivalent to

$$\begin{aligned}
&\int_\Theta 1_{[\theta,\vartheta)}(z)\int_{[0,1]} p - H_{\theta,z}(p)\ \xi(dp)\tau(dz) \geq 0 \quad \text{if} \quad \vartheta > \theta,\\
&\int_\Theta 1_{[\vartheta,\theta)}(z)\int_{[0,1]} H_{\theta,z}(p) - p\ \xi(dp)\tau(dz) \geq 0 \quad \text{if} \quad \vartheta < \theta,
\end{aligned}$$

for arbitrary finite measures τ and ξ. These inequalities hold for all such τ and ξ if and only if

$$H_{\theta,z}(p) \leq p \quad \text{if} \quad z \leq \theta \quad \text{and} \quad H_{\theta,z}(p) \geq p \quad \text{if} \quad z \geq \theta,$$

which is by definition strong unbiasedness. (ii) The proof is in essence the same as that of part (i). The requirement $\mathbf{E}L_\tau(\vartheta, Q(X_\theta)) - \mathbf{E}L_\tau(\theta, Q(X_\theta)) \geq 0$, for all $\theta, \vartheta \in \Theta$, is equivalent to

$$\begin{aligned}
&\int_\Theta 1_{[\theta,\vartheta)}(z)(1-2\mathbf{E}G_{X_\theta}(z))\tau(dz) \geq 0 \quad \text{if} \quad \vartheta > \theta,\\
&\int_\Theta 1_{[\vartheta,\theta)}(z)(2\mathbf{E}G_{X_\theta}(z)-1)\tau(dz) \geq 0 \quad \text{if} \quad \vartheta < \theta.
\end{aligned}$$

These inequalities hold if and only if Q is weakly unbiased. □

B. Proof of Lemma 6

(i) Take an arbitrary loss function $\mathrm{L}_{\tau,f}$ of the form (2), with f_0, f_1 bounded and τ finite, and write (by using Fubini) the inequality $\mathrm{R}_{\tau,f}(\theta,Q) \leq \mathrm{R}_{\tau,f}(\theta,\widetilde{Q})$ as follows

$$\begin{aligned}\int_\Theta & 1_{(-\infty,\theta)}(z)(\mathbf{E}f_0(G_{X_\theta}(z)) - \mathbf{E}f_0(\widetilde{G}_{X_\theta}(z))) \\ &+ \ 1_{[\theta,\infty)}(z)(\mathbf{E}f_1(1-G_{X_\theta}(z)) - \mathbf{E}f_1(1-\widetilde{G}_{X_\theta}(z)))\tau(dz) \leq 0.\end{aligned}$$

If this inequality is required to hold for all finite measures τ, then this is equivalent to requiring that it must hold pointwisely, i.e.

$$\begin{array}{lcll}\mathbf{E}f_0(G_{X_\theta}(z)) & \leq & \mathbf{E}f_0(\widetilde{G}_{X_\theta}(z)) & \text{if} \quad z \leq \theta \\ \mathbf{E}f_1(1-G_{X_\theta}(z)) & \leq & \mathbf{E}f_1(1-\widetilde{G}_{X_\theta}(z)) & \text{if} \quad z \geq \theta.\end{array}$$

Next, use the proof of Theorem 4 to show that the requirement that these inequalities should hold for all $\mathrm{L}_{\tau,f}$ of the form (2), with f_0, f_1 bounded and τ finite, is equivalent to the statement that

$$\mathcal{L}G_{X_\theta}(z) \leq \mathcal{L}\widetilde{G}_{X_\theta}(z) \quad \text{if} \quad z \leq \theta \quad \text{and} \quad \mathcal{L}G_{X_\theta}(z) \geq \mathcal{L}\widetilde{G}_{X_\theta}(z) \quad \text{if} \quad z \geq \psi(\theta),$$

and all $\theta \in \Theta$. (ii) The proof is the same as that of part (i) except that strict properness implies that both f_0 and f_1 are strictly increasing, and hence that ξ and τ assign positive mass to any non–degenerate interval. This changes all '$\leq$' to '$<$'. □

References

1. G. N. Wilkinson, "On Resolving the Controversy in Statistcal Inference (with discussion)," *J. R. Statist. Soc (B)*, **38**, pp. 119–171, 1977.
2. R. A. Fisher, "Inverse Probability," *Proc. Camb. Phil. Soc.*, **26**, pp. 528–535, 1930.
3. G. A. Barnard, "Pivotal Models and the Fiducial Argument," *International Statistical Review*, **63**, pp. 309–323, 1995.
4. D. A. S. Fraser, *The Structure of Inference*, Krieger, first ed., 1968.
5. D. V. Lindley, "Fiducial Inference and Bayes' Theorem," *J. R. Statist. Soc (B)*, **20**, pp. 102–107, 1958.
6. A. H. Kroese, "Distributional Inference: A Loss Function Approach." PhD Thesis Rijksuniversiteit Groningen, 1994.
7. A. H. Kroese, D. Salomé, and W. Schaafsma, "An Optimum Property for Fiducial Inference." To appear in Statistics & Decisions, 1998.
8. D. Salomé, "Statistical Inference via Fiducal Methods." PhD Thesis Rijksuniversiteit Groningen, In preparation.
9. M. J. Schervish, "A General Method for Comparing Probability Assessors," *Ann. Statistics*, **17**, pp. 1856–1879, 1989.

POSTERIOR MOMENTS OF THE CAUCHY DISTRIBUTION

DAVID R. WOLF
NASA Ames Research Center
MS 269-2 Moffet Field
CA 94035-1000[†]

Abstract. The posterior moments of parameters specifying distributions are minimum mean square Bayesian estimators for the corresponding moments of those parameters, and as such are ubiquitous in the Bayesian approach to statistical inference of distributions. The Cauchy distribution is most notable for its wide tails, decided absence of high-order moments, and non-existence of less-than-data dimension sufficient statistics. Thus it vastly differs qualitatively from the Gaussian distribution, where tails are small, moments of all orders exist, and dimension-two sufficient statistics always exist. In this paper the posterior moments of the position parameter of the Cauchy distribution are found in closed form. (Estimating the other parameter, the width or distance parameter of the Cauchy is done using the same mathematics.) The interplay between the amount of data acquired for the estimation of the position parameter and the existence of higher order moments of the inferred posterior distribution for the postition parameter is made explicit.

Key words: Cauchy distribution, Bayesian estimation, moment analysis, parameter estimation, parameter inference, sufficient statistics

1. Introduction

The Cauchy distribution [1], [2] has a clear physically motivated model as the normalized density function for reception on a line, of particles radiated omnidirectionally from a radiator. If the line is the x-axis and the radiator is at $(position, distance) = (x_0, y_0)$ then this density is proportional to the Cauchy likelihood distribution

$$P(x \mid (x_0, y_0)) = \frac{y_0}{\pi(y_0^2 + (x - x_0)^2)}. \tag{1}$$

Given N independent observations (receptions of particles) at positions $(x_1, \ldots, x_N)$ $=: \boldsymbol{x}$ and given that the prior for the true position of the radiator is given by

[†]Email: greywolf@ptolemy.arc.nasa.gov, Web: http://dino.ph.utexas.edu/ ~ wolf

W. von der Linden et al. (eds.), Maximum Entropy and Bayesian Methods, 149–156.

$P((x_0, y_0))$ then Bayes' theorem states that the probability density function for the true position of the radiator is proportional to the product of the Cauchy distribution and the prior, $P((x_0, y_0) \mid \boldsymbol{x}) \propto P(\boldsymbol{x} \mid (x_0, y_0))\, P((x_0, y_0))$. Since the observations are independent, $P(\boldsymbol{x} \mid (x_0, y_0)) = \Pi_{i=1}^{N} P(x_i \mid (x_0, y_0))$, with each term in the product having the form of (1). Taking the distance of the line from the radiator y_0 as fixed, we are left with the Cauchy posterior distribution for the position parameter x_0, $P(x_0 \mid \boldsymbol{x}, y_0)$, and this is the distribution discussed in the rest of this paper (usually dropping the redundant y_0 from the notation). However, note that the estimation of y_0 is accomplished by the exact same mathematics, as will be seen later.

Given the Cauchy posterior distribution for the position parameter x_0, we may take moments of it, and these are naturally expressed as a ratio of integrals since the proportionality constant is itself the zeroth moment of the distribution. Thus define

$$I_k[f(x) \mid \boldsymbol{x}] := \int f(x_0)^k \, P(\boldsymbol{x} \mid x_0)\, dx_0 \tag{2}$$

and satisfy oneself that for uniform $P(x_0, y_0)$

$$E[x^k \mid \boldsymbol{x}] := \int x_0^k \, P(x_0 \mid \boldsymbol{x})\, dx_0 = I_k[x \mid \boldsymbol{x}]/I_0[x \mid \boldsymbol{x}]. \tag{3}$$

The notation $I_k := I_k[x \mid \boldsymbol{x}]$ is used later. The use of the first posterior moment of the Cauchy distribution as an estimator for the true value of the position parameter is motivated by the fact that it is the Bayes' minimum mean squared error (mmse) estimator for that position parameter. Similarly, it is left as an exercise to show that the nth Cauchy posterior moment is the Bayes' mmse estimator for the nth power of the position parameter. In particular, the combination of posterior averages $E[x^2 \mid \boldsymbol{x}] - E[x \mid \boldsymbol{x}]^2$ is the mmse estimator for the variance. Thus we are immensely interested in computing these moments.

In the rest of this brief paper we show how to compute the posterior moments of the Cauchy distribution in closed form, provide explicit indications of how the existence of these moments is tied to the number of observations, discuss the use of non-uniform priors, discuss non-integer moments, and finally discuss the moment generating function. In the next section, the moments are computed when the prior is "uniform" on the *infinite* line (by treating the prior as a infinitesimal constant). Following that we discuss the computation of these moments when the prior is uniform on any *finite* interval of the line. The infinite interval solution is discussed as a two-sided limit of a large finite interval and explicit agreement with the first (infinite line) solution is shown. The final sections present comments on using non-uniform priors, and the moment generation function. For simplicity, we only consider moments of the position parameter x (moments of the distance parameter y are similar). The presentation is terse, and references to the standard mathematical literature are minimal, hopefully allowing this paper to serve as a handbook for the Cauchy distribution in Bayesian statistics.

2. Infinite integration interval

When the interval of integration is $(-\infty, \infty)$ then the problem of finding moments may be viewed as a contour integration problem. The zeros in the denominator of the Cauchy distribution are given by $x_i^{\pm} := x_i \pm iy_i$. Thus I_k is given by (same normalization for all k for fixed N)

$$I_k \propto \int_{-\infty}^{\infty} \prod_{i=1}^{N} x^k \frac{1}{y^2 + (x - x_i)^2}\, dx = \oint x^k \prod_{i=1}^{N} x^k \frac{1}{y^2 + (x - x_i)^2}\, dx = 2\pi i \sum_{i=1}^{N} R_i \tag{4}$$

with the residues given by

$$R_i = \frac{(x_i^+)^k}{x_i^+ - x_i^-} \prod_{k \neq i} \frac{1}{(x_i^+ - x_k^+)(x_i^+ - x_k^-)} \tag{5}$$

and integer k, $0 \leq k < 2N - 1$.

3. Finite integration interval

When the interval of integration is a finite (a, b) the residue theorem no longer applies. However, it is possible to expand the Cauchy likelihood via partial fractions to find

$$\prod_{i=1}^{N} \frac{1}{y^2 + (x - x_i)^2} = \sum_{i=1}^{N} \frac{C_i^+}{x - x_i^+} + \frac{C_i^-}{x - x_i^-} \tag{6}$$

where

$$C_i^{\pm} := \frac{\pm}{x_i^+ - x_i^-} \prod_{k \neq i} \frac{1}{(x_i^{\pm} - x_k^+)(x_i^{\pm} - x_k^-)}. \tag{7}$$

Now proceed with the integration as usual, noting that [1]

$$\int \frac{x^k}{x - z}\, dx = \sum_{i=1}^{k} \frac{x^i z^{k-i}}{i} + z^k log(x - z). \tag{8}$$

The result is that the desired integral is proportional to

$$I_k = \sum_{i=1}^{N} (T_i^+ + T_i^-) \mid_a^b \tag{9}$$

where

$$T_i^{\pm} := \sum_{i=1}^{N} C_i^{\pm} \left(\sum_{m=1}^{k} \frac{1}{m} x^m (x_i^{\pm})^{k-m} + (x_i^{\pm})^k log(x - x_i^{\pm}) \right) \tag{10}$$

[1]This follows immediately from the representation $\frac{x^k}{x-z} = \sum_{i=1}^{k} x^i z^{k-i-1} + \frac{z^k}{x-z}$

which simplifies to

$$I_k = \sum_{i=1}^{N} (U_i^+ + U_i^-) \mid_a^b \tag{11}$$

for [2] $0 \leq k < 2N - 1$, with

$$U_i^{\pm} := C_i^{\pm}(x_i^{\pm})^k log(x - x_i^{\pm}). \tag{12}$$

4. Large integration interval

As the interval of integration becomes larger we expect the result of the section 3 for the finite interval, to approach the infinite-line result of section 2. To show this, define $r_i^{\pm}, \theta_i^{\pm}$ via $r_i^{\pm}\theta_i^{\pm} = x - x_i^{\pm}$. Then with this notation the result (11) of section 3 is

$$I_k = \sum_{i=1}^{N} C_i^+(x_i^+)^k(log(r_i^+) + i\theta_i^+) + C_i^-(x_i^-)^k(log(r_i^-) + i\theta_i^-) \mid_a^b . \tag{13}$$

Letting $b \to \infty, a \to -\infty$ [3] and using the result of the previous footnote gives the large interval result (with the same conditions on k as in section 3)

$$I_k = 2\pi i Im[C_i^+(x_i^+)^k]. \tag{14}$$

Using definitions (5) and (7) this is immediately seen to be the result (4) of section 2.

5. Examples

Here we give three examples which i) show how a posterior moment may fail to exist for too few observations, using the context of section 2, ii) give the first posterior moments for the 2 and 3 observation cases, and iii) give the estimator for the variance in the 2 observation case.

5.1. EXAMPLE 1 - FIRST MOMENT OF THE 1 OBSERVATION POSTERIOR

Here $k = 1 = 2N - 1$ so we expect that the moment will not exist, since $k < 2N - 1$ was required, see section 3. Clearly, the numerator term for $E[x \mid \boldsymbol{x}]$ with $a \to -\infty, b \to \infty$ does not exist because the integrand of the numerator is logarithmically divergent for $k = 1$. However it is still interesting to see explicitly how this integral is ill defined in the context of the results of section 3. The expansion of the integrand of I_1 is there given by $C_1^+ x/(x - x_1^+) + C_1^- x/(x - x_1^-)$, and this integrates to $(x_1^+ log(x - x_1^+) - x_1^- log(x - x_1^-))/(2iy)$. Converting to polar

[2] Consider the leading order coefficient of x in the numerator of the expansion of $\prod_{i=1}^{N} \frac{1}{y^2+(x-x_i)^2}$. In particular, $\sum_{i=1}^{N} C_i^+(x_i^+)^k + C_i^-(x_i^-)^k = 0$ for $k = 0, 1, \ldots, 2(N-1)$.

[3] Write $log(r^{\pm}) = log(r(1 + \epsilon^{\pm}))$ and note that $\epsilon^{\pm} \to 0$ for $b \to \infty, a \to -\infty$, respectively.

form gives $(x_1^+ log(r_1^+) - x_1^- log(r_1^-) + x_1^+ i\theta_i^+ - x_1^- i\theta_i^-)/(2iy)$. As can be immediately seen, the only case when cancellation of the logarithms in the numerator is to be expected for $a \to -\infty, b \to \infty$ independently is when $y = 0$, yet we must also divide by that $y = 0$. Thus the integral is ill defined for this case, and the moment fails to exist. We may take the principle value of the integral as the object of interest. In that case, the desired moment is simply x_1.

5.2. EXAMPLE 2, FIRST MOMENTS OF THE 2 AND 3 OBSERVATION POSTERIORS

For two observations, the result is trivially given by $E[x \mid (x_1, x_2)] = (x_1 + x_2)/2$. However, the pattern set here (the average) does not continue; for 3 observations the first posterior moment is given by $E[x \mid (x_1, x_2, x_3)] =$

$$\frac{8y^2(x_1 + x_2 + x_3) + x_1^2(x_2 + x_3) + x_2^2(x_1 + x_3) + x_3^2(x_1 + x_2) - 6x_1x_2x_3}{2(12y^2 + x_1^2 + x_2^2 + x_3^2 - (x_1x_2 + x_2x_3 + x_3x_1))} \tag{15}$$

Interestingly, when $y = 0$ this is not the average. Another interesting observation is that the difference of this and the average of the three points has a numerator given by the symmetric form $(-2x_1 + x_2 + x_3)(-2x_2 + x_3 + x_1)(-2x_3 + x_1 + x_2)$. This pattern, too, does not continue. For example, for $N = 2$ the corresponding difference is zero.

5.3. EXAMPLE 3, ESTIMATOR FOR THE STANDARD DEVIATION WITH TWO OBSERVATIONS.

The second moment of the two observation posterior is given by $y^2 + (x_1^2 + x_2^2)/2$. This, along with the first moment result from example 2 above gives

$$E[(x - E[x \mid \boldsymbol{x}])^2 \mid \boldsymbol{x}] = y^2 + (x_1 - x_2)^2/4. \tag{16}$$

6. Non-uniform priors

Briefly, for an infinite interval of integration, note that when the prior is any finite sum of the form $\sum_{i=0}^{n} a_i x^i$ then moments up to order $2(N-1) - n$ exist if this quantity is non-negative. For a finite interval of integration then any series which when multiplied by the likelihood is properly integrated term-by-term may be used as a prior. Rational functions where the numerator degree minus the denominator degree is less than or equal to $2(N-1)$ may be used as priors.

7. Non-integer moments

For a finite interval of integration and uniform prior, the binomial series [4] may be used to develop an approximation in the following way. Let q be a non-integer real,

[4] The binomial series for non-integer α is given by $(1+x)^\alpha = \sum_{i=0}^{\infty} \frac{1}{i!} \frac{\Gamma(\alpha+1)}{\Gamma(\alpha-i)} x^i$, where $|x| < 1$.

and consider I_q. Make the expansion of the denominator of the Cauchy likelihood as done in section 3. In each of the terms make the substitutions $u_i^{\pm} = x - x_i^{\pm}$ respectively. Now, expand each of the resulting numerators $(u_i^{\pm} + x_i^{\pm})^q$ using the binomial series appropriate for the limit under consideration (depending upon whether $\mid u_i^{\pm} \mid > x_i^{\pm}$ or $\mid u_i^{\pm} \mid < x_i^{\pm}$ at the limits $x = a, b$ as appropriate). When it converges, then the series representing the integration may be truncated at whatever order is needed for the desired accuracy.

8. Moment generating function

The kth moment is given in terms of the kth derivative of the moment generating function by $E[x^k \mid x] = \partial_\alpha^k E[e^{\alpha x} \mid x] \mid_{\alpha=0}$. The moment generating function for the N-observation Cauchy distribution may be found in closed form. From (6)

$$I_k[e^{\alpha x} \mid x] = \int_a^b \left(\sum_{i=1}^N \frac{C_i^+ e^{\alpha x}}{x - x_i^+} + \frac{C_i^- e^{\alpha x}}{x - x_i^-} \right) dx. \tag{17}$$

After making appropriate variable changes each of the terms in the sum on the right is of the form [5]

$$Ce^{\alpha x_0} \int_{\alpha(a-x_0)}^{\alpha(b-x_0)} \frac{e^u}{u} du = Ce^{\alpha x_0}[Ei(\alpha(b - x_0)) - Ei(\alpha(a - x_0))]. \tag{18}$$

Making the substitutions for the moment generating function yields

$$E[e^{\alpha x} \mid x] = \frac{\sum_{i=1}^N Re(C_i^+ e^{\alpha x_i^+}[Ei(\alpha(b - x_i^+)) - Ei(\alpha(a - x_i^+))])}{\sum_{i=1}^N Re(C_i^+ s)} \tag{19}$$

Clearly, all moments are easily accessible since $\partial_x Ei(x) = e^x/x$. As a check of the first moment

$$\lim_{\alpha \to 0}[Ei(\alpha q) - Ei(\alpha p)] = log(p) - log(q). \tag{20}$$

9. Acknowledgements

I thank Kenneth Hanson for suggesting this problem. Part of this work was sponsored by the United States Department of Energy under contract number W-7405-ENG-36. Part of this work was sponsored by NASA.

A. Mathematica program for finding Cauchy posterior moments

This Mathematica program finds the kth posterior moment of the position of the Cauchy distribution, given the N observations $(x_1, \ldots, x_N)$, the distance $y = y_0$ and "uniform" prior on the line.

[5]The Exponential Integral is given by $Ei(z) = \int^z e^u/udu$ where the path of integration does not cut $[0, \infty)$ [3]. It may be written as $Ei(z) = \gamma + log(-z) + \sum_{i=1}^{\infty} z^i/(i!i)$ in the plane cut along the positive real axis.

```
kthMoment[k_Integer,N_Integer] := kthIntegral[k,N] / kthIntegral[0,N],
```

where

```
kthIntegral[k_Integer,N_Integer] :=
Block[ {i,j,Result},
        For[i=1,i<=N,i++,
            xp[i] = x[i] + I y;
            xm[i] = x[i] - I y;
        ];

        For[i=1,i<=N,i++,
            Cp[i] =  1/(xp[i]-xm[i]);
            Cm[i] = -1/(xp[i]-xm[i]);
            For[j=1,j<=N,j++,
                If[ j!=i,
                    Cp[i] *= 1/(xp[i]-xp[j]) 1/(xp[i]-xm[j]);
                    Cm[i] *= 1/(xm[i]-xp[j]) 1/(xm[i]-xm[j]);
                ];
            ];
        ];

        Result = 0;

        For[i=1,i<=N,i++,
            Result += Cp[i] xp[i]^k - Cm[i] xm[i]^k;
        ];

        Return[Result];
]
```

References

1. K. M. Hanson and D. R. Wolf, "Estimators for the cauchy distribution," in *Maximum Entropy and Bayesian Methods in Science and Engineering*, G. Heidbreder, ed., pp. 255–263, Kluwer Academic, Dordrecht, 1996.
2. S. F. Gull, "Bayesian inductive inference and maximum entropy," in *Maximum Entropy and Bayesian Methods in Science and Engineering*, G. Erickson and C. R. Smith, eds., pp. 53–74, Kluwer Academic, Dordrecht, 1988.
3. M. Abramowitz and I. Stegun, *Handbook of Mathematical Functions*, Dover, NY, 1972.

Ken Hanson and Giulio D'Agostini.

BAYESIAN REASONING VERSUS CONVENTIONAL STATISTICS IN HIGH ENERGY PHYSICS

G. D'AGOSTINI
Dipartimento di Fisica dell'Università "La Sapienza"
Piazzale Aldo Moro 2, I-00185 Roma (Italy)†

Abstract. The intuitive reasoning of physicists in conditions of uncertainty is closer to the Bayesian approach than to the frequentist ideas taught at University and which are considered the reference framework for handling statistical problems. The combination of intuition and conventional statistics allows practitioners to get results which are very close, both in meaning and in numerical value, to those obtainable by Bayesian methods, at least in simple routine applications. There are, however, cases in which "arbitrary" probability inversion produce unacceptable or misleading results and in these the conscious application of Bayesian reasoning becomes crucial. Starting from these considerations, I will finally comment on the often debated question: "is there any chance that all physicists will become Bayesian?"

Key words: Subjective Bayesian Theory, High Energy Physics, Measurement Uncertainty

1. Introduction

High Energy Physics (HEP) is well known for using very sophisticated detectors, status of the art computers, ultra-fast data acquisition systems, and very detailed (Monte Carlo) simulations. One might imagine that a similar level of refinement could be found in its analysis tools, on the trail of the progress in probability theory and statistics of the past half century. Quite the contrary! As pointed out by the downhearted Zech[1], "some decades ago physicists were usually well educated in basic statistics in contrast to their colleagues in social and medical sciences. Today the situation is almost reversed. Very sophisticated methods are used in these disciplines, whereas in particle physics standard analysis tools available in many program packages seem to make knowledge of statistics obsolete. This leads to strange habits, like the determination of a r.m.s. of a sample through a fit to a Gaussian. More severe are a widely spread ignorance about the (lack of) significance of χ^2 tests with a large number of bins and missing experience with unfolding

†Email: dagostini@roma1.infn.it. URL: http://www-zeus.roma1.infn.it/~agostini/

W. von der Linden et al. (eds.), Maximum Entropy and Bayesian Methods, 157–170.

methods". In my opinion, the main reason for this cultural gap is that statistics and probability are not given sufficient importance in the student curricula: *ad hoc* formulae are provided in laboratory courses to report the "errors" of measurements; the few regular lectures on "statistics" usually mix up *descriptive* statistics, *probability* theory and *inferential* statistics. This leaves a lot of freedom for personal interpretations of the subject (nothing to do with subjective probability!). Of equal significance is the fact that the *disturbing catalog of inconsistencies*[2] of "conventional"[1] statistics helps to give the impression that this subject is matter of initiates and local gurus, rather than a scientific discipline[2]. The result is that standard knowledge of statistics at the end of the University curriculum is insufficient and confused, as widely recognized. Typical effects of this (dis-)education are the *"Gaussian syndrome"*[3] (from which follows the uncritical use of the rule of combining results, weighing them with inverse of the "error" squared[4], or the habit of calling χ^2 any sum of squared differences between fitted curves and data points, and to use it as if it were a χ^2), the abused *"$\sqrt{n}$ rule"* to evaluate "errors" of counting experiments[5] and the reluctance to take into account correlations[6], just to remain at a very basic level.

I don't think that researchers in medical science or in biology have a better statistics education than physicists. On the contrary, their usually scant knowledge of the subject forces them to collaborate with professional statisticians, and this is the reason why statistics journals contain plenty of papers in which sophisticated methods are developed to solve complicated problems in the aforementioned fields. Physicists, especially in HEP, tend to be more autonomous, because of their skills in mathematics and computing, plus of a good dose of intuition. But one has to admit that it is rather unlikely that a physicist, in a constant hurry to publish results before anybody else, can reinvent methods which have been reached by

---

[1]I prefer to call the frequentist approach "conventional" rather than "classical".

[2]Who can have failed to experience endless discussions about trivial statistical problems, the solution of which was finally accepted just because of the (scientific or, more often, political) authority of somebody, rather than because of the strength of the logical arguments?

[3]I know a senior physicist who used to teach students that standard deviation is meaningful only for the Gaussian, and that it is "defined" as half of the interval around the average which contains 68 % of the events! More common is the evaluation of the standard deviation of a data sample fitting a Gaussian to the data histogram (see also the previous quotation of Zech[1]), even in those cases in which the histogram has nothing to do with a Gaussian. The most absurd case I have heard of is that of someone fitting a Gaussian to a histogram exhibiting a flat shape (and having good reasons for being considered to come from coming from a uniform distribution) to find the resolution of a silicon strip detector!.

[4]For instance, the 1998 issue of the Review of Particle Physics[3] includes an example based on this kind of mistake with the intention to show that "the Bayesian methodology ...is not appropriate for the objective presentation of experimental data" (section 29.6.2, pag. 175).

[5]Who has never come across somebody calculating the "error" on the efficiency $\epsilon = n/N$, using the standard "error propagation" starting from $\sqrt{n}$ and $\sqrt{N}$?

[6]In the past, the correlation matrix was for many HEP physicists "that mysterious list of numbers printed out by MINUIT" (MINUIT[4] is the minimization/fitting package mostly used in HEP), but at least some cared about understanding what those numbers meant and how to use them in further analysis. Currently - I agree with Zech[1] - the situation has worsened: although many people do take into account correlations, especially in combined analysis of crucial Standard Model parameters, the interactive packages, which display only standard deviations of fitted parameters, tend to "hide" the existence of correlations to average users.

others after years of work and discussion. Even those physicists who are considered experts in statistics usually read books and papers written and refereed by other physicists. The HEP community remains, therefore, isolated with respect to the mainstream of research in probability and statistics.

In this paper I will not try to review all possible methods used in HEP, nor to make a detailed comparison between conventional and Bayesian solutions to solve the same problems. Those interested in this kind of statistical and historical study are recommended to look at the HEP databases and electronic archives[5]. I think that the participants in this workshop are more interested in learning about the attitude of HEP physicists towards the fundamental aspects of probability, in which framework they make uncertainty statements, how subjective probability is perceived, and so on. The intention here will be, finally, to contribute to the debate around the question *"Why isn't everybody a Bayesian"*[2], recently turned into *"Why isn't every physicist a Bayesian"*[6]. The theses which I will try to defend are:

- there is a contradiction between a cultural background in statistics and the good sense of physicists, and physicists' intuition is closer to the Bayesian approach than one might naïvely think;
- there are cases in which good sense alone is not enough and serious mistakes can be made; it is then that the philosophical and practical advantages offered by the Bayesian approach become of crucial importance;
- there is a chance that the Bayesian approach can become widely accepted, if it is presented in a way which is close to physicists intuitions and can solve the "existential" problem of reconciling two aspects which seem irreconcilable: subjective probability and the honest ideal of objectivity which scientists have.

Some of the points sketched quickly in this paper are discussed in detail in lecture notes[7] based on several seminars and minicourses given over the past years. These notes also contain plenty of general and HEP inspired applications.

2. Measurement Uncertainty

An idea well rooted among physicists, especially nuclear and particle physicists, is that the result of a measurement must be reported with a corresponding *uncertainty*. What makes the measured values subject to a degree of uncertainty is, it is commonly said, the effect of unavoidable measurement *errors*, usually classified as *random* (or *statistical*) and *systematic*[7].

[7] This last statement may sound like a tautology, since "error" and "uncertainty" are often used as synonyms. This hints to the fact that in this subject there is neither uniformity of language, nor of methods, as is recognized by the metrological organizations, which have made great efforts to bring some order into the field[8–12]. In particular, the International Organization for Standardization (ISO) has published *"Guide to the expression of uncertainty in measurement"*[10], containing definitions, recommendations and practical examples. For example, *error* is defined as "the result of a measurement minus a true value of the measurand" *uncertainty* "a parameter, associated with the result of a measurement, that characterize the dispersion of the values that could reasonably be attributed to measurand", and, finally, *true value* "a value compatible with the definition of the given particular quantity". One can easily see that it is not just a question

Uncertainties due to statistical errors are commonly treated using the frequentist concept of confidence intervals, although the procedure is so unnatural that the interpretation of the result is unconsciously subjective (as will be shown in a while), and there are known cases (of great relevance in frontier research) in which this approach is not applicable.

As far as uncertainties due to systematics errors are concerned, there is no conventional consistent theory to handle them, as is also indirectly recognized by the ISO *Guide*[10]. The "fashion" at the moment is to add them quadratically if they are considered to be independent, or to build a covariance matrix if not. This procedure is not justified theoretically (in the frequentist approach) and I think that it is used essentially because of the reluctance of experimentalists to add linearly the dozens of contributions of a complicated HEP measurement, as the old fashioned "theory" of maximum errors suggests doing[8]. The pragmatic justification for the quadratic combination of "systematic errors" is that one is using a rule (the famous "error propagation" formula[9]) which is considered to be valid at least for "statistical errors". But, in reality, this too is not correct. The use of this formula is again arbitrary in the case of "statistical errors", if these have been evaluated from confidence intervals[10]. In fact, there is no logical reason why a probabilistic procedure proved for standard deviations of random variables (the observables) should also be valid for 68 % confidence intervals, which is considered, somehow, an uncertainty attributed to the true value.

---

of practical definitions. It seems to me that there is a well-thought-out philosophical choice behind these definitions, although it is not discussed extensively in the *Guide*. Two issues in the *Guide* that I find of particular importance are the discussion on the sources of uncertainty and the admission that all contributions to the uncertainty are of a probabilistic nature. The latter is strictly related to the subjective interpretation of probability, as admitted by the *Guide* and discussed in depth in [7]. (The reason why these comments on the ISO *Guide* have been placed in this long footnote is that, unfortunately, the *Guide* is not yet known in the HEP community and, therefore, has no influence on the behaviour of HEP physicists about which I am going to comment here. This is also the reason why I will often use in this paper typical expressions currently used in HEP and which are in disagreement with the ISO recommendations. But I will use these expressions preferably within quote marks, like "systematic error" instead of "uncertainty due to a recognized systematic error of unknown size".)

[8]In fact, one can see that when there are only 2 or 3 contributions to the "systematic error", there are still people who prefer to add them linearly.

[9]The most well-known version is that in which correlations are neglected:

$$\sigma^2(Y) = \sum_i \left(\frac{\partial Y}{\partial X_i}\right)^2 \sigma^2(X_i)\,.$$

Y stands for the quantity of interest, the value of which depends on directly measured quantities, calibration constants and other systematic effects (all terms generically indicated by X_i). This formula comes from probability theory, but it is valid if X_i and Y are random variables, $\sigma(X_i)$ are standard deviation and the linearization is reasonable. It is very interesting to look at text books to see how this formula is derived. The formula is usually initially proofed referring to random variables associated to observables and then, suddenly, it is referred to physics quantities, without any justification.

[10]As far as "systematic errors" are concerned the situation is much more problematic because the "errors" are not even operationally well defined: they may correspond to subjectivist standard deviations (what I consider to be correct, and what corresponds to the ISO *type B* standard uncertainty[10]), but they can more easily be maximum deviations, ±50 % variation on a selection cut, or the absolute difference obtained using two assumptions for the systematic effect.

These examples show quite well the contradiction between the cultural background on probability and the practical good sense of physicists. Thanks to this good sense, frequentist ideas are constantly violated, with the positive effect that at least some results are obtained[11]. It is interesting to notice that in simple routine applications these results are very close, both in value and in meaning, to those achievable starting from what I consider to be the correct point of view for handling uncertainty (subjective probability). There are, on the other hand, critical cases in which scientific conclusions may be seriously mistaken. Before discussing these cases, let us look more closely at the terms of the claimed contradiction.

3. Professed Frequentism versus Practiced Subjectivism

3.1. HEP PHYSICISTS "ARE FREQUENTIST"

If one asks HEP physicists "what is probability?", one will realize immediately that they "think they are" frequentist. The same impression is got looking at the books and lecture notes they use[13]. Particularly significant, to get an overview of ideas and methods commonly used, are the PDG[3] and other booklets[14,15] which have a kind of explicit (e.g. [3,14]) or implicit (e.g. [15]) *imprimatur* of HEP organizations.

If, instead, one asks physicists what they think about probability as *"degree of belief"* the reaction is negative and can even be violent: "science must be objective: there is no room for belief", or "I don't believe something. I assess it. This is no a matter for religion!".

3.2. HEP PHYSICISTS "ARE BAYESIAN"

On the other hand, if one requires physicists to express their opinion about practical situations in condition of uncertainty, instead of just standard examination questions, one gets a completely different impression. One realize vividly that Science is indeed based on beliefs, very solid and well grounded beliefs, but they remain beliefs "...in instrument types, in programs of experiment enquiry, in the trained, individual judgements about every local behavior of pieces of apparatus"[16].

Physicists find it absolutely natural to talk about the probability of hypotheses, a concept for which there is no room in the frequentist approach. Also the intuitive way with which they figure out the result is, in fact, a probabilistic assessment on the true value. Try to ask what is the probability that the top quark mass is between 170 and 180 GeV. No one[12] will reply that the question has no sense, since "the top quark mass is a *constant of unknown value*" (as an orthodox frequentist should complain). They will simply answer that the probability is such and such percent, using the published value and "error". They are usually sur-

[11] I am strongly convinced that a rigorous application of frequentist ideas leads nowhere.

[12] Certainly one may find people aware of the "sophistication" of the frequentist approach, but these kinds of probabilistic statements are currently heard in conferences and no frequentist guru stands up to complain that the speaker is talking nonsense.

prised if somebody tries to explain to them that they "are not allowed" to speak of probability of a true value.

Another word which physicists find scandalous is "prior" ("I don't want to be influenced by prejudices" is the usual reply). But in reality priors play a very important role in laboratory routines, as well as at the moment of deciding that a paper is ready for publication. They allow experienced physicists to realize that something is going wrong, that a student has most probably made a serious mistake, that the result has not yet been corrected by all systematic effects, and so on. Unavoidably, priors generate some subtle cross correlations among results, and there are well known cases of the values of physics quantities slowly drifting from an initial point, with all subsequent results being included in the "error bar" of the previous experiment. But I think that there no one and nothing is to blame for the fact that these things happen (unless made on purpose): a strong evidence is needed before the scientific community radically changes its mind, and such evidence is often achieved after a long series of experiments. Moreover, very subtle systematic effects may affect the data, and it is not a simple task for an experimentalist to decide when all corrections have been applied, if he has no idea what the result should be.

3.3. INTUITIVE APPLICATION OF BAYES' THEOREM

There is an example which I like to give, in order to demonstrate that the intuitive reasoning which unconsciously transforms confidence intervals into probability intervals for the true value is, in fact, very close to the Bayes' theorem. Let us imagine we see a hunting dog in a forest and have to guess where the hunter is, knowing that there is a 50 % probability that the dog is within 100 m around him. The terms of the analogy with respect to observable and true value are obvious. Everybody will answer immediately that, with 50 % probability, the hunter is within 100 m from the dog. But everybody will also agree that the solution relies on some implicit assumptions: uniform *prior* distribution (of the hunter in the forest) and symmetric *likelihood* (the dog has no preferred direction, as far as we know, when he runs away from the hunter). Any variation in the assumptions leads to a different solution. And this is also easily recognized by physicists, expecially HEP physicists, who are aware of situations in which the prior is not flat (like the cases of a bremsstrahlung photon or of a cosmic ray spectrum) or the likelihood is not symmetric (not all detectors have a nice Gaussian response). In these situations intuition may still help a qualitatively guess to be made about the direction of the effect on the value of the measurand, but a formal application of the Bayesian ideas becomes crucial in order to state a result which is consistent with what can be honestly learned from data.

3.4. BAYES VERSUS MONTE CARLO

The fact that Bayesian inference is not currently used in HEP does not imply that non-trivial inverse problems remain unsolved, or that results are usually wrong.

The solution often relies on extensive use of Monte Carlo (MC) simulation[13] and on intuition. The *inverse* problem is then treated as a *direct* one. The quantities of interest are considered as MC parameters, and are varied until the best statistical agreement between simulation output and experimental data is achieved. In principle, this is a simple numerical implementation of Maximum Likelihood, but in reality the prior distribution is also taken into account in the simulation when it is known to be non uniform (like in the aforementioned example of a cosmic ray experiment). So, in reality what it is often maximized is not the likelihood, but the Bayesian *posterior* (likelihood×prior), and, as said before, the result is intuitively considered to be a probabilistic statement for the true value. So, also in this case, the results are close to those obtainable by Bayesian inference, especially if the posterior is almost Gaussian (parabolic *negative log-likelihood*). Problems may occur, instead, when the "not used" prior is most likely not uniform, or when the posterior is very non-Gaussian. In the latter case the difference between mode and average of the distribution, and the evaluation of the uncertainty from the "Δ(log-likelihood) $= 1/2$ rule" can make quite a difference to the result.

4. Explicit use of Bayesian methods in HEP

Besides the intuitive use of Bayesian reasoning, there are, in fact, some applications in which the Bayes' theorem is explicitly applied. This happens when frequentist methods "do not work", i.e. they give manifestly absurd results, or in solving more complicated problems than just inferring the value of a quantity, like, for example, the deconvolution of a spectrum ("unfolding"). Nevertheless, these methods are mostly used with a utilitarian spirit, without having really understood the meaning of subjective probability, or even remaining skeptical about it. They are used as one uses one of the many frequentist "ad hoc-eries"[14], after it has been "proved" that they work by MC simulation[15].

Some of the cases in which the conventional methods do not work have even induced the PDG[3] to present Bayesian methods. But, according to the PDG, a paper published this year[18] finally gives a frequentist solution to the problems, and this solution is recommended for publishing the results. Let us review the situation citing directly [18]: "Classical confidence intervals are the traditional way in which high energy physicists report errors on results of experiments. ... In

[13]If there is something in which HEP physicists really believe, it is Monte Carlo simulation! It plays a crucial role in all analyses, but sometimes its use as a multipurpose brute force problem solver is really unjustified and it can, from a cultural point of view, be counterproductive. For example, I have seen it applied to solve elementary problems which could be solved analytically, like "proving" that the variance of the sum of two random numbers is the sum of the variances. I once found a sentence at the end of the solution of a standard probability problem which I consider to be symptomatic of this brute force behaviour: "if you don't trust logic, then you can make a little Monte Carlo...".

[14]For example, this was exactly the attitude which I had some years ago, when I wrote a *Bayesian unfolding* program[17], and that of the large majority my colleagues who still use the program. Now, after having attended the 1998 Valencia Meeting on Bayesian Statistics, I have realized that this pragmatic frequentist-like use of Bayesian methods is rather common.

[15]I would like to point out that sometimes the conclusions derived from MC checks of Bayesian procedures may be misleading, as discussed in detail in [7].

recent years, there has been considerable dissatisfaction ... for upper confidence limits... This dissatisfaction led the PDG to describe procedures for Bayesian interval construction in the troublesome cases: Poisson processes with background and Gaussian errors with a bounded physical region. ... In this paper, we... use (...) to obtain a unified set of classical confidence intervals for setting upper limits and quoting two-sided confidence intervals. ... We then obtain confidence intervals which are never unphysical or empty. Thus they remove an original motivation for the description of Bayesian intervals by the PDG." In fact, the 1998 issue of the *Review of Particle Physics* still exhibits the Bayesian approach (with the typical misconceptions that frequentists have about it), but then it suggests two papers by frequentists[6,2] ("a balanced discussion"[3]) to help practitioners to form their own idea on the subject, and, finally, it warmly recommends the new frequentist approach. It is easy to imagine what the reaction of the average HEP physicist will be when confronted by the authority of the PDG, unaware that "the PDG" which rules analysis methods is in reality constituted of no more than one or two persons who recommend a paper written by their friends (as is clear from the references and the cross acknowledgements). One should also notice that this paper claims important progress in statistics, but was in fact published in a physics journal (I wonder what the reaction of a referee of a statistics journal would have been...).

In conclusion, there is still a large gap between good sense and the dominating statistical culture. For this reason we must still be very careful in interpreting published results and in evaluating whether or not the conventional methods used lead to correct scientific conclusions "by chance". Some cases of misleading results will be described in the next section.

5. Examples of Misleading Results Induced by Conventional Statistics

It is well known that the frequentist denial of the concept of probability of hypotheses leads to misleading results in all cases in which the simple "dog-hunter inversion paradigm" is violated. This also happens in HEP.

As already discussed, confidence levels are intuitively thought (and usually taught) as probabilities for the true values. I must recognize that many frequentist books do insist on the fact that the probability statement is not referred to the true value. But then, when these books have to explain the "real meaning" of the result, they are forced to use ambiguous sentences which remain stamped in the memory of the reader much more than the frequentistically-correct twisted reasoning that they try to explain. For example Frodesen et al.[13] speak about "the faith we attach to this statement", as if "faith" was not the same as degree of belief...); Eadie et al.[13] introduce the argument saying that "we want to find *the range* ... which contains the true value $\theta_\circ$ with probability β"[16]; and so on.

Similarly, significance levels are usually taken as probability of the tested hypothesis. Also this non-orthodox interpretation is stimulated by sentences like "in

[16] I think that Aristoteles would have gotten mad if somebody had tried to convince him that the proposition "the range contains $\theta_\circ$ with probability β" does not imply "$\theta_\circ$ is in that range with probability β".

statistical context, the words *highly significant* mean *proved beyond a reasonable doubt*". It is also well known that the arbitrary translation of *p-values* into probability of the null hypothesis produces more severe mistakes than those concerning the use of confidence interval for uncertainty statements on true values.

Let us consider some real life examples of the misinterpretation of the two kinds just described.

5.1. CLAIMS OF NEW PHYSICS BASED ON *P*-VALUES

You may have heard in past years some rumors, or even official claims, of discoveries of "New Physics", i.e. of phenomenology which goes behind the so called Standard Model of elementary particles. Then, after some time, these announcements were systematically recognized as having been false alarms, with a consequent reduction in the credibility of the HEP community in the eyes of public opinion and payers (with easily imaginable long term aftermath for government support of this research). All these fake discoveries were based on considering low p-values as probability of the null hypothesis "no new effect". The most recent example of this kind is the so called 1997 "HERA high Q^2 events excess". The H1 and ZEUS collaborations, analyzing data collected at the HERA very high energy electron-proton collider in Hamburg (Germany), found an excess of events (with respect to expectations) in the kinematical region corresponding to very hard interactions[19]. The "combined significance"[17] of the excess was of the order of 1 %. Its interpretation as a hint of new physics was even suggested by official statement by the laboratory and by other agencies. For example the DESY official statement was *"... the joint distribution has a probability of less than one percent to come from Standard Model NC DIS processes"*[22] (then it implies "it has a > 99 % probability of not coming from the standard model!"[18]). Similarly, the Italian INFN reported that *"la probabilità che gli eventi osservati siano una fluttazione statistica è inferiore all'1 %"* (then, it implies that "with 99 % probability, the events are not a statistical fluctuation, i.e. new physics"!). This is the reason why the press reported the news as "scientists are practically sure they

[17]Physicists are not familiar with the term *p-value* (readers not familiar with this term may find a concise review in [20]). Moreover, they are usually not aware of the implications of the fact that the statistical significance takes into account also the probability of unobserved data (see, e.g., [21]).

[18]One might think that the misleading meaning of that sentence was due to unfortunate wording, but this possibility is ruled out by other statements which show clearly a quite odd point of view of probabilistic matter. In fact the DESY 1998 activity report[23] insists in saying that "the likelihood that the data produced is the result of a statistical fluctuation, ..., is equivalent to that of tossing a coin and throwing seven 'heads' or 'tails' in a row" (replacing 'probability' by 'likelihood' does not change the sense of the message). Then, trying to explain the meaning of a statistical fluctuation, the following example is given: "This process can be simulated with a die. If the number of times a die is thrown is sufficiently large, the die falls equally often on all faces, i.e. all six numbers occur equally often. The probability for each face is exactly a sixth or 16.66 % - assuming the die is not loaded. If the die is thrown less often, then the probability curve for the distribution of the six die values is no longer a straight line but has peaks and troughs. The probability distribution obtained by throwing the die varies about the theoretical value of 16.66 % depending on how many times it is thrown".

have found new physics". What I found astonishing is that most of the people I talked to had real difficulty in understanding that this probability inversion is not legitimate. Only when I forced them to state their degree of belief using the logic of the coherent bet did it emerge that most of my colleagues would not even place a 1:1 bet in favour of the new discovery. Nevertheless, they were in favour of publishing the result because the loss function was absolutely unbalanced (an indirect Nobel prize against essentially nothing).

5.2. WHAT DOES A LOWER MASS BOUND MEAN?

The second example concerns confidence intervals, and it comes from new particle search. This has always been one of the main activities of HEP. New particles are postulated by theories and experimentalists look for evidence for them in experimental data. Usually, if the particle is not "observed"[19] one says that, although the observation does not disprove the existence of the particle, this is an indication of the fact that the particle is "too heavy". The result is then quantified by a "lower bound" at a "95 % confidence level". Without entering into detail of how the limit is operationally defined (see, e.g., [24] and references therein, in particular [25], to have an idea of the level of complication reachable to solve a simple problem), I want to point out that also in this case the result can be misleading. Again I will give a real life example. A combined analysis of all the LEP experiments on the Higgs mass concluded recently that *"A 95 % confidence level lower bound of 77.5 GeV/c^2 is obtained for the mass of the Standard Model Higgs boson"*[20]. This *sounds as if* one were sure at 95 % that the mass is above the quoted bound. In fact, most of the people I interviewed about the meaning of the statement, even those belonging to the LEP experimental teams, answered "if the Higgs boson exists at all, then there is 95 % probability that its mass is above the limit". There were also a few people who answered "if I do a MC simulation of the decay of a 77.5 GeV Higgs boson, I get in only 5 % of the cases the simulation describing the data", or "if there is an Higgs boson and its mass is less than 77.5 GeV, then the observations of the search experiments have a probability of at most 5 % of being correct", or something similar. From all of which it is very difficult to understand, from a logical point of view, why one should be 95 % confident that the mass is higher than the bound[21].

The problem can be solved easily with Bayesian methods (see [7] for details). Assuming a flat prior for the mass, one finds that the value of the lower bound is

[19]This concept of "observation" is not like that of seeing a black swan, to mention a famous classical example. New particles leave signatures in the detector that on an event by event basis cannot be distinguished by other processes (background). A statistical (inferential) analysis is therefore needed.

[20]In the meanwhile new data have increased this limit, but the actual number is irrelevant for this discussion.

[21]There was also somebody who refused to answer because "your question is going to be difficult to answer", or without any justification (perhaps they realized that it was impossible to explain the statement to a scientific journalist, or to a government authority - these were the terms of my question - without using probabilistic statements which were incompatible with what they thought about probability).

more or less the published one, but only under the condition that the mass does not exceed the kinematical limit of the studied reaction. But this limit is just a few GeV above the stated lower bound. Thus in order to obtain the correct result one needs to renormalize the probability taking account of the possible range of masses above the kinematical limit and for which the experiment has no sensitivity. For this reason, in the case of [24] the probability that the mass value is above 77.5 GeV/c^2 may easily become 99.9 %, or more, depending on the order of magnitude of a possible upper bound for the mass. Then, in practice these lower bounds can be taken as certainties[22].

6. Conclusions

Although it is clear that the dominant statistical culture is still frequentism in HEP (and everywhere else), I am myself rather optimist on the possibility that the situation will change, at least in HEP, and that Bayesian reasoning will emerge from an intuitive to a conscious level. This is not a dream (although clearly several academic generations are still needed) if the theory is presented in a way that it is acceptable to an "experienced physicist".

- First, it is not difficult to get a consensus on the observation that subjective probability is the natural concept developed by the human mind to quantify the plausibility of events in conditions of uncertainty.
- Second, one should insist on the fact that Bayes' theorem is in fact a natural way of reasoning in updating probability, and not a philosophical point of view that somebody tries to apply to data analysis[23] (see [7] for details).
- Bayes' theorem is not "all". It only works in situations where the nice scheme of prior and likelihood is applicable. In many circumstances one can assess a subjective probability directly (try asking a carpenter how much he believes the result of his measurement!).
- The *coherent bet* (à la de Finetti[27]) forces people to be honest and to make the best (i.e. "most objective") assessments of probability.
- It is preferable not to mix up probability evaluation with decision problems[24]. In other words, the coherent bet should be considered hypothetical. This makes a clear distinction between our beliefs and our wishes (the example in section 5.1 should teach something in this respect).

[22] There are in fact theorists who "assume" the lower bounds as certain bounds in their considerations. Perhaps they do it intuitively, or because they have heard in the last decades of thousands of these 95 % lower bounds, and never a particle has then shown up in the 5 % side...

[23] For example frequentists completely misunderstand this points, when they state, e.g., that "Bayesian methods proceed by invoking an interpretation of Bayes' theorem, in which one deems it sensible to consider a p.d.f. for the unknown true value m_t", or that "a pragmatist can consider the utility of equations generated by the two approaches while skirting the issue of buying a whole philosophy of science"[6].

I find that also the Zellner's paper[26] demonstrating that Bayes' theorem makes the best use of the available information can help a lot to convince people.

[24] Although it may seem absurd, the Bayesian approach is recognized by "frequentists" to be "well adapted to decision-making situations"[3](see also [6,18]). I wonder what then probability is for these authors.

- One may think, naïvely, that the "objective Bayesian theory" is more suited for science than the "subjective one". Instead, it seems to me easier to convince experienced physicists that "there is nothing really that is objective", than it is to accept an objective theory containing ingredients which appear dogmatic[25] from the physicist point of view[28,7]. Any experienced physicist knows already that the only "objective" thing in science is the reading of digital scales. When we want to transform this information into scientific knowledge we have to make use of many implicit and explicit beliefs (see section 3.2). Nevertheless, the "honest" (but naïve) ideal of objectivity can be recovered if scientific knowledge is considered as a kind of very *solid networks of beliefs*, based on centuries of experimentation, with *fuzzy borders* which correspond to the areas of present research. My preferred motto is that "no one should be allowed to talk about objectivity, unless he has 10 or 20 years of experience in frontier science, economics, or any other real world application". In particular, mathematicians should refrain from using the word objectivity when talking about the physical world.
- It is very important to work on applications: the simplicity and the naturalness of the Bayesian reasoning will certainly attract people.
- Many conventional methods can be easily recovered as limit cases of the Bayesian ones, if some well defined restricting conditions are valid, as already discussed in section 3.4. For example, when I make a χ^2 fit I consider myself to be using a Bayesian method, although in a simplified form. This attitude contrasts to that of practitioners who use methods in which the Bayes' theorem is explicitly applied, but as if it were one of the many frequentist cooking recipes.
- It is important to make efforts to introduce Bayesian thinking in teaching, starting from the basic courses. I am not the first to have realized that the Bayesian approach is simple for students. The resistance comes from our colleagues, who are unwilling to renew the contents of their lectures, and who have developed a distorted way of thinking.

Finally, I would like to give a last recommendation. Don't try to convince a physicist that he already is Bayesian, or that you want to convert him to become Bayesian. A physicist feels offended if you call him "X-ian", be it Newtonian, Fermian, or Einsteinian. But, being human, he has a natural feel for probability, just like everybody else. I would like to generalize this idea and propose reducing the use of the adjective "Bayesian". I think that the important thing is to have a theory of uncertainty in which "probability" has the same meaning for everybody, precisely that meaning which the human mind has naturally developed and that

[25]Dogmatism is never desirable. It can be easily turned against the theory. For example, one criticism of [18] says, more or less, that Bayesian theory supports Jeffreys' priors, and not uniform priors, but, since Jeffreys' priors give unreasonable results in their application, then one should mistrust Bayesian methods! (see also [28].) One may object that the meaning and the role of Jeffreys' priors was misunderstood, but it seems to me difficult to control the use of *objective priors* or of *reference analysis* once they have left the community of experts aware of the "rather special nature and role of the concept of a 'minimally informative' prior specification - appropriately defined!"[29].

frequentists have tried to kill. Therefore I would rather call these methods *probabilistic methods*. And I conclude saying that, obviously, "I am not a Bayesian".

7. Acknowledgements

It is a pleasure to thank the organizers of Maxent98 for the warm hospitality in Garching, which favoured friendly and fruitful discussions among the participants.

References

1. G. Zech, "Comparing statistical data to Monte Carlo simulation - parameter fitting and unfolding", DESY 95-113, June 1995.
2. B. Efron, "Why isn't everyone a Bayesian?", Am. Stat. **40** (1986) 1.
3. PDG stands for Data Particle Group, which edits the "Review of particle physics", a very influential collection of data, formulae and methods, including sections on Probability and Statistics. The latest issues are: R.M. Barnet et al. *Phys. Rev. D* **54** (1996) 1; C. Caso et al. *Eur. Phys. J.* **C3** (1998) 1 (http://pdg.lbl.gov/).
4. F. James and M. Roos, "MINUIT - system for function minimization and analysis of the parameter errors and correlations", Comp. Phys. Comm. **10** (1975) 343.
5. http://www-slac.slac.stanford.edu/find/hep,
http://xxx.lanl.gov/,
http://alice.cern.ch/Preprints,
http://wwwas.cern.ch/library/hepdoc/hepdoc.html,
http://www-lib.kek.jp/publib.html.
6. R.D. Cousins, "Why isn't every physicist a Bayesian?", Am. J. Phys. **63** (1995) 398.
7. G. D'Agostini, "Bayesian reasoning in High Energy Physics - principles and applications", lecture notes of the Academic Training given at CERN (Geneva), May 25-29 1998. A draft is available on the web at http://www.cern.ch/Training/ACAD/reglec_E.html.
8. DIN Deutsches Institut für Normung, "Grunbegriffe der Messtechnick - Behandlung von Unsicherheiten bei der Auswertung von Messungen" (DIN 1319 Teile 1-4), Beuth Verlag GmbH, Berlin, Germany, 1985.
9. R. Kaarls, BIPM proc.-Verb. Com. Int. Poids et Mesures **49** (1981), A1-A2 (in french); P. Giacomo, *Metrologia* **17** (1981) 73 (draft of english version; for the official BIPM translation see [10] or [12]).
10. International Organization for Standardization (ISO), "Guide to the expression of uncertainty in measurement", Geneva, Switzerland, 1993.
11. International Organization for Standardization (ISO), "International vocabulary of basic and general terms in metrology", Geneva, Switzerland, 1993.
12. B.N. Taylor and C.E. Kuyatt, "Guidelines for evaluating and expressing uncertainty of NIST measurement results", NIST Technical Note 1297, September 1994; (www: http://physics.nist.gov/Pubs/guidelines/outline.html).
13. Without any attempt to be complete, I would like to cite those which are more familiar to HEP physicists (I have including also the Bayesian book by Sivia that has attracted recently the attention of my colleagues): M.C. Barford, "Experimental measurements: precision, error and truth", John Wiley & Sons, 1985; R.J. Barlow, "Statistics", John Wiley & Sons Ltd, Chichester, 1989; P.R. Bevington and D.K. Robinson, "Data Reduction and Error analysis for the Physical Sciences", McGraw-Hill, 1992; S. Brandt, Statistics and computational methods in data analysis, North-Holland, 1976; G. Cowan, "Statistical data analysis", Clarendon Press, Oxford, 1998; W. T. Eadie, D. Drijard, F. E. James, M. Roos, B. Sadoulet, "Statistical Methods in Experimental Physics", North Holland, Amsterdam, 1971; A. G. Frodeson, O. Skjeggestad, H. Tofte, "Probability and Statistics in Particle Physics", Columbia University, New York, 1979; L. Lyons, "Statistics for nuclear and particle physicists", Cambridge University Press, 1986, reprinted 1992; L. Lyons, "A practical guide to data analysis for physical science students", Cambridge University Press, 1991; A.M. Mood, F.A. Graybill and D.C. Boes, "Introduction to the theory of statistics", McGraw-Hill, 1984.

L. G. Parrat, "Probability and Experimental Errors in Science", John Wiley & Sons Ltd, 1994; S. Rabinovich, "Measurement Errors: Theory and Practice", American Institute of Physics, New York, 1993. B. P. Roe, "Probability and Statistics in Experimental Physics", Springer-Verlag New York Inc, 1992; D.S. Sivia, "Data analysis - a Bayesian tutorial", Clarendon Press, Oxford, 1996. D.L. Smith, "Probability, Statistics and Data Uncertainties in Nuclear Science and Technology", American Nuclear Society, 1991; G. L. Squires, "Practical physics", Cambridge University Press, third edition, 1985; J.R. Taylor, "An introduction for error analysis", University Science Books, 1982; H.D. Young, "Statistical analysis of experimental data", McGraw-Hill, 1962.

14. V. Blobel et al., "Formulae and Methods in Experimental Data Evaluation", European Physical Society, 1984.
15. R.K. Bock and W. Krischer, "The data analysis Briefbook", Springer, 1998 (http://www.cern.ch/Physics/DataAnalysis/BriefBook/)
16. P.L. Galison, "How experiments end", The University of Chicago Press, 1987.
17. G. D'Agostini, "A multidimensional unfolding method based on Bayes' theorem", Nucl. Instr. Meth. **A362** (1995) 487.
18. G.J. Feldman and R.D. Cousins, "Unified approach to the classical statistical analysis of small signal", Phys. Rev. **D57** (1998) 3873, April 1, 1998.
19. H1 Collaboration, C. Adloff et al., "Observation of events at very high Q^2 in ep collisions at HERA", Z. Phys., **C74** (1997) 191;
ZEUS Collaboration, J. Breitweg et al., "Comparison of ZEUS data with Standard Model predictions for $e^+p \to e + X$", Z. Phys., **C74** (1997) 207.
20. M.J. Schervish, "P values: what they are and what they are not", Am. Stat. **50** (1996) 203.
21. J.O. Berger and D.A. Berry, "Statistical analysis and the illusion of objectivity", American Scientist **76** (1988) 159.
22. "DESY Science Information on Recent HERA Results", Feb. 19, 1997, http://www.desy.de/pr-info/desy-recent-hera-results-feb97_e.html.
23. DESY '98 - Highlights from the DESY Research Center, "Throwing 'heads' seven times in a row - what if it was just a statistical fluctuation?" (report obtainable free of charge by DESY: http://www.desy.de).
24. P. Bock et al. (ALEPH, DELPHI, L3 and OPAL Collaborations), "Lower bound for the standard model Higgs boson mass from combining the results of the four LEP experiments", CERN-EP/98-046, April 1, 1998.
25. P. Janot and F. Le Diberder, "Combining 'limits' ", CERN-PPE-97-053, May 1997.
26. A. Zellner, "Optimal information processing and Bayes's theorem", Am. Stat. **42** (1988) 278 (with discussion by E.T. Jaynes, B.M. Hill, J.M. Bernardo and S. Kullback).
27. B. de Finetti, "Theory of probability", J. Wiley & Sons, 1974.
28. G. D'Agostini, "Jeffreys priors versus experienced physicist priors", contributed paper to the 6th Valencia International Meeting on Bayesian Statistics, May 30th - June 4th, 1998, Alcossebre (Spain).
29. J.M. Bernardo, A.F.M. Smith, "Bayesian theory", John Wiley & Sons Ltd, 1994.

MODEL COMPARISON IN PLASMA ENERGY CONFINEMENT SCALING

R. PREUSS, V. DOSE AND W. VON DER LINDEN
Max-Planck-Institut für Plasmaphysik, EURATOM Association
D-85748 Garching b. München, Germany †

Abstract. Energy confinement data of large fusion experiments have been analyzed in terms of dimensionless form free scaling functions. Several possible physical scenarios lead to different models with a certain number of degrees of freedom. These are used to set up the confinement function as a linear combination of dimensionless power law terms. Then one has to solve the problem, given a particular plasma model which is the optimum number of terms in the expansion, and which model is most likely in the light of the data. Based upon data containing the energy content for a wide variety of variable settings, predictions for single variable scans are made and compared to actual experiments. A very good agreement is obtained.

Key words: Model Comparison, Prior Predictive Value, Invariant Measure

1. Introduction

The plasma behavior in large fusion experiments is now described for over twenty years by energy confinement scaling functions [1]. Since an exact description of energy transport is not possible scaling laws are in common use to give an overview of the plasma machine properties. This allows inter-machine comparisons and the characterization of conditions for enhanced confinement regimes. Even more, for being quite successful in the past in predicting the performance of larger plasma devices from data of small and midsized machines, scaling laws are involved in the design of future fusion reactors.

The functional form of the scaling law is still unknown. Only for reasons of simplicity a power law dependence is assumed and has accumulated considerable credit just by experience. In order to describe the energy content W^{theo} of the confined plasma all the plasma determining quantities in the machine under consideration are fed into the scaling law: For the in this paper examined W7-AS stellarator these are the particle density n, the magnetic field B, the heating power P and the minor radius a. Additional machine parameters like the major radius of torus

†Email: preuss@ipp.mpg.de

W. von der Linden et al. (eds.), Maximum Entropy and Bayesian Methods, 171–178.

R come only into play in comparisons with machines of different size. The scaling law then reads:

$$W_i^{theo} = e^{\alpha_c} n_i^{\alpha_n} B_i^{\alpha_B} P_i^{\alpha_P} a_i^{\alpha_a} + \varepsilon_i \;, \tag{1}$$

where i labels a single measurement with unknown error ε. Also the scaling constant (e^{α_c}) has been expressed in exponential form. Now the above addressed reasons for the power law choice are obvious: Applying the logarithm to both sides of (1) an easy to solve set of linear equations in the sought α's is obtained.

2. Dimensionless Confinement Scaling

A major shortcoming of the above equation is to be solely a data fit and that it does not take care of the physical implications of the input variables. In order to obtain a dimensionally exact scaling function basic plasma models may be used. This was achieved by Connor and Taylor [2] in requiring that the invariances of basic plasma models under similarity transformations should be the same as those of the scaling function belonging to the respective model. Four kinetic models (see table 1) were obtained in considering the Boltzmann equation of motion (describing a collisional plasma) or the Vlasov equation (collisionless plasma) and discriminating further between the electrostatic limit (low-β case) or a self-consistent calculation of the electro-magnetic fields from the Maxwell equations (high-β). By the invariance requirement the plasma variables are combined in three dimensionless factors with new scaling exponents (x_1, x_2, x_3) specified according to table 1.

$$\begin{aligned} W^{theo} &\propto na^4RB^2\left(c_1\frac{P}{na^4RB^3}\right)^{x_1}\left(c_2\frac{a^3B^4}{n}\right)^{x_2}\left(c_3\frac{1}{na^2}\right)^{x_3} && (2)\\ &= c\left(\frac{P}{na^4B^3}\right)^{x_1}\left(\frac{a^3B^4}{n}\right)^{x_2}\left(\frac{1}{na^2}\right)^{x_3} && (3)\\ &= cf\left(n,B,P,a;\boldsymbol{x}\right) \;. && (4)\end{aligned}$$

The constants (c_1, c_2, c_3) in (2) consist of fundamental physical constants and carry physical units. They may be absorbed in an overall constant c in (3). Just the same may be done with the large radius R of the torus which is constant for the here considered dataset of a single machine. Finally, in (4) the scaling terms are comprised by a general function $f(n, B, P, a; \boldsymbol{x})$ where $\boldsymbol{x}$ has components x_1, x_2, x_3.

Already Connor and Taylor proposed to express a general form free energy confinement scaling function as a series of terms of the form (4) with properly chosen $\boldsymbol{x}$ and c. We want to exploit this idea and employ the following ansatz to describe the energy content $\boldsymbol{W}^{theo}$ of N different experiments:

$$\boldsymbol{W}^{theo} = \sum_{k=1}^{E} c_k \boldsymbol{f}\left(\boldsymbol{x}_k\right) = \boldsymbol{F}\boldsymbol{c} \;. \tag{5}$$

$\boldsymbol{F}$ is a $N \times E$ matrix with columns $\boldsymbol{f}(\boldsymbol{x}_k)$. In mathematical terms this is nothing but an expansion in a dimensionally exact basis. In general, N linearly independent

M_j	CT Model	x_1	x_2	x_3	N_{DOF}	$p(M_j \vert \boldsymbol{W}^{exp}, \boldsymbol{\sigma}, I)$
1	collisionless low-β	x	0	0	1	4×10^{-12}%
2	collisional low-β	x	y	0	2	99.7%
3	collisionless high-β	x	0	z	2	0.25%
4	collisional high-β	x	y	z	3	0.025%

TABLE 1. Plasma models according to Connor and Taylor [2].

vectors $\boldsymbol{f}(\boldsymbol{x}_k)$ form a complete basis in the N-dimensional data space and would therefore allow a pointwise reconstruction of the data. This is neither desirable, nor with respect to physics correct, since the corresponding vector of the measured energy content $\boldsymbol{W}^{exp}$ is corrupted by noise. What we really want is an expansion, truncated at some appropriate upper limit E describing the physics, while the residual $N - E$ terms in the expansion (5) would only fit the noise. So we have to look for the most probable expansion order. Furthermore, given the four CT models of table 1 we would like to identify the plasma physics model which describes the data best. Since the number of degrees of freedom N_{DOF} varies between one and three (see table 1) Occam's Razor will have a say in this matter. And finally we would like to estimate the energy content for certain parameter settings. An important topic here are single variable scans (e.g. the variation of the energy content as function of the density alone with all other variables fixed). Such scans are not directly accessible from published databases. On the other hand, single variable scans are experimentally cumbersome and expensive experiments have to be performed for each and every input variable of interest. It is therefore highly desirable to extract single variable scans from existing databases by employing improved data analysis techniques.

3. Bayesian Approach

The likelihood function of our problem reads:

$$p(\boldsymbol{W}^{exp}|\omega, \boldsymbol{c}, \boldsymbol{x}, E, M_j, \boldsymbol{\sigma}, I) = \left(\frac{\omega}{2\pi}\right)^{\frac{N}{2}} \prod_i^N \sigma_i^{-1}$$

$$\cdot \exp\left\{-\omega \sum_{i=1}^{N} \frac{\left(W_i^{exp} - \sum_k^E c_k f_i(\boldsymbol{x}_k)\right)^2}{2\sigma_i^2}\right\}. \quad (6)$$

M_j denotes the plasma kinetic model used to generate the expansion vectors $\boldsymbol{f}(\boldsymbol{x}_k)$. $\boldsymbol{\sigma}$ is the vector of experimental uncertainties associated with the measurement $\boldsymbol{W}^{exp}$. The uncertainty in the energy content $\boldsymbol{W}^{exp}$ contains the direct distributions from the diamagnetic measurement as well as indirect contributions from the finite precision in the input variables (n, B, P, a). Both contributions have been estimated to the best of our knowledge. However, to allow for possible deviations from the true errors we introduce an overall correction factor ω where we assume that the relative error is correctly estimated. The Bayesian analysis yields a posteriori for the most probable model and the optimum expansion order that $\boldsymbol{\sigma}$ was overestimated merely by a factor of 20%.

We are looking for the probability of a model given the data $\boldsymbol{W}^{exp}$. The odds ratio gives

$$\frac{p(M_j|\boldsymbol{W}^{exp},\boldsymbol{\sigma},I)}{p(M_k|\boldsymbol{W}^{exp},\boldsymbol{\sigma},I)} = \frac{p(M_j|\boldsymbol{\sigma},I)}{p(M_k|\boldsymbol{\sigma},I)}\frac{p(\boldsymbol{W}^{exp}|M_j,\boldsymbol{\sigma},I)}{p(\boldsymbol{W}^{exp}|M_k,\boldsymbol{\sigma},I)} . \tag{7}$$

The first ratio on the r.h.s., the so-called prior odds, is set to unity since we do not want to favor one model over another. We are left with the determination of the second ratio, the so-called Bayes factor. The global likelihood is given by a discrete sum over all expansion orders

$$p(\boldsymbol{W}^{exp}|M_j,\boldsymbol{\sigma},I) = \sum_E p(E|M_j,\boldsymbol{\sigma},I)p(\boldsymbol{W}^{exp}|E,M_j,\boldsymbol{\sigma},I) , \tag{8}$$

where $p(E|M_j,\boldsymbol{\sigma},I)$ is set constant because a priori no expansion order is to be preferred. The remaining marginal likelihood is obtained by

$$\begin{aligned} p(\boldsymbol{W}^{exp}|E,M_j,\boldsymbol{\sigma},I) &= \int p(\boldsymbol{W}^{exp}|\omega,\boldsymbol{c},\boldsymbol{x},E,M_j,\boldsymbol{\sigma},I) \\ &\cdot\ p(\omega,\boldsymbol{x},\boldsymbol{c}|E,M_j,\boldsymbol{\sigma},I)\ \mu(\omega,\boldsymbol{x},\boldsymbol{c})\ d\omega\ d\boldsymbol{c}\ d\boldsymbol{x} , \end{aligned} \tag{9}$$

where we followed the ideas brought up by Rodriguez [3] formulating an invariant measure. $\mu(\omega,\boldsymbol{x},\boldsymbol{c})$ in the marginalization integral. This stems from the requirement that the description of a problem should be the same regardless in which coordinate system it is considered. Then the posterior probability for an infinitesimal change in the parameters is given by

$$p(d\omega,d\boldsymbol{x},d\boldsymbol{c}|E,M_j,\boldsymbol{\sigma},I) = p(\omega,\boldsymbol{x},\boldsymbol{c}|E,M_j,\boldsymbol{\sigma},I)\mu(\omega,\boldsymbol{x},\boldsymbol{c})\ d\omega\ d\boldsymbol{c}\ d\boldsymbol{x} , \tag{10}$$

with $\mu(\omega,\boldsymbol{x},\boldsymbol{c})$ describing the Riemannian metric of the problem. The latter is given by

$$\mu(\omega,\boldsymbol{x},\boldsymbol{c}) = \sqrt{\det[\boldsymbol{g}]} . \tag{11}$$

$\boldsymbol{g}$ is the Fisher information matrix, which is

$$g_{ij} = -\left\langle \frac{\partial^2 \ln p(\boldsymbol{W}^{exp}|\boldsymbol{\theta},E,M_j,\boldsymbol{\sigma},I)}{\partial\theta_i\partial\theta_j} \right\rangle \tag{12}$$

and build up by the second derivative of the log-likelihood function with respect to the parameters $\boldsymbol{\theta} = (\omega,\boldsymbol{x},\boldsymbol{c})$.

The prior function decomposes into

$$p(\omega,\boldsymbol{x},\boldsymbol{c}|E,M_j,\boldsymbol{\sigma},I) = p(\omega,\boldsymbol{x}|E,M_j,I)p(\boldsymbol{c}|\boldsymbol{x},E,M_j,\boldsymbol{\sigma},I) . \tag{13}$$

While an uninformative prior is assigned to ω and $\boldsymbol{x}$ we employ what we call the Bessel prior for $p(\boldsymbol{c}|\boldsymbol{x}, E, M_j, \boldsymbol{\sigma}, I)$. This just imposes an upper boundary to the choice of possible coefficients with $\boldsymbol{c}^T\boldsymbol{F}^T\boldsymbol{F}\boldsymbol{c} \leq 2||\boldsymbol{W}^{exp}||^2$. It is equivalent to an entropic prior proposed by Rodriguez, where the hyperparameter α has been marginalized by using the Bretthorst approximation that the prior may be taken out of the integral at the maximum of the likelihood.

While the $\boldsymbol{c}$ and ω integrations may be performed analytically the remaining $\boldsymbol{x}$ integration is carried out numerically by computing the prior predictive value [4] with Markov Chain Monte Carlo (MCMC). This is given by

$$
\begin{aligned}
p(\boldsymbol{W}^{exp}|E, M_j, \boldsymbol{\sigma}, I) &= \int p(\boldsymbol{W}^{exp}|\boldsymbol{x}, E, M_j, \boldsymbol{\sigma}, I)p(\boldsymbol{x}|E, M_j, \boldsymbol{\sigma}, I)\, d\mu(\boldsymbol{x}) \quad (14)\\
&= \int \Lambda(\boldsymbol{x})\Pi(\boldsymbol{x})\, d\boldsymbol{x}\ . \quad (15)
\end{aligned}
$$

$\mu(\boldsymbol{x})$ is the remaining measure in $\boldsymbol{x}$. In (15) we combined the factors of the integral to new functions Λ and Π, where $\Lambda(\boldsymbol{x})$ contains the pure likelihood and some normalization constants, while $\Pi(\boldsymbol{x})$ comprises the prior and the terms stemming from the Riemannian measure. The latter shall be normalized to $\int \Pi(\boldsymbol{x})\, d\boldsymbol{x} = 1$. The only way to evaluate (15) is to use $\Pi(\boldsymbol{x})$ as a probability density. Since the likelihood descending term $\Lambda(\boldsymbol{x})$ is generally much more structured than $\Pi(\boldsymbol{x})$ the variance will be large and extremely long Markov chains are needed to obtain the desired accuracy. However, one can solve this problem in defining a new function

$$
Z(\beta) = \int \Lambda^\beta(\boldsymbol{x})\Pi(\boldsymbol{x})\, d\boldsymbol{x} \quad (16)
$$

with $Z(\beta = 0) = 1$ and $Z(\beta = 1)$ as the sought quantity in (15). The derivative with respect to β gives

$$
\begin{aligned}
\frac{\partial \ln Z(\beta)}{\partial \beta} &= \int \ln \Lambda(\boldsymbol{x})\rho_\beta(\boldsymbol{x})\, d\boldsymbol{x}\\
&= \langle \ln \Lambda(\boldsymbol{x})\rangle_\beta \quad (17)
\end{aligned}
$$

with

$$
\rho_\beta(\boldsymbol{x}) = \frac{\Lambda^\beta(\boldsymbol{x})\Pi(\boldsymbol{x})}{\int \Lambda^\beta(\boldsymbol{x}')\Pi(\boldsymbol{x}')\, d\boldsymbol{x}'} \quad (18)
$$

as the new sampling density. If we integrate both sides of 17,

$$
\begin{aligned}
\int_0^1 \langle \ln \Lambda(\boldsymbol{x})\rangle_\beta\, d\beta &= \int_0^1 \frac{\partial \ln Z(\beta)}{\partial \beta}\, d\beta \quad (19)\\
&= \ln Z(\beta = 1) - \ln Z(\beta = 0) \quad (20)\\
&= \ln p(\boldsymbol{W}^{exp}|E, M_j, \boldsymbol{\sigma}, I)\ , \quad (21)
\end{aligned}
$$

we are back where we started in (14). To obtain the marginal likelihood one therefore has to calculate the integral on the l.h.s. in (19) where the expectation value

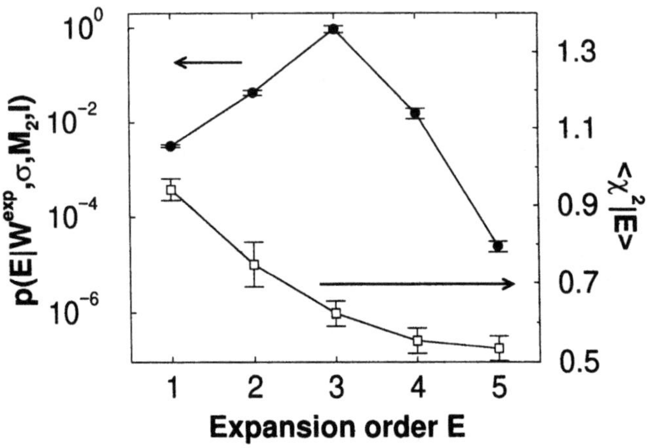

Figure 1. Upper curve: expansion order. Notice the logarithmic scale on the left. Lower curve: misfit of data and model.

$\langle \ln \Lambda(\boldsymbol{x}) \rangle_\beta$ is accessible by Markov Chain Monte Carlo techniques. The proof that this is now feasible may be found in [4].

4. Results

The data which we have used in our calculations are the 153 W7-AS data from the international stellarator energy confinement data base [5]. We have selected only data points with rotational transform of $\iota \approx 1/3$ since the single variable scans which we present further down and which we want to compare with have been performed at this value. Odds ratios obtained from (7) are converted back to probabilities using $\sum_j p(\boldsymbol{W}^{exp}|M_j, \boldsymbol{\sigma}, I) = 1$. The resulting model probabilities are depicted in the last column of table 1. We see that the $\iota \approx 1/3$ W7-AS data are best described by the collisional low-β Connor Taylor model. The high beta models follow in second and third place with much lower probability. The collisionless low-β model is clearly inappropriate to describe the transport physics in W7-AS.

The probability distribution for the expansion order given data and model is now readily obtained:

$$p(E|\boldsymbol{W}^{exp}, M_j, \boldsymbol{\sigma}, I) = \frac{p(E|M_j, \boldsymbol{\sigma}, I) p(\boldsymbol{W}^{exp}|E, M_j, \boldsymbol{\sigma}, I)}{p(\boldsymbol{W}^{exp}|M_j, \boldsymbol{\sigma}, I)} \quad . \tag{22}$$

This quantity is displayed for the most probable plasma model, the collisional low beta case, in Fig. 1 as full circles. The error bars indicate Monte Carlo integration uncertainties. The open squares, with associated error bars, depict the misfit (here square of the usual random mean square error) between data and model prediction (5) for each expansion order. We observe quite a typical behavior. Already three terms in the expansion (5) lead to a rapid decrease of the misfit in the order of 35% of the value for $E = 1$. Though it does decrease further with increasing expansion order the probability for a given E decreases rapidly so that contributions of higher expansion orders become very small demonstrating Occam's razor.

Finally we come to determination of observables. Here we want to test the predictive power of the present theory. Such a test is provided by a comparison of measured single variable scans to the predictions from the present theory. The latter is the expectation value of the energy content

$$\langle W\rangle = \frac{\int W\ p(W|\boldsymbol{W}^{exp}, \boldsymbol{v}, M_j, \boldsymbol{\sigma}, I)\ dW}{\int p(W|\boldsymbol{W}^{exp}, \boldsymbol{v}, M_j, \boldsymbol{\sigma}, I)\ dW} . \tag{23}$$

The additional condition $\boldsymbol{v}$ specifies the "input data vector" $\boldsymbol{v}^T = (n, B, P, a)$ at which the energy content measurement was performed. $p(W|\boldsymbol{W}^{exp}, \boldsymbol{v}, M_j, \boldsymbol{\sigma}, I)$ is again obtained by marginalization

$$\begin{aligned} p(W|\boldsymbol{W}^{exp}, \boldsymbol{v}, M_j, \boldsymbol{\sigma}, I) = \sum_E \int & p(W|E, \boldsymbol{x}, \boldsymbol{c}, \boldsymbol{v}, M_j, \boldsymbol{\sigma}, I) \\ & \cdot p(E, \boldsymbol{x}, \boldsymbol{c}, \omega|\boldsymbol{W}^{exp}, M_j, \boldsymbol{\sigma}, I)\ d\boldsymbol{x}\ d\boldsymbol{c}\ d\omega\ , \end{aligned} \tag{24}$$

where

$$p(W|E, \boldsymbol{x}, \boldsymbol{c}, \boldsymbol{v}, M_j, \boldsymbol{\sigma}, I) = \delta\left(W - \boldsymbol{g}^T(\boldsymbol{x}, \boldsymbol{v}) \cdot \boldsymbol{c}\right)\ , \tag{25}$$

with $\boldsymbol{g}(\boldsymbol{x}, \boldsymbol{v})$ as an E-dimensional vector with elements $f(\hat{n}, \hat{B}, \hat{P}, \hat{a}; \boldsymbol{x})$. The second factor in (24) is just

$$p(E, \boldsymbol{x}, \boldsymbol{c}, \omega|\boldsymbol{W}^{exp}, M_j, \boldsymbol{\sigma}, I) = \frac{p(\boldsymbol{W}^{exp}, E, \boldsymbol{x}, \boldsymbol{c}, \omega|M_j, \boldsymbol{\sigma}, I)}{p(\boldsymbol{W}^{exp}|M_j, \boldsymbol{\sigma}, I)} \tag{26}$$

and we get

$$\langle W\rangle = \frac{\sum_E p(\boldsymbol{W}^{exp}|E, M_j, \boldsymbol{\sigma}, I) \int \boldsymbol{g}^T(\boldsymbol{x}, \boldsymbol{v}) \boldsymbol{c}_{ML} \rho(\boldsymbol{x})\ d^E\boldsymbol{x}}{\sum_E p(\boldsymbol{W}^{exp}|E, M_j, \boldsymbol{\sigma}, I)} . \tag{27}$$

In Fig. 2 we show the result for a density scan obtained on the one hand from the present theory and on the other hand from experiments at W7-AS. Because these data are not included in the W7-AS data base the analysis is based on, this nice agreement shows the predictive power of our approach. The stellarator energy confinement data base is represented by the open circles which are spread all over since they were obtained for various settings of the variables (B, P, a). The full circles in Fig. 2 depict experimental results for a density scan. Here the variables B, P, a are held fixed and only n varies. Representative error bars signify the precision level of these data. But what would be the answer of our present semi-empirical theory based only on the data with open circles? This is shown by the continuous curve along with the confidence range indicated by the gray shaded area. Within the density range of the single variable scan the prediction of the semi-empirical theory runs straight through the data and exhibits clearly a previously supposed density saturation [5,6]. Outside this range the data set $\boldsymbol{W}^{exp}$ is too sparse, which is reflected in the rapidly widening error band and therefore indicates where the extrapolation becomes unreliable. However, it is not a shortcoming of the probabilistic approach but rather an honest outcome that

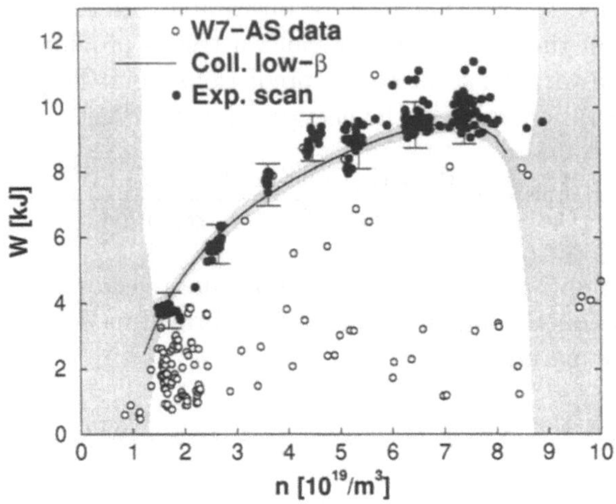

Figure 2. Experimental results of a single variable density scan (full circles) compared to the predictions of the present semi-empirical theory (continuous line, shaded area represents the error) for $B = 2.5\text{T}$, $P = 0.45\text{MW}$, $a = 0.176\text{m}$. The input data (open circles) are shown regardless of additional variations in B, P, a and are therefore spread all over. The histogram accounts for their distribution over the density axis.

an extrapolation beyond the parameter regime supported by the data base is not possible.

5. Conclusion

In summary, form free dimensionally exact energy confinement functions derived from data of the international stellarator data base and treated consistently according to the rules of Bayesian probability theory have identified the collisional low beta Connor Taylor model to be the most probable plasma physics model for W7-AS. Moreover, single variable scans were reproduced in quantitative agreement with experiments. The result of a single variable scan is therefore already hidden in the data obtained for arbitrary variable choices and can be extracted from the latter by a proper data analysis.

References

1. J. Hugill and J. Sheffield *Nucl. Fusion*, **18**, p. 15, 1978.
2. J. W. Connor and J. B. Taylor *Nucl. Fusion*, **17**, p. 1047, 1977.
3. C. Rodriguez, "From euclid to entropy," in *Maximum Entropy and Bayesian Methods*, J. W. T. Grandy, ed., Kluwer Academic, Dordrecht, 1991.
4. W. von der Linden, R. Preuss, and V. Dose, "The prior predictive value." These proceedings.
5. U. Stroth, M. Murakami, R. A. Dory, H. Yamada, S. Okamura, F. Sano, and T. Obiki *Nucl. Fusion*, **36**, p. 1063, 1996.
6. U. Stroth *Plasma Phys. Control. Fusion*, **40**, p. 9, 1998.

STRAIGHT LINES

DO KESTER
Space Research Organization Netherlands (SRON),
PO Box 800,
9700 AV GRONINGEN,
The Netherlands. †

Abstract. Bayesian evidence is used to select the best model from a large set of models. The models are straight lines which might contain glitches and sudden changes in the slope. Most of the model parameters are nuisance parameters. The standard deviation of the noise has to be estimated too. As this is bulk data processing, it has to be done within strict CPU time limitations.

Key words: Bayesian Model Selection, Infrared Detectors, Digitization noise

1. Introduction

One of the 4 instruments aboard the European satellite ISO, is the Short Wavelength Spectrometer (SWS) [1,2]. ISO has observed the infrared sky for 30 months in the period 1995 to 1998. SWS is a grating spectrometer covering the wavelength range from 2 to 45 μm. At the longer wavelengths a Fabry-Pérot (FP) could be inserted.

There are 4 × 12 photoconducting detectors in the grating section. Each of the 4 band covers a different wavelength range (see table 1). For the FP there are 2 × 2 detectors. In principle all detectors could simultaneously receive flux from an (astronomical) source, but in most settings this is not the case. Nontheless all detectors are always active.

The amount of flux seen by the detector is proportional to the current through the detector. The detectors are connected to an integrating amplifier where the charge accumulates on a capacitor. This circuit also contains a high pass filter with a known roll-off frequency. The accumulating charge on the capacitor is measured 24 times per second. After readout the samples are digitized by a 12 bit digital converter. Every 1 or 2 seconds the capacitor is short-circuited or reset, which invalidates the first 7 samples after a reset. The cycle of reset and accumulating charge results in a sawtooth pattern. The rising part of one sawtooth contains 24 or 48 samples, depending on the length of the reset interval (also called ramp).

†Email: do@sron.rug.nl

W. von der Linden et al. (eds.), Maximum Entropy and Bayesian Methods, 179–188.

As the 12 bit range is too small to accommodate the large dynamic range of astronomically interesting sources, 3 gain setting were used, with gains of 1, 4 and 16 resp.

The SWS instrument could be operated in various modes. The only distinction which is relevant here is the fact that in some modes the grating scanner moves *during* a reset interval. This is done to decrease the time needed to scan a full spectrum at the expense of spectral resolution. In the other modes the scanner is moving only at the reset itself. When the scanner is moving during the reset interval, flux of different wavelengths is falling on the detectors, which could have quite different intensities. The linear relationship between the incoming flux and the accumulating charge is no longer valid for the complete reset interval.

Some ramps are contaminated by glitches. These are caused by ionizing particles which hit the detectors or the pre-amplifiers. The particles come in at random times. The long-wavelength detectors are more sensitive to these hits than those at shorter wavelengths due mainly to their larger size. Most hits result in a positive shift of the remainder of the ramp, but sometimes negative glitches occur [3].

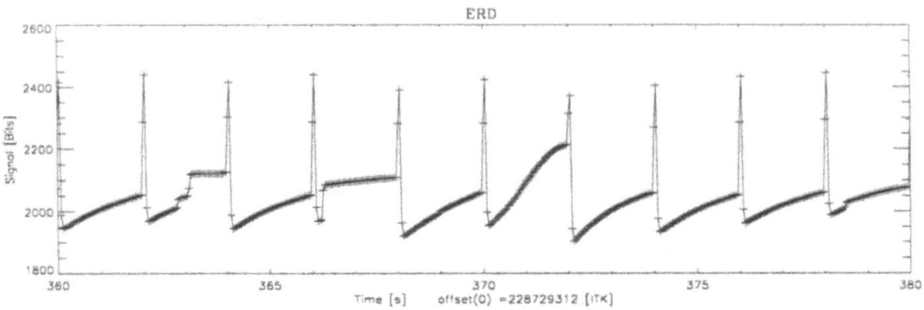

Figure 1. Ten ramps for detector 42 are displayed, each 2 seconds long. The sharp peaks at every even second is the reset pulse; 6 samples after each pulse starts a new data set. The curvature in the datasets is due to the high pass filter.

A small portion of the data for one detector (nr 42) is shown in figure 1, 10 ramps (20 seconds) in total. Detector 42 is at a long wavelength, so it is prone to particle hits: several hits in the 2nd ramp and one in the last. The 6th ramp clearly shows a different shape than the others: here we are scanning over a strong emission line in the spectrum.

One observation with the SWS instrument typically lasts an hour. The complete bulk of raw data (52 detectors with several thousands of data sets) is fed to the standard (pipeline) processing. All reset intervals for all detectors are processed whether they contain astronomically valid data or not. The pipeline flags the 7 samples just after each reset and those samples which are outside the valid bit range. The flagged samples are not used. The ramps are linearized, undoing the effects of the high pass filter. The roll-off frequency of this filter is known quite accurately. The linearized ramps are searched for glitches. The ramps are differentiated and 5σ outliers are considered glitches. To the (partial) ramps where there are no glitches a straight line is fitted, even when the scanner was moving during the reset interval. The slope of this line is considered to be proportional to the incoming flux.

There are more steps in the pipeline processing but they are not relevant here. We want to see how we can improve the pipeline processing just sketched. The product should be similar to that produced by the pipeline to enable it to pass through the other necessary processing steps. And it should all be done within a reasonable timeframe i.e. the processing should not take much more time than the observation itself lasted, i.e. 0.02 CPU seconds per detector per second.

In section 2 we introduce various models including glitches and changing slopes where we should select the best from. Section 3 assigns priors and calculates likelihoods and evidences. In Section 4 we address parameter estimation and automatic noise scaling. Strategies to search the best model are in section 5. And finally in section 6 some results and conclusions are presented.

In the following sections we are working on one reset interval of one detector, i.e. one ramp, at a time. We assume that there is no correlation between different ramps, although — it should be said in passing — this is not completely true: but that might be the subject of a future paper.

2. Building Models

The design of the instrument is to have a linear response between the incoming flux, b, and the charge build up $x(t)$.

$$x(t) = a + bt, \tag{1}$$

where a is an unknown offset. Basically we have to estimate the derivative dx/dt from a limited number of samples.

$$x_k = a + bt_k. \tag{2}$$

Here x_k is the charge at time step t_k (in seconds), a is an offset and b is the slope which is proportional to the flux. This b is the parameter we want to know, a is a nuisance parameter.

When a particle hits the detector, we can model the glitch as

$$x_k = a + bt_k + a_j H_j, \tag{3}$$

where H_j is the unit step function which steps at some (unknown) time t_j and a_j is the glitch amplitude. There might be more glitches within the same reset interval. Equation 3 can easily be expanded with more stepfunctions and amplitudes. Each glitch has 2 parameters, a glitch position and a glitch amplitude, both of which are nuisance parameters.

When the scanner steps during the reset interval, it can do so at fixed time intervals, either every 3rd sample or every 6th sample, giving rise to 8 or 4 possible 'change points' per second. At each change point the incoming flux might change and cause a change in the slope.

$$x_k = a + bt_k + a_j H_j + b_i H_i (t_k - t_i). \tag{4}$$

As before, H_i is the unit step function and b_i is the change in slope starting at time step t_i. There might more change points in a reset interval, similar to the situation of the glitches. And here too, each change point has 2 parameters, its location and its amount. But unlike the glitches, possible change point locations are restricted to every 3rd or 6th sample.

The previous models can be summarized, with a slight change in notation as

$$x_k = \sum_{j \in A} a_j H_j + \sum_{i \in B} b_i H_i (t_k - t_i), \tag{5}$$
$$A = \{0, j_1, j_2, \ldots, j_{N_A}\}, \quad B = \{0, i_1, i_2, \ldots, i_{N_B}\}.$$

A is the set of offset cq. glitch positions, of which there is always one: the offset at the start of the ramp. And B is the set of points where the slope changes. Again, there is always one slope, b_0.

The high pass filter adds curvature to the straight lines. It can be modelled as a simple differential equation.

$$\frac{dy}{dt} + \frac{y}{\tau} = \frac{dx}{dt}, \tag{6}$$

where y is the signal after the high pass filter and τ is the roll-off frequency. This frequency is well known and can be regarded as a constant. The differential equation is easily integrated and written in a discrete form, using the trapezium rule for the integration with Δt as the time step.

$$x_k = y_k + \frac{\Delta t}{2\tau} \sum_{i=1}^{k} (y_{i-1} + y_i). \tag{7}$$

All x_k are linear combinations of y_k which implies that the reverse is also true: y_k is linear in x_k. And as x_k is linear in its parameter sets $a = \{a_j\}$ and $b = \{b_i\}$, y_k is also linear in the model parameters and we can enjoy all the 'benefits of linearity' [6]. Due to the structure of equation 7, y_k cannot be written out easily, but its terms are quite simple to program.

The $\{y_k\}$ is the mock data for a model defined by the set of glitch positions and the set of change point positions (A, B). They must be describe the data $D = \{d_k\}$. Note that the d_k are integer values, digitized on board the satellite. Due to our fairly small bit range digitization is not a negligible effect. At some (low) gain settings it is the dominant source of noise while at high gain the noise in the detector chain dominates. (See section 4 for more details.)

3. Assigning Probabilities

The general model (Eq 5) allows a wild variety of scenes per reset interval, from only one constant flux to a glitch on every sample and a change point every 3rd (or 6th) sample. For a given dataset $D = \{d_k\}$ most of the possible models are just wrong. We want to select the most probable model, the one that describes best the data. Model selection is quite standard in Bayesian literature these days [4–6]. What we need is the probability for the model given the data.

The models can be indicated unambiguously by the set of glitch positions and the set of change point positions they contain: (A, B). The evidence for a model (A, B), given the data D and background information I is

$$P(A, B|D, I) = P(D|A, B, I)P(A, B|I)/P(D|I). \tag{8}$$

From this we find the odds of one model (A, B) to another (A', B'). In the odds ratio the denominator, $P(D|I)$, cancels. A and B are random processes with independent priors. These priors may be derived from simple counting arguments:

$$P(A|I) = \binom{N_A}{N}^{-1}, \quad P(B|I) = \binom{N_B}{M}^{-1}, \tag{9}$$

where M is the number of possible change points.

Neglecting the denominator in equation 8, we see that the probability of the model given the data is proportional to the prior of the model multiplied by probability of the data given the model. This latter quantity is equal to the normalization constant in the posterior of the parameters.

$$P(a, b, \sigma|D, A, B, I) = \frac{P(a, b, \sigma|A, B, I)P(D|a, b, \sigma, A, B, I)}{P(D|A, B, I)} \tag{10}$$

where σ^2 is the variance of noise.

Combining Equations 8 and 10 we find the evidence for model (A, B) by integrating over the parameters.

$$\begin{aligned} P(A, B|D, I) &\propto P(A, B|I)P(D|A, B, I) & (11) \\ &= P(A, B|I) \iiint P(a, b, \sigma|A, B, I)P(D|a, b, \sigma, A, B, I)da\, db\, d\sigma \end{aligned}$$

Assuming Gaussian noise — we postpone the discussion about uniform digitization noise until later in section 4 — the likelihood equals

$$P(D|a, b, \sigma, A, B, I) = \frac{1}{(\sigma\sqrt{2\pi})^N} \exp\left(-\sum_{k=1}^{N} \frac{(d_k - y_k)^2}{2\sigma^2}\right) \tag{12}$$

The priors can be assumed to be independent of each other. From the setup of the instrument the limited range for both a and b are easily derived: a is constrained between ± 4096 and b in a range 24 times as large. Note that b, although proportional to the incoming flux, need not always be positive. It depends on the dark current and on the bias voltage over the detector what the range of b might be. Moreover some of the b_k's are *changes* in slope, which a fortiori might be negative. For σ we take a Jeffreys' prior.

4. Parameter Estimation and Noise Scaling

As these calculations are only the beginning of a long chain of other calculations and calibrations, we don't have much use for the full probability distribution. We want to know the optimal values of b and its standard deviation, for the best model (A, B). The values of a and their positions, A, are more or less nuisance parameters. More or less because it is still interesting to know for further processing, where glitches were found.

With all probabilities in place and starting from equation 10 and 12 we integrate over σ using Jeffreys' prior $1/\sigma$.

$$P(a,b|D,A,B,I) = \frac{P(a,b|A,B,I)}{P(D|A,B,I)} \times \frac{1}{2}(N\pi)^{-\frac{1}{2}N}\Gamma(\frac{N}{2}) \left(\frac{1}{N}\sum_{k=1}^{N}(d_k - y_k)^2\right)^{-\frac{1}{2}N}. \tag{13}$$

The latter part of the equation is a Student-t distribution which equals the likelihood $P(D|a, b, A, B, I)$. Note that the constants $\frac{1}{2}(N\pi)^{-\frac{1}{2}N}\Gamma(N/2)$ only involve the number of samples, so these cancel in the odds ratio where we want to compare two model given the *same* data set.

To find the optimal values for the parameters (a, b) we differentiate equation 13 logarithmically with respect to the parameters. As y_k is linear in its parameters a set of linear equations appear which are readily solved. In the same manner we find the optimal value for σ, usually written as S, from the inversion of the second derivative [6].

$$S^2 = \frac{1}{N - N_P}\sum_{k=1}^{N}(d_k - \hat{y}_k)^2, \tag{14}$$

where $\hat{y}_k$ are the mock data using the optimal values of the parameters a and b, and N_P is the number of parameters in the model.

Using a Gaussian approximation of equation 13 we can follow MacKay's procedure [5] to calculate the evidence: Evidence $\simeq$ Best fit likelihood $\times$ Occam factor. With a few tens of samples in each data set, this approximation is very close. The 'best fit likelihood' is equation 12 where σ and y_k are replaced by the best fit values, S and $\hat{y}_k$. The Occam factor is obtained from the prior at the optimal values of the parameters and the determinant of the inverse of the Hessian, H:

$$\text{Occam factor} = P(\hat{a}, \hat{b}|A, B, I) \times (2\pi)^{\frac{1}{2}N_P}\det{}^{-\frac{1}{2}}H. \tag{15}$$

The Occam factor also needs a little adjustment with respect to MacKay's definition as the Hessian contains the likelihood, $H = -\nabla^2 \log P(D|a, b, A, B, I)$. These 2nd derivatives of the loglikelihood involve a factor $1/S^2$, so the Occam factor gets an extra factor S^{N_P} from the inverse square root of the determinant.

Up to here we have assumed that the noise was Gaussian, which is not always the case. Especially at low gain and low flux levels, the noise is dominated by uniform digitization noise. Sometimes there are over the whole reset interval only a few bitlevels. Using Gaussian noise, models would be preferred which have a 1 bit glitch at each transition from one bitlevel to the next. These transitions are not actual glitches but just the slow accumulation of charge which happens to trespass the bit boundary. In figure 2a these effects can be seen: only a few bitlevels in the data (+) and on each of the bitsteps the model (drawn line) steps too. Here, with automatic Gaussian noise scaling, erroneous models were selected with overly optimistically small noise variances of the order of 0.01.

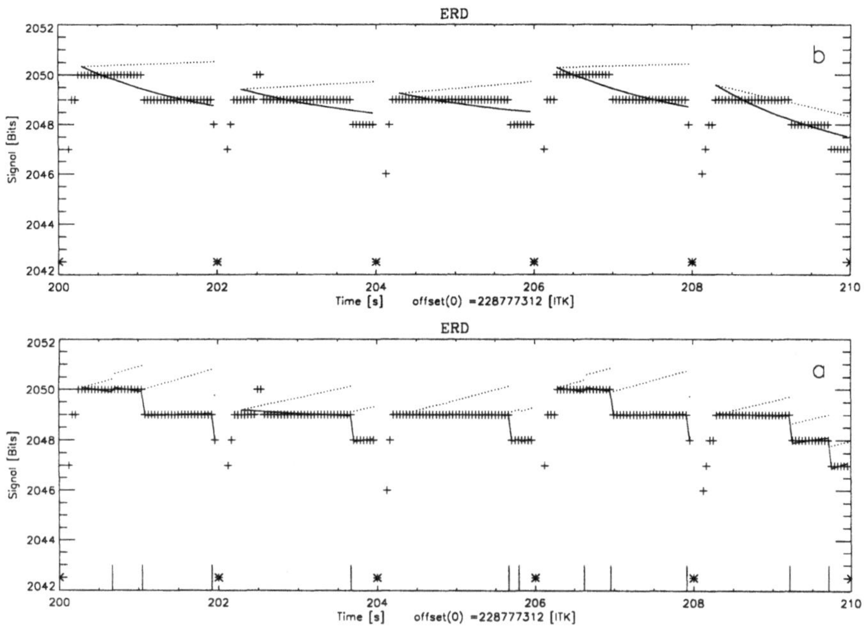

Figure 2. The effects of digitization in the data is shown. In panel (a) models with automatic noise scaling is shown; these models prefer a glitch at every bit step. They are plainly wrong. In panel (b) digitization noise is added into the likelihood. Now the models are much more believable. The data are indicated with a +. Drawn through them is the model selected. The dots display the model *before* the high pass filter. The $*$ at the bottom of the figure indicates where a new slope starts. A vertical line gives the position of a glitch.

The proper Bayesian thing to do is to incorporate the digitization into the likelihood. The error term, $d_k - y_k$, is truncated to an integer value and the probability at that integer value is the integral over unit length of the original pdf. It can be seen as adding uniformly distributed bitnoise of unit width to the error terms, to reflect our ignorance on the bitlevel. Unfortunately it bereaves us from the 'benefits of linearity', unless we approximate the uniform distribution with a Gaussian with a variance of $1/12$. We already argued that the Student-t likelihood can be approximated very well by a Gaussian likelihood. Together with Gaussian bitnoise the likelihood becomes a convolution of two Gaussians.

There is a striking similarity in the derivation of optimal values for a and b, between Gaussian distributions, where σ is known beforehand, and Student-t distributions where S is estimated from the data. When σ is replaced by S the formalism up until the second moment is identical. We use this feature replace the convolution of the two Gaussians by one Gaussian with a broader variance: the sum of a known one (digitization) and one estimated from the data as before.

$$S^2 = \frac{1}{N - N_P} \left(\sum_{k=1}^{N} (d_k - \hat{y}_k)^2 + \frac{N}{12} \right) \tag{16}$$

With this choice of S we have automatic noise scaling while at the same time preventing the system to interpret digitization as true signal. In figure 2b the effects of a minimum noise scale can be seen. Now simple models of constant flux and no glitches are preferred. Just as one would expect from such coarsely digitized data.

5. Search Strategies

There is a bewildering amount of models described in section 2. The number of possible models is in the order of $N!$, mostly glitched models. Out of this haystack we will pick the one with the highest evidence. Exhaustive search through all models for the best is simply not feasible, that would even be the case when no CPU limits were present. Some search strategies with a good chance of success must be devised. Here we depart (again) from true Bayesianism, which would have us sum over all options A and B, using the evidences as weights. As said above this is not feasible due to CPU limitations. Moreover in most cases there is only one model which is so much superior to all others that it would carry most of the weight in the final sum. We might as well neglect all the others. In the cases where two (or more) models have similar evidences obviously the models describe the data equally well and selecting one above the other(s) is not very harmfull for our purposes.

We start our search with the change points. First we calculate the evidence for a model with all possible change points present. This model has all the possibilities to follow changing flux levels. Most likely there are too many change points. One by one we omit a change point from the model and see whether the evidence increases. If so, that particular change point was not necessary — the evidence increases because a simpler model fits the data as well as a more complex one: Occam in action — and we omit the change point completely. If not, the change point at that particular location was needed, there is real evidence for it in the data and we restore the change point. In this way we check all the possible change points, omitting the ones which carry no evidence and keeping the ones which are needed. In most occasions we end up with the most simple model possible: only one offset and one slope.

This strategy works because there are always a factor 3 or 6 less change points than there are data in a reset interval. For glitches this is not the case. A glitch could occur at every data point. A different strategy is followed. If there are glitches at all in the reset interval, there is one which is the largest. We try a glitch at each location in the reset interval and see where the evidence increases most. If it does not increase at any location, we conclude there is no glitch in this ramp. Otherwise, we put a glitch at the best location and try the remaining locations for another glitch. Again searching the one with the highest increase in evidence. We continue this way until no more glitches are found. In most cases we find no glitches at all, except in band 4, the detectors at the long wavelengths.

Also whenever we find a glitch we go through the previous procedure of omitting change points, if any is present, to see whether the glitch model is better than the change point model.

So at least we have to calculate evidence for $M+N$ models of varying complexity. And for finding n glitches there are of the order of nN models of increasing complexity to be handled. As we have to go through the same models with other data, over and over again, we store each model on its first occurrence in a long linked list. If a model returns we retrieve it from the list for reuse.

6. Conclusions

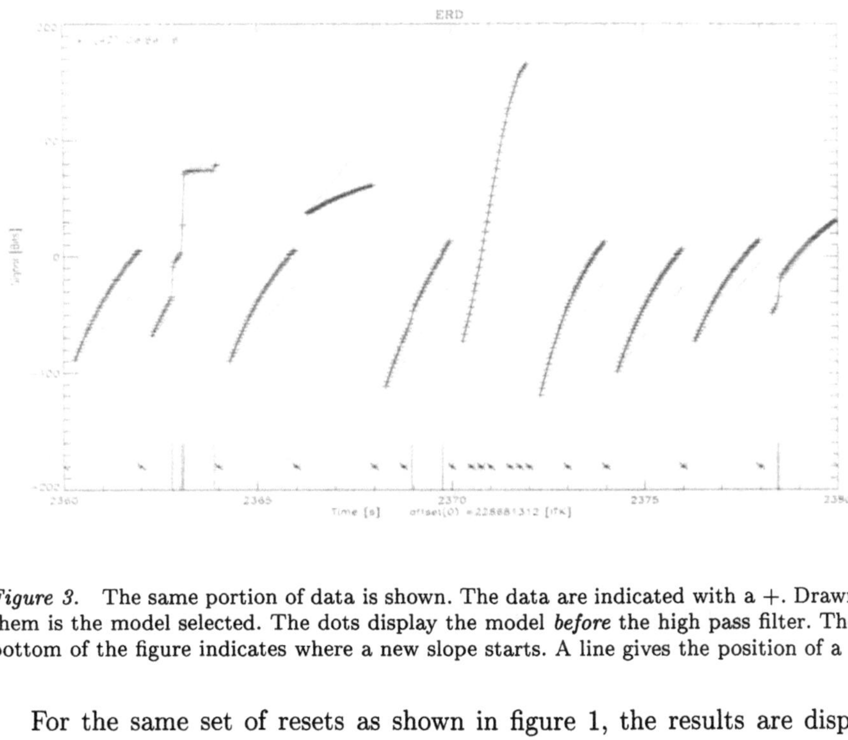

Figure 3. The same portion of data is shown. The data are indicated with a +. Drawn through them is the model selected. The dots display the model *before* the high pass filter. The * at the bottom of the figure indicates where a new slope starts. A line gives the position of a glitch.

For the same set of resets as shown in figure 1, the results are displayed in figure 3. The model is drawn as a full line through the datapoints indicated with a plus. They are hardly distinguishable. Over the whole data set, some 3700 ramps, the models fit the data to within 0.34 bits, very near the bitnoise. The gain in this detector is at the low setting. The model without the high pass filter is shown as dots. These lines are straight except for change points, indicated with an asterix at the bottom, and except for glitches, indicated at the bottom with a vertical line. The first ramp is just a straight line, the most simple model fitted best. The second has a couple of glitches, but is straight otherwise. Then two more simple models follow. Ramp number 5 has two small glitches and it starts to flex. It is the onset of an emission line in the spectrum, which is scanned over in ramp 6.

Here almost all change points are needed to model the ramp. The next ramp still sees the wing of the line and needs one change point. And in the last 3 ramps the spectrum are quiet again, with a couple of glitches in the final ramp.

TABLE 1.

Band	nr of dets	wavelength (μm)	glitchrate (%)	change points (%)	noise scale	gain	CPU per det.sec
1	12	2 - 4	1.8	0.2	0.33	4	0.034
2	12	4 - 12	3.9	0.4	0.30	1	0.036
3	12	12 - 30	4.6	1.6	0.35	1	0.037
4	12	30 - 45	29.7	2.4	0.35	1	0.061
5	2	11 - 26	28.6	4.2	0.69	16	0.067
6	2	26 - 45	25.6	1.2	0.78	16	0.060

Table 1 summarizes the results for a complete dataset. Average glitchrates are given as the percentage of ramps which are glitched. And similarly for the change points. The noise scale is the value of χ^2 divided by the degrees of freedom. If it equals 0.3 the noise is dominated by the bitnoise. The gain can either be 1 (low), 4 (intermediate) or 16 (high). The last column indicates how much CPU is needed on average to calculate one second of data. Our —selfimposed— limit was 0.02. We don't reach that but it is not uncomfortably far away. Next years computers will be faster.

The behaviour is quite typical, small glitches of a few bits are easily found and slight changes in the slopes are noticed. The measurements can be fully described with the models sketched. After processing thousands of ramps we must conclude that the system is remarkably robust and we have not seen ramps (yet) where we felt that the system had chosen a bad model.

Acknowledgements

I would like to thank the referee and Romke Bontekoe (always) for many cutting remarks and my bosses for indulging my Bayesian byways.

References

1. M. Kessler and 10 co-authors, "The Infrared Space Observatory (ISO) mission," *Astron.Astroph.*, **315**, pp. L27–L31, 1996.
2. T. de Graauw and 61 co-authors, "Observing with the ISO Short-Wavelength Spectrometer," *Astron.Astroph.*, **315**, pp. L49–L54, 1996.
3. E. Wieprecht, L. Haser and E. Wiezorrek, "SWS-Detector glitch effect correction," *submitted to Experimental Astronomy*, 1998.
4. S. Gull, "Bayesian inductive inference and maximum entropy," in *Maximum Entropy and Bayesian Methods*, G. Erickson and C. Smith, eds., Kluwer Academic, Dordrecht, 1988.
5. D. MacKay, "Bayesian interpolation," in *Maximum Entropy and Bayesian Methods*, C. Smith et al., ed., pp. 39–66, Kluwer Academic, Dordrecht, 1992.
6. D. S. Sivia, *Data Analysis, A Bayesian Tutorial*, Oxford University Press, Oxford, 1996.

BAYESIAN ANALYSIS WITH APPLICATION TO THE TIMING RESIDUALS OF A PULSAR

GILLES ZALAMANSKY
Université de Metz
Campus Bridoux, F-57000 Metz ‡

AND

CHRISTIAN P. ROBERT
CREST INSEE
Timbre J340, F-75675 Paris Cedex 14 §

Abstract. We apply Bayesian analysis to the estimation of both amplitude and exponent of a low frequency noise in the timing residuals of a millisecond pulsar, and we derive the posterior distribution for those two parameters
Key Words: Bayesian analysis, reference priors.

Because of the lack of space in this proceedings, we decided not to speak at all of the physical implications of our work. We refer the reader to [11][22][21][23] for extensive physical discussions.

1. Data and Model

Our data are $N = 165$ numbers $R = (R_1, .., R_N)$ sampled at times $(t_1, .., t_N)$ with gaussian measurement errors. The rms error at time t_k is σ_k. $(R_1, .., R_N)$, $(t_1, .., t_N)$ and $(\sigma_1, .., \sigma_N)$ are known vectors in $\Re^N$.

Our probabilistic model for R is

$$R = R_L + R_m \tag{1}$$

In (1) R_m is the measurement error. It is a gaussian vector with zero mean and covariance matrix $Diag\left(\sigma_1^2, ..., \sigma_N^2\right)$. R_L is a low frequency Gaussian noise with unknown spectral density

$$S(f) = Hu_\alpha f^{-\alpha}\ , \qquad \alpha \in [2,6] \tag{2}$$

u_α is an amplitude factor such that R_L has the same average quadratic norm as R_m. We specify its value below. In this setting, both the amplitude H and the

‡Email: zala@sciences.univ-metz.fr
§Email: robert@bayes.ensae.fr

W. von der Linden et al. (eds.), Maximum Entropy and Bayesian Methods, 189–198.

low frequency cut-off f_l of $S(f)$ (which necessarily exists, since in the other case $S(f)$ is diverging in 0) are unknown. Neither is known the exponent α of the noise. We assume f_l to be by far lower than the inverse data span. Consequently the noise is non stationary on the observation span, but it would become so if we could observe it very large spans. f_l is a true nuisance parameter in (1) since we're not interest by its value. But we are interested by infering on $\theta = (\alpha, H)$. Our problem is then a two parameters problems.

2. Inference on $\theta = (\alpha, H)$

2.1. ESTIMATORS OF θ

Our goal is now to derive estimators of θ unbiased by the nuisance term f_l. It is impossible to avoid contamination by R_m.

We introduce the orthonormal basis of $\Re^N$, $\{p_0, ..., p_{N-1}\}$, defined by

$$p_{ij} = \widetilde{p}_j(t_i)$$

In this equation p_{ij} is the ith component of vector p_j, and $\widetilde{p}_j(t)$ is a polynom with degree j, all the $\widetilde{p}_j's$ satisfying the orthonormality condition

$$\sum_{i=1}^{N} \widetilde{p}_j(t_i).\widetilde{p}_k(t_i) = \delta_{jk}$$

Note that the $\widetilde{p}_j(t_i)$ are the generalization to discrete non equispaced sampling of the Chebyshev polynoms.

For any vector a in $\Re^N$, we use the notation $a^\sigma = \left(\frac{a_1}{\sigma_1}, ..., \frac{a_N}{\sigma_N}\right)$. We introduce the $N \times N$ matrix $P = (p_{ij})$, and define the orthonormal basis of $\Re^N$ $\{q_0, ..., q_{N-1}\}$ deduced from $p_0{}^\sigma, ..., p^\sigma_{N-1}$ by the Gram-Schmidt procedure. $\{q_0, ..., q_j\}$ belongs to the vector subspace of $\Re^N$ spanned by $\{p_0, ..., p_j\}$. Define now

$$\xi_j = q_j \cdot R^\sigma \tag{3}$$

¿From (1) ξ_j can be split into two terms, $q_j \cdot R^\sigma_L$, and $q_j \cdot R^\sigma_m$. We call those terms ξ^L_j and ξ^m_j.

The covariance matrix $V_L(H, \alpha) = E\left[\xi_L \xi^t_L\right]$ of ξ_L has general term

$$\begin{aligned} E\left[\xi^L_k \xi^L_l\right] &= \sum_{r,s} \frac{q_{rk} q_{sl}}{\sigma_r \sigma_s} E\left[R_r R_s\right] \\ &= H u_\alpha \left(\int_{f_l}^{f_h} f^{-\alpha} Q_k(f).Q^*_l(f) df + \text{complex conj.} \right) \end{aligned} \tag{4}$$

with

$$\begin{aligned} Q_k(f) &= \sum_{r=1}^{N} \frac{q_{rk}}{\sigma_r} e^{i2\pi f t_r} = \sum_{n=1}^{\infty} \frac{(i2\pi f)^n}{n!} \sum_{r=1}^{N} \frac{q_{rk}}{\sigma_r} t^n_r \\ &= \sum_{n=1}^{\infty} \frac{(i2\pi f)^n}{n!} \left(q_k \cdot t_n{}^\sigma\right) \ , \ t_n = (t^n_1, .., t^n_N) . \end{aligned}$$

t_n^σ belongs to the vector $(q_0, .., q_n)$. The scalar product $q_k \cdot t_n{}^\sigma$ is then zero for every n less than k. The integrand in (4) is then dominated by $f^{-\alpha+k+l}$ near zero, and the integral converges if the exponent is larger than -1. The quantities $E[\xi_k^L \xi_l^L]$ converge when f_l goes to 0 if both k and l are larger than $j_{\min}$ satisfying[8][9]

$$j_{\min} \geq \frac{\alpha - 1}{2} \, . \tag{5}$$

If we assume $\alpha \leq 6$, this condition is fulfilled by all of our estimators $\xi_3, ..., \xi_N$. From now and then, our data set is

$$\xi = (\xi_3, .., \xi_{3+n_1}) \tag{6}$$

In the application below, we chose $n_1 = 13$, since higher values of n_1 do not lead to changes in the posterior distribution of α and H. This is related to the fact that, for large j's, the white noise dominates the low frequency noise in ξ_j. We simply call $\xi_1, .., \xi_{n_1}$ the data vector ξ defined in (6).

Even if f_l is not nearly zero, its accurate value negligibly affects the covariance matrix of ξ. Had we not discarded the first three estimators ξ_0, ξ_1, ξ_2 defined in (3) , this would not have been true. The projection we performed is a kind of regularization process.

The high frequency cut-off f_h in (4) is the average Shannon frequency but its exact value is not important here, thanks to the rapidly decreasing behavior of $S(f)$.

In our assumed probabilistic model (1), ξ is a sum of two terms, ξ^L and ξ^m

We now study the dependence of ξ with respect to $\theta = (\alpha, H)$. We write $V_L(\alpha, H) = H\, u_\alpha\, V(\alpha) = H\, Y(\alpha)$, with

$$(V(\alpha))_{kl} = 2\Re e \left(\int_{1/T}^{f_h} f^{-\alpha} U_k(f).U_l^*(f) df \right) \tag{7}$$

and the full covariance matrix of ξ is

$$E[\xi\xi^t] = H\, Y(\alpha) + I_{n_1} = W(\theta) \, . \tag{8}$$

I_{n_1} is the identity matrix in $\Re^{n_1}$.

Let $e(\alpha)$ denote the eigenvectors of $V(\alpha)$ and $\gamma_j(\alpha)$ denote its eigenvalues, $j \in \{1, ..n_1\}$. The normalization coefficient u_α can thus be defined: The average quadratic squared norm of ξ in $\Re^{n_1}$ is

$$\begin{aligned} E[||\xi||^2] &= H u_\alpha \sum_{j=1}^{n_1} \gamma_j(\alpha) + n_1 \\ &= E[||\xi_L||^2] + E[||\xi_m||^2] \, , \end{aligned}$$

$E[||\xi_L||^2]$ is the average quadratic squared norm of ξ 'due to' R_L, and $E[||\xi_m||^2]$ the part due to measurement noise. They are equal if we chose $H = 1$ and

$$u_\alpha = \frac{n_1}{\sum_{j=1}^{n_1} \gamma_j(\alpha)} \, . \tag{9}$$

Of course, u_α is a normalization coefficient internal to this paper, since it depends on the number of estimators we take. But it enables us to compare amplitudes H of noises for various exponents α, since for $H = 1$, all $\xi_L(\alpha)$'s have the same average quadratic norm, which is equal to the part of ξ due the measurement noise.

2.2. PROBABILITY DENSITY FUNCTION OF ξ GIVEN θ, AND FISHER INFORMATION MATRIX

We now have obtained a vector ξ in $\Re^{n_1}$, with a very simple probability distribution given θ :

$$p(\xi|\theta) = \frac{1}{(2\pi)^{\frac{n_1}{2}}\sqrt{|W(\theta)|}}\exp(-\frac{1}{2}\xi^t W(\theta)^{-1}\xi) \tag{10}$$

with $W(\theta)$ defined in (8). We now derive the Fisher Information Matrix $I(\theta)$. The log-likelihood $L(\theta)$ is

$$L(\theta) = -\frac{1}{2}\xi^t W(\theta)^{-1}\xi - \frac{1}{2}\ln|W(\theta)| - \frac{n_1}{2}\ln(2\pi)\,.$$

we use the identity

$$\frac{\partial}{\partial x}\left(\ln|W(x)|\right) = Tr\left(W(x)^{-1}\frac{\partial W(x)}{\partial x}\right)$$

(Tr means "trace") and find, recalling that $E[\xi\xi^t] = W(\theta)$:

$$I(\theta) = \begin{pmatrix} I_{11}(\alpha,H) & I_{12}(\alpha,H) \\ I_{21}(\alpha,H) & I_{22}(\alpha,H) \end{pmatrix}$$

with

$$\begin{aligned}
I_{11}(\alpha,\alpha) &= Tr\left(\frac{\partial^2 W(\theta)^{-1}}{\partial\alpha^2}W(\theta) + \frac{\partial W(\theta)^{-1}}{\partial\alpha}\frac{\partial W(\theta)}{\partial\alpha} + W(\theta)^{-1}\frac{\partial^2 W(\theta)}{\partial\alpha^2}\right),\\
I_{22}(\alpha,H) &= Tr\left(\frac{\partial^2 W(\theta)^{-1}}{\partial H^2}W(\theta) + \frac{\partial W(\theta)^{-1}}{\partial H}\frac{\partial W(\theta)}{\partial H} + W(\theta)^{-1}\frac{\partial^2 W(\theta)}{\partial H}\right),\\
I_{12}(\alpha,H) &= Tr\left(\frac{\partial^2 W(\theta)^{-1}}{\partial\alpha\partial H}W(\theta) + \frac{\partial W(\theta)^{-1}}{\partial\alpha}\frac{\partial W(\theta)}{\partial H} + W(\theta)^{-1}\frac{\partial^2 W(\theta)}{\partial\alpha\partial H}\right)\\
&= I_{21}(\alpha,H)\,.
\end{aligned}$$

We set $Q(\theta) = W(\theta)^{-1}$. Noting that $\frac{\partial(QW)}{\partial\alpha} = \frac{\partial(QW)}{\partial H} = 0$, and that consequently $\frac{\partial Q}{\partial\alpha} = -Q\frac{\partial W}{\partial\alpha}Q$, $\frac{\partial Q}{\partial H} = -Q\frac{\partial W}{\partial H}Q$, we derive that

$$I(\theta) = \frac{1}{2}\begin{pmatrix} H^2 Tr(Q\frac{dY}{d\alpha}Q\frac{dY}{d\alpha}) & HTr(QYQ\frac{dY}{d\alpha}) \\ HTr(QYQ\frac{dY}{d\alpha}) & Tr(QYQY) \end{pmatrix}. \tag{11}$$

The problem is now: What can we say about θ, given ξ ? To solve this problem we could use either frequentist methods, that provide confidence sets, or Bayesian analysis, which provide true posterior probabilities. We do prefer the Bayesian approach[22], but the derivation of a prior distribution for θ is seldom an easy job, and it won't be so in our problem.

3. Non informative priors

3.1. LAPLACE'S PRIOR

Laplace (1812) considered that a sensible prior π to choose when no information was available is the uniform prior: $\pi(\theta) = 1$. Unfortunately this rule of choice is not invariant under reparameterization in $\mathcal{P}$ (except when $\mathcal{P}$ is discrete), and does not always lead to a proper probability distribution.

3.2. KULLBACK-LEIBLER DISTANCE.

The only way to derive θ from x when x is random is to perform an infinite number of measurements, and find which value of θ parametrizes the histogram of x. The distance between two histograms indexed by θ_1 and θ_2 is the Kullback-Leibler distance

$$d_{KL}(p_1, p_2) = \int p_1(x) \ln \frac{p_1(x)}{p_2(x)} dx \tag{12}$$

with $p_1(x) = p(x|\theta_1)$, $p_2(x) = p(x|\theta_2)$.

And one can show that

$$d_{KL}(p_1, p_2) = \frac{1}{2}(\theta_1 - \theta_2)^t \cdot I(\theta_1) \cdot (\theta_1 - \theta_2) + \mathcal{O}(|\theta_\infty - \theta_\in|^\ni) \tag{13}$$

If we want to use a measurement x to discriminate between θ_1 and θ_2, the parametrization where $I(\theta_1)$ is uniform, if it exists, is privileged. On the average (over x), the distance between θ and neighboring values will not depend on θ.

3.3. DIFFERENTIAL GEOMETRY

If we define the distance between θ and $\theta + \delta\theta$ by

$$d(\theta, \theta + \delta\theta) = \delta\theta^t \cdot I(\theta) \cdot \delta\theta \tag{14}$$

then parameter space P, with this metric, is a Riemannian manifold, with metric tensor $I(\theta)$. If we now reparametrize P, choosing $\phi(\theta)$ as coordinate system on the manifold, instead of θ, the Fisher Information Matrix $G(\phi)$ is

$$G(\phi) = K(\theta) \cdot I(\theta) \cdot K(\theta)^t \tag{15}$$

with $K_{ij}(\theta) = \frac{\partial \theta_j}{\partial \phi_i}$.

It is a well known result in differential geometry that if the manifold is flat (i.e has vanishing Riemann tensor), then one can always find[19] a coordinate system ϕ for which $G(\phi)$ is uniform. The Kullback-Leibler distance between ϕ and $\phi + \delta\phi$ is independent on ϕ, which means that on the average over an infinite set of experiments, with all possible values of ϕ, all will be recovered at the same speed. There are no privileged values of ϕ, and the prior distribution on ϕ that respects this homogeneity is the uniform prior $\pi(\phi) = 1$. It is equivalent to the prior $\pi(\theta) = \sqrt{|I(\theta)|}$ in parametrization θ . ($|\cdot|$ denotes the matrix determinant).

As a matter of fact, the volume element filled in the manifold when θ varies by $\delta\theta$ is

$$\delta V = \sqrt{|I(\theta)|} \prod_i \delta\theta_i \tag{16}$$

which is invariant.

Unfortunately, it is also a well known result in differential geometry that when the manifold is curved (i.e has non vanishing Riemann tensor), then one can never find a parametrization $\phi(\theta)$ for which $G(\phi)$ is uniform. There will always be privileged points, for which the discriminating power of the measurement will be larger than for others.

In the case of a one dimensional parameter space, the Riemann tensor is zero, and the transformation $\theta \to \phi$ with

$$\phi(\theta) = \int \sqrt{|I(\theta)|} d\theta$$

directly gives $G(\phi) = 1$.

3.4. JEFFREYS' PRIOR

Jeffreys' prior $\pi_J(\theta)$ is the invariant volume density (16)

$$\pi_J(\theta) = \sqrt{|I(\theta)|} \tag{17}$$

In case of a flat manifold, (the most common example being the multidimensional gaussian distribution), it is the most sensible prior to select in case of no prior information on θ. Lindley showed[14] that Fisher's fiducial approach led to the same posterior distribution than Bayesian analysis with prior (17).

3.5. PARADOXES

It is tempting to use Jeffreys' prior (17) for any inference problem, unfortunately some paradoxes are met with curved manifolds. The most common example is the gaussian distribution with both mean μ and rms σ unknown, $x \in \mathcal{N}(\mu, \sigma)$ with $\theta = (\mu, \sigma)$. Jeffreys' prior is σ^{-2}, whereas σ^{-1} is quite more desirable for several reasons. I refer the reader to the review paper by Kass and Wasserman[12], but Jeffrey himself supported the prior σ^{-1} in this case, arguing that σ and μ should be estimated independently. The marginal prior for σ given μ is σ^{-1} (scaling parameter), and the marginal prior for μ is 1 (location parameter). The prior for (μ, σ) is then $\pi(\sigma|\mu)\,\pi(\mu) = \sigma^{-1}$, in contradiction with (17).

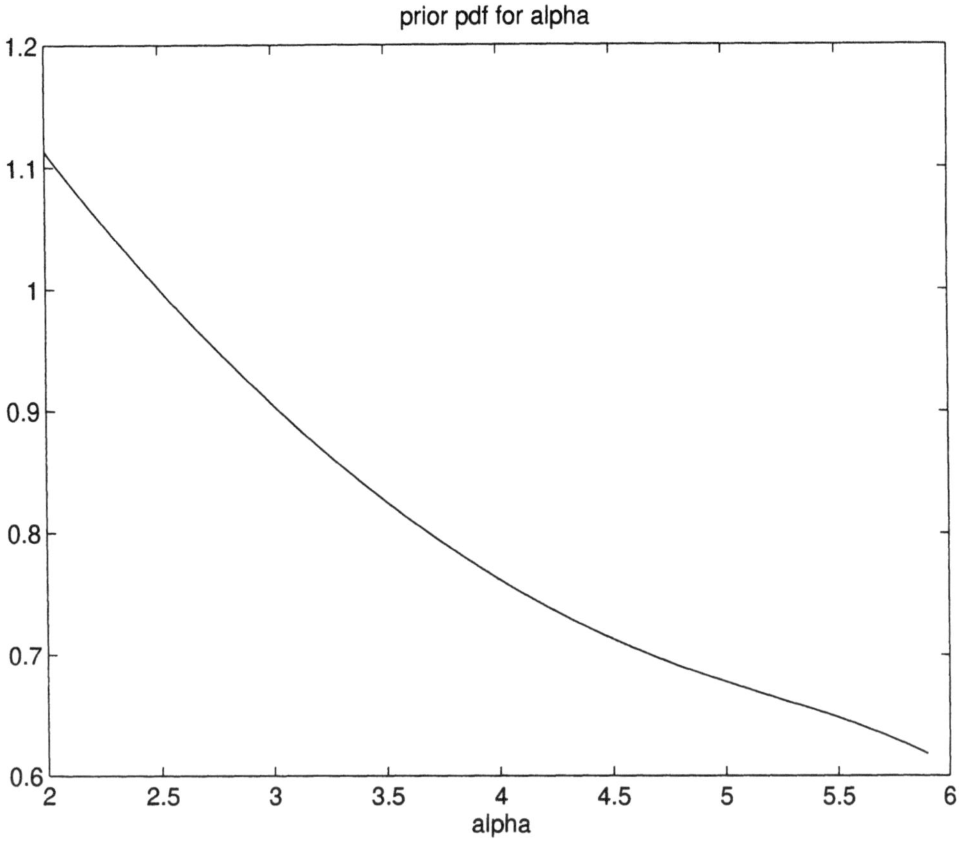

Figure 1. Reference prior for the power α.

3.6. BERNARDO-BERGER REFERENCE PRIORS

Bernardo and Berger extended this idea of step by step derivation of a non informative prior for θ. In our case, parameter space is 2 dimensional, and θ is (α, H). Basically, the procedure is the following:

1) Compute the Fisher Information Matrix. Ours is (11)
2) Compute the marginal prior for H given α, which is in our case $|I_{22}(\alpha, H)|^{\frac{1}{2}}$
3) Compute the marginal prior for α.

I refer the reader to [5] [4] [3] for the details of the computation.

3.7. RESULTS

3.7.1. *Posterior probability function for α*

We computed the posterior distribution for α, using a discrete set of values of $\theta \in \mathcal{P} \times \mathcal{H}$,with $\mathcal{P} = [2.6]$ and $\mathcal{H} = [0, 5000]$, based on the reference analysis of Bernardo Berger.

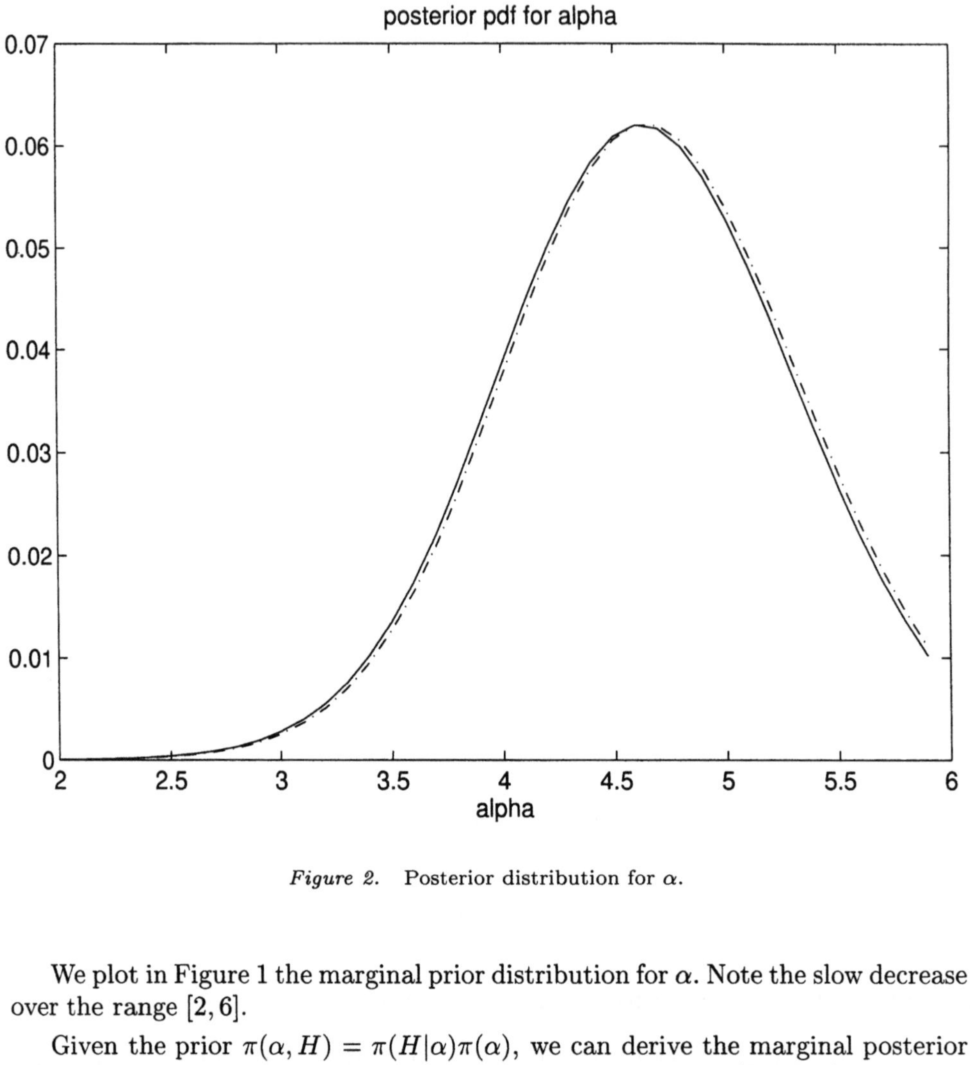

Figure 2. Posterior distribution for α.

We plot in Figure 1 the marginal prior distribution for α. Note the slow decrease over the range $[2, 6]$.

Given the prior $\pi(\alpha, H) = \pi(H|\alpha)\pi(\alpha)$, we can derive the marginal posterior probability density function for α, which is given in Figure 2 (full line).

It is peaked around $\alpha = 4.7$, and the hypothesis that ξ_L has a gravitational origin [11] is quite compatible with this curve. $p(\alpha|\xi)$ is quite close to the posterior obtained with Jeffreys' prior (dashed line).

3.7.2. *Upper limits on H*

If we assume that $\alpha = 5$, the posterior cumulative distribution function (cdf) for H can be easily derived from Jeffreys' prior for H.

To derive the marginal law of α, we should theoretically integrate over H, from 0 to infinity. In the numerical integrations, we chose instead a range $[0, H_{\max}]$ broad enough for the marginal posterior on α to be independent of $H_{\max}$. As shown by Figure 3, the marginal posterior distribution on H does not charge the larger values of H.

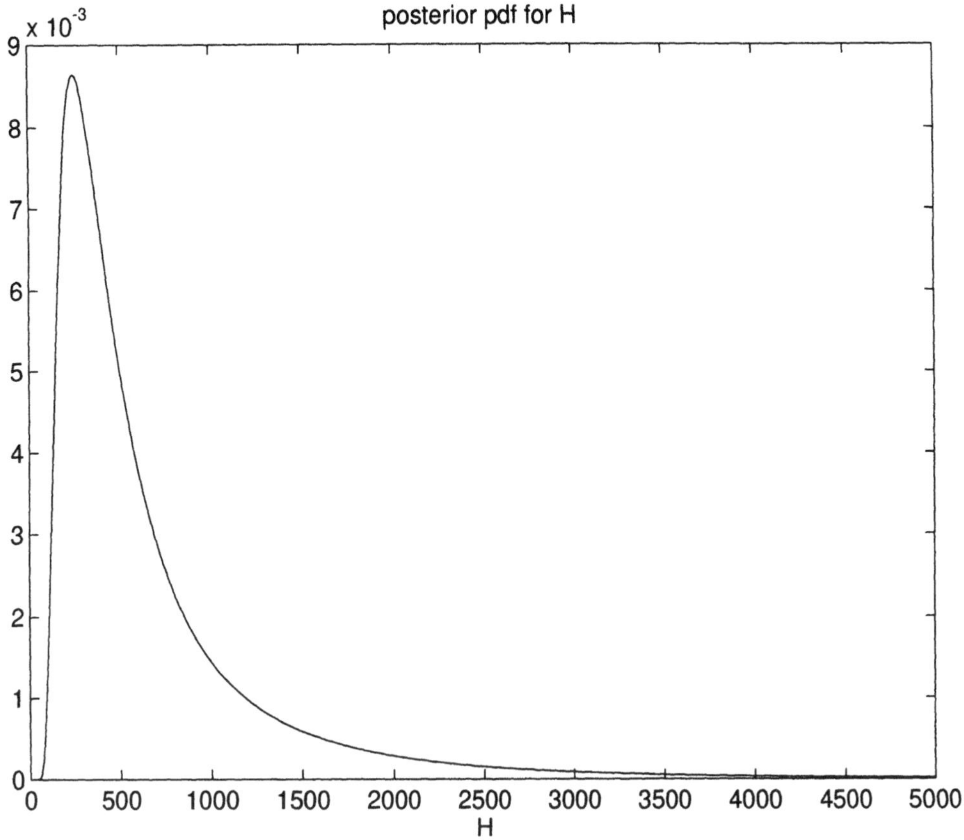

Figure 3. Posterior pdf for H.

4. Conclusion

We wanted to present here the Bayesian analysis of a rather complex 2 parameters problem of inference, with strong astrophysical interest. The difference between posterior distributions for α, obtained with Jeffreys'prior and Bernardo-Bergers' one is negligible here, but from a theoretical point of view this difference is large, and we wanted to underline it. Also did we want to underline the connection between Riemannian Geometry and Inference, as Carlos Rodrigez did in his own talks during this congress.

References

1. Bayes T.R (1763) *An Essay Towards Solving a Problem in the Doctrine of Chances*, Philosophical Transactions of the Royal Society, reprinted in Biometrika 45, 243-315, 1958
2. Berger J.(1985) *Statistical Decision Theory and Bayesian Analysis*, Springer-Verlag
3. Bernardo J., Berger J. (1989). Journal of the American Statistical Association, 84, 200-207
4. Bernardo J., A.Smith (1993) *Bayesian Theory*, Wiley&Son
5. Bernardo J. (1979), Journal of the Royal Statistical Society, ser.B, 41, 113-147

6. Blandford R., Narayan R., Romani R. (1984) J. Astrophy. Astron. 5 369
7. Cramer H (1963), *Mathematical Methods of Statistics*, Princeton University Press
8. Deeter E., Boynton E.P (1982) Astrophys. J. 261, 337
9. Groth E.J (1975) Astrophys. J. supplement series, 293, vol 29, 431-466
10. Jeffreys H. (1961) *Theory of probability* (3rd ed). Oxford Univ. Press
11. Kaspi V.,Taylor J.,Ryba M.F (1993), Astrophys. J. 428 713-728
12. Kass E., Wassermann L. (1996), Journal of the American Statistical Association, 91, no 435, 1343-1370
13. Laplace P.S (1820), *Essai Philosophique sur les probabilités*. English translation: *Philosophical essays on probabilities* (1951), New York: Dover
14. Lindley D.V (1958) Journal of the Royal Statistical Society, ser.B, 20, 102-107
15. Mc Hugh M., Zalamansky G., Vernotte F., Lantz E. (1996), Phys. Rev. D, 54, 10, 5993-6000
16. Rényi A. (1965), *Calcul des probabilités*, Ed Jacques Gabay
17. Robert C.P (1994), *The Bayesian Choice*, Springer Verlag
18. M.J Scherwish (1995), Theory of Statistics, Springer Verlag
19. Hans Stefani, *General Relativity*, Cambridge Univ. Press
20. Vachaspati T., Vilenkin A., (1985) Phys. Rev. D vol 31, 12
21. Vilenkin A., Shellard P., (1994), *Cosmic Strings and Other Topological Defects*, Cambridge Univ. Press
22. Zalamansky G.,Robert C.P,Vernotte F.,Taris F. (1997) MNRAS, 288, 533-537
23. Zalamansky G., PH-D thesis, Université Paris VII, 1996

BAYESIAN LABELING OF REMOTE SENSING IMAGE CONTENT

M. SCHRÖDER AND K. SEIDEL
Swiss Federal Institute of Technology ETH Zürich,
Communication Technology Lab, Image Science Group
Gloriastr. 35, CH-8092 Zürich[‡]

AND

M. DATCU
German Aerospace Center DLR,
German Remote Sensing Data Center DFD, Oberpfaffenhofen
D-82234 Weßling[§]

Abstract. In this paper we present a multi-level scheme for stochastic description of image content. The different levels are derived from the different degrees of abstraction. On the level of the image data, we use stochastic data models and Bayesian parameter estimation to derive low-level image features. On the next level, we derive meta features that provide both the fit of these models and the actual complexity of the data. The low-level and meta features are combined in an un-supervised clustering scheme to obtain an objective description of the image content. To obtain this objective description we use clustering by melting. The descriptions by several models are then linked to application-oriented, semantic labels using another process of Bayesian inference. We sketch in detail the various processes of inference and give an example for this kind of information on each level of abstraction using satellite images (RESURS-01 and X-SAR).

Key words: Bayesian networks, Bayesian statistics, clustering by melting, Gibbs fields, Markov random fields, remote sensing, texture

1. Introduction

Typical remote sensing images exhibit an enormous amount of information. This information is contained in possibly very many spectral bands or different polarizations. Furthermore, radiometric preprocessing and geometric correction may be necessary. Classical methods of image classification [1] mainly work with multi-spectral models, possible combined with simple textural features, in order to ob-

[‡]Email: {schro,seidel}@vision.ee.ethz.ch
[§]Email: mihai.datcu@dlr.de

W. von der Linden et al. (eds.), Maximum Entropy and Bayesian Methods, 199–206.

tain a segmentation based on given training data. The classical approach is highly application-oriented and does not provide an overall description of the image content as it is desired in future remote sensing image archives that offer the possibility of query by content [2].

In recent years, content-based query from databases of pictures and images has been subject of intensive research [3]. However, for the query from complex remote sensing image data these results are of very limited use. To close the gap between traditional methods of remote sensing image interpretation and pictorial information system we use a novel information theoretic concept of Bayesian information extraction [4,5]. We have already applied this concept in deriving a new texture descriptor in remote sensing images [6], that we shortly review in this paper. However, we also intend to apply this concept to the characterization of information on higher levels of abstraction.

The paper is organized as follows. In Sec. 2, we define the different levels of abstraction on which we characterize the image content. The probabilistic nature of each level is discussed in detail in the following sections: the Bayesian parameter estimation of low-level image features in Sec. 3, the derivation of meta features in Sec. 4 and the segmentation of the image and the possibility to incorporate geometrical models in Sec. 5. The final level of semantic content labels is reached in Sec. 6 with the description of a user-adaptive labeling scheme. Finally, we conclude with a short summary in Sec. 7.

2. Levels of Abstraction

Information in remote sensing images exhibits an enormous complexity. Our goal is to capture, describe, and assess this information using stochastic models. However, an overall, completely un-biased description of the information in remote sensing images is not possible. Therefore, we use a certain set of stochastic models and organize the obtained information according to the different levels of abstraction. In this framework, a natural grouping yields six levels of abstraction, which we summarize in Tab. 1.

level	information	elements	example
5	semantics	content labels	snow, ice, mountains
4	geometrical features	region characterization	points, lines, area
3	image segmentation	image segments	class labels
2	meta features	properties of models used	evidence, complexity
1	image features	estimations of signal models	parameters, uncertainty
0	image data	raw data	pixel values

TABLE 1. Levels of abstraction of spatial information. From bottom to top the semantic meaning becomes stronger. Each level is obtained from one or more preceding levels using a process of Bayesian inference.

Starting from the actual image data with bare pixel values (level 0) we first obtain low-level image features such as multi-spectral or textural features (level 1). These features provide additional information such as the evidence or the optimum complexity of a particular model ("information on information" or "meta information", level 2). According to these features, an un-supervised segmentation of the image into homogeneous regions is possible (level 3). These regions can be further described using geometrical objects such as points, lines, and areas (level 4). The information on levels 1–4 is—in the frame of the used models—purely objective. On the final level (level 5), semantic labels are assigned. These labels are defined by experts in particular fields of applications and linked to the features at lower levels using a formalism similar to Bayesian networks.

This scheme promises to be a powerful and robust framework for the representation of information on image content on various levels of abstraction. The processes of Bayesian inference from one level to the next are discussed in the following sections.

3. Low-level image features (level 1)

In this section, we demonstrate how to perform stochastic modeling of the observed pixels of an image, that is, how to describe image content on a level very close to the actual image data.

The stochastic model M is defined via the likelihood $p(\{x_s\}|M,\Theta)$ of the image pixels $\{x_s\}$ in terms of the parameter vector Θ. In particular, Gibbs Markov random field models are defined via

$$p(x_s|\partial x_s,\Theta) = \frac{1}{Z}\mathrm{e}^{-H(x_s;\partial x_s,\Theta)} \tag{1}$$

as the probability distribution function of the pixel x_s given its neighborhood ∂x_s. Here, $H(x_s;\partial x_s,\Theta)$ is the energy function and Z the partition sum of the Gibbs distribution.

The process of information extraction is equivalent with finding the parameter vector Θ that best explains the given data. The posterior probability of the parameter vector Θ is obtained using Bayes' equation

$$p(\Theta|\{x_s\}) = \frac{p(\{x_s\}|\Theta)\,p(\Theta)}{p(\{x_s\})}, \tag{2}$$

with the likelihood $p(\{x_s\}|\Theta)$ of the data and the prior probabilities $p(\Theta)$ and $p(\{x_s\})$. The prior $p(\Theta)$ can be derived from, e.g., group theory or maximum entropy considerations. The maximization of the posterior probability of Θ can be done in the frame of, e.g., the maximum pseudo-likelihood approximation. For certain models, computational efficient algorithms are available, such as the conditional least squares estimator of the auto-binomial model [6], which provides both the incorporation of the correct prior and a very fast calculation.

In Fig. 1, we give an example for this first level of information extraction using the auto-binomial model. In this model, the energy function is defined as

$$H(x_s;\partial x_s,\Theta) = -\ln\binom{G}{x_s} - x_s\cdot\eta, \tag{3}$$

where we denote the maximum grey value with G and the binomial coefficients with $\binom{n}{m}$. The neighboring pixels δx_s have an effective influence via the scalar quantity

$$\eta = a + \sum_{ij} b_{ij} \frac{x_{ij} + x'_{ij}}{G}. \tag{4}$$

Due to the redundancy of the parameter a [6], the auto-binomial model is characterized by the parameter vector

$$\Theta = [b_{11},\ b_{12},\ b_{21},\ b_{22},\ \ldots]. \tag{5}$$

As the example shows, there is a clear correlation between the visual appearance of the image data and the magnitude of $|\Theta|$. Depending on the choice of the model, different spatial characteristics of the image data can be captured. If a family of different models at possibly many scales of the image is applied, then we obtain a robust description of the actual image data, which incorporates all aspects the models are able to describe.

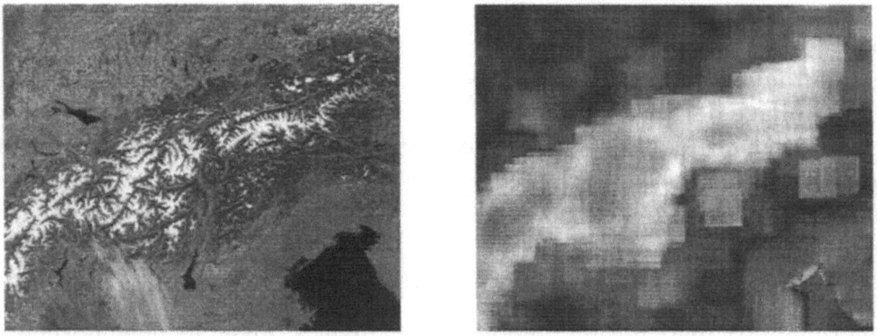

Figure 1. Bayesian parameter estimation of a Gibbs field. Remote sensing image of the alpine region (left, RESURS-01 satellite, 1km×1km per pixel) and resulting parameters Θ of a 3rd order auto-binomial model (right, 'black' to 'white' represent low to high values). Here we plot the norm $|\Theta|$ of Θ as a measure of the strength of the texture. The window size for estimation of the parameters is 32×32.

4. Meta features as image content information (level 2)

In a second level of Bayesian inference we obtain the model M_j that best describes the image data. We select this model from a family of models $\{M_i\}$ by maximizing the posterior probability

$$p(M_i|\{x_s\}) = \frac{p(\{x_s\}|M_i)\, p(M_i)}{p(\{x_s\})}, \tag{6}$$

with the prior $p(M_i)$ of the model M_i and the evidence

$$p(\{x_s\}|M_i) = \int p(\{x_s\}|\Theta, M_i)\, p(\Theta|M_i)\, d\Theta. \tag{7}$$

In Fig. 2, left, we show the evidence of the example in Fig. 1, right. Note that 'strong' textures are accompanied by a low evidence as compared to fairly homogeneous areas. This is in accordance with the fact that homogeneous areas are almost deterministic and can be described by the model with very high probability.

The use of several models with different complexity enables us to estimate the actual complexity of the image data. The decision between model M_1 and M_2 is done by considering the threshold

$$\frac{p(M_1|\{x_s\})}{p(M_2|\{x_s\})} \geq \Lambda. \tag{8}$$

The example in Fig. 2, right, depicts the result of the model selection between the 3rd and 1st order auto-binomial model with a threshold of $\Lambda = 2.3$. Note that there are regions with a better fit of the more complex model without strong texture being present in the original image (e.g., in the upper right corner). This indicates that the result of the model selection can be used as an additional characterization of image content.

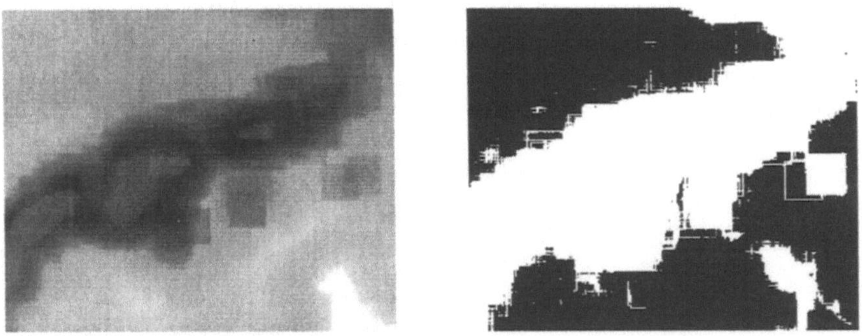

Figure 2. Example of Bayesian model selection. Evidence of the 3rd order auto-binomial model (left, 'black' to 'white' represent low to high evidence) and result of the model selection between 1st and 3rd order auto-binomial model (right). In the right example, 'white' areas correspond to areas with a significant higher evidence of the 3rd order model (threshold $\Lambda = 2.3$). Note that 'strong' texture is accompanied by low evidence as compared to homogeneous areas (left), but decides clearly for the more complex model (right). The original image data is shown in Fig. 1.

5. Image segmentation and geometrical models (level 3 and 4)

After obtaining image features on levels 1 and 2 the amount of data to describe the image is still considerably large. To further reduce the amount of data, while keeping the information on the image content, a segmentation into characteristic classes has to be performed. For this task numerous clustering algorithms are available from the fields of pattern recognition and computer vision [7].

However, to have a consistent flow of information from levels with low to high abstraction we abstain from the use of these traditional approaches and apply 'clustering by melting' [8]. This algorithms provides an un-biased segmentation, which captures the characteristics present in the data. It postulates an 'interaction energy' $E_k = (\Theta_k - \tilde{\Theta})^2$ of the parameter vector Θ_k with the cluster center $\tilde{\Theta}$. Here, the index k enumerates all estimations. Maximization of the entropy

$$S = -\sum_k p(\Theta_k|\tilde{\Theta}) \log p(\Theta_k|\tilde{\Theta}), \tag{9}$$

under the constraint of conserved expectation value of the energy, leads to the contribution $p(\Theta_k|\tilde{\Theta})$ of the kth parameter to the cluster center $\tilde{\Theta}$ of

$$p(\Theta_k|\tilde{\Theta}) = \frac{1}{Z} e^{-\frac{(\Theta_k - \tilde{\Theta})^2}{T}}, \tag{10}$$

with the partition sum Z and the temperature T. Starting at $T = 0$ with each estimation being a cluster and slowly increasing T, the algorithm produces a tree of clusters in the scale space. By selecting a final temperature the coarseness of the segmentation can be defined.

In Fig. 3, we show the result of such a clustering based on two features, one from level 1 and and one from level 2. Note that we only show a small cut of the image data of which we estimated and clustered the features. After this segmentation, the image is characterized by regions of class labels ω_i. These regions can now be further described using geometric models with basic elements such as points, lines, and polygons.

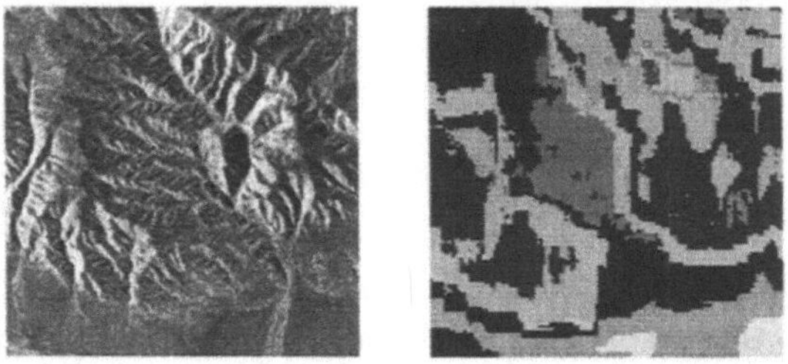

Figure 3. Un-supervised clustering using clustering by melting. Original X-SAR scene (left, scale 0.1) and result of clustering by melting (right). As features we chose the norm of the parameter vector of a 5th order Gibbs model (level 1) and the evidence of the estimation (level 2).

6. Semantic labeling (level 5)

On the highest level of abstraction, semantic labels are assigned to the image data. This is the first time in the process of information extraction where training data, that is, human knowledge, is incorporated. However, we note that through the selection of particular models and the selection of certain algorithms an implicit "human bias" has already been introduced on lower, "objective" levels.

The relationship between the elements ω_i of lower levels in the hierarchy and content labels A_ν is learned through the interaction with experts of the particular application area who sketch training areas for each cover type. This provides the likelihood $p(\omega_i|A_\nu)$ of each class label ω_i given the cover type A_ν.

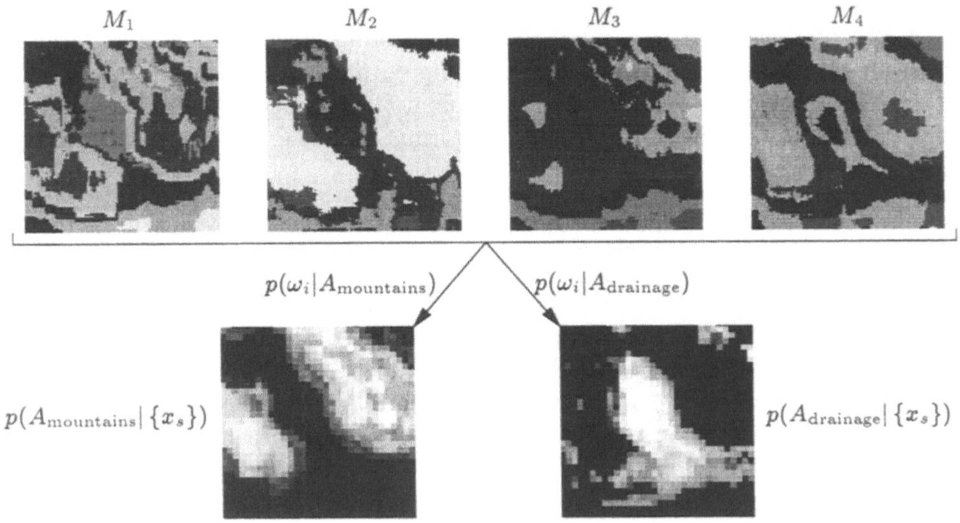

Figure 4. Semantic content labeling of the X-SAR image in Fig. 3 using four segmentations based on different Gibbs models and clustering by melting (top row, ω_i shown). The knowledge of $p(\omega_i|A_{\text{mountains}})$ and $p(\omega_i|A_{\text{drainage}})$ enables us to calculate maps of evidence of these two cover-types using Eq. (13) (bottom left and right, respectively).

Given some image data $\{x_s\}$, we now calculate $p(A_\nu|\{x_s\})$, the probability of the cover type A_ν. In terms of the objective segmentation results ω_i (level 3 and 4), we obtain

$$p(A_\nu|\{x_s\}) = \sum_i p(A_\nu|\omega_i, \{x_s\})\, p(\omega_i|\{x_s\}), \tag{11}$$

where the sum over i extends over all class labels ω_i. Using Bayes' equation yields

$$p(A_\nu|\omega_i) = \frac{p(\omega_i|A_\nu)\, p(A_\nu)}{p(\omega_i)}. \tag{12}$$

The prior probabilities $p(A_\nu)$ enable us to incorporate prior knowledge on the particular cover type, whereas the prior probabilities $p(\omega_i)$ are obtained by considering a large set of images. Altogether, we obtain

$$p(A_\nu|\{x_s\}) = p(A_\nu) \sum_i \frac{p(\omega_i|A_\nu)\, p(\omega_i|\{x_s\})}{p(\omega_i)} \tag{13}$$

as posterior probabilities of the cover-type A_ν. This way of labeling is easily extended to the case of several models. Then the descriptions on level 3 are combined to a joint set of labels ω_i. However, with many models the number of resulting elements increases very fast and the use of Bayesian prototype-trees [9] might be considered.

We give an example of semantic labeling in Fig. 4. The general information that is concentrated in the four segmentations (level 3) is transformed into maps indicating the evidence of user-specific content characteristics (in the example, mountains and drainage areas). For the application in information system indices are easily derived from such maps. Furthermore, information theoretic methods, such as the cross-entropy, can be used to evaluate the quality of these maps and in this way help to reconfigure the chain of information down to the low-level image features on level 1.

7. Summary and conclusion

We have presented an hierarchical scheme of stochastic image analysis that reflects the nature of the information content of large images. The different levels of this scheme are arranged according to their different degree of abstraction and semantics. Inference between two levels is performed in a Bayesian framework providing both a clean way of incorporating prior knowledge and measures of evidence of the features on the higher semantic level.

This scheme promises to be an important ingredient in the process of efficient extraction and storage of information from large image data sets. Extensive tests with a large number of remote sensing images of various kinds (optical, SAR, and aerial photos) are currently being prepared. However, this scheme and the concept of stochastic image analysis may also be applicable in other areas of computer vision, such as medical image analysis or astronomical image interpretation.

Acknowledgments

We thank Prof. J. Nievergelt for an inspiring discussion. The RESURS-01 scene has been provided by NPOC Switzerland, ETH Zurich, and the X-SAR scene by DFD/DLR, Oberpfaffenhofen, Germany.

References

1. J. B. Campbell, *Introduction to Remote Sensing*, Taylor and Francis, 2nd ed., 1996.
2. K. Seidel, M. Schröder, H. Rehrauer, and M. Datcu, "Query by image content from remote sensing archives," in *IGARSS'98*, 1998.
3. S.-F. Chang, "Content-based indexing and retrieval of visual information," *IEEE Signal Processing Magazine*, **14**, pp. 45–48, 7 1997.
4. R. E. Blahut, *Principles and Practice of Information Theory*, Addison-Wesley, 1987.
5. M. Datcu, K. Seidel, and M. Walessa, "Spatial information retrieval from remote sensing images: Part A. Information theoretical perspective," *IEEE Tr. on Geoscience and Remote Sensing*, September 1998. accepted for publication.
6. M. Schröder, H. Rehrauer, K. Seidel, and M. Datcu, "Spatial information retrieval from remote sensing images: Part B. Gibbs Markov random fields," *IEEE Tr. on Geoscience and Remote Sensing*, September 1998. accepted for publication.
7. R. Duda and P. Hart, *Pattern Classification and Scene Analysis*, John Wiley & Sons, 1973.
8. Y. Wong and E. C. Posner, "A new clustering algorithm applicable to multispectral and polarimetric sar images," *IEEE Tr. on Geoscience and Remote Sensing*, **31**, (3), pp. 634–644, 1993.
9. P. Myllymaeki and H. Tirri, "Constructing computationally efficient bayesian models via unsupervised clustering," in *Probabilistic Reasoning and Bayesian Belief Networks*, ch. 14, pp. 237–251, Alfred Waller Limited, 1995.

SOME RECENT DEVELOPMENTS IN BAYESIAN STATISTICS AND ECONOMETRICS

ARNOLD ZELLNER
University of Chicago
Graduate School of Business†

Abstract. After a brief review of the past and current status of the Bayesian approach to econometrics and statistics, a review of some recent results in Bayesian forecasting and their relation to time series model-building is presented. Then some remarks on the growing importance of information theory in Bayesian analysis are provided and illustrated with selected examples. It is pointed out that various information theoretic criterion functionals are being utilized to produce models for observations, prior densities, measures of the information provided by experiments, information processing rules, including Bayes' Theorem. Information processing when the form of the likelihood function is unknown via the new Bayesian method of moments and generalized maxent procedures is discussed. Last, the future of Bayesian analysis is discussed.

1. Introduction

Since the 1950s and 1960s, there has been an upswell in the use of Bayesian analysis in many sciences and applied areas. The Bayesian approach has produced exact, finite sample solutions to estimation, prediction, testing, design, control and other problems that compare very favorably with non-Bayesian solutions, when they exist. In addition, the philosophical foundations of the Bayesian approach have come to be more widely appreciated. These developments led to the creation of the International Society for Bayesian Analysis in 1992 that has sponsored world and regional meetings and has chapters in India and Chile. Also, in 1992, the American Statistical Association's Section on Bayesian Statistical Science was founded and has been active in arranging many Bayesian sessions at the annual joint statistical meetings in the U.S. with papers published in annual Proceedings of the Bayesian Section of ASA. Also, for many years, the NBER-NSF Seminar on Bayesian Inference in Econometrics and Statistics, the Valencia Conference, the Bayes-Maxent Workshop, the Workshop on Practical Applications of Bayesian Analysis and the

†Email: arnold.zellner@gsbsun.uchicago.edu. Research financed in part by the U.S. National Science Foundation and income from the H.G.B. Alexander Endowment Fund, Graduate School of Business, University of Chicago.

W. von der Linden et al. (eds.), Maximum Entropy and Bayesian Methods, 207–216.

Bayesian Decision Analysis Section of the Institute for Operations Research and Management Science (IN FORMS) have sponsored many successful research meetings, established various prizes, including the Savage Thesis Award, the Mitchell Award, etc. and published important Bayesian volumes that have had a great impact. Further, the number of published Bayesian research papers has increased amazingly over the years and some current and past editors of leading journals are Bayesians. Given these developments, it appears that we are witnessing the beginning of a Bayesian Era in statistics and econometrics, a major event in the history of science; see, *e.g.* Zellner [1].

I suggest that this upsurge in Bayesian research, applications, and publications has occurred because there is a growing awareness of the impressive foundations of Bayesian analysis and, most importantly, that Bayesian analysis has produced many solutions to theoretical and practical problems that are as good as or better than those produced by other approaches.

While a review of all areas of application is impractical, attention is focused below on some recent developments in one important area of application, namely Bayesian forecasting. Then some uses of information theory in relation to Bayesian analysis will be presented. See, *e.g.* the review articles on information theory and Bayesian analysis by Soofi [2] [3] a collection of 48 papers on recent Bayesian research in Berry, Chaloner and Geweke [4], the information theory text by Cover and Thomas [5], applications of maxent to econometric problems in Fomby and Hill [6], Golan, Judge and Miller [7], and articles on Bayesian analysis and information theory in Zellner [8]. In these works, there are many references to the pioneering research of Edwin T. Jaynes who has made many fundamental contributions to Bayesian and maximum entropy methodology and applications; see Jaynes [9]. A final section provides a brief summary and comments on prospects for the future.

2. Bayesian Modeling and Forecasting

The problem of model formulation is present in all fields of science. How to obtain models that explain the past well and predict accurately is a major issue. Prior views play an important role in model formulation in both Bayesian and non-Bayesian approaches. Generally, some, including Jeffreys, Jaynes, and many others, start with a sophisticatedly simple model and complicate it if necessary. However others, recommend starting with a complicated, general model and testing to determine whether it is adequate or whether it can be simplified. In Zellner [10], these two approaches are considered and it is pointed out that in macroeconomic modeling and forecasting, the second approach has not been successful in modeling economies. Thus in recent work some have pursued the first approach by developing relatively simple models for individual variables and determining how well they perform in forecasting using Bayesian procedures. Then, given that we have a set of forecasting equations for key variables of a system, it is possible to develop causal systems that imply algebraically these forecasting equations. See Putnam and Quintana [11] and Zellner [12] or references to such efforts in the financial economics and macroeconomics areas. An example of this approach is work to determine the forms of forecasting equations for the annual rate of growth of total output of an economy, denoted by y_{it}, the rate of growth of country i in year t. One relatively simple model for this variable is an autoregressive-leading indicator (ARLI) model,

$$\begin{aligned} y_{it} &= \beta_{0i} + \beta_{1i}y_{it-1} + \beta_{2i}y_{it-2} + \beta_{3i}y_{it-3} + \\ &\quad \beta_{4i}S_{it-1} + \beta_{5i}S_{it-2} + \beta_{6i}M_{it-1} + \beta_{7i}WR_{t-1} + u_{it}, \end{aligned} \tag{1}$$

where S_{it} and M_{it} are the rates of growth of real stock prices and real money, respectively, for the country i in year t, W_t is the world return, the median value of the countries' S values in year t, the $\beta's$ are parameters, and u_{it} is an error term.

Data for 18 countries and Bayesian methods have been employed to estimate, perform tests of forecasting performance of the above equation, and to compare and test, using posterior odds, this equation against more elaborate versions, *e.g.*, time-varying parameter, state space versions. See Garcia-Ferrer *et al.* [13], and Min and Zellner [14], for variants of the basic model and comparative forecasting results.

In this work with ARLI models, we have used Bayesian methods to compute exact posterior and predictive density functions employing normal likelihood functions and various prior densities. In recent work, Zellner, Tobias and Ryu [15], Bayesian method of moments (BMOM) techniques have been employed to derive post data densities for parameters and predictive densities without assuming the form of the likelihood function to be known and used predictive densities to make point and turning point forecasts.

Hong [16] made draws from the posterior density of the parameters of (1) to investigate dynamic properties of the models for each of the 18 countries in our sample. That is, he made draws from the marginal trivariate Student-t posterior density for the three autoregressive parameters and for each draw computed the roots of the third order process. He found that about 85 percent of the draws led to two complex conjugate roots and one real root; that is, the probability is about 0.85 that the process has an oscillatory component. From the drawn values, he computed posterior densities for the periods and amplitudes of the oscillatory components for each country's model and found amplitudes less than one and periods of about 4 to 5 years. In addition, the implications of a uniform prior density for the three AR parameters for the properties of the roots of the process were investigated and found to be reasonable for the problem at hand. These results and the exact finite sample posterior densities for coefficients of our models and their predictive densities are very useful. In particular, means of one year ahead predictive densities that are optimal point forecasts relative to quadratic loss were employed and yielded forecasts that are better in terms of root mean squared error than those of random walk, AR(3), Nelson-Plosser and other benchmark models. The simple addition of leading indicator variables to our AR (3) model led to improved point forecasts and turning point forecasts.

Another important improvement in forecasting performance was provided by use of various forms of Bayes-Stein shrinkage. That is, the country coefficient vectors were assumed drawn from a normal density with a fixed, unknown mean vector. Using this assumption, and generalizations of it, Bayesian shrinkage forecasts were computed and generally were much better than non-shrinkage point and turning point forecasts as reported in papers in Zellner [8] and also by Putnam and Quintana [11], Quintana *et al.* [17], LeSage [18], and others. In addition to shrinkage, we also introduced various time series processes for the random coefficient vectors in each of our country models, as have Putnam and Quintana [11], Quintana *et al.* [17], and others. In the simplest case, we allowed each country's coefficient vector to follow a vector random walk, that is, $\beta_{it} = \beta_{it-1} + e_{it}$. Another model involved the assumption that, $\beta_{it} = \theta_t + v_{it}$ and $\theta_t = \theta_{t-1} + w_t$, that is the country coefficient vectors are randomly distributed about a mean that is assumed generated by a random walk process. Bayesian recursive updating procedures were employed to update these and other random parameter models and to obtain forecasts that compared favorably with those of models with fixed parameters. Also, in Min and Zellner [14] posterior odds for fixed versus random parameter models have been formulated and computed using data for 18 countries. These posterior odds can be employed to choose between models or to combine models and their forecasts. Empirical results are given on the performance of individual and combined forecasts.

Further, a new decision theoretic procedure for forecasting turning points in the rates of growth has been formulated and applied; see papers reprinted in Zellner [8] for these results which involve defining a turning point, say two successive growth rates below a third in the current year and the fourth, next year's growth rate either below that of the current year, a downturn (DT) or not below that of the current year, no downturn (NDT) Given a predictive density for next year's growth rate, it is easy to compute the probability that it is below the current year's growth rate and this is the probability of a DT, say P, and 1-P is the probability of NDT. Then given a 2x2 loss structure associated with the acts, forecast DT or forecast NDT and outcomes, DT or NDT, with zero loss associated with correct forecasts and a loss c associated with incorrect forecasts, it is direct to compute expected losses for choices of the DT and NDT forecasts, compare them and choose the forecast with the lower expected loss. For this symmetric loss structure, it is optimal to forecast DT if $P > 1/2$ and NDT if $P < 1/2$, a very simple rule. Similarly, an upturn (UT) and no upturn (NUT) can be defined as two successive growth rates above a third and the next growth above the third, an UT, or not above the third, NUT. Probabilities associated with these events can be computed using the predictive density for next year's growth rate in the sequence described above and given a 2x2 loss structure an optimal UT or NUT forecast can be derived. With a symmetric loss structure, it is the case that an UT is forecasted when the probability of an UT $> 1/2$ and NUT otherwise.

Using these procedures, about 70 percent of 158 turning points were forecasted correctly in early work and in more recent work, about 70 percent of 211 were correctly forecasted. Here too, Bayesian shrinkage or pooling of information across equations for different countries produced improved turning point forecasts which are quite a bit better than those of naïve models, *e.g.* a coin flipper or an eternal optimist who always forecasts NDT at the top and UT at the bottom, etc.; see papers in Zellner [8] and Zellner and Min [14]. In Zellner, Tobias and Ryu [15] forecasts were defined to be of three types, namely minor DT, major DT and NDT at the top and minor UT, major UT and NUT at the bottom. Computing predictive probabilities for these events from predictive densities and using a 3×3 loss structure, it was possible to derive the minimum expected loss optimal turning point forecast, say minor DT.

That new Bayesian methods have proven fruitful in our and others' forecasting work is a major new development. Further, after having found an equation such as (1), or variants of it, that have been shown to work fairly well in forecasting, there is a challenge to subject matter theorists to explain this result. In this connection, Hong [16] and Min [19] have shown how equation (1) can be derived from forms of some macroeconomic theoretical models. In Zellner and Palm [20], Palm [21], and Zellner [12], the algebraic derivations of transfer function models, like that in (1), from general multivariate time series models are described and illustrated. In our work, we use information in specific simple relations like (1) that have been shown to work reasonably well in practice to formulate theoretical models that imply relations like (1) for these variables. Thus in line with the *a priori* advice of Jeffreys [22] [23], and many others, we start simply and complicate if we have to do so, in contrast to others who start with a general model and try to simplify it by testing downward. The latter approach involves a prior view that contrasts sharply with that of Jeffreys, Jaynes and many others and has yet to produce macroeconomic models that explain the past and forecast well. For further discussion of these simplicity/complexity issues and Jeffreys' measure of complexity, see Zellner [10]. The role of prior information in model construction, *e.g.* the Jeffreys-Wrinch simplicity postulate that involves associating higher prior probabilities with sophisticatedly simple models than complicated models and measurement of complexity have received increased attention in recent years, see, *e.g.*, Keuzenkamp and McAleer [24] [25] and the references cited therein. Last, data sets such as ours involving observations on a collection of individual units through time are frequently encountered in many areas, including clinical trials, government data bases, etc. and called longitudinal or panel data sets. It is the case that Bayesian shrinkage and innovative computational techniques, *e.g.* quadrature, importance function Monte Carlo, Markov Chain-Monte Carlo (MCMC) and other numerical integration techniques have been employed successfully in many different fields of application; see, *e.g.* Gelman, *et al.* [26].

3. Information Theory and Bayesian Analysis

If one views scientific induction as learning from past data and experience to produce generalizations or models that explain the past and predict as yet unobserved data, as Jeffreys [22] [23] does, it is clear that we are involved in information processing, also recognized by Jaynes [27]. In Zellner [28], it was pointed out that traditional Bayesian information processing involves two inputs, a prior density $p(\theta)$ and a data density $f(y \mid \theta)$, and two outputs, a post data density for the parameters, $p(\theta \mid D)$ where D denotes the data and prior information and a marginal density for the observations, $h(y)$. An information criterion functional, the difference between the output and input information, was minimized with respect to the output parameter density function, g subject to it being proper and the result is,

$$g(\theta \mid D) = \pi(\theta) f(y \mid \theta)/h(y). \tag{2}$$

namely Bayes' Theorem, the Bayesian learning model. Further, it was established that when (2) is employed, output information = input information, that is, no information is added or lost. That is, Bayes' Theorem in (2) is a 100 percent efficient information processing rule. Jaynes [27] commented as follows on these developments, " ...entropy has been recognized as part of probability theory since the work of Shannon (1 948)...and the usefulness of entropy maximization...is thoroughly established...This makes it seem scandalous that the exact relation of entropy to the other principles of probability theory is still rather obscure and confused. But now we see that there is, after all, a close connection between entropy and Bayes's theorem. Having seen this start, other such connections may be found, leading to a more unified theory of inference in general. Thus in my opinion, Zellner's work is probably not the end of an old story, but the beginning of a new one."

As part of the "new story," Zellner [29] has considered the prior and sample information inputs to be of differing quality in deriving an optimal information processing rule. Also, one can weight the prior information and the sample information differently in deriving an optimal IPR which results in having the optimal output density function for the parameters being proportional to the prior raised to the power w times the likelihood function raised to the power $1 - w$, with $0 < w < 1$. Then there are dynamic information processing problems to solve that involve generation and processing of sequential information given various costs of acquiring new information and of adjusting to it. As Jaynes rightly pointed out, there are indeed many new stories to be developed. See also, Cover and Thomas [5], Bernardo and Smith [30], Soofi [2] [3], and Zellner [8] for recent considerations of information processing and new and old measures of the information provided by an experiment. The new definition in Zellner [8], the information in the joint density of the observations and parameters minus the information in the prior is operational and more comprehensive than the old definition.

Another area of great activity has been research on what to do when the form of the likelihood is unknown. Of course, maxent has been and is being employed to produce models for observations in many areas, as Jaynes [27] and Zellner [29] have pointed out. Recently, Stutzer [31] has shown how maxent can be used to produce "risk neutral" densities in financial economics while Golan, Judge and Miller [7], Soofi [2], and Zellner [8] have shown how to produce many statistical and econometric models by maxent. These maxent densities involve side conditions relating to the sampling properties of data. In addition, various information criteria have been utilized to produce prior densities for parameters; see, *e.g.* Kass and Wasserman [32], Soofi [2], and Zellner [8] for reviews and comparisons of results yielded by the Jaynes, Jeffreys, Bernardo and Berger, Zellner and other approaches. Jaynes maximizes the invariant entropy of the prior while Jeffreys utilizes the Fisher information matrix to produce his invariant priors. Bernardo and Berger approximately maximize the difference between the information in the posterior and prior densities while Zellner maximizes the difference between the average information in the data density minus the information in the prior subject to side conditions. Some outstanding issues are which criterion functional to use? Should the form of the prior depend on the form of the likelihood function? How much invariance should be used? Should priors be proper or improper? How to obtain a common prior for the same parameter(s) appearing in two or more different likelihood functions? How to derive prior odds, etc.? See Kass and Wasserman [32] and Zellner [8] for discussion of these issues, examples and references. While much progress has been made in producing priors that represent the prior information that we do or do not have, there is still room for improvement.

Last, there is the problem of what to do when there is not enough information available to formulate a likelihood function. Without a likelihood function, it is impossible to employ the traditional Bayesian approach, based on Bayes' theorem. While some have attempted to obtain the likelihood function's form from the data by use of empirical likelihood and bootstrap methods, see *e.g.* Boos and Monahan [33], others have made moment assumptions about the realized error terms of a model for the given data and from these have deduced implied parameter moments. For example, if $y_i = \theta + u_i$ for $i = 1, 2, \cdots, n$, are n observed times to failure, then $\overline{y} = \theta + \overline{u}$ which is a relation for the means of the $y's$ and $u's$, and the given mean of the $y's$ satisfies $\overline{y} = E(\theta \mid D) + E(\overline{u} \mid D)$, where E is a subjective expectation operator and D denotes given data and background information about the form of the relation, namely that it is linear, etc. Then, if the measurements have been made properly with no left out variables, outliers, etc., the assumption that $E(\overline{u} \mid D) = 0$ can be made which implies $E(\theta \mid D) = \overline{y}$ and the maxent proper post data density for the parameter, subject to this mean constraint is the exponential density, $f(\theta \mid D) = (1/\overline{y}) \exp(-\theta/\overline{y})$. This is a Bayesian method of moments (BMOM) post data density based on one moment condition. See, *e.g.*, Zellner [8] [34], Green and Strawderman [35], and Tobias and Zellner [36] for applications of the approach, using additional moment restrictions on the realized errors, to a broad range of univariate and multivariate regression and other models. With additional restrictions on future error terms' moments, maxent predictive densities have been derived and used in Bayes' factors to compare alternative models in Tobias and Zellner [36]. In Golan, Judge and Miller [7] [37] the parameter and realized error term spaces are discretized and moments of realized errors and parameters are obtained and used in a maxent framework to get proper post data densities for parameters and for future observations. A number of applications are presented. Whether it is better to proceed in terms of a few parameters' moments as in the BMOM approach or many parameters associated with the discretization of the parameter space in empirical likelihood, nonparametric, and the Golan, Judge, Miller [7] [37] generalized entropy approaches is an issue that deserves attention. As Jeffreys and many others have recognized, introducing many parameters may lead to a good fit but will not generally lead to good predictions. It is possible to use Bayes' factors, see, *e.g.* Kass and Raftery [32] and Tobias and Zellner [36], and other model selection procedures and actual performance in prediction and forecasting to attempt to resolve this important range of issues.

4. Summary and Conclusions

As is evident from what has been presented above, information theory has and is playing an important role in Bayesian analysis in statistics and econometrics in producing prior densities, models for observations, information processing rules and post data densities for parameters and future observations when the form of the likelihood function is not known, etc.. As pointed out above, Bayesian analysis has been very effective in the past in solving inference and decision problems in all areas of science and application. With further attention to new developments in information theory and maxent, many of them inspired by Jaynes' work, Bayesian analysis will be applicable and provide good solutions to an even wider range of information processing and decision problems in the future.

References

1. A. Zellner, "A Bayesian era," in *Bayesian Statistics 3*, J. Bernardo, ed., Oxford University Press, London, 1988.
2. E. Soofi, "Information theory and Bayesian statistics," in *Bayesian Analysis in Statistics and Econometrics*, D. Berry, K. Chaloner, and L. Geweke, eds., pp. 179–189, Wiley, New York, 1996.
3. E. Soofi, "Information theoretic regression techniques," in *Advances in Econometrics: Applying Maximum Entropy to Econometric Problems*, T. Fomby and R. Hill, eds., pp. 25–83, Jai Press, London, 1997.
4. D. Berry, K. Chaloner, and J. Geweke, *Bayesian Analysis in Statistics and Econometrics*, Wiley, New York, 1996.
5. T. Cover and J. Thomas, *Elements of Information Theory*, Wiley, New York, 1991.
6. T. Fomby and R. Hill, *Advances in Econometrics: Applying Maximum Entropy to Econometric Problems*, Jai Press, London, 1997.
7. A. Golan, J. Judge, and D. Miller, *Maximum Entropy Econometrics*, Wiley, London, 1996.
8. A. Zellner, *Bayesian Analysis in Econometrics and Statistics: The Zellner View and Papers*, Edward Elgar Publ. Ltd., Cheltenham, UK, 1997.
9. E. Jaynes, "The intuitive inadequacy of classical statistics," *Epistemologia (Special Issue on Probability, Statistics, and Inductive Logic)*, pp. 43–74, 1984.
10. A. Zellner, "Keep it sophisticatedly simple," in *Simplicity, Inference, and Econometric Modelling*, H. Keuzenkampf, M. McAleer, and A. Zellner, eds., Cambridge University Press, Cambridge, 1998.
11. B. Putnam and J. Quintana, "The evolution of Bayesian forecasting models," in *Applying Quantative Discipline to Asset Allocation*, B. Putnam, ed., Euromoney Publications, London, 1995.
12. A. Zellner, "Time series analysis, forecasting and econometric modelling: The structural econometirc modelling, time series analysis (semtsa) approach," *Journal of Forecasting*, pp. 215–233, 1994.
13. A. Garcia-Ferrer, R. Highfield, F. Palm, and A. Zellner, "Macroeconomic forecasting using pooled international data," *Journal of Business and Economic Statistics*, pp. 53–67, 1987.
14. C.-K. Min and A. Zellner, "Bayesian and non-Bayesian methods for combining models and forecasts with applications to forecasting international growth rates," *Journal of Econometrics*, pp. 89–118, 1993.
15. A. Zellner, J. Tobias, and H. Ryu, "Bayesian method of moments analysis of time series models with an application to forecasting turning points in output growth rates," *Estadistica*, 1999.
16. C. Hong, *Forecasting Real Output Growth Rates and Cyclical Properties of Models: A Bayesian Approach*, Ph.D. thesis, Department of Economics, U. of Chicago, Chicago, IL., 1989.
17. L. Quintana, V. Chopra, and B. Putnam, "Global asset allocation: Stretching returns by

shrinking forecasts," in *Proceedings of the Section on Bayesian Statistical Science of the American Statistical Association*, 1995.

18. J. LeSage, "A comparison of techniques for forecasting turning points in regional employment activity," in *Bayesian Analysis in Statistics and Econometrics*, D. Berry, K. Chaloner, and L. Geweke, eds., pp. 41–51, Wiley, New York, 1996.
19. C.-K. Min, *Economic Analysis and Forecasting of International Growth Rates Using Bayesian Techniques*, Ph.D. thesis, Department of Economics, U. of Chicago, Chicago, IL., 1992.
20. A. Zellner and F. Palm, "Time series analysis and simultaneous equation econometric models," *Journal of Econometrics*, pp. 17–54, 1974.
21. F. Palm, "Structural econometric modelling and time series analysis: An integrated approach," in *Applied Time Series Analysis of Economic Data*, A. Zellner, ed., pp. 199–233, Bureau of the Census, U.S. Dept. of Commerce, Washington D.C., 1983.
22. H. Jeffreys, *Scientific Inference*, U. of Cambridge Press, Cambridge, 1957.
23. H. Jeffreys, *Theory of Probability*, Oxford University Press, London, 1967.
24. H. Keuzenkamp and M. McAleer, "Simplicity, scientific inference and econometric modelling," *The Economic Journal*, pp. 1–21, 1995.
25. H. Keuzenkamp and M. McAleer, "The complexity of simplicity," *Mathematics and Computers in Simulation*, pp. 553–561, 1997.
26. A. Gelman, J. Carlin, H. Stern, and D. Rubin, *Bayesian Data Analysis*, Chapman & Hall, 1994.
27. E. Jaynes, "Comment," *The American Statistician*, pp. 280–281, 1988.
28. A. Zellner, "Optimal information processing and Bayes's theorem," *The American Statistician*, pp. 278–280, 1988.
29. A. Zellner, "Bayesian methods in entropy and econometrics," in *Maximum Entropy and Bayesian Methods*, T. Grandy and L. Schick, eds., pp. 17–31, Kluwer Academic Publishers, Dordrecht, 1991.
30. J. Bernardo and A. Smith, *Bayesian Theory*, Wiley, New York, 1994.
31. M. Stutzer, "An information-theoretic index of risk in financial markets," in *Bayesian Analysis in Statistics and Econometrics*, D. Berry, K. Chaloner, and L. Geweke, eds., pp. 191–200, Wiley, New York, 1996.
32. R. Kass and L. Wasserman, "The selection of prior distributions by formal rules," *Journal of the American Statistical Association*, pp. 1343–1370, 1996.
33. D. Boos and J. Monohan, "Bootstrap methods using prior information," *Biometrika*, pp. 77–83, 1986.
34. A. Zellner, "The Bayesian method of moments (bmom): theory and applications," in *Advances in Econometrics: Applying Maximum Entropy to Econometric Problems*, T. Fomby and R. Hill, eds., pp. 85–105, Jai Press, London, 1997.
35. E. Green and W. Strawderman, "A Bayesian growth and yield model for slash pine plantations," *Journal of Applied Statistics*, pp. 285–299, 1996.
36. J. Tobias and Z. Zellner, "Further results on the Bayesian method of moments analysis of the multiple regression model," in *1997 Proceedings of the Section on Bayesian Statistical Science of the American Statistical Association*, 1997.
37. A. Golan, J. Judge, and D. Miller, "The maximum entropy approach to estimation and inference: An overview," in *Advances in Econometrics: Applying Maximum Entropy to Econometric Problems*, T. Fomby and R. Hill, eds., pp. 3–24, Jai Press, London, 1997.

John Skilling and the Erickson family.

CONVERGENT BAYESIAN FORMULATIONS OF BLIND SOURCE SEPARATION AND ELECTROMAGNETIC SOURCE ESTIMATION

K. H. KNUTH AND H. G. VAUGHAN, JR.
Department of Neuroscience
Albert Einstein College of Medicine
Bronx NY 10461 USA†

Abstract. We consider two areas of research that have been developing in parallel over the last decade: blind source separation (BSS) and electromagnetic source estimation (ESE). BSS deals with the recovery of source signals when only mixtures of signals can be obtained from an array of detectors and the only prior knowledge consists of some information about the nature of the source signals. On the other hand, ESE utilizes knowledge of the electromagnetic forward problem to assign source signals to their respective generators, while information about the signals themselves is typically ignored.

We demonstrate that these two techniques can be derived from the same starting point using the Bayesian formalism. This suggests a means by which new algorithms can be developed that utilize as much relevant information as possible. We also briefly mention some preliminary work that supports the value of integrating information used by these two techniques and review the kinds of information that may be useful in addressing the ESE problem.

1. Introduction

Determination of the spatial and temporal behavior of neural activity using non-invasive methods is critical to the study of human brain function. While there are several experimental techniques that provide information on neural activity, no single method currently provides information with the spatial and temporal resolution necessary to delineate the neural processes that underlie human cognition and sensorimotor function. Electroencephalography (EEG) and magnetoencephalography (MEG) provide sub-millisecond temporal resolution of neuroelectric current flow. However, the electric potentials and magnetic fields recorded by the detectors on the surface of the head consist of a linear superposition of the signals emitted by the active neural sources. It is well known that in the absence of additional constraints, solutions to the electromagnetic inverse problem, which consist of the determination of the underlying neural current distributions given the surface potentials

†Email: kknuth@balrog.aecom.yu.edu

W. von der Linden et al. (eds.), Maximum Entropy and Bayesian Methods, 217–226.

and fields, are not unique. The problem increases in difficulty as the number of neural sources increases. The limited ability of current methods to accurately and consistently separate and localize the neuro-electromagnetic sources has impaired confidence in the value of surface-recorded electromagnetic data for examining the dynamics of the underlying neural activity that generates them. We propose an approach to the electromagnetic inverse problem that combines information that is typically utilized in each of two methods that have been developed in parallel over the past decade: namely, blind source separation (BSS) and electromagnetic source estimation (ESE). The additional prior information that can be brought to bear through a composite approach should facilitate the development of algorithms that can achieve more accurate definition of the intracranial signal sources and the source waveforms.

Much progress has been made in the last decade to understand and attain the segregation of mixed signals using blind source separation (BSS) techniques. The success of the application of BSS toward specific problems is often surprising given the apparently minimal amount of information used to achieve separation. This minimal information usually consists of the assumptions of independence of the source signals and that the source amplitude densities can be described by a particular class of probability distributions.

Electromagnetic Source Estimation (ESE) attempts to locate and characterize the neural signals responsible for the electric potentials and magnetic fields recorded on the surface of the head during synchronized neural activity. Most ESE techniques utilize information regarding the electric properties of typical neural generators, the locations of the detectors, and the properties of propagation of electric currents and magnetic fields from the sources to the detectors. While some anatomical and physiological information may be incorporated in ESE algorithms, typically very little information about the signals themselves is included. Source signal solutions are constrained to be decorrelated or statistically independent, but no additional information about the source amplitude densities, typical amplitudes of the various sources, and stochastic variations in the source waveforms is included.

2. The Bayesian Approach

The general problems of source separation and source localization are inherently inference problems because there is not sufficient information to deduce a solution. Bayesian inference allows one to calculate the probability that a given hypothesis, or model, is correct given the data and any prior information. We demonstrate that the solutions to both of these problems can be derived from the same basic starting point, Bayes' Theorem. These derivations will diverge when we focus on particular parameters to model the physical situation, incorporate additional information, and make simplifying assumptions.

Given a model of a physical problem of interest and some additional data, we can use Bayes' Theorem to describe how this additional information affects our belief as to whether a particular model provides an accurate description:

$$P(model \mid data, I) = \frac{P(data \mid model, I)P(model \mid I)}{P(data \mid I)} \tag{1}$$

where I represents any prior information we may have.

In the case of the neural source estimation problem, there are several parameters that can be used to describe the model: the source positions $\boldsymbol{p}$ and orientations $\boldsymbol{q}$, the source signals $\boldsymbol{s}(t)$, and the mixing matrix $\boldsymbol{A}$, which describes how the source signals are mixed when recorded by the detectors. The data consist of the electromagnetic signals recorded by our detectors. Representing the recorded mixed signals as $\boldsymbol{x}(t)$, we rewrite Equation (1) as a proportionality

$$P(\boldsymbol{p}, \boldsymbol{q}, \boldsymbol{A}, \boldsymbol{s}(t) \mid \boldsymbol{x}(t), I) \propto P(\boldsymbol{x}(t) \mid \boldsymbol{p}, \boldsymbol{q}, \boldsymbol{A}, \boldsymbol{s}(t), I)\, P(\boldsymbol{p}, \boldsymbol{q}, \boldsymbol{A}, \boldsymbol{s}(t) \mid I)\,. \tag{2}$$

The prior probability on the right-hand side can be factored into several terms

$$\begin{aligned} P(\boldsymbol{p}, \boldsymbol{q}, \boldsymbol{A}, \boldsymbol{s}(t) \mid \boldsymbol{x}(t), I) \quad \propto \quad & P(\boldsymbol{x}(t) \mid \boldsymbol{A}, \boldsymbol{s}(t), I)\, P(\boldsymbol{A} \mid \boldsymbol{p}, \boldsymbol{q}, I) \\ & \times P(\boldsymbol{q} \mid \boldsymbol{p}, I)\, P(\boldsymbol{p} \mid I)\, P(\boldsymbol{s}(t) \mid I)\,. \end{aligned} \tag{3}$$

The first term on the right, the likelihood, has been simplified since the likelihood of the data $\boldsymbol{x}(t)$ depends explicitly on the matrix $\boldsymbol{A}$ and the source signals $\boldsymbol{s}(t)$ such that the mixing is linear and instantaneous, $\boldsymbol{x}(t) = \boldsymbol{A}\boldsymbol{s}(t)$. The second term, the prior $P(\boldsymbol{A} \mid \boldsymbol{p}, \boldsymbol{q}, I)$, describes the dependency of the mixing matrix on the positions and orientations of the sources. The second prior $P(\boldsymbol{q} \mid \boldsymbol{p}, I)$ describes the probability of the orientations of the sources given their positions. Prior knowledge regarding possible source positions is expressed in the third prior $P(\boldsymbol{p} \mid I)$. Since the signal propagation does not depend on the source amplitude and the source behaviors are at present not known to depend on source position, the source signal prior $P(\boldsymbol{s}(t) \mid I)$ depends only on our prior expectations of the nature of the signals themselves.

2.1. BAYESIAN FORMULATION OF BSS

BSS endeavors to separate the source signals without recourse to information on their spatial distribution and relation to the detectors. Two basic approaches are used in blind source separation: algebraic approaches, which are higher-order generalizations of singular value decomposition (SVD) or principal component analysis (PCA); and information-theoretic approaches, which utilize maximum entropy, maximum likelihood, or minimum mutual information.

To illustrate the application of the Bayesian formulation to the latter class of approaches, we focus on a popular technique, the Bell-Sejnowski Independent Component Analysis (ICA) algorithm [1]. The algorithm is cast in the form of a neural network that takes the mixed signals, performs a nonlinear transformation on them and then separates them by maximizing the entropy of the output signals so that they are statistically independent. Viewed differently, this algorithm uses information about the probability distribution of source amplitudes (via the nonlinear transformation) and the statistical independence of the source signals to separate the signals.

Derivation of the Bell-Sejnowski ICA algorithm using the Bayesian methodology has been described elsewhere [2,3] and after the initial assumptions are made, the remainder of the derivation is similar to those presented by MacKay [4] and Pearlmutter and Parra [5]. We will outline the derivation, paying special attention to the particular assumptions that distinguish the results from ESE.

BSS assumes that we know nothing about the sources or the nature of the mixing. This lack of information can be expressed by assigning uniform distributions to the priors, $P(\boldsymbol{A} \mid \boldsymbol{p}, \boldsymbol{q}, I)$, $P(\boldsymbol{q} \mid \boldsymbol{p}, I)$ and $P(\boldsymbol{p} \mid I)$ that extend over a large enough range to account for all possible values of the model parameters. We then marginalize over the model parameters $\boldsymbol{p}$ and $\boldsymbol{q}$ to obtain

$$P(\boldsymbol{A}, \boldsymbol{s}(t) \mid \boldsymbol{x}(t), I) \propto P(\boldsymbol{x}(t) \mid \boldsymbol{A}, \boldsymbol{s}(t), I)\, P(\boldsymbol{s}(t) \mid I)\,, \tag{4}$$

where all constants are absorbed by the proportionality.

If we believe the external noise level to be sufficiently small then the inverse of the mixing matrix $\boldsymbol{A}$ can be used to obtain the source signals. The source signals $\boldsymbol{s}(t)$ can be treated as nuisance parameters and removed by marginalizing over $\boldsymbol{s}(t)$

$$P(\boldsymbol{A} \mid \boldsymbol{x}(t), I) \propto \int d\boldsymbol{s} P(\boldsymbol{x}(t) \mid \boldsymbol{A}, \boldsymbol{s}(t), I)\, P(\boldsymbol{s}(t) \mid I)\,. \tag{5}$$

To reflect the statistical independence of the sources, the prior $P(\boldsymbol{s}(t) \mid I)$ is written as a product of prior probabilities for each source. Typically a hyper-Gaussian probability density, such as the derivative of a sigmoid function, is assigned to describe the source amplitude density of the source signals. Of course, if more is known about the dynamical behavior of the source signals, one can assign more detailed priors. Finally, if we assume that the mixing is linear, instantaneous and relatively noise-free, as described above, delta functions can be assigned to the likelihood. This results in

$$P(\boldsymbol{A} \mid \boldsymbol{x}(t), I) \propto \int d\boldsymbol{s} \prod_i \delta\left(x_i - A_{ik} s_k\right) \prod_l p_l(s_l) \tag{6}$$

where the Einstein summation convention is used to denote matrix multiplication. The delta functions allow one to evaluate the multi-dimensional integral. The most probable value for the matrix $\boldsymbol{A}$, is found by maximizing the logarithm of the probability

$$P(\boldsymbol{A} \mid \boldsymbol{x}(t), I) = -\log\left|\det \boldsymbol{A}\right| + \sum_l p_l\left(A_{lk}^{-1} x_k\right) + C. \tag{7}$$

The Bell-Sejnowski ICA algorithm is constructed by using a stochastic gradient search to estimate the separation matrix, $\boldsymbol{W} = \boldsymbol{A}^{-1}$, which maximizes Equation (7). This algorithm has been shown to work very well in separating mixed signals where the prior source amplitude densities provide an adequate representation of our knowledge about the source behavior.

2.2. BAYESIAN FORMULATION OF ESE

In this section we proceed with a derivation of one of the most common algorithms used in ESE. This algorithm takes a very different approach to solving the source separation problem by utilizing information ignored by ICA and ignoring information used by ICA. We focus on the model parameters describing the positions and orientations of the sources as well as the source signals. Additional information takes the form of knowledge regarding the propagation of the signals from the sources to the detectors. This consists of a description of the forward model, expressed as a function, $A_{ij} = F(\boldsymbol{d}_i, \boldsymbol{p}_j, \boldsymbol{q}_j)$ determined either analytically or numerically, that determines the matrix elements of $\boldsymbol{A}$ given the source and detector positions and orientations, and a model of the volume conductor.

We begin with Equation (3), and assign a delta function to the prior probability $P(\boldsymbol{A} \mid \boldsymbol{p}, \boldsymbol{q}, I)$,

$$P(\boldsymbol{p}, \boldsymbol{q}, \boldsymbol{A}, \boldsymbol{s}(t) \mid \boldsymbol{x}(t), I) \propto P(\boldsymbol{x}(t) \mid \boldsymbol{A}, \boldsymbol{s}(t), I)\, P(\boldsymbol{q} \mid \boldsymbol{p}, I)\, P(\boldsymbol{p} \mid I)\, P(\boldsymbol{s}(t) \mid I) \times \prod_{i,j} \delta\left(A_{ij} - F(\boldsymbol{d}_i, \boldsymbol{p}_j, \boldsymbol{q}_j)\right). \tag{8}$$

Treating the matrix $\boldsymbol{A}$ as a nuisance parameter we marginalize over $\boldsymbol{A}$,

$$P(\boldsymbol{p}, \boldsymbol{q}, \boldsymbol{s}(t) \mid \boldsymbol{x}(t), I) \propto \int d\boldsymbol{A}\, P(\boldsymbol{x}(t) \mid \boldsymbol{A}, \boldsymbol{s}(t), I) \prod_{i,j} \delta\left(A_{ij} - F(\boldsymbol{d}_i, \boldsymbol{p}_j, \boldsymbol{q}_j)\right) \times P(\boldsymbol{q} \mid \boldsymbol{p}, I)\, P(\boldsymbol{p} \mid I)\, P(\boldsymbol{s}(t) \mid I). \tag{9}$$

Unlike the previous derivation, we assign a Gaussian probability density to the likelihood function. This Gaussian density expresses that fact that we do not believe the experimental data to be precisely described by the model, and that we know the value of the expected squared deviation between the predicted and experimentally obtained data. Focusing on the integral on the right-hand side of Equation (9), we write

$$\int d\boldsymbol{A} \prod_{t,k} \exp\left[-\frac{(x_{kt} - A_{kl}s_{lt})^2}{2\sigma_k^2}\right] \prod_{i,j} \delta\left(A_{ij} - F(\boldsymbol{d}_i, \boldsymbol{p}_j, \boldsymbol{q}_j)\right). \tag{10}$$

where we explicitly express the vector of time series data $\boldsymbol{x}(t)$ as x_{kt} where k denotes the kth data channel and t denotes the tth time point. Evaluating the integral results in

$$P(\boldsymbol{p}, \boldsymbol{q}, \boldsymbol{s}(t) \mid \boldsymbol{x}(t), I) \propto P(\boldsymbol{q} \mid \boldsymbol{p}, I)\, P(\boldsymbol{p} \mid I)\, P(\boldsymbol{s}(t) \mid I) \times \prod_{t,k} \exp\left[-\frac{(x_{kt} - F_{kl}s_{lt})^2}{2\sigma_k^2}\right]. \tag{11}$$

where F_{kl} represents the function $F(\boldsymbol{d}_k, \boldsymbol{p}_l, \boldsymbol{q}_l)$. Notation can be further simplified by introducing, $\widehat{x}_{kt} = F_{kl}s_{lt}$.

The three remaining priors are all very important in that they describe our prior knowledge regarding the positions and orientations of the sources and details of the source signals. The first two priors, if not entirely neglected, are assigned functions that constrain solutions to lie on a cortical surface with a particular orientation. The prior that represents information about source signals is typically neglected. In this derivation, we will assume complete ignorance by assigning a uniform probability distribution to each of the three priors. These uniform distributions take the form of constants, which are absorbed by the proportionality. Since the ultimate goal is to find the most probable solution, we look for the parameters that maximize the logarithm of the probability

$$\log P\left(\boldsymbol{p}, \boldsymbol{q}, \boldsymbol{s}(t) \mid \boldsymbol{x}(t), I\right) = -\sum_k \sum_t \frac{(x_{kt} - \widehat{x}_{kt})^2}{2\sigma_k^2} + C, \tag{12}$$

which is equivalent to minimizing the "cost function"

$$\chi^2 = \sum_k \sum_t \frac{(x_{kt} - \widehat{x}_{kt})^2}{\sigma_k^2}. \tag{13}$$

This is the chi-squared cost function that is typically used to evaluate source models in ESE. When the variances associated with the different detector channels are identical, the expression above reduces to a minimization of the least-squared error between the data and their predicted values. It is important to remember that this result is derived from the assumptions of linear instantaneous mixing, ignorance of possible source locations and orientations, statistical independence of the source activity, and a Gaussian likelihood that the data are predicted by the model.

3. Incorporation of prior information in ESE

It is important to compare the derivations of the solutions to the problems of BSS and ESE and pay special attention to the information that is used through the assignment of the relevant prior probabilities. In most cases, the researcher has several different kinds of additional information or prior knowledge about the specific physical problem that can be naturally integrated using the Bayesian approach. In a previous paper, we derived an algorithm similar to ICA that includes information about the signal propagation, the original source positions and the signal properties [3]. Specifically, we examined an artificial problem where the signal amplitudes have a one over distance squared fall-off, and the prior source positions are described by three-dimensional Gaussian probability densities. The algorithm was shown to separate signals that are difficult to separate using ICA in conjunction with a hyper-Gaussian source density. While this preliminary work demonstrated that separation is possible using additional accurate information in conjunction with inaccurate information regarding source behavior, we stress that the information utilized should be as accurate as possible. In addition, some types of information may be more useful than others. Studies must be performed to understand the relative importance of the different types of information.

We now discuss several sources of information of specific relevance to neuro-electrophysiologic source separation. It is convenient to divide the prior information into two classes of variables, namely those associated with the source model and those related to the volume-conductive head model. The former includes information on the time varying electromagnetic signals themselves, as well as information on the number, geometry and spatial location of the signal sources. The latter includes information on the shape of the brain and its covering and the conductivities of the various tissues.

We first consider information about the source signals themselves, which are ordinarily represented by time-varying current dipole approximations of the complex patterns of synchronous neural activity within a macroscopic source. The dipolar nature of the source model allows constraints on directionality imposed by the orientation of the dipole, and on maximum amplitudes and temporal amplitude sequences of the dipole moments as well. Due to their small amplitude relative to the ongoing spontaneous brain activity, the signals of interest, so-called "event-related" potentials or fields, are enhanced relative to the background activity by signal averaging. These event-related signals are typically transients that are time-locked to an external event, usually a stimulus or motor response. Although the residual background activity is generally called "noise", its spatial and temporal properties have a statistical structure that can be empirically determined. As yet, little attention has been paid to characterizing the statistics of the residual spontaneous activity in the event-related averaged signals of interest. This residual "noise" is, however, a major source of error in ESE, and needs to be dealt with in a more direct fashion.

As previously noted, ESE approaches have rarely included information about the signals themselves. In another paper [2], we discussed the inability of ICA to separate EEG/MEG signals due to the fact that the hyper-Gaussian densities used by the algorithm do not accurately represent EEG/MEG amplitude densities. These are typically multi-modal and thus require many parameters to model. We require prior probability densities that more accurately describe what is known about the source waveforms. These priors can also describe dynamical information through the assignment of prior probabilities conditional on the derivatives of the signals.

Information regarding the possible positions and orientations of neural sources can be obtained with millimeter resolution from MR images. It is known that the equivalent dipoles that represent the net extracellular current flow within a cortical element are normal to the cortical surface due to the cellular architecture of the cortex. This provides a strong constraint on both cortical source locations and orientations [6–8]. The Bayesian formulation is particularly advantageous in providing a means by which this information can be included as accurately as possible. Thus, the cortical surface can be assigned rather precise priors that reflect our knowledge of the cortical geometry, whereas deeper, less-understood structures can be assigned uniform priors that reflect our ignorance. In this way, the entire brain is treated in a uniform manner with different probabilities reflecting knowledge specific to each region.

Determination of the number of neural sources, also known as model order, has been an extremely difficult problem in source estimation. While the number of sources is typically estimated from the number of significant singular values of the SVD of the data matrix or its covariance matrix, it is known that this provides a minimum estimate of the model order. The Bayesian approach provides a natural means by which one can perform model order comparisons. By explicitly calculating the probabilities of each model, the effect of changes in model order can be quantitatively assessed.

Turning to the head model variables, the finite conductive volume of the head constrains the equations describing the potentials and fields detected by surface recordings (i.e. the forward problem). Typically used are head models in which the brain, extra-cerebral fluid, skull and scalp are modeled as concentric spheres. These simplified volume conductor models, coupled with the current dipole model of the neural sources, permit analytic solutions that describe the potentials and fields on the surface of the head. More realistic head models take into account individual brain and skull geometry as indicated by magnetic resonance images (MRI). In these models, numerical techniques are required to estimate the surface effects of a neural source. Conductivities of the different tissues are estimated from sparse empirical data and thus are not accurately known.

While the importance of incorporating all possible sources of information to aid in solving the neuro-electromagnetic inverse problem is well appreciated, the methodology is ordinarily limited to second-order statistics. An excellent work by Dale and Sereno [8] outlines in detail the incorporation of EEG, MEG and MRI using second-order techniques. When Gaussian priors accurately describe our knowledge, these techniques are optimal. However, when this is not the case these second-order techniques are unduly restrictive.

We comment on how these ideas can be better understood and improved when examined from the Bayesian viewpoint. One of the most common misconceptions is that the Gaussian likelihood implies the existence of a physical signal consisting of independent, additive Gaussian noise. Recall that the assignment of the likelihood represents what is known about the predictive capability of a model. The assignment of the Gaussian likelihood in Equation (10) is appropriate when our prior information consists of the value of the expected squared deviation between the predicted and experimentally observed data values. In short, the variance in the Gaussian density represents our uncertainty that the model predicts the data. This uncertainty could be due to instrument noise, signals from additional or unexpected neural sources, or simplifying approximations in the description of the physical situation. Typically in the case of stimulus-elicited neuro-electromagnetic signals, instrument noise and neural background activity are estimated from the prestimulus baseline signals. While instrument noise may not exhibit correlations across channels, neural background activity often does. Detailed studies regarding the nature of the neural background will assist in a better description of the uncertainties due to these signals. More importantly, model mis-specification contributes uncertainties to the likelihood via the assignment of the delta function prior, describing a naive confidence in the approximate transfer functions (see also Dale and Sereno [8]). Research performed on the effects of model mis-specification [9,10] can aid in the assignment of a more accurate mixing matrix prior, which better describes the known limitations of our approach.

It is well known that EEG and MEG do not provide identical information about neuroelectric source currents, even though they reflect the same neural events. MEG is thought to represent primarily intracellular currents, whereas EEG is generated by extracellular volume currents. MEG is also insensitive to currents generated in radial dipolar sources, whereas EEG represents sources in all orientations. Thus, each method has its significant strengths and weakness as indices of intracranial neural activty. EEG and MEG data can be combined by augmenting the EEG mixing matrix with the MEG mixing matrix. Known differences in instrument noise and background activity recorded using these two techniques can be represented by differences in their corresponding likelihoods. In addition, information about model mis-specification for the two techniques can be accounted for in their respective mixing matrix priors.

Integration of information obtained from functional imaging, which is based on metabolic effects due to the activity of neural tissue, with ESE has received considerable recent attention. Due to the presumably better spatial accuracy of metabolic imaging methods, it has been suggested that electromagnetic sources might be constrained in location by fMRI or PET activation. These suggestions must be viewed with considerable caution inasmuch as the metabolic effects, which are visualized as differences between two states of brain activation, are by no means directly comparable to the electromagnetic indices of neural activity. Not only is the time scale vastly different - seconds versus milliseconds, but the quantitative coupling between the measures has not yet been defined. Metabolic imaging and electromagnetic source imaging techniques may index sufficiently different physiological processes, so that information obtained using one technique may not provide useful constraints on solutions based on the other technique.

In conclusion, we have laid out a Bayesian framework within which the ESE problem may be profitably explored, and provides both the means of incorporating a broad range of information with varying degrees of confidence, and for testing the probability of various alternative solutions to this inverse problem.

4. Acknowledgements

This work was supported by NIH 5P50 DC00223, NIH 5P30 HD01799, and NIH NIDCD 5 T32 DC00039-05.

References

1. A. J. Bell and T. J. Sejnowski, "An information-maximization approach to blind separation and blind deconvolution," *Neural Comp.*, **7**, pp. 1129–1159, 1995.
2. K. H. Knuth, "Difficulties applying recent blind source separation techniques to EEG and MEG," in *Maximum Entropy and Bayesian Methods*, G. J. Erickson, ed., Kluwer Academic, Dordrecht, To be published 1998.
3. K. H. Knuth, "Bayesian source separation and localization," in *SPIE Proceedings: Bayesian Inference for Inverse Problems*, A. Mohammad-Djafari, ed., vol. 3459, To be published 1998.
4. D. J. C. MacKay, "Maximum likelihood and covariant algorithms for independent component analysis." Draft Paper, http://wol.ra.phy.cam.ac.uk/mackay/, 1996.

5. B. A. Pearlmutter and L. C. Parra, "A context-sensitive generalization of ICA," in *Proceedings of the International Conference on Neural Information Processing*, 1996.
6. M. S. Hämäläinen and R. J. Ilmoniemi, "Interpreting measured magnetic fields of the brain: Minimum norm estimates," Technical Report TKK-F-A559, Helsinki University of Technology, 1984.
7. J. S. George, P. S. Lewis, D. M. Ranken, L. Kaplan, and C. C. Wood, "Anatomical constraints for neuromagnetic source models," in *SPIE Medical Imaging V: Image Physics*, vol. 1443, pp. 37 - 51, 1991.
8. A. M. Dale and M. I. Sereno, "Improved localization of cortical activity by combining EEG and MEG with MRI cortical surface reconstruction: a linear approach," *J. Cog. Neuro.*, **5**, (2), pp. 162–176, 1993.
9. B. N. Cuffin, "Effects of local variations in skull and scalp thickness on EEG's and MEG's," *IEEE Trans Biomed Eng*, **40**, (1), pp. 42–48, 1993.
10. Z. Zhang and D. L. Jewett, "Insidious errors in dipole localization parameters at a single time-point due to model misspecification of number of shells," *IEEE Trans Biomed Eng*, **40**, (1), pp. 42–48, 1993.

A MAXIMUM ENTROPY APPROACH FOR ESTIMATING NONLINEAR DYNAMIC MODELS

AMOS GOLAN
American University[‡]
4400 Massachusetts Ave. NW
Washington, DC 20016-8029
Tel: (202)885-3783; Fax: (202)885-3790[§]

Abstract. Given the objective of estimating the unknown parameters of a finite (and relatively small) data set, generated by some (possibly nonlinear) dynamic discrete time process, it is common to use a Kalman filter Maximum Likelihood (ML) approach, ML-type estimators or more recently a GMM (Imbens, Spady and Johnson, 1998) or BMOM (Zellner 1997, Tobias and Zellner 1997) estimators. All of the above ML-type methods (except the BMOM) require some distributional assumptions while the moment-type estimators require some data (or assumptions) on the moments of the underlying distribution that generated the data. Recently, Golan, Judge and Miller (1996) and Golan, Judge and Karp (1996) developed a Maximum Entropy (ME) framework for estimating dynamic models without distributional assumptions. In this paper the above ME formulation is extended for estimating nonlinear (chaotic) dynamic models. Under this new formulation, one views the errors as another set of unknown parameters to be estimated. Thus, for any data set, the estimation problem is ill-posed (under-determined) where the number of unknowns is always greater than the number of data points and therefore the ME approach is a natural way to estimate the parameters. After developing the basic primal entropy model, a computationally efficient dual, unconstrained approach is developed. Under this dual method, the optimization is done with respect to the Lagrange multipliers associated with each observation. The dual is also used to contrast this ME approach with the more traditional ML-type estimators. Statistics and inference procedures are developed as well. Monte Carlo and empirical results for estimating the parameters of noisy, chaotic systems are presented as well.

Key words: information, maximum entropy, nonlinear dynamic models

---

[‡]Visiting (1997-1999)
[§]Email: agolan@american.edu itor golan@are.berkeley.edu.

W. von der Linden et al. (eds.), Maximum Entropy and Bayesian Methods, 227–236.

1. Introduction

1.1. BACKGROUND AND OBJECTIVE

The objective of this paper is to develop and investigate a Maximum Entropy (ME) estimator for estimating the unknown parameters of a relatively small (macroeconomic) data set generated by some dynamic (possibly chaotic) discrete time process. This ME estimator extends previous work of Golan, Judge and Miller (1996) and Golan, Judge and Karp (1996). Under this new approach, no assumptions are made on the underlying distribution that generated the data. Instead of these assumptions, the errors are treated as another set of unknowns that are naturally bounded between $\pm 3\sigma_y$, where σ_y is the empirical standard deviation of the analyzed data set. Thus, both the signal and the noise (unknown error distribution) are assumed to be unknown. The objective is, then, to simultaneously recover both. The motivation for such an estimator and a brief review are discussed next.

1.2. PROBLEMS WITH MACROECONOMIC (TIME-SERIES) DATA

When data sets are (relatively) large, the design matrix (X) is well behaved (e.g., no collinear covariates), the underlying generating process is known, and we know the correct economic-theoretical model that generated the data, all traditional statistical methods perform well. However, in most economic cases (especially with macroeconomic data) data sets are non-perfect and have the following characteristics:

1. Data sets are small (only limited number of yearly/periodic observations).
2. Data are nonlinear.
3. Data are contaminated in the sense that the covariates may be autocorrelated, heteroscedastic and/or multi-collinear.
4. The initial conditions (state of the economy as is represented by the data) at the beginning of the data (observation 1) are unknown.
5. Data may exhibit unknown structural changes that may split the data into at least two subsets that are characterized differently and may involve some turning (threshold) points.

1.3. FORECASTING PROBLEMS WITH MACRO DATA

Given the above characteristics of macro data and the objective of forecasting and prediction, the main issue is what forecasting model should be used. The answer to this is, however, non trivial, and depends on which set of simplifying assumptions the researcher is willing to assume/impose on the estimation model. These restrictions include the following:

1. Distributional assumptions on the underlying mechanism that generated the data (what distribution does the random variable have? What is the relationship between present and past?, etc.).

2. Structural and economic assumptions (What is the exact functional form of the modeled economy and what variables should one use).
3. Stability issues (Is the system stable or is it unstable? Is the system stationary or non-stationary?).
4. Are the data nonlinear (or linear)?
5. Causality (what effects what).
6. Computational issues (some good theoretical modeling techniques are too complicated to compute, are non-convex or may not converge).

Given these problems, the objective is to develop an estimator that uses minimum possible distributional assumptions, uses each observation in the sample (rather than the sample moments), behaves well for small and ill-posed data and is easy to compute. The ME estimator discussed below is such an estimator.

The basic objective, and philosophy, of this estimator is similar to that of Zellner's seminal Bayesian Method of Moments (BMOM) approach (Zellner, 1997; Tobias and Zellner, 1997), the Empirical Likelihood (EL) approach of Owen (1990) and Qin and Lawless (1994), or the Generalized Method of Moments (GMM) work of Hansen (1982) and the more recent developments of Imbens, Spady and Johnson (1998). In all of these estimators the basic objective is to first search for the "natural weight" of each observation which is then used to form an empirical likelihood (discrete or continuous). However, in all of these models the empirical likelihood is constructed/estimated based on some small number of side-conditions or moments. With the estimator discussed here these side-conditions are not assumed. For a good review see, for example, Zellner (1994) and Dufour and Ghysels (1996).

The basic model is discussed in the next section. Following the basic model, a dual unconstrained estimator is formulated. This dual is used to develop inference and information measures in Section 3. Some examples are presented in Section 4. An empirical example of forecasting the U.S. unemployment rates is presented in Section 5. Finally, some concluding remarks are given in Section 6.

2. The Model

Consider the nonlinear model

$$y_t = f_t(y_{t-1}, y_{t-2}, ..., y_{t-\tau}, x_t, x_{t-1}, ..., x_{t-\delta}; \beta, \epsilon_t) = f_t(X_t; \beta, \epsilon_t) \tag{1}$$

where $\mathbf{y}$ is a T-dimensional vector of observations, X is a $(T \times K)$ design matrix that may include lag components of $\mathbf{y}$, β is a vector of unknown and unobserved parameters, ϵ is a T-dimensional error term, and f_t is some nonlinear function. To simplify notations and exposition, and without loss of generality, consider a noisy variation of the (one parameter) logistic, discrete time model

$$y_{t+1} = \beta y_t(1 - y_t) + \epsilon_t \tag{2}$$

Given the nonlinear model (1), or (2), the objective is to recover the unknown vector β with minimum assumptions. To retain this objective, *no distributional assumptions are used and no assumptions regarding an exact relationship between sample and population moments are imposed.* Thus, both the signal, β, and the noise (unknown error distribution), ϵ, are assumed to be unknown. The objective is, then, to simultaneously recover β and ϵ.

To achieve this goal we follow Golan, Judge, and Miller (1996, Chapter 6) where as a first step, the nonobserved parameters of β and ϵ are reparameterized as

$$y_{t+1} = \beta y_t(1-y_t) + \epsilon_t \equiv \sum_m \pi_m z_m y_t(1-y_t) + \sum_j w_{tj} v_j \tag{3}$$

In this reformulation, β is defined as

$$\beta \equiv \sum_m z_m \pi_m \quad \text{with} \quad \sum_m \pi_m = 1 \tag{4}$$

and where $\mathbf{z}= (z_1, z_2, ..., z_M)\prime$ is an $M \geq 2$ dimensional vector. This vector, $\mathbf{z}$, serves as a discrete support space for the unknown β, and is specified such that it will span the possible (unknown) value of β. Thus, under this formulation, the unknown β are converted from the real line to the $[0,1]$ interval with the property of a proper probability distribution π, defined over the support space $\mathbf{z}$. How should one specify $\mathbf{z}$? If the researcher posses no knowledge as to the possible value of β, $\mathbf{z}$ should be specified to be *symmetric around zero* with large negative and positive boundaries. For example, $z_{k1} = -z_{kM} = -10^6$. However, in a large number of cases the researcher has some knowledge regarding the possible values of β and may use this knowledge in specifying $\mathbf{z}$. For example, stationarity implies that for the logistic model $\beta \in [0,4]$.

In a similar way, the $\epsilon's$ may be transformed into T proper probability distributions. Define a discrete support space $\mathbf{v}$ of dimension $J \geq 2$, equally spaced and *symmetric around zero* and associate with it a set of proper weights, $\mathbf{w}_t$, such that

$$\epsilon \equiv \sum_j v_j w_{tj} \quad \text{with} \quad \sum_j w_{tj} = 1 \tag{5}$$

The end points (v_1 and v_J) are chosen to be $-3\sigma_y$ and $3\sigma_y$ where σ_y is the *empirical* (sample) standard deviation of $\mathbf{y}$.

Having converted the two sets of unknowns into probability distributions, the estimation problem becomes

$$\max_{\pi,w} H(\pi, w) = \{-\sum_m \pi_m ln\pi_m - \sum_t \sum_j w_{tj} lnw_{tj}\} \tag{6}$$

subject to

$$y_{t+1} = \sum_m \pi_m z_m y_t(1-y_t) + \sum_j w_{tj} v_j \tag{7}$$

$$\sum_m \pi_m = 1; \quad \textstyle\sum_j w_{tj} = 1 \tag{8}$$

The Lagrangian is

$$L = H(\pi, w) + \sum_t \lambda_t[y_{t+1} - \sum_m \pi_m z_m y_t(1 - y_t) - \sum_j w_{tj} v_j] + \eta[1 - \sum_m \pi_m] + \sum_t \rho_t[1 - \sum_j w_{tj}] \tag{9}$$

and the optimal solution is

$$\hat{\pi}_m = \frac{exp[-\sum_t \lambda_{t+1} z_m y_t(1 - y_t)]}{\sum_m exp[-\sum_t \lambda_{t+1} z_m y_t(1 - y_t)]} \equiv \frac{exp[-\sum_t \lambda_{t+1} z_m y_t(1 - y_t)]}{\Omega} \tag{10}$$

and

$$\hat{w}_{tj} = \frac{exp(-\lambda_{t+1} v_j)}{\sum_j exp(-\lambda_{t+1} v_j)} \equiv \frac{exp(-\lambda_{t+1} v_j)}{\Psi_t} \tag{11}$$

Finally, the point estimates $\hat{\beta}$ and $\hat{\epsilon}$ are just $\hat{\beta} \equiv \sum_m \hat{\pi}_m z_m$ and $\hat{\epsilon}_t \equiv \sum_j \hat{w}_{tj} v_j$.

Next, the dual concentrated model is

$$\begin{aligned} L(\lambda) &\equiv \sum_t y_{t+1}\lambda_{t+1} + ln\{\sum_m exp[-\sum_t \lambda_{t+1} z_m y_t(1 - y_t)]\} \\ &\quad + \sum_t ln[\sum_j exp(-\lambda_{t+1} v_j)] \\ &= \sum_t y_{t+1}\lambda_{t+1} + ln\Omega(\lambda) + \sum_t ln\Psi_t(\lambda) \end{aligned} \tag{12}$$

which is constructed by substituting the post-data $\hat{\pi}$ and $\hat{\epsilon}$ into the first element in the right hand side of (9).

Having developed the single example, the extension to more than a single parameter, and for any nonlinear function, is immediate. One starts by reparameterizing all the unobserved parameters in eq. (1). Then, the ME estimator is just a maximization of the total entropies of $\{\pi_k\}$ and $\{w_k\}$ subject to the reparameterized set of equations, and the requirements of proper distributions. This yields solutions similar to (10) and (11) which in turn yield the two vectors of estimates $\hat{\beta}$ and $\hat{\epsilon}$. Finally, following the above steps, the dual estimator can be developed as well.

3. Properties of the ME Estimator

Given four mild conditions, this ME estimator is consistent and asymptotically normal. These conditions are that (i) the errors' support v is symmetric around zero, (ii) spans the true values for each one of the unknown parameters, (iii) the errors are independently and identically distributed, and (iv) the operator X is of full rank. The proofs of consistency and asymptotic normality follow immediately from those in Golan, Judge, and Miller (1996), Golan, Judge, and Perloff (1997), and Mittelhammer and Cardell (1996). These asymptotic properties can also be established via the empirical likelihood approach (Owen, 1990, Qin and Lawless, 1994, and Golan and Judge, 1996).

On the basis of these results, the Entropy Ratio, *ER*, statistic for the different parameters of the unknown distribution generating the data have a limiting χ^2 distribution and is used to obtain confidence intervals. Let λ be the $(T-1)$ vector of Lagrange multipliers for the $T-1$ observations in (1) or (2). Let $H_u(\hat{\lambda})$ be the objective (total entropy) value for the complete ME model where *none of the parameters are constrained*, or similarly, none of the elements of λ are constrained. Thus, $H_u(\hat{\lambda})$ is just the optimal value of Equation (6). Next, let $H_M(\lambda_0)$ be the entropy value of the constrained problem where $\lambda = 0$, or equivalently all the parameters are constrained to zero (or at the *center* of their supports). Thus, $H_M(\lambda_0)$ is the maximum value of the joint entropies (objective function). It can be obtained by maximizing Equation (6) subject to no constraints (except for the requirements that all distributions are proper). Doing so yields the total entropy value of the two sets of discrete, uniform distributions π and $\{w_t\}$,

$$H_M(\lambda_0) = ln(M) + (T-1)ln(J) \tag{13}$$

where M is the dimension of the support space for β, T is the total number of observations, and J is the dimension of the support space $\mathbf{v}$. Note that if there are more unknown parameters (say, K), then the first RHS term becomes $Kln(M)$.

The entropy ratio statistic, *ER*, for testing the null hypothesis $H_0 : \beta = 0$ is

$$ER(\beta = 0) = 2H_M(\beta = 0) - 2H_u(\hat{\beta}). \tag{14}$$

Under the four mild assumptions we made above (or the assumptions of Owen, 1990 and Qin and Lawless, 1994), $ER(\beta = 0) \rightarrow \chi^2$ as $T \rightarrow \infty$ when H_0 is true. The appropriate α-level confidence intervals for β is obtained by setting $ER(\beta) \leq C_\alpha$, where C_α is chosen so that $\Pr(\chi^2 < C_\alpha) = \alpha$. Similarly, we can test any other hypothesis of the form $H_0 : \beta = \beta_0$ for all, or any subset, of the parameters.

Using the same line of reasoning as above (each constraint, or data point, represents additional potential information that may lower the value of the objective function but can never increase it) the "goodness of fit" measure for the ME estimator is

$$R^* = 1 - \frac{H_u(\hat{\lambda})}{H_M(\lambda = 0)} \tag{15}$$

where $R^* = 0$ implies no informational value of the data set, and $R^* = 1$ implies perfect certainty or perfect in-sample prediction.

The small-sample approximated variance can be computed in a number of ways. Since all are practically similar, just the simpler one is presented:

$$\hat{\sigma}^2 = \frac{1}{T-1} \sum_t \hat{\epsilon}_t^2, \tag{16}$$

where $\hat{\epsilon}_t \equiv \sum_j v_j \hat{w}_{tj}$ and $var(\hat{\beta}) \cong \hat{\sigma}^2(X'X)^{-1}$ where in the logistic example $x_t \equiv y_{t-1}$.

4. Sampling Experiments

In this section some sampling experiments are presented. The first one is based on the simple logistic (with additive noise) model discussed earlier. Table 1 summarizes the results of different experiments, each with 1000 samples of different sample size T, different level of β (in the chaotic and non-chaotic regimes) and different levels of noise. Three models are compared: the ME, the nonlinear Maximum Likelihood (ML) and the constrained, nonlinear ML (CML) where the estimate $\hat{\beta}$ (for each sample) is constrained to $[0,4]$. The statistics reported are (i) the Mean Square Error (MSE) between predicted and correct value of y_t up to six periods into the future $(MSE_1, MSE_2, ...)$, (ii) the bias of the predicted value (up to six periods into the future) and the $MSE(\hat{\beta})$ and $var(\hat{\beta})$, where $MSE_i \equiv [\sum_{g=1}^{G}(\hat{y}_{g,t+i} - y_{t+i})^2]/G$, $Bias_i \equiv [\sum_g(\hat{y}_{g,t+i} - y_{t+i})]/G$, $i = 1,2,...,6$, $MSE(\hat{\beta}) \equiv [\sum_g \sum_k(\hat{\beta}_{g,k} - \beta_k)^2]/G$ and $Var(\hat{\beta}) \equiv \sum_g[\frac{1}{T-K}\sum_k(\hat{\beta}_{g,k} - \bar{\hat{\beta}})^2]/G$, T is the number of observations and G is the number of samples (1000) in the sampling experiments.

To show the robustness of the ME estimator, the support space for β is $\mathbf{z} \equiv (-500, -250, 0, 250, 500)'$ and the error's support is $\mathbf{v} = (-3\sigma_y, 0, 3\sigma_y)'$ for all the experiments done. Thus, the β's are not restricted to the true interval $[0,4]$. The results show that the ME dominates, under all criteria, for the small samples.

In the second set of experiments, a more complex model, composed of the sum of two noisy logistic equations is investigated. Due to lack of space, table 2 summarizes only some of the results for different combinations of β's, ϵ's and T's. Again, the ME dominates for all small samples and then, for the larger samples, the ML and ME results converge. However, in all cases (small and large samples) the more nonlinear the system ($\beta_1 \to 4$ and $\beta_2 \to 4$), the better the ME performs relative to the ML estimator.

5. Forecasting the U.S. Unemployment Rate

In this section a brief comparison with the ML type estimator for forecasting the U.S. unemployment rate is done. Following the work of Montgomery et. al. (1998), the quarterly unemployment rates, u_t, for the period 1948-1993 is used. As the objective here is just to present a comparison with their results, no economic background is discussed. Given these data, a specific nonlinear Threshold Autoregressive Model (TAR) is formulated. This model is a variation of the familiar (nonlinear) Tent map. The exact model is

$$\Delta u_t = \begin{cases} \beta_{11} + \beta_{12}\Delta u_{t-1} + \beta_{13}\Delta u_{t-2} + \epsilon_{1t} & \text{if } \Delta u_{t-2} \leq C \\ \beta_{21} + \beta_{22}\Delta u_{t-1} + \beta_{23}\Delta u_{t-2} + \epsilon_{2t} & \text{otherwise} \end{cases}$$

where $\Delta u_t = u_t - u_{t-1}$.

The estimated parameters for the two methods are
$\hat{\beta}_1(ML) = (.01, 0.73, 0.10)'$ and $\hat{\beta}_2(ML) = (.018, 0.80, -0.56)'$
$\hat{\beta}_1(ME) = (.004, 0.54, 0.09)'$ and $\hat{\beta}_2(ME) = (.20, 1.02, -0.81)'$
where the critical point C is 0.1 for the ML and 0.035 for the ME model. The ML

TABLE 1. Results of sampling experiments of the logistic model

	$\beta = 3.95$; $T = 10$; $e \sim [-.1, .1]$		$\beta = 3.0$; $T = 10$; $e \sim [-.1, .1]$		$\beta = 3.95$ $T = 10$; $e \sim [-.2, .2]$		$\beta = 3.95$; $T = 100$; $e \sim [-.3, .3]$	
	ME	ML(=CML)	ME	ML	ME	ML	ME	ML
MSE1	.060	.070	.0068	.0170	.0995	.1098	.084	.0824
MSE2	.108	.112	.0095	.0110	.1270	.1246	.1225	.1212
MSE3	.156	.173	.0120	.0166	.1271	.1261	.1230	.1247
MSE4	.166	.154	.0130	.0143	.1196	.1137	.1051	.1061
MSE5	.138	.141	.0143	.0158	.1372	.1371	.0921	.0937
MSE6	.165	.160	.0149	.0168	.1330	.1300	.0959	.0968
Bias1	-.11	-.16	-.046	-.11	-.132	-.182	-.1237	-.1176
Bias2	-.07	-.06	-.04	-.056	-.034	-.047	-.0098	.0003
Bias3	-.02	-.018	-.035	-.079	.014	-.005	.1050	.1126
Bias4	.07	.07	-.048	-.072	-.020	-.035	.1024	.1088
Bias5	-.009	-.011	-.033	-.071	.064	.040	.0956	.1013
Bias6	-.04	-.03	-.060	-.087	.039	.001	.0673	.0727
MSE($\hat{\beta}$)	1.6057	1.9679	.0062	.0973	1.9746	2.5530	1.5194	1.4070
Var($\hat{\beta}$)	.599	.522	.0062	.0103	.4187	.3992	.0349	.0386

$CML \cong ML$. Note that the Constrained ML and the ML are almost identical here.

estimates are taken from Montgomery et. al. (1998). The support spaces used for the ME estimator are $z_k = (-500, -250, 0, 250, 500)'$ for all the β_k's and $\pm 3\sigma_y$ for the errors. The forecasts begin with the third quarter of 1968 and end with the third quarter of 1993.

In addition to the MSE's and bias for forecasting accuracy presented in Table 3, the confusion rate table (a table comparing the true and forecasted turning points) is also presented here. This alternative model selection criterion is particularly useful when trying to forecast the next economic turning point.

The total number of correct predictions (up and down turns) for the ME are 33 for one period forecast, 38 for two, 42 for three and 45 for four. For the ML the correct predictions are 32, 36, 41, and 48 respectively. Thus, the ME is slightly superior in predicting the future turning points.

6. Concluding Remarks

An ME estimator (based on Golan, Judge and Miller, 1996) for recovering the unknown parameters of nonlinear dynamic models was developed.

This estimator performs well for small, ill-posed and highly nonlinear dynamic data and is easy to compute. However, more work is needed in order to evaluate its small sample properties. This should include both the analytic derivation of

TABLE 2. Sampling results of the sum of two logistic equations

	β= (2.7, 3.95); $T = 10$; $e_1 \sim [-.2, .2]$; $e_2 \sim [-.1, .1]$			β= (3.95, 3.90); $T = 10$; $e_1 = e_2 \sim [-.1, .1]$		
	ME	CML	ML	ME	CML	ML
MSE1	.1271	.1331	.1460	.155	.167	.193
MSE2	.2566	.1902	.2631	.234	.202	.309
MSE3	.2092	.2044	.2403	.357	.264	.801
MSE4	.2595	.2487	.2828	.354	.338	*
MSE5	.2529	.2112	.2354	.374	.313	*
MSE6	.2399	.2602	.300	.459	.368	*
Bias1	-.12	-.23	-.25	-.145	-.241	-.270
Bias2	-.15	-.11	-.22	-.126	-.0008	-.163
Bias3	-.05	-.15	-.18	-.104	-.047	-.180
Bias4	-.10	-.14	-.19	-.079	-.081	3.511
Bias5	-.04	-.12	-.17	.035	.068	*
Bias6	-.09	-.17	-.21	-.165	-.119	*
MSE($\hat{\beta}$)	4.1225	5.0356	6.1317	4.829	4.586	8.345
Var($\hat{\beta}$)	1.979	2.178	3.540	3.420	2.178	6.077

* MSE's are above 1000 (ML does not converge)

TABLE 2. Sampling results of the sum of two logsitic equations (cont.)

	β= (2.7, 3.95); $T = 20$; $e_1 \sim [-.2, .2]$; $e_2 \sim [-.1, .1]$			β= (3.95, 3.90); $T = 20$; $e_1 = e_2 \sim [-.1, .1]$			β= (3.95, 3.90) $T = 100$; $e_1 = e_2 \sim [-.1, .1]$		
	ME	CML	ML	ME	CML	ML	ME	CML	ML
MSE1	.0913	.0766	.0866	.1432	.1138	.1240	.1079	.0983	.0982
MSE2	.1884	.1327	.1392	.2403	.2085	.2551	.1682	.1940	.1769
MSE3	.2011	.170	.1759	.2771	.2477	.2876	.2848	.2727	.2867
MSE4	.2160	.193	.1993	.3045	.3213	.3409	.3627	.3586	.3456
MSE5	.2000	.1867	.1928	.3137	.3033	.2940	.3312	.3276	.3271
MSE6	.1862	.1976	.2009	.3165	.3362	.3692	.3229	.3539	.3494
Bias1	-.05	-.11	-.12	-.15	-.17	-.19	-.05	-.05	-.05
Bias2	-.06	-.03	-.05	-.05	.11	-.01	.12	.18	.16
Bias3	-.09	.04	.02	.10	.11	.04	.23	.24	.23
Bias4	.03	.04	.02	.16	.19	.13	.18	.21	.19
Bias5	.10	.07	.06	.08	.13	.07	.21	.22	.23
Bias6	.001	-.002	-.02	.09	.12	.05	.19	.18	.16
MSE($\hat{\beta}$)	2.773	2.986	3.051	2.279	2.316	3.013	.862	.841	.874
Var($\hat{\beta}$)	.832	1.082	1.245	1.368	1.022	1.748	.286	.259	.296

TABLE 3. Comparison of the U.S. unemployment quarterly forecasts for two methods

	Relative MSE of forecasts				
Model description	One-step-ahead	Two-step-ahead	Three-step-ahead	Four-step-ahead	Five-step-ahead
TAR-ML	.078	.362	.793	1.252	1.719
TAR-ME	.087	.391	.861	1.294	1.701
	Means of forecast errors				
Model description	One-step-ahead	Two-step-ahead	Three-step-ahead	Four-step-ahead	Five-step-ahead
TAR-ML	-.032	-.096	-.122	-.124	-.135
TAR-ME	-.009	-.061	-.075	-.044	-.016

the small sample properties, as well as a well designed set of experiments that will extend the work presented here in Tables 1 and 2.

References

Dufour, J.M. and E. Ghysels, eds. (1996), Special issue of the *Journal of Econometrics*, "Recent Developments in Econometrics of Structural Change" **Vol. no. 70**,

Golan, A. and G. Judge, (1996) A Maximum Entropy Approach to Empircal Likelihood and Inference (Working Paper), UC Berkeley

Golan, A., G. Judge and L. Karp (1996) A Maximum Entropy Approach to Estimation and Inference in Dynamic Models, *Journal of Economic Dynamics and Control*, **Vol. no. 20**, pp. 559–582

Golan, A., G. Judge and D. Miller (1996), *Maximum Entropy Econometrics: Robust Estimation with Limited Data*. John Wiley.

Golan, A., G. Judge and J. Perloff (1997) Estimation and Inference with Censored and Ordered Multinomial Response Data, *Journal of Econometrics*, **Vol. no. 79**, pp. 23–51

Hansen, L. (1982) Large Sample Properties of Generalized Method of Moments Estimators, *Econometrica*, **Vol. no. 50**, pp. 1029–1054

Imbens, G.W., R. Spady and P. Johnson (1998) Information Theoretic Approaches to Inference in Moment Condition Models, *Econometrica*, **Vol. no. 66**, pp. 333–57

Mittelhammer, R.C. and N.S. Cardell (1996) On the Consistency and Asymptotic Normality of the Data-Constrained GME Estimator in the GLM (Working Paper), Washington State U.

Montgomery, A.L., V. Zarnowitz, R.S. Tsay and G.C. Tiao (1998) Foreccasting the U.S. Unemployment Rate, *Journal of the American Statistical Association*, **Vol. no. 93**, pp. 478–493

Owen, A. (1990) Empirical Likelihood Ratio Confidence Regions, *The Annals of Statistics*, **Vol. no. 18**, pp. 90–120

Qin, J. and J. Lawless (1994) Empirical Likelihood and General Estimating Equations, *The Annals of Statistics*, **Vol. no. 22**, pp. 300–325

Tobias, J. and A. Zellner (1997)*Further Results on BMOM Analysis of the Multiple Regression Analysis, H.G.B. Alexander Research Foundation*. University of Chicago

Zellner, A. (1997) The Bayesian Method of Moments (BMOM): Theory and Applications, in Advances in Econometrics, T. Fomby and R. C. Hill eds pp. 85–106

Zellner, A. (1994) Time-series Analysis, Forecasting, and Econometric Modeling: the Structural Econometric Modeling, Time-Series Analysis (SEMTSA) Approach *Journal of Forecasting*, **Vol. no. 13**, pp. 215–233

PROBABILITY AND ENTROPY IN QUANTUM THEORY

ARIEL CATICHA
Department of Physics, University at Albany-SUNY,
Albany, NY 12222 U.S.A. †

Abstract. Entropic arguments are shown to play a central role in the foundations of quantum theory. We prove that probabilities are given by the modulus squared of wave functions, and that the time evolution of states is linear and also unitary.

1. Introduction

One of the curious features of quantum mechanicsis that it is a theory in which probabilities play a most central role and yet, from a foundational point of view, the concept of entropy is conspicuously absent. Entropy appears only later as an auxiliary quantity to be used only when a problem is sufficiently complicated that clean deductive methods have failed and one is forced to use dirtier inference methods. This is curious indeed because once the use of the notion of probability has been accepted, the issue of whether or not quantum mechanics is a theory of inference has been unequivocally settled. Quantum theory should be regarded as a set of rules for reasoning in situations where even under optimal conditions the information available to predict the outcome of an experiment may still turn out to be insufficient. In such a theory entropy, as the measure of the amount of information [1][2], should play a central role. It is difficult to avoid the feeling that perhaps the use of entropic arguments has been inadvertently encoded into the usual postulates of quantum mechanics. The purpose of this paper is to show that this is in fact the case.

This paper is a continuation of previous work [3][4] in which quantum theory is formulated as the only consistent way to manipulate the amplitudes for quantum processes. The result of this consistent-amplitude approach is the standard quantum theory [5], in a form that is very close to Feynman's [6]. The first and most crucial step is a decision about the subject matter. We choose a pragmatic, operational approach: statements about a system are identified with those experimental setups designed to test them; the objective is to predict the outcomes of experiments [7]. The observation that if two setups are related in some way then information about one may be relevant to predictions about the other leads one to identify the possible relations among setups and to introduce the means for

†E-mail: ariel@cnsvax.albany.edu

W. von der Linden et al. (eds.), Maximum Entropy and Bayesian Methods, 237–246.

handling these relations quantitatively. This is the role played by the amplitudes. Amplitudes are tools for reasoning that encode information about how one builds complicated setups by combining more elementary ones. The question of how amplitudes are used to predict the outcomes of experiments is addressed through a single general interpretative rule. A brief summary is given in section 2.

It is quite remarkable that although the interpretative rule does not in itself involve probabilities it can be used to prove the Born statistical "postulate" (or, better, Born "rule") [8] provided one extra ingredient is added. The point is that the application of the interpretative rule requires a criterion to quantify the change in amplitudes when setups are modified. In ref. [4] the criterion adopted was to use the Hilbert norm as the means to measure the distance between wave functions. This may seem reasonable but such a very technical assumption without an obvious physical basis clearly detracts from the beauty of the argument. In section 3 this blemish is corrected. The realization that the components out of which setups are built, the filters, already supply us with a notion of orthogonality brings us very close to the inner product needed for the Hilbert norm. We close this remaining gap using a symmetry argument. The implication is very interesting: the opinion that quantum probabilities differ in an essential way from ordinary classical probabilities is very widespread; not only are they calculated using apparently different rules but, being given a priori by the Born rule, they also seem to depend less on information, to be more "objective". Our proof of the Born rule, tracing it to a form of the principle of insufficient reason, supports the opposite point of view, that such differences are an illusion, that there is only one kind of probability.

The constraint that amplitudes be assigned consistently leads to a time evolution that is linear [3] but remains silent about whether it should also be unitary. A common explanation is that imposing unitary time evolution guarantees that probabilities be conserved. This is true but it is also irrelevant; that probabilities should add up to one is a matter of definition [9]; the "non-conservation of probabilities" can always be "fixed" by a suitable reinterpretation. The usual symmetry arguments based on Wigner's theorem are also inadequate. Why should time evolution be a symmetry in the very technical sense of preserving inner products? Finally, arguments based on von Neumann's entropy are circular [10], they implicitly assume that one can measure observables other than position [11], an assumption which itself (see section 5) relies on unitary evolution.

The argument we offer in section 4 is based on the idea of array entropy, a concept that was briefly introduced by Jaynes [12] only to be dismissed as an inadequate candidate for the entropy of a quantum system, a quantity which he rightfully identified with von Neumann's entropy. From our point of view, however, amplitudes and wave functions are assigned not just to the system but to the whole experimental setup, and this turns the array entropy into a very useful notion. The idea is simple. In situations where the information available for the prediction of experimental outcomes is not spoiled by just waiting entropy should be conserved. Our problem is to identify the appropriate entropy (it is the array entropy). Its conservation implies the conservation of the Hilbert norm and unitary evolution. As claimed above, the postulate that time evolution is unitary is derivable from an entropic argument.

2. The consistent-amplitude approach to quantum theory

The objective of quantum theory is to predict the outcomes of experiments; statements about the quantum system are identified with the experimental setups designed to test them [3][4]. To avoid irrelevant technical distractions we consider a very simple system, a particle that lives on a discrete lattice and has no spin or other internal structure. The generalization to more complex configuration spaces should be straightforward.

The simplest experimental setup, denoted by $[x_f, x_i]$, consists of placing a source that prepares the particle at a space-time point $x_i = (\vec{x}_i, t_i)$ and placing a detector at $x_f = (\vec{x}_f, t_f)$. To test a more complex statement such as "the particle goes from x_i to x_1 and from there to x_f," denoted by $[x_f, x_1, x_i]$, requires a more complex setup involving an idealized device, a "filter" which prevents any motion from x_i to x_f except via the intermediate point x_1. This filter is some sort of obstacle or screen that exists only at time t_1, blocking the particle everywhere in space except for a small "hole" around $\vec{x}_1$. The possibility of introducing many filters each with many holes leads to allowed setups of the general form $a = [x_f, s_N, s_{N-1}, \ldots, s_2, s_1, x_i]$ where $s_n = (x_n, x_n', x_n'', \ldots)$ is a filter at time t_n, intermediate between t_i and t_f, with holes at $\vec{x}_n, \vec{x}_n', \vec{x}_n'', \ldots$

There are two basic kinds of relations among setups. The first, called *and*, arises when two setups a and b are placed in immediate succession resulting in a third setup which we denote by ab. It is necessary that the destination point of the earlier setup coincide with the source point of the later one, otherwise the combined ab is not allowed. The second relation, called *or*, arises from the possibility of opening additional holes in any given filter. Specifically, when (and *only* when) two setups a' and a'' are identical except on one single filter where none of the holes of a' overlap any of the holes of a'', then we may form a third setup a, denoted by $a' \vee a''$, which includes the holes of both a' and a''. Provided the relevant setups are allowed the basic properties of *and* and *or* are quite obvious: *or* is commutative, but *and* is not; both *and* and *or* are associative, and finally, *and* distributes over *or*.[1]

A quantitative representation of *and/or* is obtained by assigning a single complex number $\phi(a)$ to each setup a in such a way that relations among setups translate into relations among the corresponding complex numbers. What gives the theory its robustness, its uniqueness, is the requirement that the assignment be consistent: if there are two different ways to compute $\phi(a)$ the two answers must agree. The remarkable consequence of the consistency constraints is the possibility of regraduating $\phi(a)$ with a function ψ to switch to an equivalent and particularly convenient representation, $\psi(a) \equiv \psi(\phi(a))$, in which *and* and *or* are respectively represented by multiplication and addition. Explicitly, $\psi(ab) = \psi(a)\psi(b)$ and $\psi(a \vee a') = \psi(a) + \psi(a')$. Complex numbers assigned in this way are called "am-

[1]These are physical rather than logical connectives. They represent our idealized ability to construct more complex setups out of simpler ones [4] and they differ substantially from their Boolean and quantum logic counterparts. In Boolean logic not only *and* distributes over *or* but *or* also distributes over *and* while in quantum logic propositions refer to quantum properties at one time rather than to processes in time.

plitudes".For a similar (earlier) derivation of the quantum sum and product rules see ref. [13].

The observation that a single filter that is totally covered with holes is equivalent to having no filter at all leads to the fundamental equation of motion. The idea is expressed by writing the relation among setups $[x_f, x_i] = \bigvee_{\text{all } \vec{x} \text{ at } t}([x_f, x_t][x_t, x_i])$ in terms of the corresponding amplitudes [6]. Using the sum and product rules, we get $\psi(x_f, x_i) = \sum_{\text{all } \vec{x} \text{ at } t} \psi(x_f, x_t)\, \psi(x_t, x_i)$.

Following Feynman [6], we introduce the wave function$\Psi(\vec{x}, t)$ as the means to represent those features of the setup prior to t that are relevant to time evolution after t. Notice that there are many possible combinations of starting points x_i and of interactions prior to the time t that will result in identical evolution after time t. What these different possibilities have in common is that they all lead to the same numerical value for the amplitude $\psi(x_t, x_i)$. Therefore we set $\Psi(\vec{x}, t) = \psi(x_t, x_i)$ and all reference to the irrelevant starting point x_i can be omitted. The traditional language is that Ψ describes the state of the particle at time t, that the effect of the interactions was to prepare the particle in state Ψ. Now we see that the word "state" just refers to a concise means of encoding information about those aspects of the setup prior to the time t that are relevant for evolution into the future.

The equation of motion can then be written as

$$\Psi(\vec{x}_f, t_f) = \sum_{\text{all } \vec{x} \text{ at } t} \psi(\vec{x}_f, t_f; \vec{x}, t)\, \Psi(\vec{x}, t), \tag{1}$$

which is equivalent to a linear Schrödinger equationas can easily be seen [3][4] by differentiating with respect to t_f and evaluating at $t_f = t$.

This result is important because a variety of nonlinear modifications of quantum mechanicshave been proposed, either attempting to solve the problems with macroscopic quantum superpositions, or to explore the possibility that the linear theory might just be a low "intensity" limit of the true nonlinear theory [14]. Furthermore, even though experimental bounds have become increasingly stringent [15] experimentation alone cannot logically rule out small nonlinearities.

The question of amplitudes or wave functions are used to predict the outcomes of experiments is addressed through the time evolution equation (1). For example, suppose the preparation procedure is such that $\Psi(\vec{x}, t)$ vanishes at a certain point $\vec{x}_0$. Then, according to eq. (1), placing an obstacle at the single point $(\vec{x}_0, t)$ (*i.e.*, placing a filter at t with holes everywhere except at $\vec{x}_0$) should have no effect on the subsequent evolution of Ψ. Since relations among amplitudes are meant to reflect corresponding relations among setups, it seems natural to assume that the presence or absence of the filter will have no effect on whether detection at x_f occurs or not. Therefore when $\Psi(\vec{x}_0, t) = 0$ the particle will not be detected at $(\vec{x}_0, t)$.

This idea can be generalized to the following general interpretative rule: Suppose the wave function of a setup is $\Psi\,(t)$ and at time t one introduces or removes a filter that blocks out those components of Ψ characterized by a certain property $\mathcal{P}$. Suppose further that this modification of the setup has a negligible effect on the evolution of Ψ after t. Then when the wave function is $\Psi(t)$ property $\mathcal{P}$ will not be detected.

In ref. [4] we showed how this interpretative rule implies the Born statistical postulate provided one uses the Hilbert norm as the means to quantify the change in the wave function as it evolves through a filter. In the next section we show why the choice of the Hilbert norm is the natural one.

3. The Hilbert inner product and the Born rule

In order to justify the use of the Hilbert normwe show how the concepts of distance and angle among states, that is an inner product,can be physically motivated. The argument has three parts.

First, we note that wave functions form a linear space. To illustrate this point suppose that $\Psi_1(\vec{x},t) = \psi(\vec{x},t;\vec{x}_1,t_0)$ is the wave function at time t of a particle that at time t_0 was prepared at the point $\vec{x}_1$, and $\Psi_2(\vec{x},t) = \psi(\vec{x},t;\vec{x}_2,t_0)$ is the wave function at time t of a particle that at time t_0 was prepared at the point $\vec{x}_2$. One way to prepare linear superpositions of $\Psi_1(\vec{x},t)$ and $\Psi_2(\vec{x},t)$ is by placing the source at an initial point $(\vec{x}_i,t_i)$ with t_i earlier than t_0 and letting the particle evolve through a filter at t_0 with holes at $\vec{x}_1$ and $\vec{x}_2$. Then the amplitude $\psi(\vec{x},t;\vec{x}_i,t_i)$ is

$$\psi(\vec{x},t;\vec{x}_i,t_i) = \psi(\vec{x},t;\vec{x}_1,t_0)\psi(\vec{x}_1,t_0;\vec{x}_i,t_i) + \psi(\vec{x},t;\vec{x}_1,t_0)\psi(\vec{x}_1,t_0;\vec{x}_i,t_i) , \quad (2)$$

and, in an obvious notation, the wave function at time t is given by the superposition $\Psi(\vec{x},t) = \alpha\Psi_1(\vec{x},t) + \beta\Psi_2(\vec{x},t)$. Notice that the complex numbers α and β can be changed at will by changing the starting point $(\vec{x}_i,t_i)$ or by modifying the setup between t_i and t_0 in any arbitrary way.

The second part of the argument is to point out that the basic components of setups, the filters, already supply us, without any additional assumptions, with a concept of orthogonality. The action of a filter P at time t with holes at a set of points $\vec{x}_p$ is to turn the wave function $\Psi(\vec{x})$ into the wave function $P\Psi(\vec{x}) = \sum_p \delta_{\vec{x},\vec{x}_p}\Psi(\vec{x})$, and since filters P act as projectors, $P^2 = P$, any given filter defines two special classes of wave functions. One is the subspace of those wave functions such as $\Psi_P \equiv P\Psi$ that are unaffected by the filter, $P\Psi_P = \Psi_P$. The other is the subspace of those that are totally blocked by the filter, such as $\Psi_{1-P} \equiv (1-P)\Psi$, for which $P\Psi_{1-P} = 0$. We will say that these two subspaces are orthogonal to each other.

Any wave function can be decomposed into orthogonal components, $\Psi = \Psi_P + \Psi_{1-P}$. A particularly convenient expansion in orthogonal components is that defined by a complete set of elementary filters. A filter P_i is elementary if it has a single hole at $\vec{x}_i$, it acts by multiplying $\Psi(\vec{x})$ by $\delta_{\vec{x},\vec{x}_i}$; the set is complete if $\sum_i P_i = 1$. Then $\Psi(\vec{x}) = \sum_i A_i\,\delta_{\vec{x},\vec{x}_i}$, where $A_i = \Psi(\vec{x}_i)$.

In the third and last step of our argument, as a matter of convenience, we switch to the familiar Dirac notation. Instead of writing $\Psi(\vec{x})$ and $\delta_{\vec{x},\vec{x}_i}$ we shall write $|\Psi\rangle$ and $|i\rangle$, so that $|\Psi\rangle = \sum_i A_i|i\rangle$. The question is what else, in addition to the notion of orthogonality described above, is needed to determine a unique inner product. Recall that an inner product satisfies three conditions: (a) $\langle\Psi|\Psi\rangle \geqslant 0$ with $\langle\Psi|\Psi\rangle = 0$ if and only if $|\Psi\rangle = 0$, (b) linearity in the second factor $\langle\Phi|\alpha_1\Psi_1 + \alpha_2\Psi_2\rangle = \alpha_1\langle\Phi|\Psi_1\rangle + \alpha_2\langle\Phi|\Psi_2\rangle$, and (c) antilinearity in the first factor, $\langle\Phi|\Psi\rangle = \langle\Psi|\Phi\rangle^*$. Conditions (b) and (c) determine the product of the state $|\Phi\rangle = \sum_j B_j|j\rangle$ with $|\Psi\rangle = \sum_i A_i|i\rangle$ in terms of the product of $|j\rangle$ with $|i\rangle$, $\langle\Phi|\Psi\rangle = \sum_i B_j^* A_i\langle j|i\rangle$. The orthogonality of the basis functions $\delta_{\vec{x},\vec{x}_i}$ is encoded into the inner product by setting $\langle j|i\rangle = 0$ for $i \neq j$, but the case $i = j$ remains undetermined, constrained only by condition (a) to be real and positive. Clearly an additional ingredient is needed. What could be more natural than the symmetry argument that if space itself is homogeneous then there is a priori no reason to favor one location over another? We therefore choose $\langle i|i\rangle$ equal to a constant which, without losing generality, we set equal to one. Thus *the principle of insufficient reason enters quantum theory through the inner product*,$\langle i|j\rangle = \delta_{ij} \Rightarrow \langle\Phi|\Psi\rangle = \sum_i B_i^* A_i$, and this leads to the Hilbert norm $||\Psi||^2 = \sum_i |A_i|^2$.

One should emphasize that the symmetry argument invoked here differs from the usual symmetry arguments leading to conservation laws through Noether's theorem. The latter depends strongly on the particular form of the Hamiltonian; the former does not.

The deduction of the Born statistical rule now proceeds as in ref. [4]. Briefly the idea is as follows. We want to predict the outcome of an experiment in which a detector is placed at a certain $\vec{x}_k$ when the system is in state $|\Psi\rangle = \sum_i A_i|i\rangle$. In [4] we showed that the state for an ensemble of N identically prepared, independent replicas of our particle is the product $|\Psi_N\rangle = \prod_{\alpha=1}^N |\Psi_\alpha\rangle$. Now we apply the interpretative rule. Suppose that in the N-particle configuration space we place a special filter, denoted by $P_{f,\varepsilon}^k$, the action of which is to block all components of $|\Psi_N\rangle$ except those for which the fraction n/N of replicas at $\vec{x}_k$ lies in the range from $f - \varepsilon$ to $f + \varepsilon$. The difference between the states $P_{f,\varepsilon}^k|\Psi_N\rangle$ and $|\Psi_N\rangle$ is measured by the relative Hilbert distance, $||P_{f,\varepsilon}^k|\Psi_N\rangle - |\Psi_N\rangle||^2/\langle\Psi_N|\Psi_N\rangle$. The result of this calculation is [4]

$$\lim_{N\to\infty} \left\|P_{f,\varepsilon}^k\Psi_N - \Psi_N\right\|^2 = 1 - \int_{f-\varepsilon}^{f+\varepsilon} \delta\left(f' - |A_k|^2\right) df' \ , \tag{3}$$

where we have normalized $\langle\Psi|\Psi\rangle = \langle\Psi_N|\Psi_N\rangle = 1$. We see that for large N the filter $P_{f,\varepsilon}^k$ has a negligible effect on the state $|\Psi_N\rangle$ provided f lies in a range 2ε about $|A_k|^2$. Therefore the state $|\Psi_N\rangle$ does not contain any fractions outside this range. On choosing stricter filters with $\varepsilon \to 0$ we conclude that detection at $\vec{x}_k$ will certainly occur for a fraction $|A_k|^2$ and that it will not occur for a fraction $1-|A_k|^2$. For any one of the *identical* individual replicas there is no such certainty; the best one can do is to say that detection will occur with a certain probability $\Pr(k)$. In order to be consistent with the law of large numbers the assigned value must agree with the Born rule,

$$\Pr(k) = |A_k|^2 \,. \tag{4}$$

Had we weighed the $|i\rangle$'s differently and chosen a different normalization $\langle i|i\rangle = w_i$, the probability would not be given by the Born rule but rather by $\Pr(k) = w_k|A_k|^2$. It is instructive to explore further the significance of choosing $w_i = 1$ particularly in the continuum limit obtained by letting the volume $g^{1/2}\Delta x$ of a unit cell of the discrete lattice tend to 0. Thus, we replace $(g^{1/2}\Delta x)^{-1/2}|i\rangle$ by $|\vec{x}\rangle$ and the completeness condition $1 = \sum_i P_i = \sum_i |i\rangle\langle i|$ becomes $1 = \int g^{1/2}dx\,|\vec{x}\rangle\langle\vec{x}|$. Next, replace $\delta_{ij}/\Delta x$ by $\delta(\vec{x}-\vec{x}')$ and the inner product $\langle i|j\rangle = \delta_{ij}$ becomes $\langle\vec{x}|\vec{x}'\rangle = g^{-1/2}\delta(\vec{x}-\vec{x}')$. Finally, replace $(g^{1/2}\Delta x)^{-1/2}A_k$ by $A(\vec{x})$ and the state $|\Psi\rangle = \sum_i A_i|i\rangle$ becomes $|\Psi\rangle = \int g^{1/2}dx\,A(\vec{x})\,|\vec{x}\rangle$. The Born rule, eq. (4), becomes $\Pr(dx) = g^{1/2}dx\,|A(\vec{x})|^2$; $|A(\vec{x})|^2$ is the probability *density*. The extension of these results to situations in which the choice of coordinates is such that the homogeneity of space is not obvious, or even to intrinsically inhomogeneous, curved spaces is immediate.

It is sometimes argued that while there is an element of subjectivity in the nature of classical probabilities that quantum probabilities are different, that they are totally objective because they are given by $|A|^2$. We have just shown that this assignment is neither more nor less subjective than say, assigning probabilities to each face of a die. Just like one assigns probability 1/6 when there is no reason to favor one face of a die over another, the Born rule follows, even in curved spaces, from giving the same a priori weight, the same preference, to spatial volume elements that are equal. (Perhaps it should be the other way around: equally preferred spatial regions are *assigned* equal volumes. This would *explain* what a physical volume is: just a measure of a priori preference.) We have thus uncovered an interesting connection between quantum theory and the geometry of space. The full implications of this connection remain to be explored.

4. Array entropy and unitary time evolution

When we know everything that is relevant about the experimental setup prior to time $t = 0$ we know $\Psi(\vec{x},0)$; this situation is one of optimal information. But if less information is available perhaps the best we can do is conclude that the actual preparation procedure is one among several possibilities $\alpha = 1,2,3,...$ each one with probability p_α. (For simplicity we initially assume these possibilities form a discrete set.) The usual linguistic trap is to say *the system* is in state $\Psi_\alpha(\vec{x},0)$ with probability p_α, but it is better to say that *the preparation procedure* is $\Psi_\alpha(\vec{x},0)$ with probability p_α. To this state of knowledge, which one may represent as a set of weighted points in Hilbert space, and which Jaynes referred to as an array[2] [12], one may associate the entropy

$$S_A = -\sum_\alpha p_\alpha \log p_\alpha \ . \tag{5}$$

Jaynes' objection to using this quantity as the entropy of the quantum system is that if the $\Psi_\alpha(\vec{x},0)$ are not orthogonal then the p_α are not the probabilities of

[2] If the states are normalized the points of the array lie on the surface of a unit sphere, but normalization is not necessary for our argument.

mutually exclusive events. When regarded as a property or an attribute of the quantum system the various $\Psi_\alpha(\vec{x},0)$ need not, in fact, be mutually exclusive; if $\langle\Psi_\alpha|\Psi_\beta\rangle \neq 0$, knowing that the system is in $\Psi_\alpha(\vec{x},0)$ does not exclude the possibility that it will be found in $\Psi_\beta(\vec{x},0)$. However, if the $\Psi_\alpha(\vec{x},0)$ are attributes of the preparation procedure then they are mutually exclusive because the preparation devices are macroscopic! S_A is the entropy of the preparation procedure not the entropy of the quantum system.

The importance of this conceptual point cannot be overemphasized and a more explicit illustration may clarify it further. Consider a spin 1/2 particle prepared either with spin along the z direction or with spin along the x direction. These states are not orthogonal and by looking at the particle there is no sure way to tell which of the two alternatives holds, and yet the slightest glimpse at the Stern-Gerlach magnets will reveal which of the two mutually exclusive orientations was used. One can distinguish non-orthogonal states by looking at the devices that prepared the system rather than by looking at the system itself.

Turning to the issue of time evolution, we consider situations where those parts of the setup after time 0 are known and no further uncertainty is introduced. Under these conditions the points of the new array are shifted from $\Psi_\alpha(\vec{x},0)$ to $\Psi_\alpha(\vec{x},t)$ but their probabilities p_α and the corresponding array entropy S_A remain unchanged.

So far our uncertainty about the preparation procedure was of a rather simple nature, it led to a probability distribution defined over a discrete array. But in general there is no such restriction and we may deal with a continuous array. The simplest continuous array is one dimensional, a weighted curve C in Hilbert space. We could consider higher dimensional arrays but this would unnecessarily obscure the argument. The reparametrization-invariant entropy of this continuous array is

$$S_A = -\int_C d\alpha\, p(\alpha) \log \frac{p(\alpha)}{\ell(\alpha)} , \tag{6}$$

where $p(\alpha)d\alpha$ is the probability that the preparation procedure lies in the interval between α and $\alpha + d\alpha$ and $\ell(\alpha)d\alpha$ is a measure of the distance in Hilbert space between $\Psi_\alpha(\vec{x},0)$ and $\Psi_{\alpha+d\alpha}(\vec{x},0)$ [16]. As discussed in the last section the Hilbert norm is the uniquely natural choice of distance, thus $\ell(\alpha)d\alpha = \||\Psi_{\alpha+d\alpha}\rangle - |\Psi_\alpha\rangle\|$.

The possibility of continuous arrays adds a new twist to our considerations about time evolution. Again we consider setups for which no further uncertainty is introduced between times 0 and t. We find that points $\Psi_\alpha(\vec{x},0)$ of the old line array at $t = 0$ will move to points $\Psi_\alpha(\vec{x},t)$ to form a new line array at time t. Since no information was lost between times 0 and t we expect that, just as in the discrete case, the probabilities $p(\alpha)d\alpha$ remain unchanged and the corresponding array entropy S_A is conserved. But entropy conservation,

$$\frac{\partial S_A}{\partial t} = \int_C d\alpha\, \frac{p(\alpha)}{\ell(\alpha)} \frac{\partial \ell(\alpha)}{\partial t} = 0 , \tag{7}$$

should hold for any curve C and any function $p(\alpha)$, therefore $\partial\ell(\alpha)/\partial t = 0$. Thus the conservation of the array entropy leads to the conservation of Hilbert space distances. Time evolution must be unitary; the Hamiltonian must be Hermitian.

5. Observables other than position

We have only discussed the measurement of position. How about other observables,uncertainty relations, and so many other notions that are central in standard quantum theory? Our brief answer (a more detailed discussion will appear in [18]) is that other observables are useful concepts in that they facilitate the description of complex experiments but, from our point of view, they are of only secondary importance and play no role at the foundational level.

One can effectively build more complex detectors by modifying the setup (by introducing, *e.g.*, magnetic fields or diffraction gratings) just prior to the final position detection at x_f. The skill of an experimentalist consists of arranging the interactions between time t and the time of detection t_f in such a way that each state $\Phi_n(\vec{x}, t)$ of an orthogonal set evolves to a corresponding state $\phi_n(\vec{x}, t_f) = \delta_{\vec{x},\vec{x}_n}$ which also form an orthogonal set. Then, when the particle is finally detected at time t_f we say that "at time t_f the particle was found at $\vec{x}_n$," or alternatively, we convey the same information by saying, somewhat inappropriately, that "the particle was found to be in state $\Phi_n(\vec{x}, t)$ at time t".

What this particular complex detector "measures" is all observables of the form $Q = \sum_n f_n |\Phi_n\rangle \langle\Phi_n|$. It is noteworthy that the eigenvalues f_n need not be real; the observables Q are diagonalizable, *i.e.*, normal ($[Q, Q^\dagger] = 0$), but not necessarily Hermitian. Clearly, this notion of observables other than position can only be introduced after one understands that time evolution must be unitary.

6. Final remarks

For over a century now an enormous effort has been directed towards deriving the second law of thermodynamics from the laws of mechanics. The successive realization by Gibbs, by Shannon [1], and even more so by Jaynes [2] that the validity of entropic arguments rests on elements that are foreign to mechanics opens the way to inverting the logic and deriving the laws of mechanics from those same principles of inference which lie at the heart of thermodynamics. In fact, according to Jaynes' beautiful explanation [17], the validity of the second law hinges on the conservation of Gibbs' or von Neumann's entropies. Could this conservation also be used to deduce unitary time evolution? No. Such arguments would be circular because these entropies rely for their very definition on having singled out certain measures (phase space volumes, and Hilbert norms respectively) as being privileged and the only reason they are special is precisely that they are conserved under unitary time evolution.

In this work, however, we have given an argument that singles out the Hilbert norm without appealing to unitarity; this clears the road to defining an entropy, the array entropy, from the conservation of which one can deduce the unitarity of time evolution.

The mystery of why complex numbers are sufficient to encode information about relations between setups remains. It seems that one could use other mathematical objects with the required associativity and distributivity, for example matrices or other Clifford numbers [19]. The recent work by Rodríguez [20] may contain important steps in exploring this possibility from a rather different point of view. My own belief is that the connection between the quantum inner product and spatial measures of volume strongly suggests that the reason for complex numbers will be found in the geometry of space. Perhaps eventually even the geometry of space itself will be determined by entropic considerations as well.

Acknowledgments- I am indebted to C. Rodríguez and P. Zambianchi for valuable discussions and many insightful remarks. Correspondence with J. Hartle and L. Schulmann on the issue of the Hilbert norm is also gratefully acknowledged.

References

1. C. E. Shannon, Bell Systems Tech. Jour. **27**, 379, 623 (1948), reprinted in C. E. Shannon and W. Weaver "The Mathematical Theory of Communication" (Univ. of Illinois Press, Urbana, 1949); see also
2. "E. T. Jaynes: Papers on Probability, Statistics and Statistical Physics," edited by R. D. Rosenkrantz (Reidel, Dordrecht, 1983).
3. A. Caticha, Phys. Lett. **A244**, 13 (1998) (quant-ph/9803086).
4. A. Caticha, Phys. Rev. **A57**, 1572 (1998) (quant-ph/9804012).
5. P. A. M. Dirac, "The Principles of Quantum Mechanics," (Oxford, 1958).
6. R. P. Feynman, Rev. Mod. Phys. **20**, 267 (1948); R. P. Feynman and A. R. Hibbs, "Quantum Mechanics and Path Integrals," (McGraw-Hill, 1965).
7. H. Stapp, Am. J. Phys. **40**, 1098 (1972).
8. The fact that Born's postulate is actually a theorem has been independently discovered several times: A. M. Gleason, J. Rat. Mech. Anal. **6**, 885 (1957); D. Finkelstein, Trans. NY Acad. Sci. **25**, 621 (1963); J. B. Hartle, Am. J. Phys. **36**, 704 (1968); N. Graham, in "The Many-Worlds Interpretation of Quantum Mechanics" edited by B. S. DeWitt and N. Graham (Princeton, 1973). The limit $N \to \infty$ where N is the number of replicas of the system is further discussed in E. Farhi, J. Goldstone and S. Gutman, Ann. Phys. **192**, 368 (1989) and in ref. [20].
9. R. T. Cox, Am. J. Phys. **14**, 1 (1946).
10. S. Weinberg, Phys. Rev. Lett. **63**, 1115 (1989).
11. R. Blankenbecler and M. H. Partovi, Phys. Rev. Lett. **54**, 373 (1985).
12. E. T. Jaynes, Phys. Rev. **108**, 171 (1957), reprinted in [2].
13. Y. Tikochinsky, Int. J. Theor. Phys. **27**, 543 (1988) and J. Math. Phys. (1988).
14. See for example: L. de Broglie "Non-Linear Wave Mechanics-A Causal Interpretation," (Elsevier, Amsterdam, 1950); P. Pearle, Phys. Rev. **D13**, 857 (1976); I. Bialynicki-Birula and J. Mycielski, Ann. Phys. (NY) **100**, 62 (1976); A. Shimony, Phys. Rev. **A20**, 394 (1979); S. Weinberg, Phys. Rev. Let. **62**, 485 (1989), and Ann. Phys. (NY) **194**, 336 (1989); N. Gisin, Helv. Phys. Acta **62**, 363 (1989) and Phys. Lett. **143**, 1 (1990); J. Polchinski, Phys. Rev. Lett. **66**, 397 (1991); H. Scherer and P. Busch, Phys. Rev. **47**, 1647 (1993).
15. C. G. Shull, D. K. Atwood, J. Arthur, and M. A. Horne, Phys. Rev. Let. **44**, 765 (1980); R. Gahler, A. G. Klein, and A. Zeilinger, Phys. Rev. **A23**, 1611 (1981); J. J. Bollinger et al., Phys. Rev. Let. **63**, 1031 (1989).
16. E. T. Jaynes, in "Statistical Physics", Vol. **3**, K. W. Ford, ed., p. 182 (Benjamin, NY 1963) and IEEE Trans. Syst. Sci. Cybern. Vol. **SSC-4**, 227 (1968), both reprinted in [2]; J. E. Shore and R. W. Johnson, IEEE Trans. Inf. Th. Vol. **IT-26**, 26 (1980).
17. E. T. Jaynes, Am. J. Phys. **33**, 391 (1965).
18. A. Caticha, to be submitted to Phys. Rev. **A**.
19. See e.g., D. Finkelstein, J. M. Jauch, S. Schiminovich and D. Speiser, J. Math. Phys. **3**, 207 (1962) and **4**, 788 (1963); S. L. Adler, Phys. Rev. Lett. **55**, 783 (1985) and Comm. Math. Phys. **104**, 611 (1986); D. Hestenes, "Spacetime Algebra" (Gordon and Breach, 1966); W. E. Baylis (ed.) "Clifford (Geometric) Algebras" (Birkhäuser, Boston, 1996).
20. C. C. Rodríguez, "Are we cruising hypothesis space?" and "Unreal probabilities–partial truth with Clifford numbers", both in these Proceedings.

UNREAL PROBABILITIES: *PARTIAL TRUTH WITH CLIFFORD NUMBERS*

C. C. RODRIGUEZ
Department of Mathematics and Statistics
University at Albany, SUNY
Albany NY 12222, USA
carlos@math.albany.edu

Abstract. This paper introduces and studies the basic properties of Clifford algebra valued conditional measures.

Key words: Clifford algebras, probability theory, quantum mechanics, markov processes

1. Introduction

Probability theory was given a firm mathematical foundation in 1933, when Kolmogorov [1] introduced his axioms. By defining probability as an *uninterpreted* special case of a positive measure with total unit mass (plus an additional definition for independence), the subject exploded with new results and found innumerable applications. In 1946, Cox (see [2]) showed that the Kolmogorov axioms for probability are really theorems that follow from basic desiderata about the representation of partial truth with real numbers. We owe to Ed Jaynes (see [3]) the discovery of the importance of Cox's 1946 work ([2]). After Jaynes, it became clear why the calculus of probability is so successful in the real world. Probability works because its axioms axiomatize the right thing: *partial truth of a logical proposition given another*. Even more, the rules of probability are unique in the sense that any other set of consistent rules can be brought into the standard sum and product rules by a change of scale (or we may say logical gauge). This is in fact Cox's main result and it makes futile the enterprise of looking for alternatives to the calculus of normalized real valued probabilities. It is only by allowing the partial truth of a proposition to be encoded by an object other than a real number in the interval $[0,1]$ that we could find alternatives to the standard theory of probability.

We seek to find out what happens when standard probability theory is modified by relaxing the axiom that the probability of an event must be a real number in the interval $[0,1]$. We show that, by allowing the measure of a proposition to take a value in a Clifford Algebra, we automatically find the methods of standard

W. von der Linden et al. (eds.), Maximum Entropy and Bayesian Methods, 247–270.

quantum theory without ever introducing anything specifically related to nature itself.

The main motivation for this article has come from realizing that the derivations in Cox [2] still apply if real numbers are replaced by complex numbers as the encoders of partial truth. This was first mentioned by Youssef [4] and checked in more detail by Caticha [5] who also showed that non-relativistic Quantum theory, as formulated by Feynman [6], is the only consistent calculus of probability amplitudes. By measuring propositions with Clifford numbers we automatically include the reals, complex, quaternions, spinors and any combination of them (among others) as special cases.

2. The Axioms

In this section we introduce the notation and collect the simple properties about Boolean and Clifford algebras that will be needed for the definition of ψ below.

2.1. THE BOOLEAN ALGEBRA $\mathcal{A}$

Let $\mathcal{A}$ be a boolean σ-algebra of propositions $a, b, c, \ldots$. We denote by 0 the false proposition, by 1 the true proposition, by $a+b$ the logical sum, by ab the logical product and by $\bar{a}$ the negation. Each proposition $b \in \mathcal{A}$ defines the set $\mathcal{A}_b$ where,

$$\mathcal{A}_b = \{ba : a \in \mathcal{A}\} = b\mathcal{A} \tag{1}$$

Clearly, $\mathcal{A}_b$ is a subset of $\mathcal{A}$ that contains b and 0, it is closed for sums and products and thus, A_b is a sub algebra of $\mathcal{A}$ with b as the unit. From the fact that $a = ac + a\bar{c}$ it follows that

$$\mathcal{A} = \mathcal{A}_c \oplus \mathcal{A}_{\bar{c}} \tag{2}$$

Given two propositions $a, b \in \mathcal{A}$ we have,

$$\mathcal{A} \supset \mathcal{A}_a \supset \mathcal{A}_{ab} \tag{3}$$

The set $X \subset \mathcal{A}$ is called the set of elementary propositions of $\mathcal{A}$ (and we say that $\mathcal{A}$ is a σ-algebra of propositions in X) if,

1. for $x, y \in X$, $xy = 0$ whenever $x \neq y$
2. every $a \in \mathcal{A}$ is the sum of propositions in X. We write

$$a = \sum_{x \in a} x \tag{4}$$

If $\mathcal{A}$ and $\mathcal{B}$ are two Boolean σ-algebras of propositions in X and Y respectively, then $\mathcal{A} \times \mathcal{B}$ is a Boolean σ-algebra of propositions in $X \times Y$ if one defines the truth value of $(a, b) \in \mathcal{A} \times \mathcal{B}$ as the truth value of ab i.e. true only when both $a \in \mathcal{A}$ and $b \in \mathcal{B}$ are true. We denote by $\mathcal{A}^n$ the σ-algebra of n copies of $\mathcal{A}$ of propositions in X^n. We have,

$$P \in \mathcal{A}^n \Longleftrightarrow P = \sum_{x \in P} x_1 x_2 \ldots x_n \tag{5}$$

and by this we mean that P is always the sum of propositions in X^n and each $x \in X^n$ is the conjunction of n propositions, one for each copy of X. Finally we let $\mathcal{A}^* = \mathcal{A} \setminus \{0\}$.

Notice that these Boolean σ-algebras are nothing but the standard sets where general measures (in particular probability measures) are defined. We chose the notation of logical sums and products instead of the traditional set notation of unions and intersections to emphasize the fact that we are interested in the encoding of partial truth of logical propositions, but this is only a choice of notation and there is a complete one to one correspondence between the two languages. As general references see e.g. Halmos [7] or Chow and Teicher [8].

2.2. THE ALGEBRA OF CLIFFORD NUMBERS $\mathcal{G}$

Let $\mathcal{G}$ be an arbitrary finite dimensional Clifford Algebra with real scalars. We try to follow the notation in [9]. We denote the elements of $\mathcal{G}$ by capital letters like, $A, B, C, \ldots$. A general Clifford number M always expands as the sum of its scalar, vector, bivector, etc parts like:

$$\begin{aligned} M &= \langle M \rangle_0 + \langle M \rangle_1 + \langle M \rangle_2 + \ldots \\ &= \alpha + u + B + \ldots \end{aligned} \tag{6}$$

Where $\langle M \rangle_k$ denotes the k-vector part of M. If u and v are vectors in $\mathcal{G}$ then their geometric (Clifford) product uv can be decomposed into a symmetric part $u \cdot v$ and an antisymmetric part $u \wedge v$ as

$$uv = \frac{1}{2}(uv + vu) + \frac{1}{2}(uv - vu) \tag{7}$$

$$= u \cdot v + u \wedge v \tag{8}$$

The inner product between two vectors is always a scalar and their wedge product is always a bivector. The operation of *reversion* of a clifford number M is denoted by $M^\dagger$ and defined as a linear operation with the properties,

$$\alpha^\dagger = \alpha, \quad u^\dagger = u, \quad (MN)^\dagger = M^\dagger N^\dagger \tag{9}$$

where α is a scalar, u is a vector, and M and N are arbitrary Clifford numbers. The euclidean inner product on $\mathcal{G}$ is given by,

$$\langle M, N \rangle_{\mathcal{G}} = \langle M^\dagger N \rangle_0 \tag{10}$$

2.3. DEFINITION OF ψ

By a clifford algebra valued conditional measure (or simply a ψ) we mean a function,

$$\begin{aligned} \psi: \ \mathcal{A} \times \mathcal{A}^* &\longmapsto \mathcal{G} \\ (a, c) &\longmapsto \psi(a, c) \end{aligned} \tag{11}$$

satisfying the following two axioms:

(I) If $c \Rightarrow b$ then

$$\psi(a, c) = \psi(ab, c) \tag{12}$$

(II) If $\{a_1, a_2, \ldots\} \subset \mathcal{A}$, with $a_j a_k = 0$ for $j \neq k$, then

$$\psi\left(\sum_j a_j, c\right) = \sum_j \psi(a_j, c) \tag{13}$$

Since the only property a proposition in $\mathcal{A}$ always has is its truth value, we can interpret $\psi(a, c)$ as the clifford number that represents the truth in a when c is certain. Axiom (I) says that c is certain (e.g. take $a = 1$ and $b = c$) and axiom (II) says that the whole truth of a for a given c is always the sum of the truths of its separate parts.

2.3.1. *The truth of* 0

By taking each $a_j = 0$ in (13) we get,

$$\psi(0, c) = \psi(0, c) + \psi(0, c) + \ldots \tag{14}$$

and therefore, $\psi(0, c)$ is either $0 \in \mathcal{G}$ or unbounded but if it is unbounded then all the propositions will be assigned an unbounded value since $\psi(a, c) = \psi(a + 0, c) = \psi(a, c) + \psi(0, c)$. Hence,

$$\psi(0, c) = 0 \quad \text{for all } c \in \mathcal{A}^* \tag{15}$$

3. The spaces H_c

The functions ψ, as defined by (12) and (13), are specified independently at each $c \in \mathcal{A}^*$. So far, there is no link between the ψ in the domain of discourse of c, i.e. $\psi(\cdot, c)$ and ψ in the more specialized domain of discourse of bc, i.e. $\psi(\cdot, bc)$. We shall talk about changing domains of discourse in the next section but in this section we describe the important properties that the functions $\psi(\cdot, c)$ have as functions of their first argument only, for fixed $c \in \mathcal{A}^*$. To simplify the notation simply write $\psi(a)$ instead of $\psi(a, c)$ in the formulas below. Thus, whenever the background proposition c is not subject to change we take ψ as any σ-additive function defined on $\mathcal{A}_c$ with values in $\mathcal{G}$. The condition (12) is automatically satisfied since c is the true proposition in $\mathcal{A}_c$.

Let H_c be the set of all σ-additive functions defined on $\mathcal{A}_c$ with values in $\mathcal{G}$.

3.1. THE H_C ARE HILBERT SPACES

Since the sum of two σ-additive functions and the product of a σ-additive function by a scalar are still σ-additive functions we have that the H_c are vector spaces. The scalars are the scalars in $\mathcal{G}$. In principle the field of scalars could be taken as the reals or the complex numbers but it seems that the reals is all that is needed in most applications.

3.1.1. *The inner product in H_c*

For, $\varphi, \psi \in H_c$ define the real inner product between them by:

$$< \varphi, \psi > \; = \; \sum_{x \in X} < \varphi(x), \psi(x) >_{\mathcal{G}} \qquad (16)$$

$$= \; \sum_{x \in X} < \varphi(x)^{\dagger} \psi(x) >_0 \qquad (17)$$

By considering only ψs with finite norm we make H_c a real Hilbert space. From now on we assume the finite norm to be part of the definition of H_c itself, i.e.

$$H_c = \{\psi : \psi \text{ is } \sigma - \text{additive on } \mathcal{A}_c \text{ and } \sum_{x \in X} < \psi^{\dagger}(x)\psi(x) >_0 \; < \infty\} \qquad (18)$$

Notice that the spaces H_c are complete for the inner product (17) since $\mathcal{G}$ with the scalar product $< .,. >_{\mathcal{G}}$ is complete. When X is a finite set (i.e. when it contains only a finite number of propositions) the proof is trivial, just use the fact that if $\{\phi_n\}_{n \in \mathbb{N}} \subset H_c$ is a Cauchy sequence then for each $x \in X$ the sequence $\{\phi_n(x)\}_{n \in \mathbb{N}}$ is also a Cauchy sequence of elements of $\mathcal{G}$ and thus it converges to some $\phi(x) \in \mathcal{G}$ and therefore $\phi \in H_c$ is the limit of the original sequence in H_c. When X is infinite we need to reinterpret the sums as integrals, (for which we need a measure in X), and we also need to reinterpret the ψs as $\mathcal{A}$ -measurable densities, but after that, the proof is essentially the standard proof that L^2 is complete. An important example of an infinite X occurs when the propositions in X are labeled with the vectors in $\mathcal{G}$. In this case the sum in (17) is replaced by the integral with respect to the standard Lebesgue measure in X.

3.2. THE ISOMORPHIC SPACES: $H(\mathcal{A}_C) \simeq H_C(\mathcal{A})$

In order to be able to understand the differences between the current approach and ordinary probability theory it is convenient to introduce two other spaces closely related to H_c. These are, the space of all σ-additive functions on $\mathcal{A}_c$ with values on $\mathcal{G}$,

$$H(\mathcal{A}_c) = \{\psi_c : \mathcal{A}_c \longrightarrow \mathcal{G} \; | \psi_c \; \sigma\text{-additive on } \mathcal{A}_c\} \qquad (19)$$

and the space,

$$H_c(\mathcal{A}) = \{\psi : \mathcal{A} \longrightarrow \mathcal{G} \; | \psi \; \sigma\text{-additive on } \mathcal{A} \text{ AND if } c \Rightarrow b, \psi(ab) = \psi(a)\} \quad (20)$$

Both are Hilbert with the inner product (17) and considering only elements of finite norm.

Notice that if $\psi \in H_c(\mathcal{A})$ then its restriction to $\mathcal{A}_c$ belongs to $H(\mathcal{A}_c)$, i.e.

$$\psi|_{\mathcal{A}_c} = \psi_c \in H(\mathcal{A}_c)$$

and conversely, if $\psi_c \in H(\mathcal{A}_c)$ then the function ψ defined by:

$$\psi(a) = \psi_c(ac) \; \forall a \in \mathcal{A}$$

belongs to $H_c(\mathcal{A})$ since it is clearly σ-additive and if $c \Rightarrow b$ then $\bar{c} + b = 1$ and multiplying both sides by c we get, $bc = c$ from where,

$$\psi(ab) = \psi_c(abc) = \psi_c(ac) = \psi(a)$$

The map $\psi \rightarrow \psi_c$ is obviously linear one to one and onto so it makes the two spaces isomorphic.

Consider now two propositions b and c such that $c, bc \in \mathcal{A}^*$. Then we can write:

$$H(\mathcal{A}_{bc}) \simeq H(b\mathcal{A}_c) \simeq H_b(\mathcal{A}_c) \tag{21}$$

In other words each $\psi(\cdot, bc) \in H(\mathcal{A}_{bc})$ uniquely defines a function $\varphi(\cdot b, c) \in H_b(\mathcal{A}_c)$ and that is all we can say. Since the ψs are unnormalized we can not write a general product rule as in normalized standard probability theory. Nevertheless, it is possible to justify a restricted product rule for independence as we do in section 5 below. When $c = 1 \in \mathcal{A}$ we simply write $H(\mathcal{A})$ instead of $H_1(\mathcal{A})$ or $H(\mathcal{A}_1)$.

4. The truth with ψ

The remarkable fact about the functions ψ is that without committing to a particular value for $\psi(1)$ in $\mathcal{G}$, they still allow to tell what propositions are true. We show in this section that, $b \in \mathcal{A}$ is considered to be true by ψ when $\psi(\bar{b}a) = 0\ \forall a \in \mathcal{A}$. By liberating ordinary probabilities from the constrain that the probability of the whole space must always be fixed at one, we make the space of all possible assignments of partial truth into a Hilbert space without losing the ability to identify truth.

4.1. PROPOSITIONS AS OPERATORS

Each proposition $b \in \mathcal{A}$ defines two complementary linear operators on H_c by multiplication, $\hat{b}$, and by addition, $\check{b}$ to the first argument of ψ. In symbols

$$\hat{b}\psi(a, c) = \psi(ab, c)\ \forall a \in \mathcal{A} \tag{22}$$
$$\check{b}\psi(a, c) = \psi(a + b, c)\ \forall a \in \mathcal{A} \tag{23}$$

To simplify the notation we often omit the hats and simply write $b\psi$ instead of $\hat{b}\psi$. From $bb = b$ and $b + b = b$ it follows that $\hat{b}$ and $\check{b}$ are projectors and therefore they are self-adjoint with eigen values either 0 or 1. We can write,

Theorem 1 *The following two complementary statements are true.*

1. *If $\psi \in H_c$ is an eigen vector of the operator $\hat{b}$ with eigen value 1 then $\psi(b) = \psi(1)$ and we say that ψ makes b true conditional on c. Conversely, if $c \Rightarrow b$ then every $\psi \in H_c$ is an eigen vector of the operator $\hat{b}$ with eigen value 1.*
2. *If $\psi \in H_c$ is an eigen vector of the operator $\check{b}$ with eigen value 1 then $\psi(b) = 0$ and we say that ψ makes b false conditional on c. Conversely, if $c \Rightarrow \bar{b}$ then every $\psi \in H_c$ is an eigen vector of the operator $\check{b}$ with eigen value 1.*

Proof

1.

$$\hat{b}\psi = \psi \Rightarrow \psi(ab, c) = \psi(a, c)\ \forall a \in \mathcal{A}\ \Rightarrow \psi(b, c) = \psi(1, c)$$

where the last implication follows by taking $a = 1$. Conversely, if $c \Rightarrow b$ then from (12) we have,

$$\psi(ab, c) = \psi(a, c)\ \forall a \in \mathcal{A}$$

Thus, $\hat{b}\psi = \psi$.

2.

$$\check{b}\psi = \psi \Rightarrow \psi(a + b, c) = \psi(a, c)\ \forall a \in \mathcal{A}\ \Rightarrow \psi(b, c) = \psi(0, c)$$

where the last implication follows by taking $a = 0$. Conversely,

$$(c \Rightarrow \bar{b}) \Rightarrow \bar{c} + \bar{b} = 1 \Rightarrow a + bc = a\ \forall a \in \mathcal{A}$$

Now from the fact that ψ is a function and applying (12) twice we have,

$$\begin{aligned} \psi(a + bc, c) &= \psi(a, c)\ \forall a \in \mathcal{A} \\ \psi(ac + bc, c) &= \psi(a, c)\ \forall a \in \mathcal{A} \\ \psi(a + b, c) &= \psi(a, c)\ \forall a \in \mathcal{A} \end{aligned}$$

Thus, $\check{b}\psi = \psi$ •

The following theorem elaborates on the same theme.

Theorem 2 *Let $b \in \mathcal{A}$ be an arbitrary proposition in a σ-algebra of propositions in X and let $\psi \in H(\mathcal{A})$. The following are all equivalent:*

1. *$b\psi = \psi$ i.e., ψ makes b true.*
2. *$\bar{b}\psi = 0$ i.e., ψ makes $\bar{b}$ false.*
3. $||\bar{b}\psi|| = 0$
4. $||\psi|| = ||b\psi||$

Proof:
We show that $1 \Leftrightarrow 2 \Leftrightarrow 3 \Leftrightarrow 4$. First equivalence follows from $\psi = b\psi + \bar{b}\psi$, the second equivalence is a property of the norm and the third equivalence is Pythagoras theorem since $(b\psi) \perp (\bar{b}\psi)$ •

It is evident from this last theorem that the norm in the Hilbert spaces H_c provides a mechanism for translating the clifford numbers $\psi(b)$ assigned to the propositions in $\mathcal{A}$ by a function $\psi \in H(\mathcal{A})$ into positive real numbers

$$\begin{aligned} ||\bar{b}\psi||^2 &= ||(1 - b)\psi||^2 = ||\psi - b\psi||^2 &\quad (24) \\ &= ||\psi||^2 - ||b\psi||^2 &\quad (25) \end{aligned}$$

measuring how close is ψ from making the proposition b true. It is also clear from (24) and (25) that it is the square of the norm and not just the norm what is needed. It is only with the square of the norms that we can say that the amount of truth of $\bar{b}$ (measured by $||\bar{b}\psi||^2$) equals the amount of truth assigned to the true proposition (measured by $||1\psi||^2$) minus the amount of truth assigned to b (measured by $||b\psi||^2$).

4.2. COMMUTATIVITY, ORTHOGONALITY AND A CLIFFORD NUMBER TIMES A PROPOSITION

Propositional operators can be composed to form other operators. Thus, if $a, b \in \mathcal{A}$ and $\psi \in H(\mathcal{A})$ we have,

$$
\begin{aligned}
(\hat{a}\hat{b})\psi(x) &= \hat{a}(\hat{b}\psi(x)) = \psi(abx) = (\widehat{ab})\psi(x) && (26)\\
(\hat{a}\check{b})\psi(x) &= \hat{a}\psi(x+b) = \psi(ax+ab) && (27)\\
(\check{b}\hat{a})\psi(x) &= \check{b}\psi(ax) = \psi(ax+b) && (28)
\end{aligned}
$$

and we can see that checks commute with other checks and hats commute with other hats, but in general, hats don't commute with checks.

If $A \in \mathcal{G}$ and $b \in \mathcal{A}$ we can define the operator $A\hat{b}$ by,

$$(A\hat{b})\psi(x) = A(\hat{b}\psi(x)) = A\psi(bx) \tag{29}$$

and similarly for $A\check{b}$. These definitions allow a very rich algebra of operators that mix boolean and clifford algebra properties in new ways. One particularly interesting example of this kind of mix is given by the following statement: *mutually exclusive propositions are orthogonal.* More explicitly, if $a, b \in \mathcal{A}$ and $\psi_1, \psi_2 \in H(\mathcal{A})$ then, $ab = 0 \Rightarrow < a\psi_1, b\psi_2 >= 0$ and pythagoras theorem holds,

$$||a\psi_1 + b\psi_2||^2 = ||a\psi_1||^2 + ||b\psi_2||^2 \tag{30}$$

5. Independence

If the clifford number $\psi(a, c)$ is interpreted as a representation of the partial truth of a when we assume c to be certain then there is only one rational way to define independence namely:

Preliminary Definition: We say that ψ makes propositions a and b in $\mathcal{A}$ *logically independent* conditionally on $c \in \mathcal{A}^*$ if the additional knowledge of one of them does not affect the value of ψ for the other. i.e,

$$
\begin{aligned}
\psi(a, bc) &= \psi(a, c)\\
&AND \\
\psi(b, ac) &= \psi(b, c)
\end{aligned}
\tag{31}
$$

whenever the conditional ψs exist.

5.1. A RESTRICTED PRODUCT RULE

If we try to find the value of $\psi(ab,c)$ in terms of the partial truths that ψ assigns to a and b, then the most general relation is,

$$\psi(ab,c) = F(\psi(a,c),\psi(a,bc),\psi(b,c),\psi(b,ac)) \tag{32}$$

where F is an arbitrary function of its arguments. If we assume further that a and b are logically independent conditionally on c then using (31) the most general relation becomes,

$$\psi(ab,c) = F(\psi(a,c),\psi(b,c)) \tag{33}$$

Let $u = \psi(a,c), v = \psi(b,c)$ and $w = \psi(d,c)$ and use the commutativity and associativity properties of the logical product to get the following two properties for the function F:

$$F(u,v) = F(v,u) \tag{34}$$
$$F(F(u,v),w) = F(u,F(v,w)) \tag{35}$$

In other words, F must be symmetric and it must satisfy the usual associativity equation. If the ψs take values only on a commutative subspace of $\mathcal{G}$ (e.g. reals, complex or pseudo scalars) then the only solution is $F(u,v) = uv$ (see [5]) but this can not be the solution if $uv \neq vu$. Given that $F(u,v)$ must be symmetric, and that it must reduce to uv when u and v commute, and obvious solution is given by the symmetrization of the product, i.e. $(uv+vu)/2$. In principle, it seems feasible that a modification of the standard argument of Aczel (see [10] or [11]) may yield the symmetrized product as the unique solution of (35) and (34) for $u,v \in \mathcal{G}$ at least for u,v in some subset of $\mathcal{G}$ for which (35) is still true. At the present time there is no such proof. In any case the lack of a proof for the uniqueness is not a deterrent to turn the formula into the definition for independence. If it turns out that there are multiple solutions (which seems highly unlikely) the results obtained from this particular solution will still be valid. Thus, from now on we say that ψ makes a and b (logically) independent given c if,

$$\psi(ab,c) = \frac{1}{2}[\psi(a,c)\psi(b,c) + \psi(b,c)\psi(a,c)] \tag{36}$$

More generally we have,

Definition: We say that ψ makes $a_1,a_2,\ldots,a_n$ logically independent given c if, for $k = 1,2,\ldots,n$ and $1 \leq i_1 < i_2 < \ldots < i_k \leq n$

$$\psi(\prod_{j=1}^{k} a_{i_j},c) = \frac{1}{k!}\sum_{\sigma}\psi(a_{\sigma(i_1)},c)\psi(a_{\sigma(i_2)},c)\ldots\psi(a_{\sigma(i_k)},c) \tag{37}$$

where the sum runs over all the permutations, σ of $(i_1,i_2,\ldots,i_k)$.

The associativity equation (35) imposes a heavy restriction on the possible ψ assignments for independent propositions. In fact we have,

Theorem 3 *If ψ makes three or more propositions $a, b, d, \ldots$ independent conditionally on c then the clifford numbers $u = \psi(a,c), v = \psi(b,c), w = \psi(d,c), \ldots$ are such that each of them commutes with the anticommutator of any other two.*

Proof:
From (37) it suffices to show that v must commute with $F(u,w) = (uw + wu)/2$ when a, b, d are independent given c.

The right hand side of (35) simplifies to,

$$F(u, F(v,w)) = \frac{1}{4}\{uvw - uwv + vwu - wvu\} \tag{38}$$

Similarly the left hand side of (35) is given by,

$$F(F(u,v),w) = \frac{1}{4}\{uvw - vuw + wuv - wvu\} \tag{39}$$

Equating (38) to (39) and simplifying we get,

$$[v, F(u,w)] = 0 \tag{40}$$

where $[u,v] = uv - vu$ denotes the usual commutator product •

Notice that when the clifford numbers $u, v, w, \ldots$ either commute or anticommute with each other then (40) is true. But there are many other solutions. For example (40) is also true when $u, v, w, \ldots$ are arbitrary vectors.

5.2. INDEPENDENCE AND ORTHOGONALITY

The above definition for independence makes the following statement true:

Theorem 4 *If $\psi(ab,c) = 0$ then $\psi(a,c)$ anticommutes with $\psi(b,c)$ when and only when ψ makes a and b independent given c.*

There is nothing like this in standard probability theory where mutually exclusive events that are possible (i.e. that have positive probability) are never independent. We are used to think that this makes sense, for if we know that one of the events happens then we also know that the other couldn't happen. The events are totally linked so they can't be independent.

This is fine for real numbers, that are commutative, but not with clifford numbers. There is however, a extreme case where the above theorem is true even in standard probability theory. Suppose that a, b and c are three mutually exclusive propositions. Then, $\psi(ab,c) = \psi(a,c) = \psi(b,c) = \psi(a,c)\psi(b,c) = 0$ and we would have to say that a and b are independent given c even though neither a nor b are possible given c. Anticommutativity allows this to happen even when $\psi(a,c)$ and $\psi(b,c)$ are not zero. Two events can be completely linked (i.e. mutually exclusive) and at the same time be logically independent from each other! This is as weird as entanglement in quantum mechanics.

6. Flipping n coins

If $\mathcal{A}$ is a σ-algebra of propositions in X then, by the σ-additivity property, every $\psi \in H(\mathcal{A})$ is completely specified on $\mathcal{A}$ by just giving $\psi(x)$ for all $x \in X$, i.e.,

$$\psi(x) = \sum_{y \in X} \psi(y)\delta_y(x) \tag{41}$$

where for $x, y \in X$, $\delta_y(x)$ is $1 \in \mathcal{G}$ if $x = y$ and $0 \in \mathcal{G}$ otherwise.

We consider the following special case.

6.1. THE BINOMIAL EXPERIMENT WITH ψS

Let a be an arbitrary proposition and let $X = \{a, \bar{a}\}$ and $\mathcal{A} = \{1, 0, a, \bar{a}\}$. Clearly $\mathcal{A}$ is a boolean algebra of propositions in X. From (41) we have

$$\psi(x) = A\delta_a(x) + B\delta_{\bar{a}}(x) \tag{42}$$

where $\psi \in H(\mathcal{A})$ and $A, B \in \mathcal{G}$. This is the canonical Bernoulli experiment. There are only two possible outcomes a and $\bar{a}$ with partial truths encoded by the clifford numbers $A = \psi(a)$ and $B = \psi(\bar{a})$. As in standard probability theory, consider now n independent repetitions of the Bernoulli experiment. i.e., consider X^n with its corresponding boolean algebra $\mathcal{A}^n$ of elements in X^n (see (5)). From (41) a general $\psi_n \in H(\mathcal{A}^n)$ is given by,

$$\psi_n(x) = \sum_{y \in X^n} \psi_n(y)\delta_y(x) \tag{43}$$

From the assumption that ψ_n make the different repetitions independent, we obtain, using (37) that

$$\psi_n(x) = \psi_n(x_1, \ldots, x_n) = M_n(m(x)) \tag{44}$$

where for each integer k with $0 \leq k \leq n$, $M_n(k) \in \mathcal{G}$ is the symmetrization of the product $A^k B^{n-k}$ and $m(x)$ is the number of a's in x. Here are some examples for $n = 2$ and $n = 3$,

$$\begin{aligned} \psi_2(a, a) &= A^2 = M_2(2), \quad \psi_2(a, \bar{a}) = \psi_2(\bar{a}, a) = \frac{1}{2}[AB + BA] = M_2(1) \\ \psi_3(a, a, \bar{a}) &= \psi_3(a, \bar{a}, a) = \psi_3(\bar{a}, a, a) = \frac{1}{3}[A^2B + ABA + BA^2] = M_3(2). \end{aligned}$$

Now define the proposition $P_n^k \in \mathcal{A}^n$ by,

$$P_n^k = \text{“exactly } k \text{ of the } n \text{ repetitions is an } a\text{”} \tag{45}$$

Recall that by (22) we have,

$$P_n^k \psi_n(x) = \psi_n(P_n^k x) = \begin{cases} \psi_n(x) & \text{if } P_n^k x \neq 0 \\ 0 & \text{otherwise} \end{cases} \tag{46}$$

By the first part of theorem (1) we have that ψ_n makes P^k_n true when $P^k_n\psi_n = \psi_n$. So the question is: *How far is ψ_n from making P^k_n true?*. Answer: $||\psi_n - P^k_n\psi_n||^2$.

6.2. COMPUTATION OF $||\psi_N - P^K_N\psi_N||^2$

To compute this distance we use the fact that P^k_n, and its negation in $\mathcal{A}^n$, $1-P^k_n$, are mutually exclusive propositions hence orthogonal (see (30)) and by pythagoras,

$$||\psi_n - P^k_n\psi_n||^2 = ||\psi_n||^2 - ||P^k_n\psi_n||^2 \tag{47}$$

Let us compute each of these terms. From (17),

$$||\psi_n||^2 = \sum_{x\in X^n} \langle\psi^\dagger_n(x)\psi_n(x)\rangle_0. \tag{48}$$

and using (43) and (44) we can write,

$$\psi_n(x) = \sum_{y\in X^n} M_n(m(y))\delta_y(x) \tag{49}$$

from where we obtain,

$$\begin{aligned}\psi^\dagger_n(x)\psi_n(x) &= \sum_{y_1,y_2\in X^n} M^\dagger_n(m(y_1)M_n(m(y_2))\delta_{y_1}(x)\delta_{y_2}(x)\\ &= \sum_{x\in X^n} M^\dagger_n(m(x))M_n(m(x))\end{aligned}$$

and replacing in (48) we get,

$$\begin{aligned}||\psi_n||^2 &= \sum_{x\in X^n} |M_n(m(x))|^2\\ ||\psi_n||^2 &= \sum_{j=0}^{n}\binom{n}{j}|M_n(j)|^2\end{aligned} \tag{50}$$

the last equation followed from the fact that there are $\binom{n}{j}$ propositions in X^n with exactly j components equal to a. We use the same fact again to compute the other norm in (47),

$$||P^k_n\psi_n||^2 = \sum_{x\in X^n} \langle\psi^\dagger_n(P^k_n x)\psi_n(P^k_n x)\rangle_0 \tag{51}$$

to obtain,

$$||P^k_n\psi_n||^2 = \binom{n}{k}|M_n(k)|^2 \tag{52}$$

Replacing (50) and (52) in (47) we get,

$$||\psi_n - P^k_n\psi_n||^2 = \sum_{j=0}^{n}\binom{n}{j}|M_n(j)|^2 - \binom{n}{k}|M_n(k)|^2 \tag{53}$$

Let us consider the proposition, $P^f_{n,\epsilon} \in \mathcal{A}^{\,n}$ defined by,

$$P^f_{n,\epsilon} \;=\; \text{"The observed frequency of } a\text{'s in } n \text{ independent repetitions is } k/n \text{ with } f-\epsilon \leq \tfrac{k}{n} \leq f+\epsilon\text{"} \tag{54}$$

in other words for $x \in X^n$, $P^f_{n,\epsilon}x \neq 0$ when and only when the proportion of a's in $x = (x_1, \ldots, x_n)$ is within ϵ from the specified frequency f. The proposition $P^f_{n,\epsilon}$ is equal to the following disjunction of $2n\epsilon + 1$ mutually exclusive propositions P^k_n:

$$P^f_{n,\epsilon} = \sum_{k=n(f-\epsilon)}^{n(f+\epsilon)} P^k_n \tag{55}$$

hence, from (30) we get,

$$||P^f_{n,\epsilon}\psi_n||^2 = \sum_{k=n(f-\epsilon)}^{n(f+\epsilon)} ||P^k_n\psi_n||^2 \tag{56}$$

and from (47) and (53) we can write,

$$||\psi_n - P^f_{n,\epsilon}\psi_n||^2 = \sum_{j=0}^{n} \binom{n}{j} |M_n(j)|^2 - \sum_{k=n(f-\epsilon)}^{n(f+\epsilon)} \binom{n}{k} |M_n(k)|^2 \tag{57}$$

In general this distance increases without limit as $n \to \infty$ but it can converge relative to the size of ψ_n. Let us define the relative error by,

$$\Delta^f_{n,\epsilon} = \frac{||\psi_n - P^f_{n,\epsilon}\psi_n||^2}{||\psi_n||^2} \tag{58}$$

Using (57) and (50) we have,

$$\Delta^f_{n,\epsilon} = 1 - \frac{\displaystyle\sum_{k=n(f-\epsilon)}^{n(f+\epsilon)} \binom{n}{k} |M_n(k)|^2}{\displaystyle\sum_{k=0}^{n} \binom{n}{k} |M_n(k)|^2} \tag{59}$$

We separate the computation of $\Delta^f_{n,\epsilon}$ into three different cases.

6.3. CASE: $AB = BA$

From (59) we can write the following,

Theorem 5 *If $AB \neq 0$, $AB = BA$ and $|A^k B^{n-k}| = |A|^k |B|^{n-k}$ then,*

$$\Delta^f_{n,\epsilon} = 1 - \sum_{k=n(f-\epsilon)}^{n(f+\epsilon)} \binom{n}{k} p^k (1-p)^{n-k} \tag{60}$$

where,

$$p = \frac{|A|^2}{|A|^2 + |B|^2} \tag{61}$$

Proof:
Under the conditions of the theorem we have,

$$|M_n(k)|^2 = |A|^{2k}|B|^{2(n-k)}$$

replacing this last equation in (59) and noticing that,

$$\sum_{k=0}^{n} \binom{n}{k} |A|^{2k}|B|^{2(n-k)} = \left(|A|^2 + |B|^2\right)^n$$

we immediately obtain (60) and (61) •

It is not always true that for $A, B \in \mathcal{G}$, $|AB| = |A||B|$ even when $AB = BA$ (take for example $A = 1 + \alpha u, B = 1 - \beta u$ for a unit vector u and scalars α and β) so the extra condition besides commutativity is needed for the theorem to be true.

6.4. CASE: $AB = 0$

Unlike the real (or complex) numbers, the product of non zero clifford numbers can be zero (e.g. take $\alpha = \beta = 1$ in the example above) so this case is not trivial. When $AB = 0$ the following is true,

Theorem 6 *If $AB = 0$ then,*

$$\Delta^f_{n,\epsilon} = \begin{cases} \dfrac{|A|^{2n}}{|A|^{2n} + |B|^{2n}} & \text{if } f = 0 \\ 1 & \text{if } 0 < f < 1 \\ \dfrac{|B|^{2n}}{|A|^{2n} + |B|^{2n}} & \text{if } f = 1 \end{cases} \tag{62}$$

Proof:
Notice that when $AB = 0$ then all the symmetrized products, except the two extremes are zero, i.e., $M_n(k) = 0$ for all $0 < k < n$ and $M_n(n) = A^n$ and $M_n(0) = B^n$. Substituting these values into (59) we obtain (62) •

6.5. CASE: $AB = -BA$

When A and B anticommute we have,

Theorem 7 *If $AB \neq 0$, $AB = -BA$ and $|A^k B^{n-k}| = |A|^k |B|^{n-k}$ then,*

$$\Delta^f_{n,\epsilon} = 1 - \frac{\sum_{k=n(f-\epsilon)}^{n(f+\epsilon)} b_n(k)(1 - 2\lambda_n(k))^2}{\sum_{k=0}^{n} b_n(k)(1 - 2\lambda_n(k))^2} \tag{63}$$

where $b_n(k)$ are the binomial probabilities,

$$b_n(k) = \binom{n}{k} p^k (1-p)^{n-k}, \text{ with } p \text{ as before. i.e., } p = \frac{|A|^2}{|A|^2 + |B|^2} \tag{64}$$

and the numbers $\lambda_n(k)$ satisfy $\lambda_n(k) = \lambda_n(n-k)$ and for $k \leq n/2$, $\lambda_n(k)$ is the chance of drawing and odd number of RED balls out of k draws without replacement from a box containing either: $n/2$ REDS and $n/2$ BLUES if n is even or $(n+1)/2$ REDS and $(n-1)/2$ BLUES if n is odd.

Proof:
Recall that $M_n(k)$ is the symmetrization of $A^k B^{n-k}$, i.e., the average over all the permutations of $A^k B^{n-k}$. There are $\binom{n}{k}$ permutations and, by the assumed anticommutativity of A with B, each permutation is either $A^k B^{n-k}$ or $-A^k B^{n-k}$ so we have,

$$M_n(k) = \frac{\rho(n,k) A^k B^{n-k}}{\binom{n}{k}} \tag{65}$$

where $\rho(n,k)$ is an integer. From the fact that $|M_n(k)|$ is invariant under the transformation: $A \to B$, $B \to A$, and $k \to (n-k)$ it follows that $|\rho(n,k)| = |\rho(n, n-k)|$. In order to prove the theorem it is sufficient to show that,

$$\frac{|\rho(n,k)|}{\binom{n}{k}} = |1 - 2\lambda_n(k)| \tag{66}$$

since if (66) is true, by using the conditions of the theorem we have,

$$|M_n(k)|^2 = (1 - 2\lambda_n(k))^2 |A|^{2k} |B|^{2(n-k)} \tag{67}$$

and dividing the numerator and the denominator of (58) by $(|A|^2 + |B|^2)^n$ we obtain (63).

Let us show that (66) is true by giving an explicit formula for $|\rho(n,k)|$ when $k \leq n/2$. To do this, represent each permutation of $A^k B^{n-k}$ by the k integers $(j_1 j_2 \ldots j_k)$ that correspond to the positions of the A's in increasing order. For example, for $n = 6$ and $k = 3$, the permutation $ABABBA$ is represented by (136), since the As are found at positions 1, 3 and 6. The permutation $AABBBA$ is represented by (126) etc. Define the parity of $(j_1 \ldots j_k)$ as

$$\text{parity of } (j_1 j_2 \ldots j_k) = (-1)^{j_1 + j_2 \ldots + j_k} = (-1)^{j_1} (-1)^{j_2} \ldots (-1)^{j_k} \tag{68}$$

Note that the transposition of an A with a B, located next to it, changes by one the position of that A in the permutation and hence, the parity of the permutation obtained after the transposition is always the reverse of the parity of the original permutation. From this and the fact that we can transform any permutation into any other by a sequence of transpositions it follows that two permutations have the same parity if and only if the number of flips (transpositions) necessary for transforming one permutation into the other is even.

The permutation $A^k B^{n-k}$ always corresponds to $(12 \ldots k)$ and therefore an arbitrary permutation $(j_1 j_2 \ldots j_k)$ will have the same parity as $A^k B^{n-k}$ if the parity of the number of odd integers in the set $\{j_1, j_2, \ldots, j_k\}$ is the same as the parity of the number of odd integers in the set $\{1, 2, \ldots, k\}$. In other words, if there are an even number of odd integers in the set $\{1, 2, \ldots, k\}$ then every permutation $(j_1 j_2 \ldots j_k)$ which also contains an even number of odd integers can be reorder into $A^k B^{n-k}$ but if the number of odd integers in $\{j_1, \ldots, j_k\}$ is odd then the permutation reorders into $-A^k B^{n-k}$. Therefore, we can write

$$|\rho(n,k)| = | \sum_{1 \le j_1 < j_2 \ldots < j_k \le n} (-1)^{j_1 + j_2 + \ldots + j_k} | \tag{69}$$

Thus, if we call N_e the number of permutations with an even number of odd integers among $\{j_1 \ldots, j_k\}$ and we call N_o the number of permutations with an odd number of odds, then,

$$|\rho(n,k)| = |N_e - N_o| \tag{70}$$

using the fact that $N_e + N_o = \binom{n}{k}$ we also have that,

$$|\rho(n,k)| = |\binom{n}{k} - 2N_o| \tag{71}$$

We now turn to the computation of N_o. Let $N_o(m)$ be the total number of permutations $(j_1 j_2 \ldots j_k)$ with exactly m of the positions of the A's being odd. We have,

$$N_o = \begin{cases} \sum_{t=0}^{\frac{k}{2}-1} N_o(2t+1) & \text{if } k \text{ is even} \\ \\ \sum_{t=0}^{\frac{k-1}{2}} N_o(2t+1) & \text{if } k \text{ is odd} \end{cases} \tag{72}$$

where, for $0 \le m \le k \le n/2$

$$N_o(m) = \begin{cases} \binom{n/2}{m}\binom{n/2}{k-m} & \text{if } n \text{ is even} \\ \\ \binom{(n+1)/2}{m}\binom{(n-1)/2}{k-m} & \text{if } n \text{ is odd} \end{cases} \tag{73}$$

this is because the set $\{1, 2, \ldots, n\}$ contains an equal number of odd and even numbers when n is even but the number of odds is one more than the number of even when n is odd. So dividing (71) by $\binom{n}{k}$ and using (72) and (73) we obtain (66) with $\lambda_n(k)$ defined as the theorem says. There are four different formulas for $\lambda_n(k)$ depending on the parities of n and k.Let us check one of them. When n and k are both even and $k \le n/2$ we have,

$$\lambda_n(k) = \sum_{t=0}^{\frac{k}{2}-1} \frac{\binom{n/2}{2t+1}\binom{n/2}{k-2t-1}}{\binom{n}{k}} \tag{74}$$

and we can see that (74) is the chance of drawing an odd number of red balls when drawing at random k balls, without replacement, from a box containing $n/2$ red balls and $n/2$ blue balls. This completes the proof of the theorem •

7. The weak law of large numbers

7.1. TAKING LIMITS AS $N \to \infty$

In this section we compute

$$\lim_{n\to\infty} \Delta^f_{n,\epsilon}$$

for the three cases considered in the previous section.

Theorem 8 *If $AB \neq 0$, $|A^k B^{n-k}| = |A|^k |B|^{n-k}$ and either $AB = BA$ or $AB = -BA$ then, for all sufficiently small $\epsilon > 0$,*

$$\lim_{n\to\infty} \Delta^f_{n,\epsilon} = \begin{cases} 0 & \text{if } f = p \\ 1 & \text{if } f \neq p \end{cases} \tag{75}$$

where as before, $p = \frac{|A|^2}{|A|^2+|B|^2}$. Moreover, if $AB = 0$, then $\forall \epsilon > 0$,

$$\lim_{n\to\infty} \Delta^f_{n,\epsilon} = \begin{cases} 0 & \text{if } (f = 0 \text{ and } |A| < |B|) \text{ or } (f = 1 \text{ and } |A| > |B|) \\ 1/2 & \text{if } |A| = |B| \text{ and either } f = 0 \text{ or } f = 1 \\ 1 & \text{otherwise} \end{cases} \tag{76}$$

Proof

For the first part we use equations, (60) and (63). By the usual gaussian approximation for the binomial probabilities (e.g. see [12] p.59) we have that for any integers $0 \leq k_1 \leq k_2 \leq n$ and any function g_n with finite expectation with respect to the standard gaussian,

$$\sum_{k=k_1}^{k_2} b_n(k) g_n(k) = \int_{\frac{k1-np}{\sqrt{npq}}}^{\frac{k2-np}{\sqrt{npq}}} g_n(np + x\sqrt{npq}) \frac{1}{\sqrt{2\pi}} e^{\frac{-x^2}{2}} dx \left(1 + o(n^0)\right) \tag{77}$$

thus, taking $k_1 = n(f - \epsilon)$, $k_2 = n(f + \epsilon)$, $g_n(y) = 1$ for $0 \leq y \leq n$ and $g_n(y) = 0$ outside $[0, n]$ we obtain from equation (60) that,

$$\lim_{n\to\infty} \Delta^f_{n,\epsilon} = 1 - \lim_{n\to\infty} \int_{\frac{n(f-p-\epsilon)}{\sqrt{npq}}}^{\frac{n(f-p+\epsilon)}{\sqrt{npq}}} \frac{1}{\sqrt{2\pi}} e^{\frac{-x^2}{2}} dx \tag{78}$$

hence, when $f \neq p$ for any $0 < \epsilon < |f - p|$ the limits of the integral in equation (78) are both positive or both negative and both going to ∞ as $n \to \infty$ so the desired limit is $1 - 0 = 0$. On the other hand when $f = p$ for any $\epsilon > 0$ the desired limit is $1 - 1 = 0$ and this shows that (75) is true for the commutative case. To show (75) for the anticommutative case we take

$$g_n(y) = \frac{1}{4} y^2 (y - 1)^2 \tag{79}$$

which increases like y^6 and therefore it has finite expectation with respect to the standard gaussian. If we show that

$$g_n(k) = n^2(1-2\lambda_n(k))^2 + o(n^0) \tag{80}$$

then it will follow from (80), (77) and (63) that,

$$\lim_{n\to\infty} \Delta^f_{n,\epsilon} = 1 - \lim_{n\to\infty} \frac{\int_{\frac{n(f-p-\epsilon)}{\sqrt{npq}}}^{\frac{n(f-p+\epsilon)}{\sqrt{npq}}} g_n(np+x\sqrt{npq})\frac{1}{\sqrt{2\pi}}e^{\frac{-x^2}{2}}\,dx}{\int_{\frac{-np}{\sqrt{npq}}}^{\frac{nq}{\sqrt{npq}}} g_n(np+x\sqrt{npq})\frac{1}{\sqrt{2\pi}}e^{\frac{-x^2}{2}}\,dx} \tag{81}$$

and by the same reasoning as in the commutative case we obtain (75) for the anticommutative case. Let us then show (80). Notice that from (74) we can write,

$$\lambda_n(k) = \sum_{t=0}^{\frac{k}{2}-1} W(n,2t+1,k) \tag{82}$$

where the hypergeometric probabilities,

$$\begin{aligned} W(n,m,k) &= \frac{\binom{n/2}{m}\binom{n/2}{k-m}}{\binom{n}{k}} \\ &= \frac{\left[\frac{1}{n^m}\binom{n/2}{m}\right]\left[\frac{1}{n^{k-m}}\binom{n/2}{k-m}\right]}{\left[\frac{1}{n^k}\binom{n}{k}\right]} \\ &= \frac{\left[\frac{1}{m!}\frac{1}{2}(\frac{1}{2}-\frac{1}{n})\cdots(\frac{1}{2}-\frac{m-1}{n})\right]\left[\frac{1}{(k-m)!}\frac{1}{2}(\frac{1}{2}-\frac{1}{n})\cdots(\frac{1}{2}-\frac{k-m-1}{n})\right]}{\left[\frac{1}{k!}1(1-\frac{1}{n})(1-\frac{2}{n})\cdots(1-\frac{k-1}{n})\right]} \end{aligned} \tag{83}$$

expanding the products up to terms of order $(1/n)$ and letting $W = W(n,m,k)$ we have,

$$\begin{aligned} W &= \binom{k}{m}\frac{\left[2^{-m}\{1-\frac{m(m-1)}{n}+o(n^{-1})\}\right]\left[2^{m-k}\{1-\frac{(k-m)(k-m-1)}{n}+o(n^{-1})\}\right]}{1-\frac{k(k-1)}{n}+o(n^{-1})} \\ &= \binom{k}{m}\left(\frac{1}{2}\right)^k\left\{1-\frac{1}{n}[m^2+(k-m)^2-k]+o(n^{-1})\right\}\left\{1+\frac{k(k-1)}{n}+o(n^{-1})\right\} \\ &= \binom{k}{m}\left(\frac{1}{2}\right)^k\left\{1+\frac{2}{n}m(k-m)+o(n^{-1})\right\} \end{aligned} \tag{84}$$

We can readily check that,

$$\sum_{t=0}^{\frac{k}{2}-1}\binom{k}{2t+1}\left(\frac{1}{2}\right)^k = \frac{1}{2} \tag{85}$$

and that,

$$\sum_{t=0}^{\frac{k}{2}-1}(2t+1)(k-2t-1)\binom{k}{2t+1} = \frac{1}{4}k(k-1)2^{k-1} \tag{86}$$

From (85), (86), (84) and (82) we have,

$$\lambda_n(k) - \frac{1}{2} = \frac{k(k-1)}{4n} + o(n^{-1}). \tag{87}$$

Squaring both sides of (87) and multiplying through by $4n^2$ we obtain,

$$n^2(1-2\lambda_n(k))^2 = \frac{1}{4}k^2(k-1)^2 + o(n^0) \tag{88}$$

which is exactly (80). This ends the proof for the anticommutative case. The second part of the theorem i.e. (76) follows directly from (62) by taking limits as $n \to \infty$•

7.2. FLIPPING AN INFINITE NUMBER OF COINS

As in standard probability theory there is a subtle nuisance with limits such as (75) and (76) that needs to be faced in order to have a straight probabilistic interpretation for laws of large numbers. The problem with (75) and (76) is that it is not clear how to paste all the ψ_n together into one global ψ_∞. It was due to these kind of problems that modern measure-theoretic probability theory was born.

To be able to make statements about infinite sequences of bernoulli trials we need to specify a boolean σ-algebra, $\mathcal{A}^{\infty}$, that contains at least those statements. This can be done as in standard probability theory (e.g. see [8]), i.e. $\mathcal{A}^{\infty}$ is defined as the smallest σ-algebra containing the cylinder sets, in particular it contains the propositions P_n^k defined in (45) but now n refers to the first n repetitions in an infinite sequence of bernoulli trials. Having constructed $\mathcal{A}^{\infty}$ we also need to construct the Hilbert space, $H(\mathcal{A}^{\infty})$, containing the functions $\psi = \psi_\infty$. Again, the construction is not trivial but well known in functional analysis as the standard construction of an infinite tensor product of Hilbert spaces (e.g. see [13]). These standard constructions allow us to write,

$$P_n^k\psi = P_n^k\psi_n \tag{89}$$

where $\psi = \psi_\infty \in H(\mathcal{A}^{\infty})$. Equation (89) can be used to re-write the statements (75) and (76) as,

Theorem 9 *Let X^∞ be the space of infinite sequences of independent tosses of a coin and let $\mathcal{A}^{\infty}$ be the smallest σ-algebra containing all the propositions P_n^k about elements in X^∞. If for each toss the ψ values for falling heads and tails are the clifford numbers A and B satisfying,*

1. $|A|^2 + |B|^2 = 1$
2. $AB \neq 0$
3. *either* $AB = BA$ *or* $AB = -AB$

4. $|A^k B^{n-k}| = |A|^k |B|^{n-k} \quad \forall n \in \mathbb{N}, \forall 0 \leq k \leq n.$

Then, for all sufficiently small $\epsilon > 0$ *the propositions,*

$$P^{|A|^2}_{\infty,\epsilon} \in \mathcal{A}^{\infty}$$

are true.

Proof

Under the conditions of the theorem we have from (89) and (75) that when the ψ_n are normalized i.e. when $||\psi_n|| = 1$ for all n then,

$$||\psi - P^{|A|^2}_{n,\epsilon} \psi|| \to 0 \text{ as } n \to \infty$$

or equivalently,

$$\lim_{n\to\infty} P^{|A|^2}_{n,\epsilon} \psi = P^{|A|^2}_{\infty,\epsilon} \psi = \psi \tag{90}$$

so that ψ is an eigen vector of the operator $P^{|A|^2}_{\infty,\epsilon}$ with eigen value 1 and thus, it makes the proposition true •

We also have,

Theorem 10 *Let* X^∞ *and* $\mathcal{A}^{\infty}$ *be as in the previous theorem but now suppose that the clifford numbers* A *and* B *satisfy,*

1. $AB = 0$

2. $|A| > |B|$

Then for all $\epsilon > 0$ *the propositions,*

$$P^{1}_{\infty,\epsilon} \in \mathcal{A}^{\infty}$$

are true.

Proof

Under the conditions of the theorem we have from (89) and (76) that when the ψ_n are all of unit norm then

$$||\psi - P^{1}_{n,\epsilon}\psi|| \to 0 \text{ as } n \to \infty$$

or equivalently,

$$\lim_{n\to\infty} P^{1}_{n,\epsilon}\psi = P^{1}_{\infty,\epsilon}\psi = \psi \tag{91}$$

so that ψ is an eigen vector of the operator $P^{1}_{\infty,\epsilon}$ with eigen value 1 and thus, it makes the proposition true •

7.3. INTERPRETATION AND EXAMPLES

The previous two theorems can be interpreted as in standard probability theory. They say that an infinite sequence of independent tosses of a coin with $\psi($ heads $) = A$ and $\psi($ tails $) = B$ will have for sure (relative to ψ) a frequency of heads within ϵ from $|A|^2$ in the first case and within ϵ from 1 in the $AB = 0$ case. When

$AB = 0$ the theorem assures us that (again relative to ψ) the coin will show up heads with frequency 100% whenever $|A| > |B|$!

The four conditions on A and B that are needed for the $AB \neq 0$ case, impose heavy restrictions on the possible values that A and B can take but there are lots of examples. Let p be a real number in the interval $[0, 1]$ and consider,

Example 1

$$A = \sqrt{p} \qquad B = \sqrt{1-p} \tag{92}$$

Example 2

$$A = \sqrt{p} \qquad B = \sqrt{1-p}\hat{B} \tag{93}$$

where $\hat{B} = \sigma_1\sigma_2\ldots\sigma_r$ is a unit blade, i.e. it can be factorized into a product of orthogonal (anticommuting) unit vectors σ_j.

Example 3

$$A = \sqrt{p}\hat{A} \qquad B = \sqrt{1-p}\hat{B} \tag{94}$$

where $\hat{A}$ and $\hat{B}$ are both unit blades possibly of different dimensions.

Example 4

$$A = \sqrt{p}\, e^{\mathbf{i}\alpha}\hat{A} \qquad B = \sqrt{1-p}\, e^{\mathbf{i}\beta}\hat{B} \tag{95}$$

where α and β are scalars, $\hat{A}$ and $\hat{B}$ are both unit blades and $\mathbf{i}$ is any multivector such that $\mathbf{i}^2 = -1$ and $\mathbf{i}$ commutes or anticommutes with both $\hat{A}$ and $\hat{B}$ i.e. $\mathbf{i}\hat{A} = \pm\hat{A}\mathbf{i}$ and $\mathbf{i}\hat{B} = \pm\hat{B}\mathbf{i}$

It can be readily check that all these examples satisfy the four conditions of the theorem and hence, coin tosses with these ψs will show up heads with probability p.

7.4. WHY ISN'T EVERY ONE A FREQUENTIST?

For the same reason as in probability theory these laws of large numbers can not be used to define what we mean by *the partial truth that the coin will show up heads in the next toss* since the theorem only says that the propositions $P^p_{\infty,\epsilon}$ are made true *by* ψ. So any attempt to use the law of large numbers as the definition of what ψ is, or means, is therefore circular.

8. The Boolean algebra of Caticha's temporal filters

Let X be a set and let $\mathcal{B}$ be a σ-algebra of subsets of X. Notice that we are using the standard set notation for the elements of $\mathcal{B}$ instead of the logical notation used in the rest of the paper. The reason for changing the notation is that the boolean σ-algebra that we are trying to define is not $\mathcal{B}$ itself but only based on $\mathcal{B}$. Think of X as the set of possible locations for a point particle and define the elementary propositions $e(x,t)$ by the statement: *the particle is at location x at time t*. As in [5], $e(x,t)$ is a pure hypothesis not the result of a measurement. The truth value of $e(x,t)$ can be obtained, at least in principle, by imagining a filter that covers all of X except at location x where it has an infinitesimal hole. This magical filter materializes only for an instant at time t and then disappears leaving no trace of its existence. If after time t we still find the particle somewhere then we conclude that $e(x,t)$ is true. These filters form a boolean algebra with the definitions below.

Let T be a subset of the real line and define for $t \in T$ and $B \in \mathcal{B}$ the proposition $e(B,t)$ as: *an elementary filter at time t with B open.* Thus, $e(B,t)$ is true if and only if the statement: *the particle is somewhere in B at time t* is true. We define the logical product of two elementary filters as the operation of putting one on top of the other and we define the negation of an elementary filter as the filter that closes the holes and opens the rest. In symbols:

$$e(B_1,t)e(B_2,t) = e(B_1 \cap B_2,t) \tag{96}$$

$$\overline{e(B,t)} = e(\overline{B},t) \tag{97}$$

$$e(B_1,t) + e(B_2,t) = e(B_1 \cup B_2,t) \tag{98}$$

where $\overline{B} = X \setminus B$ is the complement of B with respect to X. Notice that (98) follows from (96) and (97) by using De'Morgan's law i.e.,

$$\begin{aligned} e(B_1,t) + e(B_2,t) &= \overline{e(\overline{B_1},t)e(\overline{B_2},t)} \\ &= \overline{e(\overline{B_1} \cap \overline{B_2},t)} \\ &= e(B_1 \cup B_2,t) \end{aligned}$$

We also have that for all $s,t \in T$,

$$e(B_1,s)e(B_2,t) = \text{“Filter at time s followed (or on top of) filter at time t”}$$

$$e(B_1,s) + e(B_2,t) = \text{“Filter at time s } \mathbf{OR} \text{ filter at time t”}$$

$$e(\phi,t) = \text{“Barrier (nothing open) at time } t\text{”} = 0 \tag{99}$$

$$e(X,t) = \text{“Absence of filter (all open) at time } t\text{”} = 1 \tag{100}$$

We define $\mathcal{F}$ as the smallest σ-algebra containing the elementary filters $e(B,t)$ i.e.,

$$\mathcal{F} = \sigma\{e(B,t) : B \in \mathcal{B}, t \in T\} \tag{101}$$

The boolean algebra of temporal filters $\mathcal{F}$ is a spell out of the usual algebra of events of a stochastic process with state space X.

8.1. THE MARKOV PROPERTY

Due to the fact that there is no product rule for the unnormalized ψs we cannot make use of the standard Markov property of probability theory directly. The following definition is all that is needed to recover non relativistic quantum mechanics,

Definition: $\psi \in H(\mathcal{F})$ is said to have *independent segments given* $c \in \mathcal{F}$ if for all $n = 1,2,\ldots$, all times $t_0 < t_1 < t_2 < \ldots < t_n$ in T and all locations $x_0, x_1, x_2, \ldots, x_n$ in X the propositions

$$e(x_0,t_0)e(x_1,t_1),\ e(x_1,t_1)e(x_2,t_2),\ldots,e(x_{n-1},t_{n-1})e(x_n,t_n)$$

are independent given c.

8.2. TIME EVOLUTION AND THE SCHRÖDINGER EQUATION

When $\psi \in H(\mathcal{F})$ has independent segments, it evolves according to the Schrödringer equation. The usual jargon of quantum mechanics is recovered with the notation,

Probability Amplitude: $\psi(e(x,s)e(y,t), e(x_0,t_0))$ is the amplitude for the particle to go from location x at time s to location y at time $t > s$ given that it was initially prepared at location x_0 at time t_0. We denote this amplitude by $K(y,t;x,s)$.

Wave Function: $\psi(e(x,t), e(x_0,t_0))$ is the amplitude of going from the initial position to location x at time t. It is often denoted by just $\Psi(x,t)$.

Thus, with this notation, a particle which is prepared by $e(x_0,t_0)$ and for which $\psi \in H(\mathcal{F})$ has independent segments conditionally on this preparation, will satisfy,

$$\Psi(x,t) = \sum_{y\in X} \frac{1}{2}\left[K(x,t;y,s)\Psi(y,s) + \Psi(y,s)K(x,t;y,s)\right] \tag{102}$$

since

$$\begin{aligned}\Psi(x,t) &= \sum_{y\in X} \psi(e(x,t)e(y,s), e(x_0,t_0)) \\ &= \sum_{y\in X} \psi\left([e(x_0,t_0)e(y,s)]\,[e(y,s)e(x,t)], e(x_0,t_0)\right)\end{aligned}$$

taking derivatives in (102) with respect to t and evaluating at $t = s$ we obtain,

$$\left.\frac{\partial\Psi(x,t)}{\partial t}\right|_{t=s} = \sum_{y\in X} \frac{1}{2}\left[\left.\frac{\partial K(x,t;y,s)}{\partial t}\right|_{t=s} \Psi(y,s) + \Psi(y,s)\left.\frac{\partial K(x,t;y,s)}{\partial t}\right|_{t=s}\right]$$

Defining the Hamiltonian H by,

$$\left.\frac{\partial K(x,t;y,s)}{\partial t}\right|_{t=s} = -\frac{i}{\hbar}H(x,y,s) \tag{103}$$

where i is any multivector that squares to -1 and that it commutes with all the ψs. Relabeling s with t we can write Schrödinger equation for possible non-commuting ψs as,

$$i\hbar\frac{\partial\Psi(x,t)}{\partial t} = \sum_{y\in X} \frac{1}{2}\left[H(x,y,t)\Psi(y,t) + \Psi(y,t)H(y,x,t)\right] \tag{104}$$

when the wave functions Ψ commute with the Hamiltonian, (e.g. when all the ψs take values in a commutative subspace of $\mathcal{G}$) (104) reduces to the usual Schrödinger equation.

9. Next:

Using the Spacetime algebra How to connect the above with the Dirac-Hestenes equation.

ψ assignments in the real continuous case Minimum Fisher information and the Huber-Frieden derivation of the time independent Schrödinger equation.

ψ and Brownian motion Nagasawa's diffusion model.

Comments and conclusion What the hell is this all about and what it may be likely to become....

References

1. A. N. Kolmogorov, *Foundations of the Theory of Probability, 1933*, Chelsea, New York, 1950.
2. R. T. Cox, "Probability, frequency and reasonable expectation," *American Journal of Physics*, **14**, pp. 1–13, 1946.
3. E. T. Jaynes, "Probability theory: The logic of science." http://omega.albany.edu:8008/JaynesBook.html.
4. S. Youssef *Mod. Phys. Lett.*, **6**, p. 225, 1991.
5. A. Caticha, "Consistency, amplitudes and probabilities in quantum theory," *Phys. Rev.*, 1998.
6. R. P. Feynman *Rev. Mod. Phys.*, **20**, p. 267, 1948.
7. P. R. Halmos, *Measure Theory*, vol. 18 of *GTM*, Springer-Verlag, 1974.
8. Y. S. Chow and H. Teicher, *Probability Theory: Independence, Interchangeability, Martingales*, Springer Texts in Statistics, Springer-Verlag, second ed., 1988.
9. D. Hestenes and G. Sobczyk, *Clifford Algebra to Geometric Calculus*, D. Reidel, 1984. For links to online resources on Clifford algebra check, http://www.mrao.cam.ac.uk/~clifford/.
10. J. Aczel, *Lectures on Functional equations and their appliations*, Academic Press, New York., 1966.
11. C. R. Smith and G. J. Erickson, "Probability theory and the associativity equation," in *Maximum Entropy and Bayesian Methods*, P. F. Fougere, ed., pp. 17–30, Dartmouth, USA, Kluwer, Academic publishers, 1990.
12. A. N. Shiryayev, *Probability*, vol. 95 of *Graduate Texts in Mathematics.* Springer Verlag, 1984.
13. J. B. Hartle *Am. J. Phys.*, **36**, p. 704, 1968.

YET ANOTHER DERIVATION OF THE PRINCIPLE OF MAXIMUM ENTROPY

ANTHONY J.M. GARRETT
Byron's Lodge, 63 High Street
Grantchester
Cambridge CB3 9NF
United Kingdom

Abstract. A new derivation is presented of the principle of maximum entropy, for assigning probability distributions given equality constraints linear in the probabilities. In common with the Shannon–Jaynes derivation the idea involves an invariance under arbitrary partitioning of the points on which probabilities are defined. However, the Shannon–Jaynes derivation is a two-stage process: assert that a particular functional of the probabilities, that has a certain interpretation, should be maximised; then use that interpretation to generate an equation for the functional itself. This equation is routinely solved to give the usual $p \log p$ form. By contrast the derivation presented here is a one-stage process which does not rest on any interpretation.

Key words: Maximum entropy, Cox, Shannon, Tsallis entropy

1. Introduction

Our problem is how to assign a probability distribution p_i on a set of discrete points i, given 'testable' information about that distribution, like "$p_1 + p_3 = 0.5$" or "$p_2^3 + 7p_4 \geq 3$", not just samples from it. This is important when, for example, a parameter in a distribution has physical significance, and its value is known or is to be estimated from samples. Only the values of expectations, which are linear in the probabilities, will be treated here (apart from the non-negativity inequality $p_i \geq 0$). Generalisations will be mentioned at the end.

Our strategy for finding the probabilities p_i will be to extremise some functional of the p_i's. We can always take this operation to be a maximising process, if necessary by flipping the sign of the functional. Denote the unknown functional $G(\{p_i\})$. Then the probability distribution p_i, $i = 1, 2, \ldots$ is assigned by maximising $G(\{p_i\})$ subject to normalisation,

$$Q \equiv \sum_i p_i = 1 \tag{1}$$

W. von der Linden et al. (eds.), Maximum Entropy and Bayesian Methods, 271–288.

and the constraints supplied by the testable information,

$$\mathbf{T} \equiv \sum_i p_i \mathbf{f}_i = \mathbf{F}. \tag{2}$$

(The constraint of non-negativity will be discussed later.) These constraints are written in vector form to emphasise that there may be several. Each has the form of an expectation of some function f_i defined on the points i, taking expectation value F. If there are N points, so that i runs from 1 to N, then there must not be more than $N-1$ linearly independent constraints, or else – with the normalisation constraint – the problem is over-determined and there is no solution for the probabilities. Even with $N-1$ constraints the maximisation process is redundant and the probabilities can be calculated as the solution of N algebraic equations. For fewer constraints maximisation *is* necessary to pick out a solution. It is supposed that the constraints are sensible, in that the hyperplanes representing them in probability space do intersect in the positive orthant (all probabilities positive). This necessarily requires that

$$\min_i \mathbf{f}_i \leq \mathbf{F} \leq \max_i \mathbf{f}_i \tag{3}$$

where the inequality applies to every component of the vector.

This is a problem to which the answer is not in doubt. The functional is

$$G(\{p_i\}) = -\sum_i p_i \ln p_i\ . \tag{4}$$

Many derivations of this form exist: these include the Shannon–Jaynes argument, discussed below; the combinatorial argument, which uses Stirling's approximation in the limit of a large, possibly imaginary, ensemble [1]; and the Shore–Johnson argument that a joint distribution to be reconstructed from its single-variable marginals should not exhibit correlation between the variables, since no evidence is given for it (or, admittedly, against it) [2]. All these arguments give the same result but begin from different axioms, so that which derivation you prefer depends simply on which starting axioms you prefer. Although we have great confidence in the answer, since so many distinct arguments lead to it, the proliferation of methods indicates a measure of dissatisfaction over the starting axioms. Each fresh derivation is a response to unhappiness with all preceding ones. The present derivation is no different; it proceeds from a new axiom, which I shall argue is more compelling than previous axioms.

2. The Shannon–Jaynes Derivation

Shannon's founding work of information theory [3] derived the $p \log p$ form and argued for its maximisation in certain problems concerning communication down a channel. Jaynes [4] then asserted that the same mathematics is involved in distributing probability given testable information. Shannon argued that it is a measure of the *uncertainty* in a distribution that is required – there is more uncertainty in a broad distribution than a sharply peaked one – and in Jaynes' context this means that the most uncertain distribution should be selected that is compatible with the constraints; to do otherwise is to overcommit. Shannon then argued that the measure of uncertainty should obey a certain relation: partition the N points into subsets, each subset having probability equal to the sum of the probabilities of its points; and define a subdistribution for each subset by rescaling the probabilities of its points by their sum, the probability of that subset. Now consider a person who initially knows only about the subsets and the probabilities of each, but who then looks inside the subsets to learn the full picture. The change in uncertainty associated with this learning process should be the expectation value of the uncertainties of the subdistributions, weighted by the probabilities of the subsets. This was the principal axiom from which Shannon derived the $p \log p$ form (4), which became known as the *information entropy* . It has many elegant properties: with a trivial change of scale on taking logarithms to base 2 it is the expectation of the smallest number of binary questions whose answer takes us to certainty (*i.e.*, one probability is unity and the rest zero). By 'smallest number of questions' is meant an optimal question-posing strategy, involving interval halving.

Shannon's arguments for his axiom did not add much; and in view of the obvious correctness of $p \log p$, they seem to have a whiff of *a posteriorism* about them. Others evidently agree, or the various post-Shannon derivations would not exist.

3. The New Axiom

The idea to be presented here is 'one-step', in not interpreting the functional to be maximised – the maximand – in advance. Like Shannon's axiom, it exploits partitioning. It is best presented as a sequence of steps:

- Assign $\{p_i\}$ on the full set of points by maximising the (unknown) functional $G(\{p_i\})$ subject to the constraints $\sum_i p_i \mathbf{f}_i = \mathbf{F}$ and normalisation.
- Take a subset of points, S, and calculate probabilities p_i^S on it by rescaling to unity (re-normalising) those probabilities just found that lie in S.
- Calculate, from this subdistribution p_i^S, the expectation values in S of the constraint functions: $\mathbf{F}^S = \sum_{i \in S} p_i^S \mathbf{f}_i$.
- Assign probabilities p_i' in S by maximising G subject to constraint values $\mathbf{F}^S$ and normalisation, and require that the resulting distribution be equal to p_i^S so that $p_i' = p_i^S$.

This requirement induces a functional equation for the unknown functional G which, taken in conjunction with some uncontroversial minor axioms, permits solution.

The justification for this procedure is consistency. Given constraint functions $\mathbf{f}_i$, their expectation variables $\mathbf{F}$ imply a distribution via the maximisation principle. Conversely, a distribution implies constraint variables $\mathbf{F}$ as the expectation values of $\mathbf{f}_i$. Given the $\mathbf{f}_i$, we are requiring this mutual implication of constraint values and distributions to hold firm under the taking of arbitrary subdistributions.

An example should make the motivation more transparent. Suppose we are to assign probabilities to the faces of a six-sided die, in terms of just the mean of the number of spots that show. We obtain the distribution by maximising a chosen functional G given the variable representing the value of the mean. Now consider somebody who is only aware of the first three faces. This person should end up with a distribution in which the ratios $p_1 : p_2 : p_3$ are the same as in the full distribution, and a mean variable which is the expectation of this subdistribution over faces 1–3. That mean should generate the same subdistribution when G is maximised subject to it.

A further justification of this axiom will be given in the next section.

4. Functional Equation for the Maximand from the New Axiom

We now derive the form of the maximand $G(\{p_i\})$ from the new axiom. The first step is to prove that G is of additive form $\sum_i \Phi(p_i)$ for some function Φ. Consider a problem in which one constraint specifies the value of p_1 (call it P_1), and substitute this value wherever p_1 appears in the other constraints. The resulting optimisation of $G(p_1, p_2, p_3, \ldots)$ over all probabilities, with these constraints including $p_1 = P_1$ and normalisation, is equivalent to the problem of maximising $G(p_1 = P_1, p_2, p_3, \ldots)$ over p_2, $p_3 \ldots$ with the same constraints and with normalisation replaced by $p_2 + p_3 + \ldots = 1 - P_1$. Evidently the same functional form of p_2, $p_3 \ldots$ should be maximised in each case, so that

$$G(p_1, p_2, p_3, \ldots) = G(p_1 = P_1, p_2, p_3, \ldots) + \Phi \tag{5}$$

where Φ is independent of p_2, $p_3 \ldots$; for this equation to make sense Φ must be a function of p_1 alone. Now we can play a similar trick with p_2, and so on. The result is the additive form

$$G(p_1, p_2, p_3, \ldots) = \sum_i \Phi_i(p_i). \tag{6}$$

Now consider a problem in which the constraints are invariant under arbitrary interchanges of the labels i. It follows that G must be exchangeable with respect to the p_i, so that all the functions Φ_i are the same, and

$$G(\{p_i\}) = \sum_i \Phi(p_i). \tag{7}$$

Another proof of this result is given in an appendix of a tutorial derivation of $p \log p$ by Gull and Skilling [5], who use the Shore–Johnson method.

We now suppose that the maximum of G can be found by the method of Lagrange multipliers; this point will be taken up later. At the maximum we have $dG = 0$, or

$$\sum_i \Phi'(p_i) dp_i = 0 \tag{8}$$

where the variations dp_i are not all independent, but obey the relations

$$dQ = \sum_i dp_i = 0 \tag{9}$$

(from normalisation (1)) and

$$\mathbf{dT} = \sum_i \mathbf{f}_i dp_i = \mathbf{0} \tag{10}$$

from constraints (2). Upon multiplying the normalisation condition by a Lagrange multiplier $-\eta$ and the constraint condition by a vector-valued multiplier $-\boldsymbol{\lambda}$, and adding, we have for a maximum of G:

$$dG = d(G - \eta Q - \boldsymbol{\lambda}.\mathbf{T}) = \sum_i \left[\Phi'(p_i) - \eta - \boldsymbol{\lambda}.\mathbf{f}_i\right] dp_i = 0. \tag{11}$$

In this form we can take the dp_i all to vary independently, removing freedom from the values of the Lagrange multipliers in compensation. The coefficients of dp_i must all vanish, so that

$$p_i = \Psi(\eta + \boldsymbol{\lambda}.\mathbf{f}_i) \tag{12}$$

where, functionally, $\Psi = \Phi'^{-1}$, which we suppose is well-defined. The Lagrange multipliers now satisfy the equations obtained by substituting this result into the normalisation condition and the constraints, which become

$$\sum_i \Psi(\eta + \boldsymbol{\lambda}.\mathbf{f}_i) = 1 \ , \qquad \sum_i \Psi(\eta + \boldsymbol{\lambda}.\mathbf{f}_i)\, \mathbf{f}_i = \mathbf{F}. \tag{13}$$

These should be seen as implicit equations for η, $\boldsymbol{\lambda}$ whose solutions, substituted into (12), give the probability distribution in terms of the constraint values.

We now follow the next step of our axiom and select a subspace S, calculating the subdistribution in S by rescaling the probabilities of those points that fall in S:

$$p_i^S = \frac{1}{K}\Psi(\eta + \boldsymbol{\lambda}.\mathbf{f}_i), \qquad K \equiv \sum_{i \in S} \Psi(\eta + \boldsymbol{\lambda}.\mathbf{f}_i). \tag{14}$$

The constraint value in S is:

$$\mathbf{F}^S = \sum_{i \in S} p_i^S \mathbf{f}_i = \frac{1}{K} \sum_{i \in S} \Psi(\eta + \boldsymbol{\lambda}.\mathbf{f}_i)\, \mathbf{f}_i. \tag{15}$$

The next step in the axiom is to assign probabilities p_i' in S using this constraint value. The result is

$$p_i' = \Psi(\alpha + \boldsymbol{\beta}.\mathbf{f}_i) \tag{16}$$

where the Lagrange multipliers α, $\boldsymbol{\beta}$ are given by

$$\sum_{i \in S} \Psi(\alpha + \boldsymbol{\beta}.\mathbf{f}_i) = 1 \tag{17}$$

(normalisation) and

$$\sum_{i \in S} \Psi(\alpha + \boldsymbol{\beta}.\mathbf{f}_i)\, \mathbf{f}_i = \mathbf{F}^S = \frac{1}{K} \sum_{i \in S} \Psi(\eta + \boldsymbol{\lambda}.\mathbf{f}_i)\, \mathbf{f}_i. \tag{18}$$

Finally we require that these probabilities p_i' be equal to the rescaled probabilities p_i^S in S. This condition is

$$\Psi(\alpha + \boldsymbol{\beta}.\mathbf{f}_i) = \frac{1}{K}\Psi(\eta + \boldsymbol{\lambda}.\mathbf{f}_i). \tag{19}$$

where all the Lagrange multipliers are given by the various equations of constraint above, and the rescaling constant by (14). Equation (19) is a functional equation for Ψ, with the f_i as independent variables.

Before tackling this equation, we observe that it is also generated if, instead of partitioning a set of points into subsets, we go the other way and assemble a set from smaller sets, given the probability for each of those subsets. This way round, the axiom may have greater appeal. In spaces S_1 and S_2 we have independent distributions

$$p_i^{(1)} = \Psi(\alpha_1 + \boldsymbol{\beta}_1.\mathbf{f}_i^{(1)}), \qquad p_i^{(2)} = \Psi(\alpha_2 + \boldsymbol{\beta}_2.\mathbf{f}_i^{(2)}) \tag{20}$$

where the Lagrange multipliers are given via constraints in S_1, S_2 as usual, and the constraint functions $\mathbf{f}_i^{(1)}$, $\mathbf{f}_i^{(2)}$ are defined strictly on each of S_1 and S_2, so that they include Kronecker factors $\delta_{i\in S_1}$ and $\delta_{i\in S_2}$ respectively. Then we require the distribution in the combined space to be

$$p_i^{(12)} = K_1 p_i^{(1)}\delta_{i\in S_1} + K_2 p_i^{(2)}\delta_{i\in S_2} \tag{21}$$

where K_1 is the probability for space S_1 in the combined space, so that $K_1 = \sum_{i\in S_1} p_i^{(12)}$, and likewise for $K_2 = 1 - K_1$. We apply the same constraints on the expectations of $\mathbf{f}_i^{(1)}$ and $\mathbf{f}_i^{(2)}$, only now in the combined space so that the probabilities in subspaces S_1 and S_2 are scaled by factors K_1 and K_2; this is most simply handled by rescaling the expectation values to be $K_1\mathbf{F}^{(1)}$ and $K_2\mathbf{F}^{(2)}$ in the combined space. We also replace the normalisation constraint with the values of K_1 and K_2, to obtain

$$p_i^{(12)} = \Psi\Big((\eta_1 + \boldsymbol{\lambda}_1.\mathbf{f}_i^{(1)})\delta_{i\in S_1} + (\eta_2 + \boldsymbol{\lambda}_2.\mathbf{f}_i^{(2)})\delta_{i\in S_2}\Big). \tag{22}$$

Upon equating the last two expressions for $p_i^{(12)}$ we obtain the same functional equation as before, both in S_1 and S_2.

Now we begin to tackle the functional equation (19). A necessary condition for the existence of non-zero solutions is that $\boldsymbol{\beta} = \nu\boldsymbol{\lambda}$. We can make a linear transformation of variables, from the components of the vector $\mathbf{f}_i$, to a new set in which the first component is $\boldsymbol{\lambda}.\mathbf{f}_i$. The right-hand side of (19) depends only on this one component, but unless $\boldsymbol{\beta} = \nu\boldsymbol{\lambda}$ the left-hand side depends on further components. That can be true only if both sides of (19) are constant, and since $K \neq 1$ the constant would have to be zero, making normalisation impossible. Hence we require that $\boldsymbol{\beta} = \nu\boldsymbol{\lambda}$.

We now solve equation (19) under the assumption that $\boldsymbol{\beta} = \nu\boldsymbol{\lambda}$, checking *a posteriori* that solutions correspond to this condition and rejecting any that do not. Begin by taking logarithms of both sides, by defining as independent variable $x = \boldsymbol{\lambda}.\mathbf{f}_i$, and also

$$\Gamma = \ln \Psi, \qquad C = -\ln K. \tag{23}$$

Then (19) becomes

$$\Gamma(\alpha + \nu x) = \Gamma(\eta + x) + C. \tag{24}$$

This is a linear functional equation with linear arguments in the independent variable. Though we anticipate a linear solution, systematic techniques reveal that there is more to it than that. Begin by defining a shifted function:

$$\Gamma(z) = \Gamma_q(z + q) \tag{25}$$

with the aim of choosing the shift such that the arguments on both sides of (24) have the same shift: with independent variable $z \equiv x + \eta$ we have

$$\Gamma_q\Big(\nu\big(z + \frac{q + \alpha - \nu\eta}{\nu}\big)\Big) = \Gamma_q(z + q) + C. \tag{26}$$

Therefore, if we choose the shift q such that

$$\frac{q + \alpha - \nu\eta}{\nu} = q, \tag{27}$$

or $q = (\alpha - \nu\eta)/(\nu - 1)$, the equation becomes

$$\Gamma_q(\nu y) = \Gamma_q(y) + C. \tag{28}$$

At this point it is clear that we have to treat the case $\nu = 1$ separately, since it generates a contradiction in this equation and a divergent solution for q. We do this in the next section.

5. Solution for the Maximand when $\nu = 1$

We return to general ν later. For $\nu = 1$ equation (24) reads

$$\Gamma(\alpha + x) = \Gamma(\eta + x) + C. \tag{29}$$

Define a new independent variable $y = x + \alpha$, so that

$$\Gamma(y) = \Gamma(y + \eta - \alpha) + C. \tag{30}$$

Differentiation with respect to y removes the constant C to give $\Gamma'(y) = \Gamma'(y+\eta-\alpha)$, the periodicity equation, so that Γ' is a periodic function with period $\eta-\alpha$. The general solution is almost everywhere given, subject to convergence, by a Fourier series:

$$\Gamma'(y) = A_0 + \sum_{n=1}^{\infty} A_n \cos\big(\frac{2\pi n}{\eta - \alpha} y - \phi_n\big). \tag{31}$$

By integrating this and substituting the result into (30), which fixes the constant term A_0, we obtain the solution

$$\Gamma(y) = \frac{C}{\alpha - \eta} y + B + \sum_{n=1}^{\infty} A_n \cos\left(\frac{2\pi n}{\eta - \alpha} y - \phi_n\right), \tag{32}$$

where the coefficients A_n of the oscillatory terms have been redefined. The assumption of differentiability can be removed by using a trial solution $\Gamma(y) = Ay + \Lambda(y)$ in (30); we find the coefficient of the inhomogeneous term $A = C/(\alpha - \eta)$, and that Λ satisfies the periodicity equation, so that we recover (32) without differentiation.

From (23) the function Ψ representing probabilities is $\exp(\Gamma)$, and the function Φ defining the maximand (7) can be deduced from $\Psi = \Phi'^{-1}$ or, in terms of Γ,

$$\Phi(y) = \int dy\, \Gamma^{-1}(\ln y). \tag{33}$$

The general form for Ψ is therefore

$$\Psi(y) = \exp\Big[Ay + B + \sum_{n=1}^{\infty} A_n \cos(2\pi nky - \phi_n)\Big] \tag{34}$$

where A, B, k, A_n, ϕ_n are constant parameters which appear in the maximand function Φ. Upon substituting this form into the functional equation we know that a necessary condition for solutions is $k = 1/(\eta - \alpha)$. But η and α are Lagrange multipliers whose values depend on details of the problem, and we are seeking a universal functional valid for all problems. Therefore k cannot appear, implying that the A_n are zero and all the oscillatory terms vanish. (We have not in any case verified that $\boldsymbol{\beta} = \nu\boldsymbol{\lambda}$ with $\nu = 1$ for the oscillatory solutions, or that Γ^{-1} and hence Φ is well-defined.) Consequently

$$\Gamma(y) = Ay + B, \qquad \Psi(y) = \exp(Ay + B) \tag{35}$$

and

$$\Phi(y) = A^{-1} y \ln y - A^{-1}(1 + B)y + \text{const.} \tag{36}$$

This is what we have been aiming for – the $y \ln y$ term. In the maximand $G = \sum_i \Phi(p_i)$ the term in Φ proportional to y simply shifts the value of the Lagrange multiplier corresponding to normalisation, while the constant term is irrelevant to maximisation. However it remains to verify that A and B are not forced to take problem-dependent values and the maximand is universal; it seems this is not so, since by looking at (32) we have $A = C/(\alpha - \eta)$, and the Lagrange multipliers are problem-dependent. We must also verify that $\boldsymbol{\beta} = \boldsymbol{\lambda}$ so that $\nu = 1$. Only then can we say we have derived the standard entropic form; and even then we have still to deal with any solutions of the functional equation arising from the case $\nu \neq 1$.

Rather than verify that A and B can be chosen arbitrarily, we fix their values and show that the resulting formalism for an arbitrary problem does not grind to an inconsistent halt. We choose $A = B = -1$, so that $\Phi(y) = -y \ln y$ and we have the standard maximum entropy procedure for assigning probabilities [4]:

$$p_i = \exp(-1 - \eta - \boldsymbol{\lambda}.\mathbf{f}_i), \tag{37}$$

where normalisation is

$$\sum_i \exp(-1-\eta-\boldsymbol{\lambda}.\mathbf{f}_i) = 1 \tag{38}$$

so that

$$\exp(1+\eta) = \sum_i \exp(-\boldsymbol{\lambda}.\mathbf{f}_i) \tag{39}$$

(the partition function), and the constraints are

$$\sum_i \exp(-1-\eta-\boldsymbol{\lambda}.\mathbf{f}_i)\,\mathbf{f}_i = \mathbf{F}. \tag{40}$$

Let us verify existence and uniqueness of the maximum entropy solution to this constrained problem. In probability space, without any constraint (even normalisation), we have a unique maximum of $G = -\sum_i p_i \ln p_i$ at the point where each probability has value e^{-1}. Hypersurfaces of constant entropy are therefore nested round this point, with entropy diminishing monotonically away from it; there is no 'volcano rim'. In the positive orthant such a hypersurface passes through every point, since entropy is well-defined there. Because of the logarithmic term, it is easily verified that these hypersurfaces approach the walls bounding the positive orthant at a zero angle, and do not continue on the other side of the walls; the walls are asymptotic to these hypersurfaces.

We next show that every hypervolume bounded by an isentropic hypersurface is convex. (This convex hypervolume is often incorrectly referred to as a convex hypersurface.) At every point inside a hypersurface on which entropy takes value G_0 we have $G(\{p_i\}) \geq G_0$, while outside the hypersurface $G(\{p_i\}) \leq G_0$. Convexity means, reasonably, that every point on the straight line segment between two points, each inside the hypersurface, is also inside the hypersurface. Let these two points in probability space be $\{p_i^{(1)}\}$ and $\{p_i^{(2)}\}$, with entropies $G^{(1)} \geq G_0$ and $G^{(2)} \geq G_0$. Then a point on the straight line segment between them is $\{\rho p_i^{(1)} + (1-\rho)p_i^{(2)}\}$ where ρ lies between 0 and 1, with entropy

$$G^{(\rho)} = -\sum_i [\rho p_i^{(1)} + (1-\rho)p_i^{(2)}] \ln[\rho p_i^{(1)} + (1-\rho)p_i^{(2)}]. \tag{41}$$

We aim to prove $G^{(\rho)} \geq G_0$. Now the function $\Phi(y) = -y \ln y$ is *concave*, meaning that

$$\Phi\big(\rho y + (1-\rho)z\big) \geq \rho\Phi(y) + (1-\rho)\Phi(z). \tag{42}$$

To prove this, put $z = y + \epsilon$ and expand both sides in powers of ϵ; zero and first powers of ϵ cancel and negativity of the second derivative of Φ implies that Φ is locally concave, and is locally concave everywhere and so globally concave. It follows that

$$G^{(\rho)} \geq \rho G^{(1)} + (1-\rho)G^{(2)} \tag{43}$$

and, since $G^{(1)} = G_0 + \Delta^{(1)}$ and $G^{(2)} = G_0 + \Delta^{(2)}$ where the Δ terms are non-negative, we have

$$G^{(\rho)} \geq G_0 + \rho\Delta^{(1)} + (1-\rho)\Delta^{(2)} \tag{44}$$

so that $G^{(\rho)} \geq G_0$; the chosen point on the line segment between $\{p_i^{(1)}\}$ and $\{p_i^{(2)}\}$ also lies inside the hypersurface of entropy S_0, verifying convexity.

Our existence proof is completed by the observation that this convexity also implies convexity of the hypervolume of intersection of isentropic surfaces with any hyperplane. (The corresponding algebraic theorem is that the intersection of two convex sets of points is itself convex; hyperplanes are trivially convex.) Linear constraints and normalisation correspond to hyperplanes, so that provided the constraints are well-conditioned there is always a unique solution. Moreover, the fact that isentropic hypersurfaces do not intersect the walls bounding the positive orthant means that the solution lies inside that orthant, not on any vertex; the solution for the Lagrange multipliers is regular. Consequently, even though $A = -(\alpha - \eta)^{-1} \ln K$ in (36) *seems* to be problem-dependent, the formalism works whatever A and B; this relation must also emerge from the subspace axiom.

To verify that $\boldsymbol{\beta} = \boldsymbol{\lambda}$ and $\nu = 1$, as required, we must use the subspace axiom. The subdistribution obtained by rescaling is

$$p_i^S = \frac{1}{K} \exp(-1 - \eta - \boldsymbol{\lambda}.\mathbf{f}_i) \tag{45}$$

where

$$K = \sum_{i \in S} \exp(-1 - \eta - \boldsymbol{\lambda}.\mathbf{f}_i), \tag{46}$$

so that the expectation value in S of the constraint function is

$$\mathbf{F}^S = \sum_{i \in S} p_i^S = \frac{\sum_{i \in S} \exp(-1 - \eta - \boldsymbol{\lambda}.\mathbf{f}_i)\, \mathbf{f}_i}{\sum_{i \in S} \exp(-1 - \eta - \boldsymbol{\lambda}.\mathbf{f}_i)} \tag{47}$$

$$= \frac{\sum_{i \in S} \exp(-\boldsymbol{\lambda}.\mathbf{f}_i)\, \mathbf{f}_i}{\sum_{i \in S} \exp(-\boldsymbol{\lambda}.\mathbf{f}_i)}. \tag{48}$$

The subdistribution obtained by maximising entropy over the subspace S using this constraint value is

$$p_i' = \exp(-1 - \alpha - \boldsymbol{\beta}.\mathbf{f}_i) \tag{49}$$

where normalisation implies

$$\sum_{i \in S} \exp(-1 - \alpha - \boldsymbol{\beta}.\mathbf{f}_i) = 1 \tag{50}$$

and the constraint equation becomes

$$\sum_{i \in S} \exp(-1 - \alpha - \boldsymbol{\beta}.\mathbf{f}_i)\, \mathbf{f}_i = \mathbf{F}^S. \tag{51}$$

We have sought solutions $\Psi(\bullet)$ to the equation $p_i^S = p_i'$ (equation (19), with the $\mathbf{f}_i$ independent) for which $\boldsymbol{\beta} = \nu\boldsymbol{\lambda}$ and $\nu = 1$, so that $\boldsymbol{\beta}.\mathbf{f}_i = \boldsymbol{\lambda}.\mathbf{f}_i$. By comparing $p_i^S \propto \exp(-\boldsymbol{\lambda}.\mathbf{f}_i)$ with $p_i' \propto \exp(-\boldsymbol{\beta}.\mathbf{f}_i)$ we see immediately how our solution corresponds to this condition. However it remains to verify from the axiom that the Lagrange multipliers satisfy $\boldsymbol{\beta} = \boldsymbol{\lambda}$, as consistency requires. Divide the left-hand side of (51) by the left-hand side of (50), which is unity, and cancel the α terms. Upon substituting for $\mathbf{F}^S$ from (48), the result is

$$\frac{\sum_{i\in S}\exp(-\boldsymbol{\beta}.\mathbf{f}_i)\,\mathbf{f}_i}{\sum_{i\in S}\exp(-\boldsymbol{\beta}.\mathbf{f}_i)} = \frac{\sum_{i\in S}\exp(-\boldsymbol{\lambda}.\mathbf{f}_i)\,\mathbf{f}_i}{\sum_{i\in S}\exp(-\boldsymbol{\lambda}.\mathbf{f}_i)}\,, \tag{52}$$

which by comparison of either side tells us that $\boldsymbol{\beta} = \boldsymbol{\lambda}$, as hoped for. Uniqueness of maximum entropy solutions guarantees that there is no other solution of (52) for $\boldsymbol{\beta}$.

We have therefore established that the solution of the basic functional equation (24) for $\nu = 1$ implies the usual entropic form, with all its desirable features. But are there further solutions with $\nu \neq 1$? We answer this in the next section.

6. Solution for the Maximand when $\nu \neq 1$

Let us go back to the functional equation in the form (28),

$$\Gamma_q(\nu y) = \Gamma_q(y) + C \tag{53}$$

where $\Gamma(y) = \Gamma_q(y+q)$ and $q = (\alpha - \nu\eta)/(\nu - 1)$. This form is well behaved when $\nu \neq 1$. Change the independent variable to z where $y = e^z$, so that the factor ν becomes additive, and define the function Θ such that

$$\Gamma_q(e^z) = \Theta(z). \tag{54}$$

The transformed equation is

$$\Theta(z + \ln\nu) = \Theta(z) + C. \tag{55}$$

Equation (30), which we have just solved, is of precisely this form, so that Θ' is periodic with period $\ln\nu$ and

$$\Theta(z) = \frac{C}{\ln\nu}z + B + \sum_{n=1}^{\infty} A_n \cos\Big(\frac{2\pi n}{\ln\nu}z - \phi_n\Big). \tag{56}$$

We now undo all transformations to recover the function Ψ that represents probabilities, and the function Φ that specifies the maximand G. The solution for $\Gamma = \ln\Psi$ is

$$\Gamma(y) = \frac{C}{\ln\nu}\ln\Big(y + \frac{\alpha - \nu\eta}{\nu - 1}\Big) + B + \sum_{n=1}^{\infty} A_n \cos\Big[\frac{2\pi n}{\ln\nu}\ln\Big(y + \frac{\alpha-\nu\eta}{\nu-1}\Big) - \phi_n\Big]. \tag{57}$$

(We expect that the $\nu = 1$ solution (32) can be derived as a limit of this.) Write the solution for Ψ in the form

$$\Psi(y) = B\Big(y + \frac{\alpha - \nu\eta}{\nu - 1}\Big)^{\frac{C}{\ln\nu}} \exp\Big\{\sum_{n=1}^{\infty} A_n \cos\Big[\frac{2\pi n}{\ln\nu}\ln\Big(y + \frac{\alpha-\nu\eta}{\nu-1}\Big) - \phi_n\Big]\Big\} \tag{58}$$

(with a redefinition of B), while Φ is, implicitly,

$$\Phi(y) = -(\frac{\alpha - \nu\eta}{\nu - 1})y + \int dy \exp\left[\Theta^{-1}(\ln y)\right]. \tag{59}$$

We rule out the oscillatory terms as before: the general form of solution for Ψ is

$$\Psi(y) = B(y+d)^m \exp\Big\{\sum_{n=1}^{\infty} A_n \cos\left[2\pi nk \ln(y+d) - \phi_n\right]\Big\} \tag{60}$$

where all parameters appear in the expression for the maximand function Φ. Yet, by comparison with (59), k must depend on details of the problem. Since we seek a universal principle this is not acceptable: k cannot appear, implying that all the A_n are zero. It then follows by substituting (56) into (59) that

$$\Phi(y) = By^{(1+\frac{\ln \nu}{C})} - \frac{\alpha - \nu\eta}{\nu - 1}y + \text{const.} \tag{61}$$

(Again B has been redefined.) As before the term linear in y just shifts the normalisation Lagrange multiplier, and the constant term is irrelevant. Therefore the crucial term is that by which $\Phi(y)$ is proportional to some power of y. We now find the value of ν in $\boldsymbol{\beta} = \nu\boldsymbol{\lambda}$ to which this solution corresponds, and investigate whether the equations of constraint giving the multipliers then rule out that solution to leave just the entropy.

If $\Psi(y) \propto y^m$, and $\Phi(y) \propto y^{1+1/m}$ (except for $m = -1$ when $\Phi(y) \propto \ln y$), the basic functional equation (19) reads

$$(\alpha + \boldsymbol{\beta}.\mathbf{f}_i)^m = \frac{1}{K}(\eta + \boldsymbol{\lambda}.\mathbf{f}_i)^m, \tag{62}$$

which is satisfied with $\alpha = \eta/K^{1/m}$ and $\boldsymbol{\beta} = \boldsymbol{\lambda}/K^{1/m}$, so that $\nu = K^{-1/m}$. Following this, it is routine to verify that all equations of constraint generated in section 4 by the basic axiom *are* satisfied. The multipliers η and $\boldsymbol{\lambda}$ are given via normalisation and constraints in the problem in the full space; with $\Psi(y) = y^m$,

$$\sum_i (\eta + \boldsymbol{\lambda}.\mathbf{f}_i)^m = 1, \qquad \sum_i (\eta + \boldsymbol{\lambda}.\mathbf{f}_i)^m \mathbf{f}_i = \mathbf{F}. \tag{63}$$

It is convenient, in view of our purpose to single out the entropy form, to treat existence for power-law solutions later. At this stage we have therefore narrowed the maximand down to entropic form or power-law form (except for $m = -1$ as above). These can be combined in the single expression

$$G(\{p_i\}) = m \sum_i p_i(1 - p_i^{1/m}), \tag{64}$$

since by normalisation the first term is unity, and, by writing $p^{1/m}$ as $\exp(m^{-1}\ln p)$ and expanding in inverse powers of m, we recover the entropic form in the limit

$m \to \infty$. Expression (64) is the generalised entropy studied by Tsallis in a series of papers beginning with [6].

Finally in this section, we calculate the extremal value of the functional G. This value is obtained by substituting the solution $p_i = \Phi'^{-1}(\eta + \boldsymbol{\lambda}.\mathbf{f}_i)$ for the probabilities into the functional $G = \sum_i \Phi(p_i)$. For the Tsallis form $\Phi(y) = my(1-y^{1/m})$, the extremal value is readily shown to be

$$\sum_i \frac{m}{m+1}(1+\eta+\boldsymbol{\lambda}.\mathbf{f}_i)p_i \tag{65}$$

$$= \frac{m}{m+1}(1+\eta+\boldsymbol{\lambda}.\mathbf{F}). \tag{66}$$

For other forms of the function Φ the result involves more complicated moments of the constraint function $\mathbf{f}_i$, and the maximum value cannot be written in terms of $\mathbf{F}$ alone. Tsallis' expression is essentially the most general allowing this: we require $\Phi(\Psi(y)) = (A+By)\Psi(y)$ where $\Psi = \Phi'^{-1}$ and, by changing variables to $z = \Psi(y)$ we have $\Phi(z) = Az + Bz\Phi'(z)$, a differential equation with solution $\Phi(z) = (-Az + Dz^{1/B})/(B-1)$, which apart from a trivial shift in the normalisation multiplier is the Tsallis form.

7. Singling out the Entropy

It is clear that the present derivation can immediately be generalised to multivariate distributions $p_{jk\ldots}$; our suffix i above simply runs through all possible values of $jk\ldots$. Generalisation is also immediate to the continuum, and (by rewriting the continuous variable, say x, as $M(x)$ and defining $m(x) = M'(x)$) with allowance for a measure $m(x)$.

We now use a specific multivariate problem, together with the requirement of universality, to take us further and rule out all positive powers of m. (We cannot use arguments that examine the value of G, since we are refraining from interpreting that quantity; a way to proceed is to seek an unpalatable probability assignment.) It is required that the formalism work in every case in which the problem is well-posed: that is, it should yield real non-negative probabilities in all problems where the constraints are compatible. Consider now the problem of reconstructing the 2×2 joint distribution of two variables from its marginals, in the case where one marginal distribution has probabilities zero and unity. This problem is depicted in Table 1, where p_1 to p_4 are to be assigned and there are in fact two independent

	0	1
W	p_1	p_2
$1-W$	p_3	p_4

TABLE 1. The 2×2 reconstruction problem from the marginals.

constraints in addition to normalisation, which we choose as

$$p_1 + p_3 = 0, \qquad p_1 + p_2 = W. \tag{67}$$

(It is harmless to include further linearly dependent constraints such as $p_3 + p_4 = 1 - W$; doing so is equivalent to altering the values of the Lagrange multipliers for whatever linearly independent set has been chosen.) Solutions now form a one-parameter family, out of which the variational procedure singles out the required distribution. For power-law maximand the result is

$$p_i = \left[\eta + \lambda^{[1]}(\delta_{i1} + \delta_{i3}) + \lambda^{[2]}(\delta_{i1} + \delta_{i2})\right]^m \tag{68}$$

or

$$\begin{aligned} p_1 &= (\eta + \lambda^{[1]} + \lambda^{[2]})^m \\ p_2 &= (\eta + \lambda^{[2]})^m \\ p_3 &= (\eta + \lambda^{[1]})^m \\ p_4 &= \eta^m \end{aligned} \tag{69}$$

where η is, as usual, the Lagrange multiplier corresponding to normalisation, $\lambda^{[1]}$ is the multiplier for the constraint $p_1 + p_3 = 0$ and $\lambda^{[2]}$ the multiplier for $p_1 + p_2 = W$; clearly $p_1^{1/m} + p_4^{1/m} = p_2^{1/m} + p_3^{1/m}$. The solution should have $p_1 = p_3 = 0$, since otherwise the constraint $p_1 + p_3 = 0$ gives either complex probabilities, or one of p_1, p_3 positive and the other negative, and complex or negative probabilities are not acceptable. If $p_3 = 0$ then, provided m is positive, (69) tells us that $\eta + \lambda^{[1]} = 0$, following which $p_1 = 0$ tells us that $\lambda^{[2]} = 0$. But if $\lambda^{[2]} = 0$ then (69) immediately implies that $p_2 = p_4$; yet $p_2 = W$ and $p_4 = 1 - W$ where W is arbitrary in $[0, 1]$. This contradiction means our assumption that the solution has $p_1 = p_3 = 0$ is wrong; and since $p_1 + p_3 = 0$ it must have either p_1, p_3 complex or one of these negative. Either case is unacceptable, and our requirement that the formalism work for every well-posed problem now rules out positive values of m. Geometrically, this is because the hypersurfaces of constant G intersect the walls of the first, positive-probability orthant at a non-zero angle and may continue into neighbouring orthants (when $G = -\sum_i p_i^2$, for example), into which the solution may fall. Of course, any optimisation problem may still possess a solution when the search is restricted to the positive-probability orthant; we are really adding the requirement that this solution should be accessible by the Lagrange multiplier method and not lie at a vertex. If any of these 'regular' solutions fall in other orthants, however, the derivation of the functional equation via arbitrary partitioning is wrecked, since for some partitions the functional equation will arise as usual, while for others something more irregular, having a different solution, will take its place. Yet we cannot have two answers to the same question.

By contrast, for the entropic form of the maximand we have

$$\begin{aligned} p_1 &= \exp(-\eta - \lambda^{[1]} - \lambda^{[2]}) \\ p_2 &= \exp(-\eta - \lambda^{[2]}) \\ p_3 &= \exp(-\eta - \lambda^{[1]}) \\ p_4 &= \exp(-\eta) \end{aligned} \tag{70}$$

so that $p_1 p_4 = p_2 p_3$, which in combination with normalisation and the constraints immediately gives $p_1 = p_3 = 0$ as required, with the Lagrange multiplier $\lambda^{[1]}$

corresponding to $p_1 + p_3 = 0$ upward divergent. It is readily shown from (70) that the reconstructed distribution is always just the product of the marginal distributions.

What of negative values of the power index m? As with entropy the condition $p_1 + p_3 = 0$, a single equation which forces two separate equations $p_1 = p_3 = 0$, no longer generates two distinct equations in Lagrange multiplier space: the single solution $\lambda^{[1]} = +\infty$ implies both of $p_1 = p_3 = 0$. For $-1 < m < 0$ we maximise $-\sum_i p_i^{1+1/m}$, where $1 + 1/m$ is negative (with the sign reversed there would be an unbounded maximum where any probability is zero); for $m = -1$ we maximise $\sum_i \ln p_i$; while for $m < -1$ we maximise $\sum_i p_i^{1+1/m}$, where $1 + 1/m$ lies between 0 and 1. In each case the convexity property studied above still holds, and the walls of the positive probability orthant are still asymptotic to the hypersurfaces of constant G. Hence, given sensible constraints, a sensible and unique solution always exists, with no probability zero unless the constraints force it directly.

To rule out power solutions with $m < 0$ there is no alternative now to seeking an unpalatable probability assignment. We could require zero correlation among the variables in a joint distribution reconstructed from single marginals, for example, in a specific model problem. But because entropy can be derived from uncorrelation in the *general* case [2,5], this would virtually render the present development superfluous. Such a derivation supposes that, where there is no evidence for or against correlation, none is allowed. (The reasoning then proceeds that marginal distributions are multiplied but Lagrange multipliers are additive, so that the gradient of the maximand must be logarithmic.) Here a less presumptuous argument is given, still based on the relation between joint and marginal distributions.

The idea is that we have two sets of constraints, relating to two distinct distributions. We then wish to reconstruct the joint distribution by applying those constraints in the joint space. The marginals of that joint distribution should be the same as the two distributions assigned independently in each subspace. This requirement proves more than strong enough to rule out powers $m < 0$. With a notation borrowed from the similar (but logically distinct) partitioning argument of section 4, the distribution in the first subspace is

$$p_i^{(1)} = \Psi(\eta_1 + \boldsymbol{\lambda}_1.\mathbf{f}_i^{(1)}) \tag{71}$$

where the multipliers are given by

$$\sum_i \Psi(\eta_1 + \boldsymbol{\lambda}_1.\mathbf{f}_i^{(1)}) = 1 \ , \qquad \sum_i \Psi(\eta_1 + \boldsymbol{\lambda}_1.\mathbf{f}_i^{(1)})\, \mathbf{f}_i^{(1)} = \mathbf{F}_1, \tag{72}$$

and likewise for the second distribution $p_j^{(2)}$. The joint distribution, reinterpreting the constraints to apply in the joint space, is

$$p_{ij}^{(12)} = \Psi(\eta_{12} + \boldsymbol{\lambda}_{12}.\mathbf{f}_i^{(1)} + \boldsymbol{\lambda}_{21}.\mathbf{f}_j^{(2)}) \tag{73}$$

where

$$\sum_{ij} \Psi(\eta_{12} + \boldsymbol{\lambda}_{12}.\mathbf{f}_i^{(1)} + \boldsymbol{\lambda}_{21}.\mathbf{f}_j^{(2)}) = 1 \tag{74}$$

and

$$\sum_{ij} \Psi(\eta_{12} + \boldsymbol{\lambda}_{12}.\mathbf{f}_i^{(1)} + \boldsymbol{\lambda}_{21}.\mathbf{f}_j^{(2)})\, \mathbf{f}_i^{(1)} = \mathbf{F}_1 \; ,$$
$$\sum_{ij} \Psi(\eta_{12} + \boldsymbol{\lambda}_{12}.\mathbf{f}_i^{(1)} + \boldsymbol{\lambda}_{21}.\mathbf{f}_j^{(2)})\, \mathbf{f}_j^{(2)} = \mathbf{F}_2 \; . \tag{75}$$

We now require the marginals of $p_{ij}^{(12)}$ to be $p_i^{(1)}$ and $p_j^{(2)}$, so that

$$\sum_j p_{ij}^{(12)} = p_i^{(1)} \; , \qquad \sum_i p_{ij}^{(12)} = p_j^{(2)} \tag{76}$$

or

$$\sum_j \Psi(\eta_{12} + \boldsymbol{\lambda}_{12}.\mathbf{f}_i^{(1)} + \boldsymbol{\lambda}_{21}.\mathbf{f}_j^{(2)}) \;=\; \Psi(\eta_1 + \boldsymbol{\lambda}_1.\mathbf{f}_i^{(1)}) \; ,$$
$$\sum_i \Psi(\eta_{12} + \boldsymbol{\lambda}_{12}.\mathbf{f}_i^{(1)} + \boldsymbol{\lambda}_{21}.\mathbf{f}_j^{(2)}) \;=\; \Psi(\eta_2 + \boldsymbol{\lambda}_2.\mathbf{f}_j^{(2)}) \; . \tag{77}$$

Though these functional equations are worth tackling in their own right, we wish only to substitute the entropic and power-law forms for Ψ into them and see what survives.

For entropy we have $\Psi(y) = \exp(-1-y)$ and the joint distribution $p_{ij}^{(12)}$ separates in i and j:

$$p_{ij}^{(12)} \;\propto\; \exp(-\boldsymbol{\lambda}_{12}.\mathbf{f}_i^{(1)}) \exp(-\boldsymbol{\lambda}_{21}.\mathbf{f}_j^{(2)}) \; ,$$
$$p_i^{(1)} \;\propto\; \exp(-\boldsymbol{\lambda}_1.\mathbf{f}_i^{(1)}) \; , \tag{78}$$
$$p_j^{(2)} \;\propto\; \exp(-\boldsymbol{\lambda}_2.\mathbf{f}_j^{(2)}).$$

Since normalisation takes care of the constants of proportionality, our functional equations are satisfied provided that $\boldsymbol{\lambda}_{12} = \boldsymbol{\lambda}_1$ and $\boldsymbol{\lambda}_{21} = \boldsymbol{\lambda}_2$. To investigate whether these relations hold, we eliminate η_{12} from the constraint equations (75) employing the normalisation condition (74); the first equation of (75) becomes

$$\sum_{ij} \exp(-\boldsymbol{\lambda}_{12}.\mathbf{f}_i^{(1)} - \boldsymbol{\lambda}_{21}.\mathbf{f}_j^{(2)})\, \mathbf{f}_i^{(1)} = \mathbf{F}_1 \sum_{ij} \exp(-\boldsymbol{\lambda}_{12}.\mathbf{f}_i^{(1)} - \boldsymbol{\lambda}_{21}.\mathbf{f}_j^{(2)}), \tag{79}$$

and analogously for the second equation. The sum over j separates and cancels on either side, leaving an equation for $\boldsymbol{\lambda}_{12}$. This equation proves identical to the corresponding equation for $\boldsymbol{\lambda}_1$ in the first subspace; hence $\boldsymbol{\lambda}_{12} = \boldsymbol{\lambda}_1$, and likewise $\boldsymbol{\lambda}_{21} = \boldsymbol{\lambda}_2$, completing our verification that the entropy satisfies the new relations (77) and associated constraints.

What of the power-law solutions? The first equation in (77) is

$$\sum_j [\eta_{12} + \boldsymbol{\lambda}_{12}.\mathbf{f}_i^{(1)} + \boldsymbol{\lambda}_{21}.\mathbf{f}_j^{(2)}]^m = [\eta_1 + \boldsymbol{\lambda}_1.\mathbf{f}_i^{(1)}]^m. \tag{80}$$

Analogously with our earlier analysis, a necessary condition for solution is that $\boldsymbol{\lambda}_{12} = \nu'\boldsymbol{\lambda}_1$; with $x \equiv \boldsymbol{\lambda}_1.\mathbf{f}_i^{(1)}$ the equation reads

$$\sum_j [\nu' x + \eta_{12} + \boldsymbol{\lambda}_{21}.\mathbf{f}_j^{(2)}]^m = [x + \eta_1]^m \tag{81}$$

or, with $y \equiv x + \eta_1$,

$$\sum_j \left[\nu' + \frac{\gamma_j}{y}\right]^m = 1 \tag{82}$$

where $\gamma_j = \eta_{12} - \nu'\eta_1 + \boldsymbol{\lambda}_{21}.\mathbf{f}_j^{(2)}$. By making a binomial expansion of the left-hand side in inverse powers of y we have

$$\sum_j \left(\nu'^m + m\frac{\nu'^{m-1}}{y}\gamma_j + \frac{m(m-1)}{2!}\frac{\nu'^{m-2}}{y^2}\gamma_j^2 + \cdots\right) = 1. \tag{83}$$

By comparing coefficients of inverse powers of y it follows from the zeroth power that $\nu' = J^{-1/m}$, where j runs from 1 to J. From the first inverse power we have $\sum_j \gamma_j = 0$ (supposing that $m \neq 0$), while from the second inverse power we have either $m = 1$ or $\sum_j \gamma_j^2 = 0$. But the γ_j are functions of j and cannot generally be zero; hence $m = 1$, and it must be verified that $\sum_j \gamma_j = 0$. Routine but lengthy analysis confirms that the $m = 1$ solution $\Psi(y) \propto y$ (corresponding to $G(\{p_i\}) = -\sum_i p_i^2$) satisfies the functional equations and the constraints. However, since $m > 0$, it is ruled out by the 2×2 marginal model problem treated previously. Therefore no power-law solution survives, and we have finally singled out the entropy.

(For those who wish to investigate the functional equations further, the general version of (82) is

$$\sum_j \Psi(\nu' y + \gamma_j) = \Psi(y). \tag{84}$$

Inspection reveals a linear solution $\Psi(y) \propto y - (J-1)^{-1}\sum_j \gamma_j$ provided that $\nu' = J^{-1}$; this corresponds to $m = 1$. If $\nu' = 1$ there is an exponential solution corresponding to the entropy; we expect this solution to be a special case, as in section 5. Further solutions of (84) are not investigated here.)

8. Conclusions

We have derived the entropy functional

$$G(\{p_i\}) = -\sum_i p_i \ln p_i \tag{85}$$

as the unique maximand for assigning probability given linear constraints. The derivation was inspired by Shannon's use of partitioning arguments, but no advance interpretation of the maximand G has been made here. Of course, having derived it, the conventional interpretation of information entropy can be used as set out in section 2.

This difference in philosophy is operationally irrelevant for problems with constraints linear in the probabilities, but for nonlinear constraints a difference emerges. According to conventional maximum entropy the meaning of $-\sum_i p_i \ln p_i$ implies it should be maximised whatever the constraints, but in the present derivation the functional equation that gave rise to the entropic form was dependent on linearity of the constraints. For other types of constraint a different functional equation, giving a different form for G, would result. For probabilities this difference is not presently important, since only linear constraints – expectation values – have arisen in real problems. However, arguments yet unknown might give rise to nonlinear constraints in future; also the same mathematical analysis might be used to find a maximand in problems where *proportions*, not probabilities, are to be assigned. Image processing is an example, in which the intensity over an image is required [5]. Though constraints are often linear there, a nonlinear problem also arises – the phaseless Fourier problem , typically in crystallography where intensities of X-ray reflections measure Fourier components of the electron density of the material sample [5]. Also, in solid state physics the occupation probability of an energy level may depend on the number of electrons already there. Nonlinear optimisation ('nonlinear programming') is a vast field, and its application in appropriate generalisations of the procedure described here might generate better crystallographic reconstructions and improve experiment-theory accord in solid state physics.

References

1. E.T. Jaynes, "Information theory and statistical mechanics," in: *Statistical Physics* (1962 Brandeis Lectures), ed. K.W. Ford, W.A. Benjamin Publishing, New York, USA, 1963. Reprinted in: *E.T. Jaynes: Papers on Probability, Statistics and Statistical Mechanics*, ed. R.D. Rosenkrantz, Reidel, Dordrecht, Netherlands, 1983.
2. J.E. Shore and R.W. Johnson, "Axiomatic derivation of maximum entropy and the principle of minimum cross-entropy," *IEEE Trans.*, **IT-26**, p. 26–37, 1980.
3. C.E. Shannon, "A mathematical theory of communication," *Bell Syst. Tech. J.*, **27**,p. 379–423 and p. 623–659, 1948. Reprinted in: *The Mathematical Theory of Communication*, C.E. Shannon and W.W. Weaver, University of Illinois Press, Urbana, USA, 1949.
4. E.T. Jaynes, "Information theory and statistical mechanics," *Phys. Rev.*, **106**, p. 620–630, 1957.
5. S.F. Gull and J. Skilling, "Maximum entropy method in image processing," *IEE Procs.*, **131F**,p. 646–659, 1984.
6. C. Tsallis, "Possible generalization of Boltzmann–Gibbs statistics," *J. Stat. Phys.*, **52**,p. 479–487, 1988.

EFFICIENT ESTIMATION OF MISSING INFORMATION IN MULTIVALUED SINGLY CONNECTED NETWORKS USING MAXIMUM ENTROPY

D. HOLMES

School of Computing and Mathematics,
University of Bradford, Bradford, BD7 1DP, U.K. †

Abstract. Although it has been shown that in the general case solving causal networks using maximum entropy techniques is *NP*-complete, P.C. Rhodes, G.R. Garside and the present author have previously provided correct and efficient algorithms for calculating minimally prejudiced estimates for missing conditional probabilities in causal trees and causal inverted trees with multivalued events. This paper considers multivalued singly connected causal networks (polytrees) and shows that in this case, maximum entropy can be used to find minimally prejudiced estimates for missing information efficiently.

Key words: Distributional Inference, Decision Theory, Foundations of Statistics, Incomplete Information.

1. Introduction

Resistance to Bayesian (causal) networks as a problem solving strategy usually involves concern about how the required probabilities are found. This concern highlights some of the limitations of causal networks with complete information as a vehicle for modelling problem domains. Jensen [1] points out that there is often no theoretical way of determining all the required probabilities and gives examples of how they are ascertained in practice. Sometimes they are guessed, sometimes a complex and subjective procedure is gone through in order to produce an approximate and necessarily biased value. When multivalued events are to be modelled, the situation becomes complex. In some situations there are inevitably too many conditional probabilities for an expert to reliably estimate. Thus the need for a theoretically sound technique for estimating them in a minimally

†Email: D.Holmes@bradford.ac.uk

W. von der Linden et al. (eds.), Maximum Entropy and Bayesian Methods, 289–300.

prejudiced fashion becomes apparent. The maximum entropy formalism provides just such a technique. However, Maung and Paris [2] have shown that the general problem of finding the maximum entropy solution in probabilistic systems is *NP*-complete; and so, if we are to apply this methodology to causal networks, we must show that the estimates for missing information can be found efficiently. P.C. Rhodes, G.R. Garside and the present author have previously provided correct and efficient algorithms for calculating minimally prejudiced estimates for missing conditional probabilities in causal trees [3] and causal inverted trees [4] with multivalued events.

2. Causal Networks

Following the style adopted by Jensen [1], Neapolitan [5], Paris [6] and others a causal network can be defined as follows. Let

(i) **V** be a finite set of vertices
(ii) **B** be a set of directed edges (branches) between vertices with no feedback loops. The vertices together with the directed edges form a directed acyclic graph $\mathbf{G} =< \mathbf{V}, \mathbf{B} >$
(iii) a set of events be depicted by the vertices of **G** and hence also represented by **V**, each event having a finite set of mutually exclusive outcomes
(iv) E_i be a variable which can take any of the outcomes e_i^j of the event i, $j = 1...n_i$
(v) **P** be a probability distribution over the combinations of events, i.e. **P** consists of all possible $P(\bigwedge_{i \in V} E_i)$.

Let **C** be the following set of constraints:

(2i) The elements of **P** sum to unity.
(2ii) for each event i with a set of parents Mi there are associated conditional probabilities $P(E_i | \bigwedge_{j \in M_i} E_j)$ for each possible outcome that can be assigned to E_i and E_j.
(2iii) those independence relationships implied by *d*-separation in the directed acyclic graph [5].

Then $\mathbf{N} =< \mathbf{G}, \mathbf{P}, \mathbf{C} >$ is a causal network if **P** satisfies **C**. In the present paper causal networks are restricted to multivalued singly connected causal networks (polytrees). Polytrees are a causal network where each vertex in **V** may have many parents and many children but where there is no more than one path between any two nodes. When the tree is depicted by a graph, the nodes of the tree are the vertices of the graph and the branches of the tree are the edges of the graph. The causal information in (2ii) above is always of the form $P(E_{c_1}|M_{c_1})$ except at a source node, s, where it is of the form $P(E_s)$. Then $\mathbf{G} =< \mathbf{V}, \mathbf{B} >$ is a directed acyclic graph which represents a polytree. The causal information required by the causal network described above is

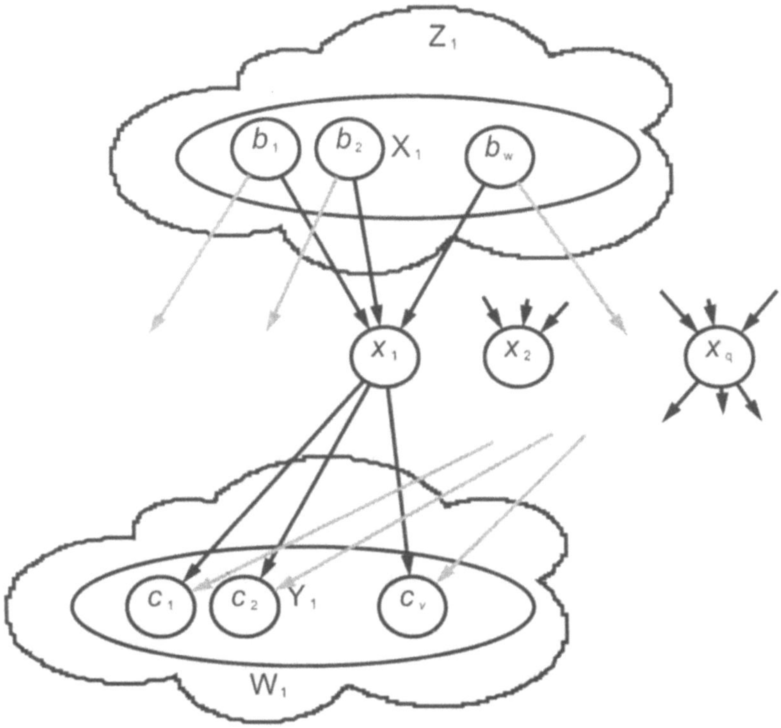

Figure 1. Representation of the causal polytree **N**.

(3i) the marginal probabilities of the sources nodes, i.e. $P(e_i^j) = a_i^j,\ j = 1...n_i - 1$ when $\mathbf{X}_i = \{\}$

(3ii) the conditional probability of each non-source node c_i, given the state of its parents, i.e. $P(e_{c_i}^m | M_{c_1}) = \beta(c_i^m, M_{c_1})$, for $m = 1...n_{c1} - 1$ when $\mathbf{X}_{c_1} \neq \{\}$, where α_i^j and $\beta(c_i^m, M_{c_1})$ are constants.

3. Solution Using Bayesian Techniques

The general state S of the causal tree is the conjunction $\underset{i \in V}{\wedge} E_i$. A particular state is obtained by assigning some e_i^j to each E_i. Subtrees can have their states defined in a similar manner. Hence, any tree or subtree can only be in one of a finite set of states and the state of the tree is the conjunction of the states of any set of subtrees which constitutes a partition of the tree.

Consider the general causal polytree N, as represented by Figure 1. Let

$\mathbf{X_i}$ denote the parents of x_i, $i = 1...q$
$\mathbf{Z_i}$ denote the ancestors of x_i, $i = 1...q$
$\boldsymbol{X}_i$ denote a state of $\mathbf{X_i}$, $i = 1...q$
$\boldsymbol{X}'_i$ denote a state of $(\mathbf{Z_i} - \mathbf{X_i})$, $i = 1...q$

$\mathbf{Y_i}$ denote the children of x_i, $i = 1...q$
$\mathbf{W_i}$ denote the descendants of x_i
$\boldsymbol{Y}_i$ denote a state of $\mathbf{Y_i}$, $i = 1...q$
$\boldsymbol{Y}'_i$ denote a state of $(\mathbf{W_i} - \mathbf{Y_i})$, $i = 1...q$

$\mathbf{U_i}$ denote the parents of c_i, $i = 1..v$.
$\boldsymbol{U}_i$ denote a state of $\mathbf{U_i}$, $i = 1..v$.
$\bar{\boldsymbol{U}}_1$ denote a state of $(\mathbf{U_i} - x_1)$, $i = 1..v$.

Since the parents of c_i always contain x_1, then $x_1\bar{\boldsymbol{U}}_1 = \mathbf{U}_1$. Then the probability of any state S of $\boldsymbol{N}$ is given by

$$P(S) = P(\boldsymbol{X'}_1...\boldsymbol{X'}_q\boldsymbol{X}_1...\boldsymbol{X}_q E_{x_1}...E_{x_q}\boldsymbol{Y}_1...\boldsymbol{Y}_q\boldsymbol{Y'}_1..\boldsymbol{Y'}_q) \ .$$

Expanding out using the chain rule and simplifying using those independence relationships implied by d-separation (see Neapolitan[5]) we get

$$P(S) = P(\boldsymbol{X}_1\boldsymbol{X'}_1)...P(\boldsymbol{X}_q\boldsymbol{X'}_q)P(E_{x_1}|E_{b_1}...E_{b_w})P(E_{x_2}|\boldsymbol{X}_2)...P(E_{x_q}|\boldsymbol{X}_q)$$
$$P(E_{c_1}|x_1\bar{\boldsymbol{U}}_1)...P(E_{c_v}|x_1\bar{\boldsymbol{U}}_v)P(\boldsymbol{Y}_1\boldsymbol{Y}'_1)...P(\boldsymbol{Y}_q\boldsymbol{Y}'_q)$$

Each event can take on many values. In (iv) above we let event i have n_i values. In what follows we denote $n_{x_1}...n_{x_q}$ by $k_1...k_q$ respectively. Choose x_1 arbitrarily to be of special interest. Let $m_1...m_v$ be any one of the possible outcomes $(1...n_{c_1})$, ..., $(1...n_{c_v})$ of events $c_1...c_v$ respectively. Denote by $\bar{\boldsymbol{U}}_i^{\boldsymbol{K}_i}$ and $\bar{\boldsymbol{U}}_i^{\boldsymbol{J}_i}$ that each event in $\bar{\boldsymbol{U}}_i$ is instantiated by its k th and j th value respectively.
Consider the states (4i) to (4iv) below where $c_1...c_v$ are instantiated by $m_1...m_v$ respectively and

(4i) events $x_1...x_q$ are instantiated by arbitrary values $e_{x_1}^{j_1}...e_{x_q}^{j_q}$ respectively, denoted by $S_{x_1...x_q:j_1...j_q}$
(4ii) event x_1 is instantiated by $e_{x_1}^{k_1}$ and events $x_2...x_q$ are instantiated by arbitrary values $e_{x_2}^{j_2}...e_{x_q}^{j_q}$ respectively, denoted by $S_{x_1...x_q:j_1...j_q}$
(4iii) events $x_1...x_q$ are instantiated by arbitrary values $e_{x_1}^{k_1}...e_{x_q}^{k_q}$ respectively, denoted by $S_{x_1...x_q:k_1...k_q}$
(4iv) event x_1 is instantiated by $e_{x_1}^{j_1}$ and events $x_2...x_q$ are instantiated by arbitrary values $e_{x_2}^{k_2}...e_{x_q}^{k_q}$ respectively, denoted by $S_{x_1,j_1:x_2...x_q:k_2...k_q}$

Where an event is not explicit in states (4i)-(4iv) it must be instantiated by the same arbitrary value throughout. Forming the quotient of probabilities of states from (4iii) and (4iv) we get

$$\frac{P(S_{x_1\ldots x_q,k_1\ldots k_q})}{P(S_{x_1,j_1:x_2\ldots x_q,k_2\ldots k_q})} = \frac{\beta(x_1^{k_1},b_1^{r_1}\ldots b_w^{r_w})\prod_{i=1}^{v}\beta(c_i^{m_i},x_1^{k_1}\bar{\boldsymbol{U}}_i^{\boldsymbol{K}_i})}{\beta(x_1^{j_1},b_1^{r_1}\ldots b_w^{r_w})\prod_{i=1}^{v}\beta(c_i^{m_i},x_1^{j_1}\bar{\boldsymbol{U}}_i^{\boldsymbol{K}_i})} \tag{1}$$

Similarly, forming the quotient Q, of probabilities of states with (4i) and (4iii) as numerators and (4ii) and (4iv) as denominators, we get after cancelling

$$\begin{aligned} Q &= \frac{P(S_{x_1\ldots x_q,j_1\ldots j_q})}{P(S_{x_1,k_1:x_2\ldots x_q,j_2\ldots j_q})}\frac{P(S_{x_1\ldots x_q,k_1\ldots k_q})}{P(S_{x_1,j_1:x_2\ldots x_q,k_2\ldots k_q})} \\ &= \frac{\prod_{i=1}^{v}\beta(c_i^{m_i},x_1^{j_1}\bar{\boldsymbol{U}}_i^{\boldsymbol{J}_i})\prod_{i=1}^{v}\beta(c_i^{m_i},x_1^{k_1}\bar{\boldsymbol{U}}_i^{\boldsymbol{K}_i})}{\prod_{i=1}^{v}\beta(c_i^{m_1},x_1^{k_1}\bar{\boldsymbol{U}}_i^{\boldsymbol{J}_i})\prod_{i=1}^{v}\beta(c_i^{m_i},x_1^{j_1}\bar{\boldsymbol{U}}_i^{\boldsymbol{K}_i})}. \end{aligned} \tag{2}$$

These quotients are interesting in that most of the terms have cancelled and this will subsequently prove to be useful.

4. Solution Using Maximum Entropy Techniques

An account of the principle in general is given by Jaynes [7]. The conditional probabilities already known for a given tree are treated as constraints by the maximum entropy formalism. Given the constraints imposed upon the tree the maximum entropy method will estimate the least prejudicial probability distribution. When there are no constraints other than that of being a probability distribution, a uniform distribution gives the highest entropy, thus complying with the principle of insufficient reason. When complete causal and independence information is given, the maximum entropy solution is fully constrained, which results in the same probability distribution as that achieved by the Bayesian method used above. Consider again the knowledge domain represented by the polytree $\boldsymbol{N}$ in Figure 1. The number of states N_S in a causal tree is given by $N_S = \prod_{i\in V} n_i$ where n_i is the number of values possessed by the event i. For the first part of this section it will be helpful if we consider the states numbered from $1\ldots N_S$ and denoted by S_i, $i = 1\ldots N_S$.

The maximum entropy formalism requires us to maximise

$$H = -\sum_{i=1}^{N_s} P(S_i)\log P(S_i)$$

whilst conforming with the necessary constraints. The Lagrange multiplier technique, as first described by Jaynes [8], will be used. Rhodes and Garside [9] have shown that if the work done by Jaynes and Tribus is modified to handle non-linear constraints arising from independence, the generalised form of **P** is

$$P(S_i) = \prod_{C_r \in C} \exp\left((-\lambda_r)\left(\frac{\partial C_r}{\partial P(S_i)}\right)\right) \tag{3}$$

where λ_r is the Lagrange multiplier associated with constraint C_r and C is the set of constraints described in (2i)-(2iii). An important constraint is (2i), the requirement that a probability distribution sums to unity, since the Lagrange multiplier technique will not automatically assume this. Since the contribution from this constraint is $\exp(-\lambda_0)$ in all expressions for the probabilities of states, these cancel in Q. Other constraints include the causal information (2ii) and the independence information (2iii) required by the polytree. We first consider the contributions of the causal information to the quotient under consideration. The notation of (2ii) has been introduced in (3i) and (3ii),of the Bayesian model. Since, in Q, only values of $x_1 ... x_q$ change, all terms arising from constraint equations involving only nodes other than these will cancel. The causal constraints (3ii) can be re-expressed in terms of state probabilities and written as

$$C(c_1^m, x_1^{j_1}\bar{\boldsymbol{U}}_i^{\boldsymbol{J}_i}) \quad : \quad \frac{P(e_{c_1}^m e_{x_1}^{j_1}\bar{\boldsymbol{U}}_i^{\boldsymbol{J}_i})}{P(e_{x_1}^{j_1}\bar{\boldsymbol{U}}_i^{\boldsymbol{J}_i})} = \beta(c_1^m, x_1^{j_1}\bar{\boldsymbol{U}}_i^{\boldsymbol{J}_i}) \,. \tag{4}$$

Expanding (4) and writing it as sums of state probabilities we get

$$\left(1 - \beta(c_1^m, x_1^{j_1}\bar{\boldsymbol{U}}_i^{\boldsymbol{J}_i})\right)\sum_{y \in Y} P(S_y) - \beta(c_1^m, x_1^{j_1}\bar{\boldsymbol{U}}_i^{\boldsymbol{J}_i})\sum_{z \in Z} P(S_z) = 0 \tag{5}$$

where $Y = \{y | S_y \text{ contains } e_{c_1}^m e_{x_1}^{j_1}\bar{\boldsymbol{U}}_i^{\boldsymbol{J}_i}\}$ and $Z = \{z | S_z \text{ contains one of } e_{c_1}^{n_{c_1}} e_{x_1}^{j_1}\bar{\boldsymbol{U}}_i^{\boldsymbol{J}_i}\}$. This determines a family of constraint equations associated with c_1 and its parents. When information is complete, every such set of nodes will have a family of constraint equations associated with it. (A source node is a special case). Each constraint in (5) has associated with it a Lagrange multiplier λ, these being labelled in a manner similar to the corresponding β, i.e. $\lambda(c_1^1, x_1^1\bar{\boldsymbol{U}}_1^1) ... \lambda(c_1^{n_{c_1}}, x_1^{k_1}\bar{\boldsymbol{U}}_i^{\boldsymbol{K}_i})$. We now find the same quotient of probabilities of states, Q, as in the solution using Bayesian techniques but this time using maximum entropy techniques Consider the causal information required by Q. The probability of a state $P(S_{x_1 ... x_q, j_1 ... j_q})$ contains the following Lagrangian term from (3)

$$\prod_{x=1}^{v}\left\{\exp\left[-\lambda(c_x^{m_x}, x_1^{j_1}\bar{\boldsymbol{U}}_i^{\boldsymbol{J}_i})\right] i \prod_{y=1}^{n_{c_x}-1} \exp\left[-\lambda(c_x^y, x_1^{j_1}\bar{\boldsymbol{U}}_i^{\boldsymbol{J}_i})\left(-\beta\left(c_x^y, x_1^{j_1}\bar{\boldsymbol{U}}_i^{\boldsymbol{J}_i}\right)\right)\right]\right\} \tag{6}$$

Similarly $P(S_{x_1, j_1 : x_2 ... x_q, k_2 ... k_q})$ contains the term

$$\prod_{x=1}^{v}\left\{\exp\left[-\lambda(c_x^{m_x}, x_1^{j_1}\bar{\boldsymbol{U}}_i^{\boldsymbol{K}_i})\right] \prod_{y=1}^{n_{c_x}-1} \exp\left[-\lambda(c_x^y, x_1^{j_1}\bar{\boldsymbol{U}}_i^{\boldsymbol{K}_i})\left(-\beta\left(c_x^y, x_1^{j_1}\bar{\boldsymbol{U}}_i^{\boldsymbol{K}_i}\right)\right)\right]\right\} \tag{7}$$

$P(S_{x_1...x_q,k_1...k_q})$ contains the term

$$\prod_{x=1}^{v}\left\{\exp\left[-\lambda(c_x^{m_x},x_1^{k_1}\bar{\boldsymbol{U}}_i^{\boldsymbol{K}_i})\right]\prod_{y=1}^{n_{x_x}-1}\exp\left[-\lambda(c_x^{y},x_1^{k_1}\bar{\boldsymbol{U}}_i^{\boldsymbol{K}_i})\left(-\beta\left(c_x^{y},x_1^{k_1}\bar{\boldsymbol{U}}_i^{\boldsymbol{K}_i}\right)\right)\right]\right\} \tag{8}$$

and $P(S_{x_1,k_1:x_2...x_q,j_2...j_q})$ contains the term

$$\prod_{x=1}^{v}\left\{\exp\left[-\lambda(c_x^{m_x},x_1^{k_1}\bar{\boldsymbol{U}}_i^{\boldsymbol{J}_i})\right]\prod_{y=1}^{n_{c_x}-1}\exp\left[-\lambda(c_x^{y},x_1^{k_1}\bar{\boldsymbol{U}}_i^{\boldsymbol{J}_i})\left(-\beta\left(c_x^{y},x_1^{k_1}\bar{\boldsymbol{U}}_i^{\boldsymbol{J}_i}\right)\right)\right]\right\} \tag{9}$$

The above analysis is now applied to each of $x_1...x_q$ and its respective parents. Each of these has the same form so without loss of generality we consider x_1 and its parents, $b_1...b_w$. The form of the constraints is again dictated by the causally related constraint equations but with different event labels. We recall that $r_1...r_w$ are the arbitrary values of events $b_1...b_w$ respectively. Then $P(S_{x_1...x_q,j_1...j_q})$ and $P(S_{x_1,j_1:x_2...x_q,k_2...k_q})$ both contain the term

$$\exp\left[-\lambda(x_1^{j_1},b_1^{r_1}...b_w^{r_w})\right]\prod_{y=1}^{k_1-1}\exp\left[-\lambda(x_1^{y},b_1^{r_1}...b_w^{r_w})\left(-\beta\left(x_1^{y},b_1^{r_1}...b_w^{r_w}\right)\right)\right] \tag{10}$$

Similarly $P(S_{x_1...x_q,k_1...k_q})$ and $P(S_{x_1,k_1:x_2...x_q,j_2...j_q})$ both contain the term

$$\prod_{y=1}^{k_1-1}\exp\left[-\lambda(x_1^{y},b_1^{r_1}...b_w^{r_w})\left(-\beta\left(x_1^{y},b_1^{r_1}...b_w^{r_w}\right)\right)\right] . \tag{11}$$

Hence terms arising from $x_1...x_q$ and their parents, cancel in Q. Since the assignments to all events in the other causal constraints are identical in all four probabilities in Q, the contributions from these constraints to the quotient also cancel.

We now consider the non-linear constraints, which arise from information concerning independence. Verma and Pearl [10] have identified all such relationships implicit within general causal networks. The number of independence relationships is exponential in the number of events in the system and we must show that this does not imply an exponential amount of computational work when calculating estimates of missing information based on the maximum entropy model. The independence relationships (2iii) implied by d-separation in $\mathbf{G}$ are far greater in number than actually required by the maximum entropy model. It has been shown in Holmes, Rhodes and Garside [4] that in Q, the independence constraints $C\left(\begin{smallmatrix}j_1\\x_1\end{smallmatrix}...\begin{smallmatrix}j_q\\x_q\end{smallmatrix}\right)$ that need explicitly stating are

$$C\left(\begin{smallmatrix}j_1\\x_1\end{smallmatrix}...\begin{smallmatrix}j_q\\x_q\end{smallmatrix}\right) \quad : \quad P(e_{x_1}^{j_1}e_{x_2}^{j_2}...e_{x_q}^{j_q})-P(e_{x_1}^{j_1})P(e_{x_2}^{j_2})...P(e_{x_q}^{j_q})=0 \tag{12}$$

with associated Lagrange multiplier $\lambda\binom{j_1 \ldots j_q}{x_1 \ldots x_q}$. The contribution to Q from these terms has been shown in Holmes [4] to be

$$\frac{\exp\left(-\lambda\binom{j_1 \ldots j_q}{x_1 \ldots x_q}\right)\exp\left(-\lambda\binom{k_1 \ldots k_q}{x_1 \ldots x_q}\right)}{\exp\left(-\lambda\binom{j_1\ k_2 \ldots k_q}{x_1 \cdot x_q \ldots x_q}\right)\exp\left(-\lambda\binom{k_1\ j_2 \ldots j_q}{x_1 \cdot x_q \ldots x_q}\right)}. \tag{13}$$

It has also been shown, in Holmes [4], that the contribution to the left hand side of (1) contains the term

$$\frac{\exp\left(-\lambda\binom{j_1 \ldots j_q}{x_1 \ldots x_q}\left(1 - P(e_{x_2}^{j_2})\ldots P(e_{x_q}^{j_q})\right)\right)}{\exp\left(-\lambda\binom{k_1 j_2 \ldots j_q}{x_1 x_2 \ldots x_q}\left(1 - P(e_{x_2}^{j_2})\ldots P(e_{x_q}^{j_q})\right)\right)} \times \frac{\prod_{<s_1 \ldots s_q> \neq <j_1 \ldots j_q>} \exp\left(-\lambda\binom{j_1 s_2 \ldots s_q}{x_1 x_2 \ldots x_q}\left(-P(e_{x_2}^{s_2})\ldots P(e_{x_q}^{s_q})\right)\right)}{\prod_{<k_1 s_2 \ldots s_q> \neq <k_1 j_2 \ldots j_q>} \exp\left(-\lambda\binom{k_1 s_2 \ldots s_q}{x_1 x_2 \ldots x_q}\left(-P(e_{x_2}^{s_2})\ldots P(e_{x_q}^{s_q})\right)\right)}. \tag{14}$$

Equation (14) will be used in the next section to find an expression for an arbitrary causally related Lagrange multiplier.

5. Solving For The Lagrange Multipliers

Using maximum entropy techniques Q consists of a combination of the causally related terms and independence terms which we have found in the previous section. To simplify the algebra let

$$\mathrm{B}(\bar{\boldsymbol{U}}_i^{\boldsymbol{J}_i}) = \prod_{x=1}^{v} \frac{\beta(c_x^{m_x}, x_1^{j_1}\bar{\boldsymbol{U}}_i^{\boldsymbol{J}_i})}{\beta(c_x^{m_x}, x_1^{k_1}\bar{\boldsymbol{U}}_i^{\boldsymbol{J}_i})} \times$$

$$\prod_{x=1}^{v} \frac{\left\{\exp\left[-\lambda(c_x^{m_x}, x_1^{k_1}\bar{\boldsymbol{U}}_i^{\boldsymbol{J}_i})\right] \prod_{y=1}^{n_{c_x}-1} \exp\left[-\lambda(c_x^{y}, x_1^{k_1}\bar{\boldsymbol{U}}_i^{\boldsymbol{J}_i})\left(-\beta\left(c_x^{y}, x_1^{k_1}\bar{\boldsymbol{U}}_i^{\boldsymbol{J}_i}\right)\right)\right]\right\}}{\left\{\exp\left[-\lambda(c_x^{m_x}, x_1^{j_1}\bar{\boldsymbol{U}}_i^{\boldsymbol{J}_i})\right] \prod_{y=1}^{n_{c_x}-1} \exp\left[-\lambda(c_x^{y}, x_1^{j_1}\bar{\boldsymbol{U}}_i^{\boldsymbol{J}_i})\left(-\beta\left(c_x^{y}, x_1^{j_1}\bar{\boldsymbol{U}}_i^{\boldsymbol{J}_i}\right)\right)\right]\right\}} \tag{15}$$

and let $\mathrm{B}(\bar{U}_i^{K_i})$ be similarly defined. Equating the Bayesian expression, and the maximum entropy expression for Q, by putting together (2), (6), (7), (8), (9) and (13) substituting $\mathrm{B}(\bar{\boldsymbol{U}}_i^{\boldsymbol{J}_i})$ and $\mathrm{B}(\bar{\boldsymbol{U}}_i^{\boldsymbol{K}_i})$ we get

$$\frac{\exp\left(-\lambda\binom{j_1 \ldots j_q}{x_1 \ldots x_q}\right)}{\exp\left(-\lambda\binom{k_1 j_2 \ldots j_q}{x_1 x_2 \ldots x_q}\right)} \frac{\mathrm{B}(\bar{\boldsymbol{U}}_i^{\boldsymbol{K}_i})}{\mathrm{B}(\bar{\boldsymbol{U}}_i^{\boldsymbol{J}_i})} = \frac{\exp\left(-\lambda\binom{j_1 k_2 \ldots k_q}{x_1 x_2 \ldots x_q}\right)}{\exp\left(-\lambda\binom{k_1 \ldots k_q}{x_1 \ldots x_q}\right)}. \tag{16}$$

Applying the above analysis again but using Q where $< s_2 ... s_q > \neq < k_2 ... k_q >$, it has been shown by Holmes [4] that

$$\frac{\exp\left(-\lambda\left(\begin{smallmatrix} j_1 & & j_q \\ x_1 & \cdots & x_q \end{smallmatrix}\right)\right)}{\exp\left(-\lambda\left(\begin{smallmatrix} k_1 j_2 & & j_q \\ x_1 x_2 & \cdots & x_q \end{smallmatrix}\right)\right)} \frac{\mathrm{B}(\bar{U}_i^{S_i})}{\mathrm{B}(\bar{U}_i^{J_i})} = \frac{\exp\left(-\lambda\left(\begin{smallmatrix} j_1 s_2 & & s_q \\ x_1 x_2 & \cdots & x_q \end{smallmatrix}\right)\right)}{\exp\left(-\lambda\left(\begin{smallmatrix} k_1 s_2 & & s_q \\ x_1 x_2 & \cdots & x_q \end{smallmatrix}\right)\right)} . \tag{17}$$

We now find an expression for an arbitrary Lagrange multiplier associated with causal information by considering (1). From the solution using Bayesian techniques (2), the causally related terms (6), (9), (10) and (11), and the independence related terms (14) we get, after simplification

$$\begin{aligned} \exp\left(-\lambda(x_1^{j_1}, b_1^{r_1} ... b_w^{r_w})\right) &= \mathrm{B}(\bar{\boldsymbol{U}}_i^{\boldsymbol{J}_i}) \frac{\beta(x_1^{j_1}, b_1^{r_1} ... b_w^{r_w})}{\beta(x_1^{k_1}, b_1^{r_1} ... b_w^{r_w})} \\ &\times \frac{\exp\left(-\lambda\left(\begin{smallmatrix} k_1 j_2 & & j_q \\ x_1 x_2 & \cdots & x_q \end{smallmatrix}\right)\left(1 - P(e_{x_2}^{j_2}) ... P(e_{x_q}^{j_q})\right)\right)}{\exp\left(-\lambda\left(\begin{smallmatrix} j_1 & & j_q \\ x_1 & \cdots & x_q \end{smallmatrix}\right)\left(1 - P(e_{x_2}^{j_2}) ... P(e_{x_q}^{j_q})\right)\right)} \\ &\times \frac{\prod_{<s_2 ... s_q> \neq <j_2 ... j_q>} \exp\left(-\lambda\left(\begin{smallmatrix} k_1 s_2 & & s_q \\ x_1 x_2 & \cdots & x_q \end{smallmatrix}\right)\left(-P(e_{x_2}^{s_2}) ... P(e_{x_q}^{s_q})\right)\right)}{\prod_{<s_1 ... s_q> \neq <j_1 ... j_q>} \exp\left(-\lambda\left(\begin{smallmatrix} j_1 s_2 & & s_q \\ x_1 x_2 & \cdots & x_q \end{smallmatrix}\right)\left(-P(e_{x_2}^{s_2}) ... P(e_{x_q}^{s_q})\right)\right)} . \end{aligned} \tag{18}$$

Substituting the left-hand side of (17) in (18) and simplifying gives

$$\begin{aligned} \exp\left(-\lambda(x_1^{j_1}, b_1^{r_1} ... b_w^{r_w})\right) &= \mathrm{B}(\bar{\boldsymbol{U}}_i^{\boldsymbol{J}_i}) \frac{\beta(x_1^{j_1}, b_1^{r_1} ... b_w^{r_w})}{\beta(x_1^{k_1}, b_1^{r_1} ... b_w^{r_w})} \\ &\times \frac{\exp\left(-\lambda\left(\begin{smallmatrix} k_1 j_2 & & j_q \\ x_1 x_2 & \cdots & x_q \end{smallmatrix}\right)\left(1 - P(e_{x_2}^{j_2}) ... P(e_{x_q}^{j_q})\right)\right)}{\exp\left(-\lambda\left(\begin{smallmatrix} j_1 & & j_q \\ x_1 & \cdots & x_q \end{smallmatrix}\right)\left(1 - P(e_{x_2}^{j_2}) ... P(e_{x_q}^{j_q})\right)\right)} \\ &\times \prod_{<s_2 ... s_q> \neq <j_2 ... j_q>} \left(\frac{\exp\left(-\lambda\left(\begin{smallmatrix} k_1 j_2 & & j_q \\ x_1 x_2 & \cdots & x_q \end{smallmatrix}\right)\right)}{\exp\left(-\lambda\left(\begin{smallmatrix} j_1 & & j_q \\ x_1 & \cdots & x_q \end{smallmatrix}\right)\right)} \frac{\mathrm{B}(\bar{\boldsymbol{U}}_i^{\boldsymbol{J}_i})}{\mathrm{B}(\bar{\boldsymbol{U}}_i^{\boldsymbol{S}_i})}\right)^{-\left(P(e_{x_2}^{s_2}) ... P(e_{x_q}^{s_q})\right)} \end{aligned} \tag{19}$$

Since exponential terms on the right hand side of (19) cancel and terms in $\mathrm{B}(\bar{\boldsymbol{U}}_i^{\boldsymbol{J}_i})$ combine to give a single term we obtain

$$\exp\left(-\lambda(x_1^{j_1}, b_1^{r_1} ... b_w^{r_w})\right) = \frac{\beta(x_1^{j_1}, b_1^{r_1} ... b_w^{r_w})}{\beta(x_1^{k_1}, b_1^{r_1} ... b_w^{r_w})} \prod \mathrm{B}(\bar{U}_i^{S_i})^{P(e_{x_2}^{s_2}) ... P(e_{x_q}^{s_q})} . \tag{20}$$

We thus have an expression for $\exp\left(-\lambda(x_1^{j_1}, b_1^{r_1} ... b_w^{r_w})\right)$ in terms of known information. The sink node and the source nodes are special cases of (20). The solution given above is for a completely specified system. The next section explores how the analysis above may be utilised to provide minimally prejudiced estimates for any missing information and thus allow standard methods to be implemented.

6. Estimating Missing Information

Suppose $\beta(x_1^{j_1}, b_1^{r_1}...b_w^{r_w})$ is missing. Then the maximum entropy solution no longer generates the term $\exp\left(-\lambda(x_1^{j_1}, b_1^{r_1}...b_w^{r_w})\right)$. So putting

$\Omega_{j_1} = \prod \mathrm{B}(\bar{\boldsymbol{U}}_i^{\boldsymbol{S}_i})^{-P(e_{x_2}^{s_2})...P(e_{x_q}^{s_q})}$ and re-arranging terms (20) becomes

$$\Omega_{j_1} = \frac{{}_{mp}\beta(x_1^{j_1}, b_1^{r_1}...b_w^{r_w})}{\beta(x_1^{k_1}, b_1^{r_1}...b_w^{r_w})} \tag{21}$$

where ${}_{mp}\beta(x_1^{j_1}, b_1^{r_1}...b_w^{r_w})$ denotes a minimally prejudiced estimate of $\beta(x_1^{j_1}, b_1^{r_1}...b_w^{r_w})$. Since

$$\beta(x_1^{k_1}, b_1^{r_1}...b_w^{r_w}) = 1 -_{mp} \beta(x_1^{j_1}, b_1^{r_1}...b_w^{r_w}) - \sum_{\substack{y=1 \\ y \neq j_1}}^{k_1-1} \beta(x_1^{y}, b_1^{r_1}...b_w^{r_w})$$

equation (21) becomes

$${}_{mp}\beta(x_1^{j_1}, b_1^{r_1}...b_w^{r_w}) = \left(1 -_{mp} \beta(x_1^{j_1}, b_1^{r_1}...b_w^{r_w}) - \sum_{\substack{y=1 \\ y \neq j}}^{k_1-1} \beta(x_1^{y}, b_1^{r_1}...b_w^{r_w})\right)\Omega_{j_1} \tag{22}$$

Hence

$${}_{mp}\beta(x_1^{j_1}, b_1^{r_1}...b_w^{r_w}) = \frac{\Omega_{j_1}}{1+\Omega_{j_1}}\left(1 - \sum_{\substack{y=1 \\ y \neq j_1}}^{k_1-1} \beta(x_1^{y}, b_1^{r_1}...b_w^{r_w})\right) \tag{23}$$

is a minimally prejudiced estimate of $\beta(x_1^{j_1}, b_1^{r_1}...b_w^{r_w})$ when one piece of information is missing.

Suppose that many $\beta(x_1^{j_1}, b_1^{r_1}...b_w^{r_w})$ terms are missing. Without loss of generality we can assume that $\beta(x_1^{j_1}, b_1^{r_1}...b_w^{r_w})$, $j_1 = 1...g$ are missing. Let minimally prejudiced estimates of g pieces of missing information be

$${}_{mp}\beta(x_1^{1}, b_1^{r_1}...b_w^{r_w}),\ {}_{mp}\beta(x_1^{2}; b_1^{r_1}...b_w^{r_w}),\ ...,\ {}_{mp}\beta(x_1^{g}, b_1^{r_1}...b_w^{r_w}).$$

Then, re-arranging (21) we get

$$\frac{{}_{mp}\beta(x_1^{1_1}, b_1^{r_1}...b_w^{r_w})}{\Omega_1} = \frac{{}_{mp}\beta(x_1^{2}, b_1^{r_1}...b_w^{r_w})}{\Omega_2} = ... = \frac{{}_{mp}\beta(x_1^{g}, b_1^{r_1}...b_w^{r_w})}{\Omega_g} = {}_{mp}\beta(x_1^{k_1}, b_1^{r_1}...b_w^{r_w}) \tag{24}$$

Substituting ${}_{mp}\beta(x_1^{k_1}, b_1^{r_1}...b_w^{r_w}) = 1 - \sum_{y=1}^{g} {}_{mp}\beta(x_1^y, b_1^{r_1}...b_w^{r_w}) - \sum_{y=g+1}^{k_1-1} \beta(x_1^y, b_1^{r_1}...b_w^{r_w})$ into (23) and factorising gives

$$ {}_{mp}\beta(x_1^{k_1}, b_1^{r_1}...b_w^{r_w}) \left(1 + \sum_{y=1}^{g} \Omega_y\right) = 1 - \sum_{y=g+1}^{k_1-1} \beta(x_1^y, b_1^{r_1}...b_w^{r_w}) \ , $$

and hence

$$ {}_{mp}\beta(x_1^{k_1}, b_1^{r_1}...b_w^{r_w}) = \frac{1 - \sum_{y=g+1}^{k_1-1} \beta(x_1^y, b_1^{r_1}...b_w^{r_w})}{\left(1 + \sum_{y=1}^{g} \Omega_y\right)} \ . \tag{25} $$

Thus, substituting (25) in (24) gives

$$ \frac{{}_{mp}\beta(x_1^1, b_1^{r_1}...b_w^{r_w})}{\Omega_1} = \frac{{}_{mp}\beta(x_1^2, b_1^{r_1}...b_w^{r_w})}{\Omega_2} = ... = \frac{{}_{mp}\beta(x_1^g, b_1^{r_1}...b_w^{r_w})}{\Omega_g} $$
$$ = \frac{1 - \sum_{y=g+1}^{k_1-1} \beta(x_1^y, b_1^{r_1}...b_w^{r_w})}{\left(1 + \sum_{y=1}^{g} \Omega_y\right)} \tag{26} $$

that is

$$ {}_{mp}\beta(x_1^{j_1}, b_1^{r_1}...b_w^{r_w}) = \left(\frac{\Omega_{j_1}}{1 + \sum_{y=1}^{g} \Omega_y}\right) \left(1 - \sum_{y=g+1}^{k_1-1} \beta(x_1^y, b_1^{r_1}...b_w^{r_w})\right) \tag{27} $$

for $j_1 = 1...g$, which are minimally prejudiced estimates of the missing information. When all information between a given node and its parents is missing; i.e. when $g = k_1 - 1$ equation (26) becomes

$$ \frac{{}_{mp}\beta(x_1^1, b_1^{r_1}...b_w^{r_w})}{\Omega_1} = \frac{{}_{mp}\beta(x_1^2, b_1^{r_1}...b_w^{r_w})}{\Omega_2} = ... = \frac{{}_{mp}\beta(x_1^{k_1-1}, b_1^{r_1}...b_w^{r_w})}{\Omega_{k_1-1}} $$
$$ = \frac{1}{\left(1 + \sum_{y=1}^{k_1-1} \Omega_y\right)} \tag{28} $$

from which the minimally prejudiced estimates of the missing information can be easily ascertained. Estimating missing information at source nodes and sink nodes are special cases of the above and are not reproduced here due to lack of space. The techniques described above can be brought together in an iterative algorithm which converges to the estimates for missing information. The local nature of the causal information required by the computational process together with the relative simplicity of the graphical structure lead to a relatively straightforward algorithm.

7. Conclusion

The work described above is part of a larger project being undertaken by Rhodes P.C., Garside G.R. and the present author, in which work already completed has provided correct and efficient algorithms for calculating minimally prejudiced estimates of missing information in causal trees and inverted causal trees. This paper has extended these results to polytrees. The results obtained so far lead the team to feel optimistic that they will be able to tackle general causal networks with a view to providing a front-end to be used in conjunction with existing methods of updating information in causal networks when some of the information is missing.

References

1. F. V. Jensen, *An Introduction to Bayesian Networks*, UCL Press, 1996.
2. I. M. I. and J. B. Paris, "Note on the infeasibility of some inference processes," *International Journal of Intelligent Systems*, **5**, p. 595, 1990.
3. D. E. Holmes and P. C. Rhodes, "Reasoning with incomplete information in a multivalued causal tree using the maximum entropy formalism," *International Journal of Intelligent Systems*, 1998. To appear.
4. D. E. Holmes, P. C. Rhodes, and G. R. Garside, "Efficient computation of marginal probabilities in multivalued causal inverted multiway trees given incomplete information," *International Journal of Intelligent Systems*, 1998. Submitted.
5. R. E. Neapolitan, *Probabilistic Reasoning in Expert Systems*, John Wiley, 1990.
6. J. B. Paris, *The Uncertain Reasoner's Companion: A Mathematical Perspective*, Cambridge University Press, 1994.
7. E. T. Jaynes, "Where do we stand on maximum entropy?," in *The Maximum Entropy Formalism*, R. D. Levine and M. Tribus, eds., p. 15, MIT Press, Cambridge, Mass., 1979.
8. E. T. Jaynes, "Notes on probability theory in science and engineering," tech. rep., Physics Department, Washington University,St.Louis, 1969.
9. P. C. Rhodes and G. R. Garside, "Computing marginal probabilities in causal inverted binary trees given incomplete information," *Knowledge Based Systems*, **10**, p. 213, 1998.
10. T. Verma and J. Pearl, "Causal networks: Semantics and expressiveness," in *Proceedings of the Fourth AAAI Workshop on Uncertainty in Artificial Intelligence*, University of Minnesota, Minneapolis, Minnesota, 1998.

OVERVIEW AND BAYESIAN PERSPECTIVE OF PHASE UNWRAPPIN

G. NICO, L. GUERRIERO
Dipartimento Interateneo di Fisica and
Istituto Nazionale di Fisica della Materia (INFM)
I-70126 Bari ‡

AND

G. PALUBINSKAS, M. DATCU
Deutsches Zentrum für Luft und Raumfahrt (DLR) e.V.
Deutsches Fernerkundungsdatenzentrum (DFD), Oberpfaffenhofen
D-82234 Weßling §

Abstract. The problem of phase unwrapping of two-dimensional phase signals has gained a considerable interest in recent years. The areas of applications include speckle imaging, magnetic resonance imaging and Synthetic Aperture Radar (SAR) interferometry. Phase unwrapping deals with the problem of estimating an absolute phase f from the observation of its noisy principal (wrapped) values g. This is an ill-posed problem since many possible solutions correspond to a given observation. We give an overview of several phase unwrapping algorithms and analyze them in the Bayesian framework, discussing the quality of models used to encapsulate the *a priori* knowledge.

1. Introduction

In many fields in physics and engineering, the interferometric phase is used to retrieve a physical information. In this paper we will focus on Synthetic Aperture Radar Interferometry (InSAR) [1] due to the growing importance it has assumed in the last years. InSAR is a remote sensing technique able to perform global and continous monitoring of the environment [1–3]. An interferogram - a regular array of interferometric phase values - is formed by appropriately combining the electromagnetic fields gathered by two spatially and/or temporally separated SAR antennas. The interferogram quality depends on the correlation between the two received electromagnetic fields. Sources of decorrelation are: thermal noise, temporal changes of the scene, spatial diversity. We will refer to such a decorrelation as interferogram noise [4]. The interferogram gives phase values in the principal inter-

‡Email: {Giovanni.Nico,Luciano.Guerriero}@ba.infn.it
§Email: {Gintautas.Palubinskas,Mihai.Datcu}@dlr.de

W. von der Linden et al. (eds.), Maximum Entropy and Bayesian Methods, 301–308.

val $(-\pi, \pi]$, whereas the physical information is related to the absolute phase. The problem of phase unwrapping (PU), consisting in estimating the absolute phase from the observed interferogram, is a key problem in all quantitative applications of InSAR.

2. Phase unwrapping: notation and definitions

We introduce a regular grid I, the phase fields: $\{f(i,j)\}$ – the absolute phase – and $\{g(i,j)\}$ – the wrapped observed phase, both defined on I. Moreover, we define the wrapping operator $\mathcal{W}$ as:

$$\mathcal{W} : f(i,j) \in R \rightarrow g(i,j) = \arctan\left[\frac{\Im f(i,j)}{\Re f(i,j)}\right] \in (-\pi, \pi]. \tag{1}$$

PU consists in estimating $\{f(i,j)\}$ from the observed noisy interferogram $\{g(i,j)\}$

$$g(i,j) = \mathcal{W}[f(i,j) + n(i,j)], \tag{2}$$

where $\{n(i,j)\}$ is the field describing the noise on the interferogram [5]. This is an *ill-posed* problem, in the meaning, that to a given interferogram many solutions correspond. The concept of residue is a peculiarity of two-dimensional PU. A residue occurs where the sum of following quantities

$$\Delta_{x(y)}(i,j) = \mathcal{W}[\nabla_{x(y)}g(i,j)] \tag{3}$$

calculated along the smallest closed path, a square of 2×2 pixels, is different from zero. Residues, due to steep absolute phase gradients verifying the condition $|\nabla f(i,j)| > \pi$, are used to have an idea of the difficulty in unwrapping an interferogram (see figure 1). In the most general sense, it is impossible to unambiguously recover the unknown phase f corrupted by noise from its wrapping into the interval $(-\pi, \pi]$. However, certain assumptions on the underlying sampling process or the nature of the desired solution can make PU tractable.

3. Least Squares Estimation

The Least Squares Estimation (LSE) algorithm takes into consideration the first phase derivatives fields $\{\nabla f(i,j)\}$ and $\{\nabla g(i,j)\}$. The most common assumption on the sampling process is that the desired unwrapped phase verify the condition

$$|\nabla f(i,j)| < \pi \tag{4}$$

everywhere on the grid I. This is equivalent to the requirement that the sampled signal from which g is extracted is not aliased. In this case the relation

$$\nabla f(i,j) = \mathcal{W}[\nabla g(i,j)]. \tag{5}$$

is verified. The LSE algorithm gives an estimate of $\nabla f(i,j)$ fitting the observed quantities $\mathcal{W}[g(i,j)]$. We refer to table 1 for the functional which is minimized.

(a) (b)

Figure 1. (a) Absolute phase surface; (b) Residues map of the interferogram obtained from the absolute phase surface depicted in (a): residues occur where the sum, along the smallest closed path consisting of a square of 2×2 adjacent pixels, of the quantities defined in (3) is different from zero (see section 2). The positive (black) and negative (white) residues are displayed. As an example, we consider the following absolute phase values $(0, 1.6\pi, 0.9\pi, 0.3\pi)$ defined on a square of 2×2 adjacent pixels. By using (1) we obtain the corresponding interferogram values $(0, -0.4\pi, 0.9\pi, 0.3\pi)$ and the quantities $(-0.4\pi, -0.7\pi, -0.6\pi, -0.3\pi)$ introduced in (3) which sum is -2π. This results in a negative residue. The residue occurs because one of the absolute phase derivatives $(1.6\pi, -0.7\pi, -0.6\pi, -0.3\pi)$ verifies the condition $|\nabla f| > \pi$ due to undersampling of the phase signal caused by steep absolute phase gradients or noise.

The sets I_x and I_y are subsets of I such that the first derivatives can be defined. This solution has two drawbacks. The first one is that the LSE does not take into account the ill–posedness of the problem: there are many possible solutions $\{\nabla f(i,j)\}$ minimizing the functional. The second drawback is that the assumption $|\nabla f(i,j)| < \pi$, equivalent to stating that the interferogram has not residues, is not always verified. When the LSE algorithm is applied to a situation, such as that depicted in figure 1, the solution, reported in figure 2-b, is *biased*, with this we mean that every true phase slope $\nabla f(i,j)$, which value is outside $(-\pi, \pi]$, cannot be correctly estimated.

4. Bayesian inference

We need a way to estimate a $\{f(i,j)\}$, which well explains $\{g(i,j)\}$ and at the same time fulfills some regularity criterion, which allows us to choose among the many possible solutions the most probable one according to our *a priori* knowledge. The Bayesian approach allows one to take into account both requirements simultaneously. The MAP estimate of $\{f(i,j)\}$ is obtained by maximizing the *a posteriori* probability distribution $P_{f|g}$

$$P_{f|g} = \frac{P_{g|f} \cdot P_f}{P_g}, \tag{6}$$

where $P_{g|f}$ describes the fidelity to the data, P_f gives the *prior* and P_g is a normalization factor. Assuming a χ^2-measure of the error between the reconstructed and the observed data, $P_{f|g}$ assumes a Gibbsian form with a functional consisting of a regularity and a fidelity terms. In the following, algorithms will be introduced directly using the corresponding functionals, summarized in table 1. In the text we will refer to this table for the form of regularization functionals.

5. Tikhonov and Gull-Skilling regularization of PU

In this section two algorithms are described, which try to solve PU applying the classical recipe of Bayesian estimation. Both algorithms give an estimate of the first derivative $\nabla f(i,j)$. The functional used in Marroquin et al. [6] consists of the classical regularity and fidelity terms. This algorithm constitutes a generalization of the LSE algorithm: the two functionals coincide, when we do not take into account the first terms in each sum (see table 1), embedding the *a priori* knowledge. A similar algorithm gives an estimate of the unwrapped phase using the *maximum entropy principle* to regularize the PU [7] [1] . The fidelity is still expressed through the LSE functional. The regularity is posed in a statistical manner: one states that in absence of the constraint due to observations, the absolute phase would assume everywhere the same value, because this is the configuration having the minimum information and so the maximum entropy. Instead, the knowledge we have from data reduce the entropy of the absolute phase surface and give it a shape, that we are looking for. Even if both functionals are able to inject in the solution some *a priori* knowledge about its regularity, they have the same limits essentially due to the fidelity term. In fact, impose that the absolute phase gradient be as likely as possible to the wrapping of the wrapped phase gradient, is not a useful manner to pose the fidelity between the reconstructed and the observed interferogram. The true phase slopes could be in module greater than π and then such a fidelity causes distortions when applied to real interferograms as described in section 3. In figure 2-c the result obtained applying such algorithms, with $\lambda = 0.1$ to the interferogram of the gaussian depicted in figure 2-a is reported. Moreover, these algorithms do not give a solution but a family of solutions each of one depends on the parameter λ. The choice of such a parameter gives an arbitrary flavor to the PU solution.

6. Bayesian estimation through residues analysis

After the first attempts to regularize PU, new algorithms were proposed trying to release the classical recipe of Bayesian estimation. In table 1 is reported the functional used by Costantini [8] in his regularization of PU through residues analysis and integer variables. The quantities $\{k_x(i,j)\}_{(i,j)\in I_x}$, $\{k_y(i,j)\}_{(i,j)\in I_y}$ are integers and give the number of 2π cycles to sum the wrapped phase gradients $\{\mathcal{W}[\nabla_x g(i,j)]\}_{(i,j)\in I_x}$, $\{\mathcal{W}[\nabla_y g(i,j)]\}_{(i,j)\in I_y}$ in order to have an estimate of the true phase gradients $\{\nabla_x f(i,j)\}_{(i,j)\in I_x}$, $\{\nabla_y f(i,j)\}_{(i,j)\in I_y}$. The constraint equation (see table 1) states that the sum of the quantities $\nabla_{x(y)} f(i,j) = \nabla_{x(y)} g(i,j) + 2\pi k_{x(y)}(i,j)$ along the smallest closed paths is zero, that is no "residues" are on the unwrapped phase. This is the peculiarity of the algorithm wich differ it from the LSE. The solution is obtained exploiting the integer nature of PU. It is worth

[1]Both f and g have to be properly normalized in order to fulfill the condition $\sum f = 1$.

to note, that in this case, it is not possible easily to recognize the regularity term. The *a priori* knowledge enters through the residues analysis, which states that the *a priori* probability distribution is zero everywhere except for the configurations satisfying the constraint. The functional is similar to the term minimized by the LSE algorithm. So, even if the result is better than LSE it evidences the same trouble characterizing the LSE solution as can been seen from figure 2-d.

7. Bayesian estimation through residues analysis and hidden data

Recently, a new algorithm has been developped for solving PU in the framework of Bayesian inference [9]. The functional which is minimized is nothing but a Laplacian operator expressed in terms of differences between first derivatives. These are written using the terms $\{\mathcal{W}[\nabla_x g(i,j)]\}_{(i,j)\in I_x}$, $\{\mathcal{W}[\nabla_y g(i,j)]\}_{(i,j)\in I_y}$ measured on the interferogram and the quantities $\{k_x(i,j)\}_{(i,j)\in I_x}$, $\{k_y(i,j)\}_{(i,j)\in I_y}$ to be estimated as described in section 6. The used sets are subsets of I defined in such a way that all second derivatives can be defined. The constraint is the same as used in [8] and expresses the condition that all residues are "absorbed" by the solution $\{k_x^*(i,j)\}_{(i,j)\in I_x}$, $\{k_y^*(i,j)\}_{(i,j)\in I_y}$. This means that we avoid to fit $\{\nabla f(i,j)\}$ to $\{\mathcal{W}[\nabla g(i,j)]\}$, so releasing the condition (4). In doing this we take more *a priori* information on the nature of the sampling process. This information comes from the residues map which highlightes the fact that the sampled phase signal from which the interferogram is extracted is aliased. In these conditions the use of (5) is not justified. All distortions in solutions obtained from the previous algorithms come from tha abuse of (5). The condition (4) is substituted by the simple assumption on the regularity of the first derivatives $\{\nabla f(i,j)\}$. This is expressed through the minimization of the phase Laplacian. The *a priori* knowledge, usually embedded in the regularity term, is both in the functional, which is minimized, and in the constraint. Also the fidelity to the interferogram $\{g(i,j)\}$ is in both terms, even if in a hidden manner. The functional expresses the regularity of the phase surface expressed in terms of the measured quantities $\{\mathcal{W}[\nabla_x g(i,j)]\}_{(i,j)\in I_x}$, $\{\mathcal{W}[\nabla_y g(i,j)]\}_{(i,j)\in I_y}$ and the unknown integer fields $\{k_x(i,j)\}_{(i,j)\in I_x}$ and $\{k_y(i,j)\}_{(i,j)\in I_y}$. The roughness of the unwrapped phase surface depends on the estimated quantities $\{k_x(i,j)\}_{(i,j)\in I_x}$, $\{k_y(i,j)\}_{(i,j)\in I_y}$ and on the fidelity to the data. The same observation is valid for the constraint term, which imposes the condition the unwrapped phase surface be consistent. The constraint takes into account both fidelity to the data and the estimated integer 2π–cycles to adjoin to the data (in order to calculate the first derivatives to be summed along the smallest closed path). This new way of writing the regularization functional avoids the problem of biased solutions encountered in all previous algorithms. In figure 2-e the result obtained applying this algorithm to the interferogram corresponding to the absolute phase surface in figure 2-a is reported. In principle, this algorithm allows the reconstruction of phase surfaces having an arbitrary slope, releasing the condition (4).

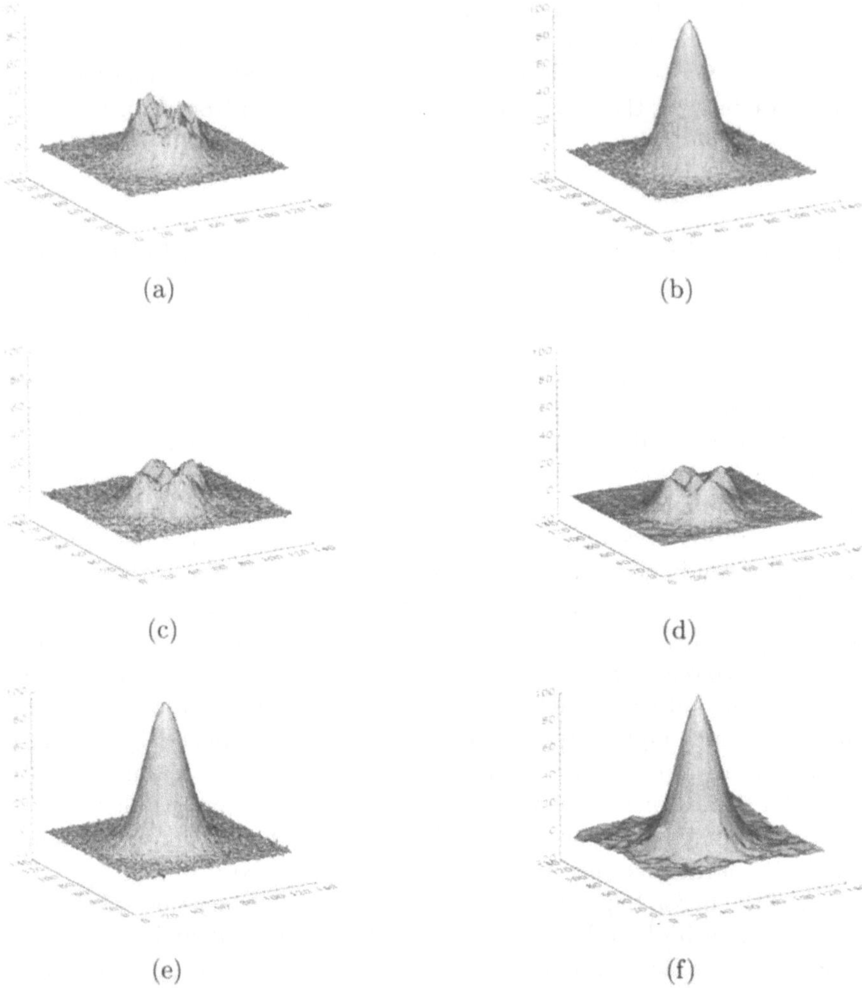

Figure 2. (a) Original absolute phase surface with noise corresponding to a coherence value $\gamma = 0.9$; Results obtained unwrapping the interferogram of this surface are reported using (b) LSE algorithm (section 3); (c) Tikhonov and Gull-Skilling regularization (section 5); (d) regularization through residues analysis, (section 6) (e) regularization through residues analysis and hidden data (section 7), (f) fractal model for topography (section 8).

8. Bayesian estimation using a terrain model

In any of the previous algorithms the attention was on the real absolute phase field $\{f(i,j)\}$ to be reconstructed after an estimate of the absolute phase derivatives $\nabla f(i,j)$. In this section an approach is described, which addresses the solution of PU paying attention on the topography, the physical quantity to be reconstructed. A forward model describing the InSAR process, giving the interferogram, is used. The *a priori* knowledge is given in terms of a fractional Brownian motion (fBm) used to model the topography. A complex interferogram $\{J[h(i,j)]\} = \{\exp(-if[h(i,j)])\}$ is generated using the forward model. This simulated interferogram is compared with the observed one $\{\tilde{J}\}$. In the regularization functional the dependence of absolute phase f on terrain height h is evidenced in the Fourier $\mathcal{F}$ transformed domain. With Ω we denote the interferogram support in the frequency domain, whereas η is a parabolic window. D is the fractal dimension. This algorithm gives another way to release the condition (4), because the fidelity is not measured using the reconstructed absolute phase slopes and the corresponding quantities from the observed interferogram.

9. Conclusions

In this paper several PU algorithms have been described in the Bayesian perspective. We have shown the need of the Bayesian inference in solving the PU. In particular we showed, that the PU requires a careful use of the Bayesian theory in order to take more *a priori* knowledge about the sampling process of the phase signal. We showed that the better result is obtained by the algorithm which take into account the integer nature of PU and assume a less restrictive condition on the phase gradient. This allows the reconstruction of aliased phase signals. Moreover, an algorithm has been described which solves PU using a different kind of *a priori* information on the physical quantity to be measured instead of on the sampling process.

References

1. H. A. Zebker, R. M. Goldstein, "Topographic mapping from interferometric aperture radar observations", Journal of Geophysical Research, **91(B5)**, 4993-4999, 1986.
2. R. M. Goldstein, H. A. Zebker, "Interferometric radar mapping of ocean currents", Nature, **328**, 707-709, 1987.
3. D. Massonnet, M. Rossi, C. Carmona, F. Adragna, G. Peltzer, K. Feigl, T. Rabaute, "The displacement field of the Landers earthquake mapped by radar interferometry", Nature, **364**, 138-142, 1993.
4. E. Rodriguez, J. M. Martin, "Theory and design of interferometric synthetic aperture radars", IEE Proc., **F(139)**, 147-159, 1992.
5. R. Bamler, N. Adam, G. W. Davidson, D. Just, "Noise–induced slope distortion in 2–d phase unwrapping by linear estimators with application to SAR interferometry", IEEE Transactions on Geoscience and Remote Sensing, **36(3)**, 913-921, 1998.
6. J. Marroquin, M. Rivera, "Quadratic regularization functionals for phase unwrapping", Journal of Optical Society of America A, **12(11)**, 2393-2400, 1995.
7. M. Datcu, "Maximum entropy solution for InSAR phase unwrapping", Proceedings of IGARSS'96, 310-314, 1996.
8. M. Costantini, "A novel phase unwrapping method based on network programming", IEEE Transactions on Geoscience and Remote Sensing **36(3)**, 813-821, 1998
9. L. Guerriero, G. Nico, G. Pasquariello, S. Stramaglia, "A new regularization scheme for phase unwrapping", Applied Optics, **37(14)**, 3053-3058, 1998.
10. M. Datcu, G. Palubinskas, "Multiscale Bayesian terrain height estimation from SAR interferometry", Proceedings of IGARSS'98, 1998.

TABLE 1. Comparison of different regularization functionals to be minimized, for estimating the absolute phase surface by unwrapping an observed interferogram.

Algorithm	Regularization functionals
LSE	$\sum_{(i,j)\in I_x}(\nabla_x f(i,j) - \mathcal{W}[\nabla_x g(i,j)])^2 + \sum_{(i,j)\in I_y}(\nabla_y f(i,j) - \mathcal{W}[\nabla_y g(i,j)])^2$
Tikhonov regularization [6]	$\left\{\sum_{(i,j)\in I_x}(\nabla_x f(i,j))^2 + \sum_{(i,j)\in I_y}(\nabla_y f(i,j))^2\right\} +$ $\lambda\left\{\sum_{(i,j)\in I_x}(\nabla_x f(i,j) - \mathcal{W}[\nabla_x g(i,j)])^2 + \sum_{(i,j)\in I_y}(\nabla_y f(i,j) - \mathcal{W}[\nabla_y g(i,j)])^2\right\}$
Gull-Skilling regularization [7]	$\sum_{(i,j)\in I} f(i,j)\ln f(i,j) +$ $\lambda\left\{\sum_{(i,j)\in I_x}(\nabla_x f(i,j) - \mathcal{W}[\nabla_x g(i,j)])^2 + \sum_{(i,j)\in I_y}(\nabla_y f(i,j) - \mathcal{W}[\nabla_y g(i,j)])^2\right\}$
Regularization with residues analysis and integer variables [8]	$\sum_{(i,j)\in I_x}\lvert k_x(i,j)\rvert + \sum_{(i,j)\in I_y}\lvert k_y(i,j)\rvert$ with the constraint: $\sum_{i,j}\{\mathcal{W}[\nabla_x g(i,j)] + 2\pi k_x(i,j) + \mathcal{W}[\nabla_y g(i+1,j)] + 2\pi k_y(i+1,j) -$ $\mathcal{W}[\nabla_x g(i,j+1)] - 2\pi k_x(i,j+1) - \mathcal{W}[\nabla_y g(i,j)] - 2\pi k_y(i,j)\}^2 = 0$
Regularization with residues analysis, hidden data and integer variables [9]	$\sum_{(i,j)\in I_{xx}}\{\mathcal{W}[\nabla_x g(i+1,j)] + 2\pi k_x(i+1,j) - \mathcal{W}[\nabla_x g(i,j)] - 2\pi k_x(i,j)\}^2 +$ $\sum_{(i,j)\in I_{xy}}\{\mathcal{W}[\nabla_y g(i+1,j)] + 2\pi k_y(i+1,j) - \mathcal{W}[\nabla_y g(i,j)] - 2\pi k_y(i,j)\}^2 +$ $\sum_{(i,j)\in I_{yx}}\{\mathcal{W}[\nabla_x g(i,j+1)] + 2\pi k_x(i,j+1) - \mathcal{W}[\nabla_y g(i,j)] - 2\pi k_y(i,j)\}^2 +$ $\sum_{(i,j)\in I_{yy}}\{\mathcal{W}[\nabla_y g(i,j+1)] + 2\pi k_y(i,j+1) - \mathcal{W}[\nabla_y g(i,j)] - 2\pi k_y(i,j)\}^2$ with the constraint: $\sum_{i,j}\{\mathcal{W}[\nabla_x g(i,j)] + 2\pi k_x(i,j) + \mathcal{W}[\nabla_y g(i+1,j)] + 2\pi k_y(i+1,j) -$ $\mathcal{W}[\nabla_x g(i,j+1)] - 2\pi k_x(i,j+1) - \mathcal{W}[\nabla_y g(i,j)] - 2\pi k_y(i,j)\}^2 = 0$
Fractal Model for topography [10]	$\sum_{(\alpha,\beta)\in\Omega}\eta(\omega)\lVert\mathcal{F}\{\exp(ig - if[h])\}(\omega)\rVert^2$ with $h \in \{fBm \mid D = D_0\}$

OPERATION OF THE BAYES INFERENCE ENGINE

K.M. HANSON AND G.S. CUNNINGHAM
Los Alamos National Laboratory, MS P940
Los Alamos, New Mexico 87545, USA†

Abstract. We have developed a computer application, called the Bayes Inference Engine, to enable us to make inferences about models of a physical object from radiographs taken of it. In the BIE calculational models are represented by a data-flow diagram that can be manipulated by the analyst in a graphical-programming environment. We demonstrate the operation of the BIE in terms of examples of two-dimensional tomographic reconstruction including uncertainty estimation.

Key words: Bayesian analysis, MAP estimator, uncertainty estimation, radiographic modeling, Markov chain Monte Carlo, adjoint differentiation, object oriented design

1. Introduction

We developed the Bayes Inference Engine (BIE) as a modeling tool for Bayesian data analysis. While its intended application is analysis of radiographic data in terms of geometric models of the objects being radiographed, we developed it with an eye toward generality. Thus, the design approach is readily extendible to other kinds of data analysis. Significant aspects of the BIE include the great flexibility with which object models and measurement models can be configured, the high degree of interactivity with those models, and the visualization of the overall modeling process.

The BIE represents a computational approach to Bayesian inference, as opposed to the traditional analytical approach [1]. The computational approach affords great flexibility in modeling, which facilitates the construction of complex models for the objects under study. The BIE easily deals with data that are nonlinearly dependent on the model parameters. For example, radiographic data are not linearly related to material densities [2]. Furthermore, the computational approach allows one to use nonGaussian probability distributions, such as likelihood functions based on Poisson distributions.

This paper is devoted to demonstrating the operation of the BIE. We will mention enough of the underlying technology to make the basis of operation un-

†Email: kmh@lanl.gov, cunning@lanl.gov

W. von der Linden et al. (eds.), Maximum Entropy and Bayesian Methods, 309–318.

derstandable. Many of the details behind the remarkable capabilities of the BIE can be found in our previous publications [1,3–6], most of which are available on our web site: `http://www.lanl.gov/home/kmh`. Our demonstration will center on several simple 2D tomographic reconstruction examples.

2. Historical Background

From the beginning of our development effort, we intended that BIE should be easy to use and should provide a high degree of interactivity with good visualization of the models involved and of the inference process. Additionally, we wanted the application to provide the analyst with a great deal of flexibility in configuring object models and measurement models. After about ten man-years of development, we believe that we have largely met these goals. While the BIE is intended to solve a limited class of problems, that is, radiographic modeling, its underlying design is eminently suited to many other applications.

The basic approach of the BIE is to employ a forward model to calculate the measurements that would be obtained for a given realization of a hypothesized object. Comparison of the calculated and observed measurements is done in terms of the likelihood function, which probabilistically summarizes the differences relative to the statistical uncertainties in the actual measurements. The posterior of the model parameters is obtained by combining the likelihood with a prior probability function that accounts for what is known beforehand.

We started developing the BIE in 1993. We embraced from the outset the principles of object-oriented (OO) design, which we felt would result in an application with great flexibility and interactivity. As we have commented before [5], the OO approach to software design has greatly aided and influenced our development. It has had a very positive effect on our approach to algorithm design.

Our assessment was that the OO language SmallTalk80 supplied in the development environment called VisualWorks, supplied by ParcPlace Systems (now ObjectShare[1]) provided the most advanced software-development environment available at that time. Unfortunately, for computationally intensive calculations, SmallTalk is as much as an order of magnitude slower than code written in complied languages, such as FORTRAN and C. Therefore, SmallTalk is used in the BIE to provide the OO structure essential for the data-flow diagram interface and the data structures. The underlying calculations are coded in FORTRAN and C. Another down side of SmallTalk is that most programmers lack experience with it.

A major thrust in our work has been to learn to cope in a general way with large numbers of parameters, e.g., pixel values, geometrical description, etc. Thus we have avoided numerical methods that are not extendible to numerous variables, which excludes many methods that deal directly with matrices, for example. Our approach is based on dealing directly with numerical valuation of the posterior and its gradient or potentially second-order derivatives. This tack is perhaps less efficient than analytic methods often appropriate for solving small, simple problems,

[1] ObjectShare, 16811 Hale Ave., Suite A, Irvine, CA 92606; Web: www.objectshare.com

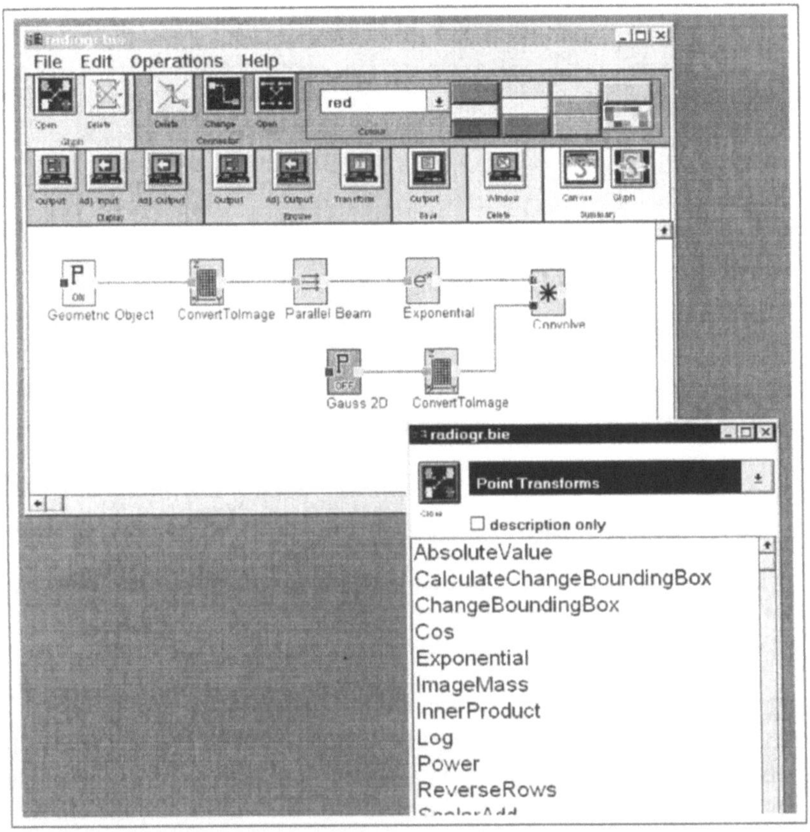

Figure 1. The users interacts with the Bayes Inference Engine through its canvas shown here. The set of transforms shown here represent a forward model for a pair of radiographs taken of a physical object that is modeled in terms of its shape. The Glyph Manager window, also shown here at the bottom, provides the selection of transforms that can be placed on the canvas.

but it is able to handle complex models and nonlinear measurement situations, which can not be solved analytically.

3. Data Flow Diagram

Models are created in the Bayes Inference Engine through the graphical-programming interface shown in Fig. 1. This means of interaction facilitates understanding of the analysis model. A data-flow diagram is constructed by connecting together transforms, represented visually by rectangles on the canvas [7]. We call these rectangles glyphs, after the nomenclature of the well-known signal-processing application Khoros[2]. While the BIE may appear to be similar to Khoros, its aim and underlying operation differ significantly from Khoros. In fact, we realized early on

[2]Khoral Research, Inc., 6200 Uptown Blvd. NE, Albuquerque, NM 87110-4142; URL: www.khoral.com

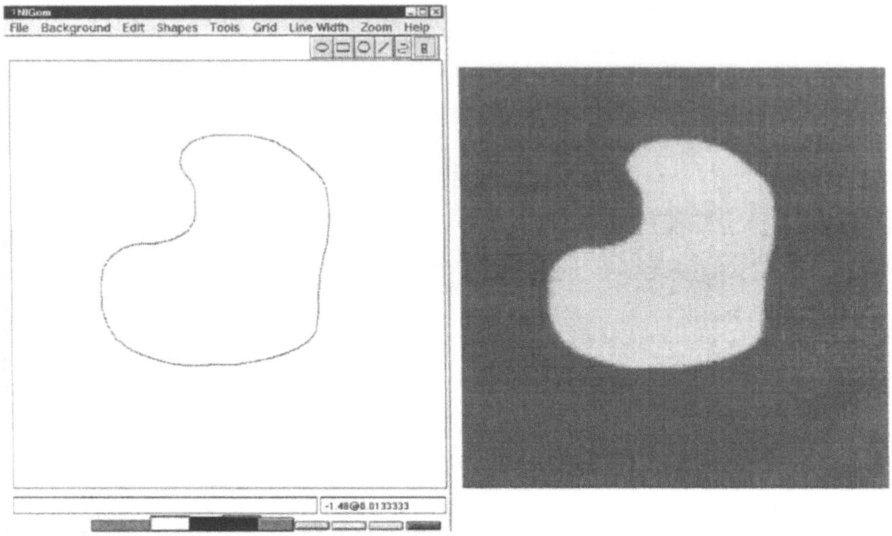

Figure 2. The object created in Fig. 1 as seen a) at the output of the first glyph showing it in terms of its geometrical description, and b) at the output of the second glyph as a gray-scale image, 128 pixels square.

that the Khoros environment did not meet our needs, which meant that we had to design the BIE from scratch. Some of the important features embodied in the BIE that are not contained in Khoros (at the time we started developing the BIE) include: an object-oriented (OO) programming environment, direct user interaction with the modules represented on the canvas, interaction between the modules that are linked together in the data-flow diagram, and the ability to reverse the flow of data relative to the forward data-flow diagram, which is necessary for adjoint differentiation described below.

When the BIE application is started, a blank canvas is brought up on the computer screen. The analyst creates a data-flow diagram by selecting transforms from a pull-down menu on the Glyph Manager window such as that shown in Fig. 1. The transforms are arranged in categories such as Models and Parameters, Basic Math Transforms, Point Transforms, Priors, Posteriors, and Likelihoods, etc. When a transform is selected, a glyph appears on the canvas. A panel can be brought up to allow the analyst to specify the salient attributes of the transform. Glyphs are linked together by clicking next to the output of one and then next to the input of another, or visa-versa. The connection is visualized as a line drawn between the two glyphs.

The data-flow diagram shown in Fig. 1 represents the calculation of three radiographs of a 2D object. This calculation is based on a forward model of the radiographic process. The leftmost two glyphs calculate a density image of an object in two steps. The first step is to generate a geometrical description of the object boundary. The output of this glyph is shown in Fig. 2a. In this case the object is represented by a polygon, which is drawn freehand. The second glyph converts the output of the geometrical description to a density image, shown in Fig. 2b.

The remaining glyphs in Fig. 1 correspond to the calculation of three projections of the density image at specified angles, taking the negative exponential to represent x-ray attenuation, and convolution to represent detector blur. In the following analysis, we will neglect the last three steps and work directly with just the unblurred projections. The flexibility of the BIE readily allows extensions to include further details in the radiographic model, including detector sensitivity function, inclusion of known material such as collimators, spectral effects, scattered radiation, beam-source blur, etc. It is possible to save a BIE canvas and subsequently read it back in to replay a computation or resume an analysis.

The output of each glyph can be seen by any one of several modes of interaction. One can double-click on the glyph and the output of the glyph appears. Alternatively, one can see the output by selecting the glyph by clicking on it once and then clicking on the leftmost button in the Display group of hot buttons above the canvas window or by selecting the "display output" option under the Operations button on the window tool bar.

4. Object Modeling and Parameter Estimation

Figure 3 shows the data-flow diagram used to reconstruct an object from the projection data generated by Fig. 1 to which is added uncorrelated Gaussian noise (5% relative to the maximum projection value). The object is modeled in terms of a geometric description of its boundary consisting of a 60-sided polygon, initially in the shape of a circle. We have found that using deformable geometric models [8,9] greatly improves tomographic reconstructions of objects from just a few views [10–12].

In Fig. 3 we use Bayes law in its logarithmic form

$$\varphi = -\log[p(\mathbf{x}|\mathbf{d})] = -\log[p(\mathbf{d}|\mathbf{x})] - \log[p(\mathbf{x})] + constant \ , \tag{1}$$

where $p(\mathbf{x}|\mathbf{d})$ is the posterior for the parameters $\mathbf{x}$ given the data $\mathbf{d}$, $p(\mathbf{d}|\mathbf{x})$ is the likelihood, and $p(\mathbf{x})$ is the prior on the parameters. The likelihood involves calculating the measurements for the given object model, which is handled in Fig. 3 by the same set of glyphs used in Fig. 1 to calculate the projections. The minus-log-likelihood is calculated as $\frac{1}{2}\chi^2 = \frac{1}{2}\sum_i (y_i - y_i*)^2/\sigma^2$, where the y_i are the actual measurements and y_i* are the corresponding values calculated from the object model, and σ is the estimated rms deviation of the measurement noise. The flexibility of the polygon is controlled by placing a prior on the boundary. The prior used here is an approximation to the integral around the boundary of the square of curvature of the boundary [13], which is calculated in the leftmost glyph along the bottom row of the canvas.

The shape of the object that minimizes the minus-log-posterior is found by the optimizer (the output of the upper-right glyph) by adjusting the vertices of the polynomial boundary of the object. The optimizer operates using the technique described in the next section. The geometry-based reconstruction is shown

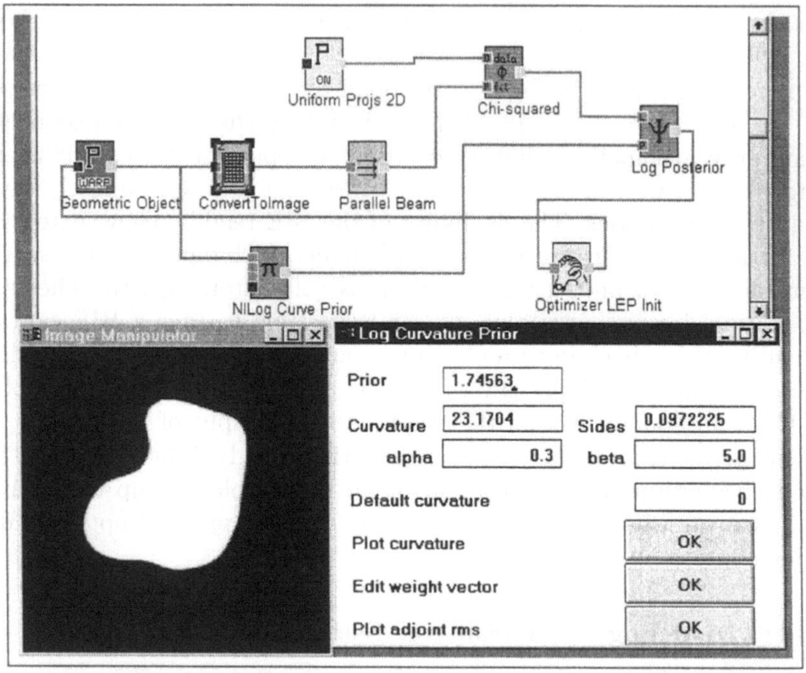

Figure 3. The data-flow diagram to estimate the geometric model parameters, namely the vertices of the polygon, that minimize the minus-log-posterior. The input measurements in the top glyph (Uniform Projs 2D) represent noisy projections of the unknown object from three directions. Also shown on the lower right is the panel specifying the details of the LogCurvaturePrior glyph and, on the lower left, the reconstructed gray-scale image, which is the output of the ConvertToImage glyph.

in Fig. 4a. In the context of the BIE, alternative means of representing the object are easy to try, for example, using Bézier curves instead of polygons, or simply employing a pixelated image, in which case one obtains a standard linear solution to the reconstruction problem. The latter approach results in the reconstruction shown in Fig. 4b. The poor quality of this reconstruction is caused by the extremely limited nature of the data compared to the large number of parameters in the image model, namely $128^2 = 16384$ pixel values. The generally noisy nature of maximum-likelihood reconstructions is evident, despite imposing a nonnegativity constraint on the image model and the limited number of iterations (20). The noise in this reconstruction only gets worse with more iterations.

4.1. OPTIMIZATION

To obtain the maximum a posteriori (MAP) estimate, we need to minimize the scalar function φ, given in Eq. (1), by varying the parameters that specify the object model. This optimization problem would be intractable without knowing the gradient of φ, or sensitivities, with respect to the many parameters on which it de-

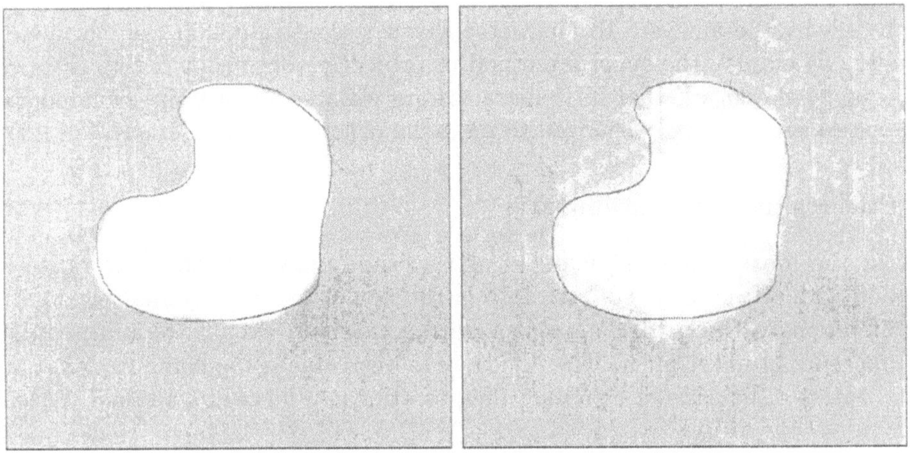

Figure 4. Gray-scale renditions of reconstructions from three noisy projections obtained using two different models for the reconstructed object: a) on the left, a geometric model for the object boundary, subject to a prior on the curvature, and b) on the right, a 128×128-pixel image, subject to a nonnegativity constraint on the pixel values. For comparison, the dark curves show the boundary of the original object.

pends. We employ a technique to calculate these crucial sensitivities, called adjoint differentiation [14]. Using the adjoint differentiation technique, the calculation of all these derivatives can be done in a computational time that is comparable to the forward calculation through the data-flow diagram. The adjoint sensitivity technique is crucial to the efficient operation of the BIE, especially for optimization.

Adjoint differentiation amounts to implementing the chain rule of differentiation for the computer code. It requires a backward flow of the derivatives of φ, which is easily accomplished in the OO framework [3]. In the BIE, each transform module propagates the adjoint derivative from its output side to its input side. In the code-based approach that we advocate, the adjoint differentiation code is based on the code for the forward calculation. Importantly, the CPU time needed to calculate the derivatives with respect to all the variables in the model is comparable to forward calculation of that model.

We and our collaborators have shown that adjoint differentiation can be beneficially employed in many types of forward modeling codes, for example in a finite-difference code to find the inverse of a time-dependent diffusion problem [15,16] and hydrodynamic codes [17]. See [18] for more details.

With the gradients available, we can perform the optimization using efficient gradient-based algorithms. The algorithm that we use most often is the DFP (Davidon-Fletcher-Powell) quasi-Newton method in which an estimate of the inverse curvature matrix is used to take a Newton step. The inverse curvature matrix is obtained by starting with a diagonal matrix and updating it with each step taken, based on the change in gradient. In the BIE the inverse curvature matrix is not explicitly stored; the accumulated update expression is evaluated each time one needs to multiply a vector by the matrix. This approach maintains our ability to cope with large numbers of variables. The diagonal terms of the starting matrix can be set by the analyst. To guarantee that the constraints on the parameters are met, we employ the general method of projection onto convex sets (POCS). Each model checks whether its constraints are violated. If they are violated, the parameters are minimally changed to meet the constraints [4].

5. Uncertainty Estimation

One of the most important features of Bayesian analysis is that the posterior characterizes of the degree of certainty in the models used in an analysis. One of our intentions with the BIE is to provide the means to exploit the full posterior as a measure of uncertainty. The difficulty to overcome is the computational and visualization difficulties of exploring the posterior in a large dimensional space.

5.1. MARKOV CHAIN MONTE CARLO

One way to visualize the reliability of an inferred model is to display a sequence of solutions that are randomly chosen from the posterior probability distribution, as proposed by Skilling et al. [19]. The sequence of images, typically calculated off line, is presented as a video loop. By showing a representative range of plausible solutions, the degree of variability of this presentation provides the viewer with a visual impression of the degree of uncertainty in the inferred model. The present emphasis in Bayesian research on Markov chain Monte Carlo methods [20,21] should yield improved methods to generate random samples of the posterior. Of course, MCMC is more than a tool for visualizing uncertainties; it provides quantitative estimates of the uncertainty in the inferred models. The uncertainty in any aspect of the model may be estimated with respect to any type of uncertainty measure desired, for example, in terms of variance. A notable advantage of MCMC is that the results are obtained with marginalization with respect to any nuisance parameters. It can also be used to estimate the posterior mean (as an alternative to the posterior mode).

To assess the uncertainty for our present problem, the only change required in the optimization canvas, Fig. 3, is to replace the optimizer glyph with an MCMC glyph. The MCMC module in the BIE permits one to save every nth sample from a sequence generated in a long MCMC run. One can then replay the subsampled MCMC sequence at a later time to scrutinize the model variations. Further details can be found in [13].

5.2. PROBING THE MODEL STIFFNESS

A useful alternative to MCMC is based on drawing an analogy between φ and a physical potential [6,22]. Then the gradient of φ is analogous to a force. To explore the reliability of a particular feature of a MAP solution in the BIE, the user specifies it by directly perturbing the selected combination of parameters that characterize that feature. Then, all the parameters are reoptimized to minimize the new φ. The uncertainty in the parameters is indicated by the amount that they move away from their MAP values relative to the strength of the applied external force. The correlations between parameters experiencing the external force and the others is demonstrated by how much and in what direction the parameters change. We have shown that this approach leads to a quantitative estimate for an appropriate part of the covariance matrix for problems in which the posterior is approximately Gaussian and the parameters are unconstrained [6,22].

6. Future Directions

Many extensions to the present capabilities of the BIE are planned. We are developing 3D models and already have very encouraging results [23,24]. A number of our underlying algorithms need improvement. The optimization algorithm is always under development to make it more robust. Recent work [25] has shown the limitations in efficiency of the Metropolis MCMC algorithm. We will explore alternatives that promise improved efficiency. So-called "systematic uncertainties" have always been a bugaboo in the uncertainty analysis of experiments. We think that we can address these in the BIE as highly-correlated uncertainties in the measurements that contribute appropriately to the posterior.

Acknowledgements

This work has been supported by the United States Department of Energy under contract number W-7405-ENG-36. Many have contributed to the development of the BIE over the years, including Xavier Battle, Roger Bilisoly, George Jennings, Jr., Igor Koyfman, Robert McKee, John Pang, David Wolf, Marion Yapuncich, and Hanji Zhao. We would like to acknowledge the helpful conversations we have had with John Skilling, Steve Gull, Kyle Myers, Robert Wagner, Julian Besag, James Gee, Jim Gubernatis, and Richard Silver.

References

1. K. M. Hanson and G. S. Cunningham, "A computational approach to Bayesian inference," in *Computing Science and Statistics* **27**, M. M. Meyer and J. L. Rosenberger, eds., pp. 202–211, Interface Foundation, Fairfax Station, VA 22039-7460, 1996.
2. K. M. Hanson, "A Bayesian approach to nonlinear inversion: Abel inversion from x-ray data," in Transport Theory, Invariant Imbedding, and Integral Equations, *Lect. Notes in Pure and Appl. Math.* **115**, P. Nelson, V. Faber, D. L. Seth, and A. B. White, Jr., eds., pp. 363–368, Marcel Dekker, New York, 1989.
3. G. S. Cunningham, K. M. Hanson, G. R. Jennings, Jr., and D. R. Wolf, "An interactive tool for Bayesian inference," in *Review of Progress in Quantitative Nondestructive Evaluation*, D. O. Thompson and D. E. Chimenti, eds., vol. 14A, pp. 747–754, Plenum, New York, 1995.
4. G. S. Cunningham, K. M. Hanson, G. R. Jennings, Jr., and D. R. Wolf, "An object-oriented optimization system," in *Proc. IEEE Int. Conf. Image Processing, vol. III*, pp. 826–830, IEEE, 1994.
5. K. M. Hanson and G. S. Cunningham, "The Bayes inference engine," in *Maximum Entropy and Bayesian Methods*, K. M. Hanson and R. N. Silver, eds., pp. 125–134, Kluwer Academic, Dordrecht, 1996.
6. K. M. Hanson and G. S. Cunningham, "The hard truth," in *Maximum Entropy and Bayesian Methods*, J. Skilling and S. Sibisi, eds., pp. 157–164, Kluwer Academic, Dordrecht, 1996.

7. G. S. Cunningham, K. M. Hanson, G. R. Jennings, Jr., and D. R. Wolf, "An object-oriented implementation of a graphical-programming system," in *Image Processing*, M. H. Loew, ed., *Proc. SPIE*, **2167**, pp. 914–923, 1994.
8. M. Kass, A. Witkin, and D. Terzopoulos, "Snakes: active contour models," *Inter. J. Comp. Vision*, **1**, pp. 321–331, 1988.
9. R. Szeliski, "Probabilistic modeling of surfaces," *Proc. SPIE*, **1570**, pp. 154–165, 1991.
10. K. M. Hanson, "Flexible prior models in Bayesian image analysis," in *Maximum Entropy and Bayesian Methods*, A. Mohammad-Djafari and G. Demoment, eds., pp. 399–406, Kluwer Academic, Dordrecht, 1993.
11. K. M. Hanson, "Bayesian reconstruction based on flexible prior models," *J. Opt. Soc. Amer. A*, **10**, pp. 997–1004, 1993.
12. K. M. Hanson, G. S. Cunningham, G. R. Jennings, Jr., and D. R. Wolf, "Tomographic reconstruction based on flexible geometric models," in *Proc. IEEE Int. Conf. Image Processing, vol. II*, pp. 145–147, IEEE, 1994.
13. K. M. Hanson, G. S. Cunningham, and R. J. McKee, "Uncertainty assessment for reconstructions based on deformable models," *Int. J. Imaging Systems and Technology*, **8**, pp. 506–512, 1997.
14. W. C. Thacker, "Automatic differentiation from an oceanographer's perspective," in *Automatic Differentiation of Algorithms: Theory, Implementation, and Application*, A. Griewank and G. F. Corliss, eds., pp. 191–201, SIAM, Philadelphia, 1991.
15. S. S. Saquib, K. M. Hanson, and G. S. Cunningham, "Model-based image reconstruction from time-resolved diffusion data," in *Medical Imaging: Image Processing*, K. M. Hanson, ed., *Proc. SPIE*, **3034**, pp. 369–380, 1997.
16. A. H. Hielscher, A. Klose, and K. M. Hanson, "Model-based iterative image reconstruction schemes for time-resolved optical tomography," submitted to *IEEE Trans. Med. Imaging*, 1998.
17. M. L. J. Rightley, R. J. Henninger, and K. M. Hanson, "Adjoint differentiation of hydrodynamic codes," in *CNLS Research Highlights*, Center for Nonlinear Studies, Los Alamos National Laboratory, April, 1998 (WWW: http://cnls.lanl.gov/Publications/highlights.html).
18. K. M. Hanson, G. S. Cunningham, and S. S. Saquib, "Inversion based on computational simulations," in *Maximum Entropy and Bayesian Methods*, G. Erickson, ed., Kluwer Academic, Dordrecht, 1998 (to be published).
19. J. Skilling, D. R. T. Robinson, and S. F. Gull, "Probabilistic displays," in *Maximum Entropy and Bayesian Methods*, W. T. Grandy, Jr. and L. H. Shick, eds., pp. 365–368, Kluwer Academic, Dordrecht, 1991.
20. J. Besag, P. Green, D. Higdon, and K. Mengersen, "Bayesian computation and stochastic systems," *Stat. Sci.*, **10**, pp. 3–66, 1995.
21. W. R. Gilks, S. Richardson, and D. J. Spiegelhalter, *Markov Chain Monte Carlo in Practice*, Chapman and Hall, London, 1996.
22. K. M. Hanson and G. S. Cunningham, "Exploring the reliability of Bayesian reconstructions," in *Image Processing*, M. H. Loew, ed., *Proc. SPIE*, **2434**, pp. 416–423, 1995.
23. G. S. Cunningham, K. M. Hanson, and X. L. Battle, "Three-dimensional reconstructions from low-count SPECT data using deformable models," *Opt. Express*, **2**, pp. 227–236, 1998.
24. X. L. Battle, G. S. Cunningham, and K. M. Hanson, "3D tomographic reconstruction using geometrical models," in *Medical Imaging: Image Processing*, K. M. Hanson, ed., *Proc. SPIE*, **3034**, pp. 346–357, 1997.
25. K. M. Hanson and G. S. Cunningham, "Posterior sampling with improved efficiency," in *Medical Imaging: Image Processing*, K. M. Hanson, ed., *Proc. SPIE*, **3338**, pp. 371–382, 1998.

THE PRIOR-PREDICTIVE VALUE:

A PARADIGM OF NASTY MULTI-DIMENSIONAL INTEGRALS

W. VON DER LINDEN, R. PREUSS, AND V. DOSE
Max-Planck-Institut für Plasmaphysik, EURATOM Association
D-85740 Garching b. München, Germany ¶

Abstract. At the heart of Bayesian model comparison lies the so-called prior-predictive value. In the important class of Quantified-MaxEnt applications analytic approximations are routinely used which often give rise to noise-fitting and ringing. We present an improved analytic expression which overcomes these shortcomings. In most interesting real-world problems, however, standard approximations and straight forward application of Markov-Chain Monte Carlo are hampered by the complicated structure of the likelihood in parameter space. At the Maxent workshop 1997 in Boise John Skilling suggested to employ a formalism, borrowed from statistical physics, to compute the prior-predictive value. We have scrutinized his suggestion: IT WORKS!

Key words: Prior-predictive value, Bayes factor, MCMC, Model comparison

1. Introduction

Bayesian probability theory provides a general frame for model evaluation and hypothesis testing [1] irrespective of the complexity of the models involved, the structure of the underlying parameter space and the number of data points or model parameters. Bayes theorem relates the probability for model/hypothesis/theory M in the light of the data and background information $\mathcal{I}$ to probabilities which are at our disposal

$$P(M|D,\mathcal{I}) = \frac{P(M|\mathcal{I})p(D|M,\mathcal{I})}{\mathcal{Z}} . \tag{1}$$

The quantity $\mathcal{Z}$, which ensures normalization, is irrelevant as it drops out in model comparison. $P(M|\mathcal{I})$ stands for the prior probability, i.e. the probability we assign to M prior to our knowing the data D. The centerpiece of the Bayesian approach is the so-called prior-predictive value (ppv)

$$p(D|M,\mathcal{I}) = \int_V p(D|x,M,\mathcal{I})p(x|M,\mathcal{I})d\mu(x) , \tag{2}$$

¶e-mail: wvl@ipp-garching.mpg.de

W. von der Linden et al. (eds.), Maximum Entropy and Bayesian Methods, 319–326.

i.e. the evidence for the data under model M given all background information $\mathcal{I}$. The terms under the integral are likelihood, prior and measure $\mu(x)$ in the parameter space spanned by the allowed model parameters x in volume V. Quite generally, the ppv is a multi-dimensional integral of the form $\int f(x)\pi(x)d\mu(x)$, where $\pi(x)$ is a probability density and $f(x)$ an arbitrary function. This integral plays an important role in many areas of science. Several techniques have been proposed to compute the ppv, including Laplace-type of approximations which are asymptotically exact [2].

QUANTIFIED MAXIMUM ENTROPY (QME) [3,4] applications are an important class of problems for which the Laplace approximation, also known as *steepest descent* approximation, is routinely used [3,4]. It is required to determine the regularization parameter α in the so-called evidence approximation [3,4]. The probability for α given the data is proportional to the ppv which has the specific form

$$p(D|\alpha,\mathcal{I}) \propto \int_0^\infty e^{-\frac{1}{2}\chi^2+\alpha S} \prod_i x_i^{-1/2} d^N x \; . \tag{3}$$

For concreteness we have chosen a Gaussian likelihood although the following considerations are valid for arbitrary argument of the exponential $\phi = -\frac{1}{2}\chi^2 + \alpha S$. In the Laplace approximation, in which ϕ is expanded to second order in x_i and the lower integration limits are moved to $-\infty$, the ppv reads

$$p(D|\alpha,\mathcal{I}) \propto e^{\phi^*} |H^0|^{-\frac{1}{2}} \prod_i \hat{x}_i^{-1/2} \; . \tag{4}$$

Here ϕ^* stands for the maximum value of ϕ and $H^0 = \nabla\nabla^T\phi|_{\hat{x}}$ is the hessian of ϕ at the maximum a-posteriori (MAP) value $\hat{x}$. In the case of an ill-posed inversion problem, e.g. deconvolution of a broad point spread function, the Laplace approximation causes the following problem: For small α, i.e. weak regularization, the MAP-image $\hat{x}$ exhibits pronounced scatter (ringing) which along with the positivity constraint produces many image values close to zero. But particularly for $\hat{x}_i \to 0$ the Laplace approximation strongly overestimates the integral (3) due to the modification of the lower integration limits. Consequently, the probability for α, which is proportional to $p(D|\alpha,\mathcal{I})$, will be largely overestimated for small α which results in noise-fitting and ringing. It is important to note that the origin for the problems is not the Gaussian approximation of e^ϕ but rather the modification of the integration volume [5].

We have shown in previous publications that the QME results improve significantly if either [6] the α-marginalization is performed exactly – hence avoiding the Laplace approximation altogether – or [5] if the Laplace approximation is improved such as to allow for the correct integration limits. The latter had the advantage that existing QME programs could be used with only minor modifications but it had the disadvantage that the integral correction was still approximate. There is, however, a simple way to avoid additional approximations by using the representation $x_i = \exp(\rho_i)$. The integration limits for ρ_i are now really $\pm\infty$. On top of that this is the natural representation of the MaxEnt solution, see e.g. [7]. The ppv transforms into

$$p(D|\alpha,\mathcal{I}) \propto \int_{-\infty}^{\infty} e^{\phi+\frac{1}{2}\sum_i \rho_i}\, d^N\rho \tag{5}$$

As pointed out above, it is now legitimate to employ the Laplace approximation

$$p(D|\alpha,\mathcal{I}) \propto e^{\phi^*}|H^0 + \frac{1}{2}\mathrm{diag}(\hat{x}_i^{-2})|^{-\frac{1}{2}} \prod_i \hat{x}_i^{-1/2} \tag{6}$$

Our experience to date with this expression for the ppv or rather α-evidence is very positive: less ringing and the results are comparable to what one gets in the case of α-marginalization.

For PARAMETER-ESTIMATION problems, however, the approximate schemes are often to crude since the posterior probability is far from being a Gaussian and Monte Carlo integration methods are the only approaches furnishing reliable results. Straight forward Markov-Chain Monte Carlo (MCMC) [8,9] is, however, ineligible since sampling from the prior $\pi(x|\mathcal{I})$[1] would in general result in huge variances and extremely long Markov chains. The reason is that the likelihood is generally much more structured and localized than the prior. At the Maxent workshop 1997 in Boise John Skilling suggested to employ the thermodynamic integration method[10] (thin-MCMC), borrowed from statistical physics, to compute the ppv. There is a wide variety of different MCMC methods for calculating Bayes factors which are discussed by Neal [10]. Some of them seem promising but have not been applied in really demanding problems yet. We have analyzed thin-MCMC in great detail and found that it works also in challenging problems, though it still is computationally demanding. Here we will present some general considerations why and when it works. In another paper of these proceedings [11] it is applied to a challenging real-world application.

2. The thermodynamic integration scheme

In order to make contact to statistical physics, we consider $p(D|M,\mathcal{I})$ as the partition function $Z(\beta)$ of a canonical ensemble with a fictitious energy function equal to the log-likelihood $E(x) = -\ln(p(D|M,x,\mathcal{I}))$ and a fictitious inverse temperature β. The probability for the system being in micro-state volume dx corresponds to the posterior probability

$$P(dx|M,D,\beta,\mathcal{I}) = \frac{1}{Z(\beta)}e^{-\beta E(x)}\pi(x)d\mu(x) = p(x|M,D,\beta,\mathcal{I})d\mu(x) \tag{7}$$

$$Z(\beta) = \int e^{-\beta E(x)}\pi(x)d\mu(x)\ . \tag{8}$$

[1] The prior is the only part for which the normalization is known.

Having established the correspondence to statistical physics we can borrow ideas to compute the partition function or equivalently the free energy F which is defined via $Z(\beta) = \exp(-\beta F)$. One of these ideas is the thermodynamic integration method. It proceeds from the derivative of $Z(\beta)$with respect to β

$$\frac{d\ln Z(\beta)}{d\beta} = \frac{1}{Z(\beta)}\frac{dZ(\beta)}{d\beta} = -\int E(x)p(x|M,D,\beta,\mathcal{I})d\mu(x) = -\langle E(\check{x})\rangle_\beta \ , \qquad (9)$$

which corresponds to the negative internal energy. We recall, $Z(0) = 1$ by virtue of the normalization of the prior and $Z(1)$ is the desired ppv. In order to avoid confusion, we mark the variable, which has been marginalized (averaged) over, x in the present case, by a check $\check{x}$. The solution of the differential equation (9) readily furnishes the ppv as the thermodynamic integral of the internal energy

$$\ln[Z(1)] = -\int_0^1 \langle E(\check{x})\rangle_\beta d\beta = \ln[p(D|M,\mathcal{I})] \ . \qquad (10)$$

Now $\langle E(\check{x})\rangle_\beta$ can be computed for a selection of inverse temperatures $\beta \in [0,1]$ by MCMC using the sampling density $p(x|M,D,\beta,\mathcal{I})$. The problem appears to be more tractable as before since we need to compute the expectation value of the energy (log-likelihood) only and the probability density contains the structure of the likelihood as well. Both facts reduce the variance significantly. In the worst case $\beta \to 0$ the sampling density is the bare prior but the variation of the likelihood enters merely on a logarithmic scale. Provided, $\langle E(\check{x})\rangle_\beta$ is a reasonably smooth function, the β-integral can easily be performed by standard numerical integration schemes.

3. Test Example

In order to assess the thermodynamic integration scheme we consider a simple but fairly realistic example. For likelihood and prior we use Gaussians

$$p(D|M,x,\omega,\mathcal{I}) = (\frac{\omega}{2\pi})^{\frac{N}{2}} e^{-\frac{\omega}{2}\|d-Ux\|^2} \ ; \qquad \pi(x) = (\frac{\theta}{2\pi})^{\frac{n}{2}} e^{-\frac{\theta}{2}\|x\|^2} \ , \qquad (11)$$

with $x \in \mathbb{R}^n$ and $d \in \mathbb{R}^N$. U is a $N \times n$ matrix which represents the expansion in orthonormal functions, i.e. $U^T U = \mathbb{1}$. This is not really an example for which MC-techniques are required but it is fairly realistic as far as the MC-performance is concerned if we assume that we are able to generate uncorrelated samples from the posterior distribution. If this is the case, no matter how bizarre the structure of the posterior is it can always be re-arranged into a uni-modal convex function without modifying the MC performance. Of course, there is still the important topic of how to generate uncorrelated and ergodic samples from a multi-modal distribution. Neal [12] suggested a promising method based on periodically heating and annealing the system. Extensions of this idea are described by Skilling in another paper of these proceedings. Our experience to date suggests that moderate multi-modality can be overcome by a modified move-generation and acceptance schedules, like those based on Cauchy distributions [13] or Tsallis statistics [14,15]. For the Gaussian example the ppv can be determined analytically

$$p(D|M,\mathcal{I}) = \langle p(D|M,\check{x},\mathcal{I})\rangle_{\beta=0} = (\frac{\omega}{2\pi})^{\frac{N}{2}}(\frac{\theta}{\omega+\theta})^{\frac{n}{2}} e^{-\frac{\omega}{2}d^T(1\!\!1-\frac{\omega}{\omega+\theta}UU^T)d} \,. \quad (12)$$

In order to assess the performance of straight forward MCMC based on the sampling density $\pi(x)$ we need

$$\langle p(D|M,\check{x},\mathcal{I})^2\rangle_{\beta=0} \;=\; (\frac{\omega}{2\pi})^{N}(\frac{\theta}{2\omega+\theta})^{\frac{n}{2}} e^{-\omega d^T(1\!\!1-\frac{2\omega}{2\omega+\theta}UU^T)d} \,. \quad (13)$$

In the worst case, $n = N$ and $\omega \gg \theta$ the relative Monte Carlo error is to leading order in θ/ω

$$\epsilon_{\mathrm{MC}} := \sqrt{\frac{\langle\Delta^2 p(D|M,\check{x},\mathcal{I})\rangle_{\beta=0}}{N_{\mathrm{MC}}}}\Big/p(D|M,\mathcal{I}) = (\frac{\omega}{2\theta})^{\frac{N}{4}} e^{\frac{\theta}{4\omega}\omega N\overline{d^2}}/\sqrt{N_{\mathrm{MC}}} \,, \quad (14)$$

where N_{MC} is the number of independent points in the Markov chain. For a set of realistic but still moderate parameters: signal-to-noise ratio $\sqrt{\omega\overline{d^2}} = 10$, prior width to likelihood width ratio $\omega/\theta = 100$, and $n = N = 128$ the relative MC error is

$$\epsilon_{\mathrm{MC}} \approx 10^{68}/\sqrt{N_{\mathrm{MC}}} \,. \quad (15)$$

This quantifies our statement that straight-forward MCMC is simply not eligible in the case of weak prior-knowledge. We would need more than 10^{138} independent MC moves for 10% accuracy.

Next we will see whether the thermodynamic integration scheme can cope with this order of magnitude. To this end we determine the internal energy using (12)

$$\langle E(\check{x})\rangle_\beta = -\frac{N}{2}\ln[\frac{\omega}{2\pi}] + \frac{n\omega}{2(\beta\omega+\theta)} + \frac{\omega}{2}d^T(1\!\!1 - UU^T(1-\frac{\theta^2}{(\beta\omega+\theta)^2}))d \quad . \quad (16)$$

In Fig.1 $\langle E(\check{x})\rangle_\beta$ is depicted. We see that the integrand is indeed a good-natured smooth function which can be integrated by standard numerical techniques with only a small number of inverse temperatures $\beta \in [0,1]$. From Fig.1 and from (16) it is obvious that the temperature scale is set by ω/θ. We therefore expect a more structured β-dependence if the likelihood has several scales. Next we determine the statistical error associated with the MC evaluation of $\langle E(\check{x})\rangle_\beta$ for which we need the variance

$$\langle\Delta^2 E(\check{x})\rangle_\beta = -\frac{d}{d\beta}\langle E(\check{x})\rangle_\beta = \frac{n\omega^2}{2(\beta\omega+\theta)^2} + \frac{\omega^2\theta^2}{(\beta\omega+\theta)^3}d^TUU^Td \,. \quad (17)$$

The β-dependence of the variance (17) is plotted in Fig.1. Fortunately, the variance is good-natured even for $\beta \to 0$. The total variance is obtained upon integrating (17) with respect to β

$$-\int_0^1 \frac{d}{d\beta}\langle E(\check{x})\rangle_\beta d\beta = \langle E(\check{x})\rangle_\beta\Big|_0^1 = \frac{n\omega^2}{2\theta(\omega+\theta)} + \frac{\omega}{2}(1-\frac{\theta^2}{(\omega+\theta)^2})d^TUU^Td. \quad (18)$$

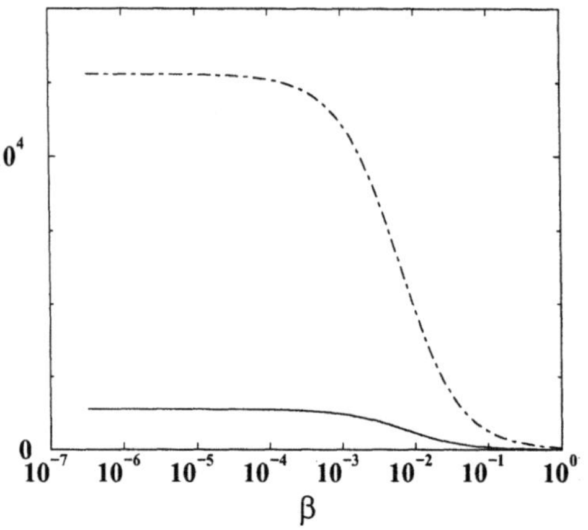

Figure 1. *β-dependence on a log-linear scale of the internal energy $\langle E(\check{x})\rangle_\beta - \langle E(\check{x})\rangle_\infty$ (dashed line) and the variance $\sqrt{\langle \Delta^2 E(\check{x})\rangle_\beta}$ (solid line) for $\omega\overline{d^2} = 100$, $\omega/\theta = 100$, and $N = 128$.*

We consider only the interesting and realistic case in which the likelihood is significantly more informative (localized) than the prior, i.e. $\omega \gg \theta$. In addition, the model under consideration is only sensible if the basis U is large enough to ensure $d^T U U^T d \approx ||d||^2$ and we have the estimate

$$\Delta^2_{\mathrm{MC}} \approx \left(n\omega/\theta + N\omega\overline{d^2}\right)/2N_{\mathrm{MC}} . \tag{19}$$

The MC error Δ_{MC} in the log-ppv corresponds to a relative error in the ppv, provided $\Delta_{\mathrm{MC}} < 1$. In this case Δ_{MC} corresponds roughly to the relative error in $p(D|M,\mathcal{I})$. To achieve a relative error $\epsilon = 10\%$ the number of uncorrelated samples for the previous set of parameters is dramatically reduced to $\underline{\boldsymbol{N_{\mathrm{MC}} > 10^6}}$!. By virtue of the smoothness of $\langle E(\check{x})\rangle_\beta$ as function of β, it takes roughly 10 separate MC computations to do the β-integration by standard numerical integration techniques.

4. Direct Computation of Bayes Factors

Instead of computing the absolute value of the ppv for each model separately, it would be expedient to head for the Bayes factor directly. This is indeed feasible if the two models cover the same state-space (micro-states) [10]. Generalization to models with different state-space are conceivable upon introduction of dummy variables. Let the un-normalized (posterior) probability for model M_α, $\alpha \in \{1,2\}$ be $f_\alpha(x) = p(D|x, M_\alpha, \mathcal{I})p(x|M_\alpha, \mathcal{I})\mu(x,\alpha)$. We define a mixture probability

$$p(x,\beta) = f_1(x)^{1-\beta} f_2(x)^\beta / Z(\beta) \tag{20}$$

$$Z(\beta) = \int f_1(x)^{1-\beta} f_2(x)^\beta \, dx . \tag{21}$$

with mixing coefficient β and mixture-partition function $Z(\beta)$. Obviously, the individual partition functions for the two models are recovered for $\beta = 0$ and $\beta = 1$, respectively. As before we obtain

$$\frac{d\ln Z(\beta)}{d\beta} = \int \ln(\frac{f_2(x)}{f_1(x)})p(x|M,\beta)dx = \langle \ln(\frac{f_2(\check{x})}{f_1(\check{x})})\rangle_\beta \ . \tag{22}$$

Finally, the Bayes factor of the two models M_1 and M_2, respectively, reads

$$\ln(\frac{P(M_2|D,\mathcal{I})}{P(M_1|D,\mathcal{I})}) = \ln(\frac{Z(1)}{Z(0)}) = \int_0^1 \langle \ln(\frac{f_2(\check{x})}{f_1(\check{x})})\rangle_\beta d\beta \ . \tag{23}$$

Like before, we want to assess this approach guided by the example given in (11). The two models M_α differ in the sets of orthonormal basis vectors U_α. This results in two Gaussian posterior distributions of identical width $\sigma^2 = 1/(\omega+\theta)$ located at $x_\alpha = \frac{\omega}{\omega+\theta}U_\alpha^T d$ and with amplitudes $A_\alpha \propto \exp(x_\alpha^2/2\sigma^2)$:

$$\begin{aligned} f_\alpha(x) &= A_\alpha e^{-(x-x_\alpha)^2/2\sigma^2} \\ p(x|\beta) &= (2\pi\sigma^2)^{-n/2} e^{-(x-[(1-\beta)x_1+\beta x_2])^2/2\sigma^2} \ . \end{aligned} \tag{24}$$

The integrand of (23) has a trivial β-dependence

$$\langle \ln(f_2(\check{x})/f_1(\check{x}))\rangle_\beta = (\beta - 1/2)(x_2 - x_1)^2/\sigma^2 \tag{25}$$

which presumably will not get much more complicated in real-world problems. Also the MC error has a good-natured parameter dependence

$$\Delta_{\mathrm{MC}}^2 = n(x_2 - x_1)^2/(N_{\mathrm{MC}}\sigma^2) \ . \tag{26}$$

In order to achieve a relative error for the Bayes factor of ϵ the number of independent MC steps has to obey $N_{\mathrm{MC}} \geq n(x_2 - x_1)^2/(\sigma^2\epsilon^2)$. A pleasant feature is the weak (quadratic) dependence on the distance $(x_2 - x_1)$ between the posterior modes of the 2 models. To give some numbers we consider the previously discussed example of two linear models. The $N = 128$ dimensional data vector d should be described by the two models U_α with $n = 10$ expansion coefficients x for each model. Since the x_α-dependence of the Bayes factors are approximately given by $\exp(-(x_2 - x_1)^2/2\sigma^2)$ high accurate results for the Bayes factors are only needed for $(x_2 - x_1)^2/2\sigma^2 \lesssim 1$. The number of MC-steps for a 10% accurate result is therefore $N_{\mathrm{MC}} = 10^5$, which is easily doable on present-days computers. It is also reassuring that the prior does not enter the expressions explicitly and weak priors do not deteriorate the performance.

5. Concluding remarks

We have demonstrated that the prior-predictive value can be determined by thermodynamic integration MCMC. We studied an example where the required sample size could be reduced from more than 10^{138} down to $\approx 10^6$, which is a manageable order of magnitude. Moreover, we have shown that the numerical effort can be reduced even further if we directly determine Bayes factors.

There is another class of MCMC approaches to determine the free energy, which are closely related to the thin-MCMC, the so-called 'method of expanded ensembles' [16] or 'acceptance ratio method' [10]. These methods are described by Skilling in these proceedings. Besides the applications given there, a similar method has been applied successfully to an electrolyte model in physical chemistry [16]. In general one would expect thin-MCMC to be superior, since the number of intermediate 'temperatures' (β) is small and does not depend on the dimension of the problem, while in the other schemes the number of intermediate 'temperatures' will increase with the system size, since the free energy is an extensive quantity. On the other hand, these methods might be superior, if the posterior distribution has a pronounced multi-modal structure with isolated modes. In this case thin-MCMC might have ergodicity-problems at low temperatures.

References

1. H. Jeffreys, *Theory of Probability*, Oxford University Press, London, 1939.
2. R. Kass and A. E. Raftery, "Bayes factors and model uncertainty," *J. Amer. Statist. Assoc.*, **90**, pp. 773–795, 1995.
3. S. Gull, "Developments in maximum entropy data analysis," in *Maximum Entropy and Bayesian Methods*, J. Skilling, ed., p. 53, Kluwer Academic Publishers, Dordrecht, 1989.
4. J. Skilling, "Quantified maximum entropy," in *Maximum Entropy and Bayesian Methods*, P. Fougère, ed., p. 341, Kluwer Academic Publishers, Dordrecht, 1990.
5. W. von der Linden, R. Fischer, and V. Dose, "Evidence integrals," in *Maximum Entropy and Bayesian Methods*, R. Silver and K. Hanson, eds., p. 443, Kluwer Academic Publishers, Dordrecht, 1996.
6. R. Fischer, W. von der Linden, and V. Dose, "On the importance of α marginalization in maximum entropy," in *Maximum Entropy and Bayesian Methods*, R. Silver and K. Hanson, eds., p. 229, Kluwer Academic Publishers, Dordrecht, 1996.
7. W. von der Linden, "Maximum-entropy data analysis," *Appl. Phys. A*, **60**, p. 155, 1995.
8. J. Hammersley and D. Handscomb, eds., *Monte Carlo Methods*, Chapman and Hall Ltd, London, 1964.
9. J. Ruanaidh and W. Fitzgerald, *Numerical Bayesian Methods Applied to Signal Processing*, Springer, New York, 1996.
10. R. Neal, "Probabilistic inference using markov chain monte carlo methods," Dept. of Computer Science, University Toronto, 1993.
11. R. Preuss, V. Dose, and W.v.d.Linden, "Model comparison in plasma energy confinement scaling," Kluwer Akademic, 1999.
12. R. Neal, "Sampling from multimodal distributions using tempered transitions," *Statistics and Computing*, **6**, pp. 353–366, 1996.
13. H. Szu and R. Hartley, "Fast simulated annealing," *Physics Letters A*, **122**, pp. 157–162, 1987.
14. C. Tsallis and D. Stariolo, "Generalized simulated annealing," cond-matter/9501047Jan95, 1995.
15. T. Penna, "Traveling salesman problem and tsallis statistics," *Phys.Rev.E*, **51**, pp. R1–3, 1995.
16. A. Lyubartsev, A. Martsinovski, S. Shevkunov, and P. Vorontsov-Velyaminov *J. Chem.Phys.*, **96**, p. 1776, 1992.

AN UNSUPERVISED CLUSTERING METHOD BY ENTROPY MINIMIZATION

G. PALUBINSKAS
Deutsches Zentrum für Luft und Raumfahrt (DLR) e.V.
Deutsches Fernerkundungsdatenzentrum (DFD), Oberpfaffenhofen
D-82234 Weßling, Germany †

Abstract. We address the problem of unsupervised clustering using a Bayesian inference. The entropy is considered to define a prior and enables us to overcome the problem of defining *a priori* the number of clusters and an initialisation of their centers. Under assumption of gaussian distribution of data the proposed clustering method reduces to a deterministic algorithm which appears to be an extension of the standard k-means algorithm. The robustness of the proposed method is shown on a magnetic resonance (MR) images database containing 65 volumetric (3D) images.

Key words: Clustering, k-means, ISODATA, entropy, Bayesian inference

1. Introduction

Clustering or classification of data is one of the first and essential steps in pattern recognition, image processing, computer vision. The quality of further processing steps like image analysis and interpretation rely on the clustering accuracy. Usually clustering methods are divided into supervised and unsupervised. Supervised methods require operator input for selecting training pixels in the image. So the result is operator dependent and may require actualisation of training data for each new image. Unsupervised methods are automatic, in the sense that operator intervention is reduced to initialising the process (define the number of clusters) and completing the process (interpret the result), but the result is here operator independent.

Unsupervised clustering methods include popular ones as ISODATA [1], or often called as k-means or nearest mean reclassification, their fuzzy equivalent, fuzzy c-means [2], and the maximum likelihood with expectation maximisation [3] or sometimes called as maximum likelihood estimation. The latter can be interpreted as an extension of the fuzzy c-means algorithm and is based on a mixture of Gaussian distributions model.

†Email: Gintautas.Palubinskas@dlr.de

W. von der Linden et al. (eds.), Maximum Entropy and Bayesian Methods, 327–334.

All these unsupervised clustering algorithms require an initialisation of the number of clusters and of the cluster centers (cluster validity problem). Various measures were proposed to find out automatically the number of clusters in a data. All of them are based on the statistical characteristics of clusters (variance, *a priori* probabilities and the difference of cluster centers) and the results are not always as expected and data dependent. Some criteria issued from the information theory have been proposed. The Minimum Description Length (MDL) criterion evaluate a compromise between the likelihood of the classification and the complexity of the model [4]. The prior concerning the complexity of the model is not really adapted to image modelling and the theoretical values for the hyper-parameter do not provide satisfactory results [5]. Empirical values for the hyper-parameter are also data dependent.

Herein we propose to embed the clustering problem into a Bayesian inference to automatically detect the number of clusters. The prior of the proposed model is derived from the entropy. Some automatic thresholding methods have been proposed using entropy either by maximising the information between two clusters derived from Renyi's entropy [6] or by minimising the cross entropy [7]. We consider the clustering problem where we have to reduce the complexity of the grey level description. We therefore minimise the entropy associated with the clustering histogram. To this prior a Gaussian likelihood term is added.

This paper is organised as follows. We derive the proposed model (section 2) which leads to an extension of the k-means algorithm (section 3). The results of both algorithms are then discussed and concluding remarks are derived.

2. A Bayesian Approach with Entropy as a Prior

Denote an image by $X = \{x_1, ..., x_j, ...x_N\}$ where the subscripts j refer to coordinates of the lattice L and the x_j to as the grey values. A clustering is defined by a partitioning the grey level set $\mathcal{G} = \{g_i, i = 1, ..., K\}$. A partition of the image corresponds to a cluster of the image (classification) defined by the region $S_i = \{j \in L | x_j = i\}$.

Usual clustering algorithms search for the partition which minimise a distance between the data and the classification. However, the number of clusters is fixed in the optimisation function and is therefore required. In this paper, we propose to add a prior on the minimised function to include the number of clusters as a variable in the minimised function and to estimate it. The Bayesian approach is a flexible framework to minimise an energy (or cost function) taking into account both a distance between the data and the result and a prior on the result. Denote by $Y = \{y_1, ..., y_N\}$ the classified image. A classified image is obtained by maximising the posterior probability $P(Y|X)$. Using the Bayes rule, we have the following result:

$$P(Y|X) = \frac{P(X|Y)P(Y)}{P(X)} \propto P(X|Y)P(Y), \tag{1}$$

where $P(X|Y)$ refers to as the likelihood, data attachment or goodness of fit term (GOF) and $P(Y)$ is the prior or *a priori* model.

We assume that the data are independent conditionally to Y, that is we consider correlation in the classification but the noise is supposed to be independent. Therefore we have the following relation:

$$P(X|Y) = \prod_{j\in L} p(x_j|y_j) = \exp\left(\sum_{j\in L} \ln p(x_j|y_j)\right). \qquad (2)$$

In the proposed approach we consider that the number of clusters is unknown. Therefore we first consider one cluster for each grey level. To reduce this number of clusters we have to sharpen the histogram associated with the clustering. We propose to minimise the entropy of the classified image histogram. We first initialise the classified image with the data. The minimum of the entropy is achieved for uniform images and, more generally, the entropy decreases as the number of levels with probability 0 increases. Therefore when adding the GOF to an entropy prior we reach a compromise between the likelihood of the classification and the simplicity of the description (a few number of clusters). Referring to Shannon [8], the information (or entropy) is defined as follows:

$$H = -\sum_{i=1}^{K} p_i \ln p_i, \qquad (3)$$

where K is the number of grey levels (*a priori* clusters). Note that if some clusters are empty ($p_i = 0$) the value of the entropy is not affected by removing the corresponding term. Therefore, if we consider a clustering containing k clusters, we have:

$$H = -\sum_{i=1}^{K} p_i \ln p_i = -\sum_{i=1}^{k} p_i \ln p_i. \qquad (4)$$

We define the prior as an exponentially shape probability given by:

$$P(Y) \propto \exp\left(\alpha_E \sum_{i=1}^{K} p_i \ln p_i\right), \qquad (5)$$

where $p_i = \#S_i/N$ is the prior probability of cluster i.

The I-model (Information model) [9] is defined by the posterior probability:

$$P(Y|X) \propto \exp\left(\sum_{j\in L} \ln p(x_j|y_j)\right) \exp\left(\alpha_E \sum_{i=1}^{K} p_i \ln p_i\right) \propto \exp(-U), \qquad (6)$$

where U refers to as the energy and is written as follows:

$$U = -\sum_{j\in L} \ln p(x_j|y_j) - \alpha_E \sum_{i=1}^{K} p_i \ln p_i, \qquad (7)$$

where α_E refers to as the hyper-parameter and weights the GOF and the *a priori* knowledge.

3. K-means algorithm

In this section, we derive the extended k-means algorithm to minimise the energy associated with the I-model. We assume that the x_j are Gaussian distributed with mean values $y_i, i = 1, ..., k$ and constant variation for clusters:

$$p(x_j|y_i) = \frac{1}{\sqrt{2\pi\sigma^2}} \exp\left(-\frac{(x_j - y_i)^2}{2\sigma^2}\right). \tag{8}$$

Omitting the constants we have:

$$\ln p(x_j|y_i) = -\frac{(x_j - y_i)^2}{2\sigma^2}. \tag{9}$$

From (7) we have:

$$U = \sum_{i=1}^{k} \sum_{j \in S_i} \frac{(x_j - y_i)^2}{2\sigma^2} - \alpha_E \sum_{i=1}^{k} p_i \ln p_i, \tag{10}$$

or

$$U = \sum_{i=1}^{k} \sum_{j \in S_i} \left[\frac{(x_j - y_i)^2}{2\sigma^2} - \frac{\alpha_E}{N} \ln p_i\right]. \tag{11}$$

When $\alpha_E = 0$, U is the energy of k-means (KM) clustering algorithm. So the standard k-means algorithm is a particular case of the extended k-means (KME) which uses the entropy and minimises the energy from equation (11).
The iterative procedure for k-means (KM) consists of the following 5 steps:

1. Select the number of clusters $k \geq 2$ and a small value ϵ to stop iterating.
2. Initialise the cluster centers y_i^m and *a priori* probabilities $p_i^m, i = 1, 2, ..., k$ and the number of iterations $m = 0$.
3. Classify the data x_j (as a result we get regions S_i)

$$x_j \in r \Leftrightarrow d(x_j, y_r^m, p_r^m) = \min_i d(x_j, y_i^m, p_i^m), \tag{12}$$

where d is defined as

$$d(x_j, y_i^m, p_i^m) = \frac{(x_j - y_i)^2}{2\sigma^2} - \frac{\alpha_E}{N} \ln p_i. \tag{13}$$

4. Update cluster centers

$$y_i^{m+1} = \frac{1}{\#S_i} \sum_{j \in S_i} x_j \tag{14}$$

and *a priori* probabilities of clusters

$$p_i^{m+1} = \frac{\#S_i}{N}. \tag{15}$$

5. Check convergence criterion

$$\text{if} \quad D = \max_i |y_i^{m+1} - y_i^m| > \epsilon \quad \text{then } m = m + 1, \text{ goto 3 else stop} \tag{16}$$

Usually less than 20 iterations are required for convergence.

The extended k-means (KME) clustering algorithm allows us to avoid the initialisation of the number of clusters and cluster centers. We start the iterative procedure with the number of clusters equal to the number of grey values in an image and with the cluster means equal to corresponding grey values. Initial values for *a priori* probabilities of clusters are computed from the image histogram. We therefore overcome the problems of selection of the number of clusters and initialisation (steps 1 and 2 in the iterative procedure). During iterations some clusters vanish ($p_i = 0$) due to the entropy term. The final number of clusters depends on the α_E parameter value. For small values of α_E we get a lot of clusters whereas for large values we get very few clusters. In extreme case only one cluster containing all the pixels of an image is found. The value of α_E depends on the resolution of the clustering we are seeking for, i.e. the complexity of the description. The obtained clustered image matches the data with an accuracy defined by the parameter α_E, whereas fixing the number of clusters is arbitrary with respect to the data. For example, if there is a tumour in a brain image, this can be automatically detected with the KME approach. When dealing with unpeeled image the optimal number of clusters highly depend on the scanner and on the data itself whereas a given value of α_E provides more stable results. We therefore claim that the α_E parameter is more reliable than the number of clusters as it depends on the desired classification but is also data independent. Herein we propose a heuristical argument to estimate a value for α_E parameter which provides satisfactory results. We estimate the entropy factor in the following way. We first rewrite (10) as follows:

$$U = \sum_{i=1}^{k} \frac{p_i N \sigma_i^2}{2\sigma^2} - \alpha_E \sum_{i=1}^{k} p_i \ln p_i, \tag{17}$$

where σ_i^2 is the variance of a cluster. If we want a good equilibrium between the entropy and the GOF, the prior value in equation (17) must be of the same order as the data attachment sum, so that we can assume the following:

$$\frac{N\sigma_i^2}{2\sigma^2} \propto -\alpha_E \ln p_i, \tag{18}$$

or

$$\alpha_E = -A \frac{N\sigma_i^2}{2\sigma^2 \ln p_i} = -A \frac{N}{2 \ln p_i}, \tag{19}$$

when simplifying equation (19) by assuming that $\sigma_i^2 = \sigma^2$ is the variance of a noise in the image and $p_i = 1/M$, where M is the expected number of clusters. The noise value in an image can be estimated using a very fast noise estimation method [10]. So the entropy constant is normalised to the data. The proportionality constant A in (19) is used to control the number of the clusters. For $A = 1$ quite a good

clustering can be achieved for most of the images. Smaller value of A result in a finer clustering, larger value - coarser clustering.
This iterative clustering procedure is very fast as it works on the image histogram (grey level set partition).

4. Results: KME versus KM

The clustering algorithms k-means (KM) and (KME) were tested on magnetic resonance (MR) images of a human head.

Visual evaluation of the results of clustering was performed on 65 high-resolution volumetric (3D) MR brain data sets, which were measured on 8 different MR scanners at 5 different locations (see Acknowledgements). Two slices of the 3D sample image are presented in Figure 1.a and 1.b. This image was clustered with k-means (KM) into 3 clusters, with k-means (KME) $A = 1.5$, and with k-means (KM) into 6 clusters. The results are presented in Figure 1. When comparing with the hand segmentation, KM with 3 clusters underestimates WM (white matter) because of the intensity inhomogeneities in an image (Figure 1c,d). KME produces 6 clusters: 2 for CSF (cerebrospinal fluid), 1 for GM (grey matter) and 3 for WM (Figure 1e,f). So it respects intensity inhomogeneity in an 3D image. A similar result is received with KM and a known number of clusters equal to 6 (Figure 1.g and 1.h). Tests on the 65 images showed that KME is a robust procedure as we use the same value for $A = 1.5$ whereas the number of clusters found varies from 4 to 7, mostly depending on noise and intensity inhomogeneities in an image. If we require the same number of clusters for each image in a KM algorithm, the results are not good for the whole data sets.

We have also performed a quantitative analysis of clustering algorithms for manually segmented image [11] into 4 classes: background, cerebrospinal fluid (CSF), grey matter (GM) and white matter (WM). This allowed to perform quantitative comparison of clustering algorithms by calculating the kappa coefficient - clustering accuracy measure - from the confusion matrix [12]. Peeled (non-brain areas removed) image was clustered with k-means (KM) into 3 clusters, with k-means (KME) using various $A = 1, 1.5, 6.6$, and finally again with k-means (KM) into 13 clusters which corresponds to the number of clusters obtained with $A = 1.5$. For that image, the best result is obtained with $A = 6.6$. This can be explained by analysing the histograms. We tend to detect several clusters for the CSF. This is due to different data normalisation which have been used for the studied database and for this manually segmented image. In order to compute the confusion matrix the received clusters (if more than three) were merged to get the main three clusters. The results are presented in Table 4. K-means (KME) with the proposed estimate for entropy constant in equation (19) produces about 2.5% higher accuracy than standard k-means (KM) with 3 clusters.

5. Conclusion

Usual clustering algorithm even when referred to as unsupervised requires the number of clusters and an initialisation of their centers. Proposed methods to *a*

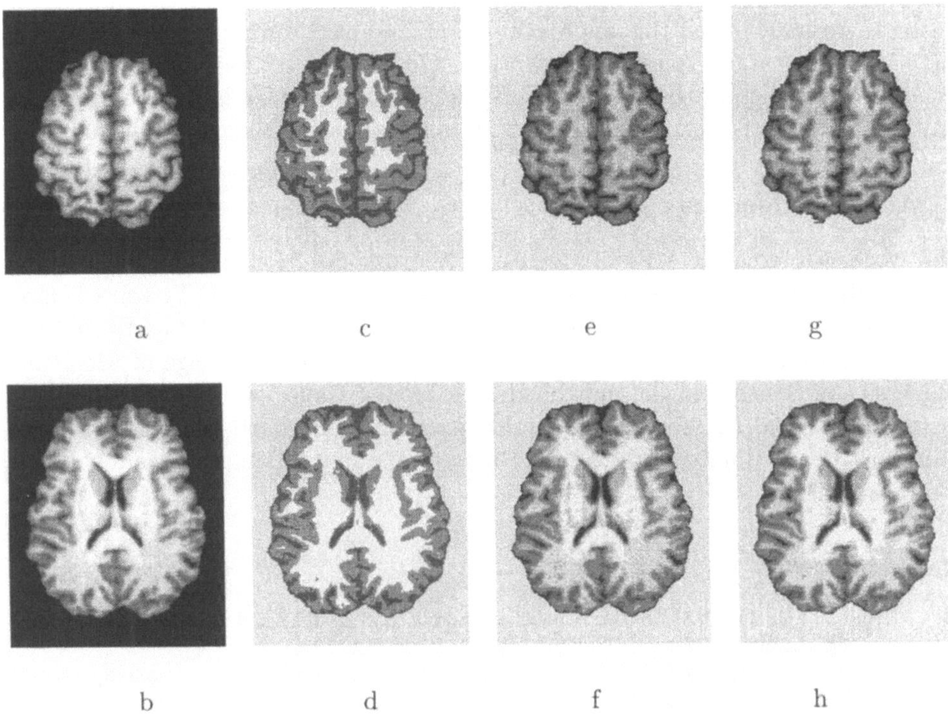

Figure 1. MR sample image: T_1 slice 1 (a) and 2 (b), k-means (MD) with 3 clusters (c,d), k-means (MDE) with entropy constant equal to 1.5 (e,f) and k-means (MD) with 6 clusters (g,h).

TABLE 1. Comparison of KM and KME algorithms: number of clusters, value of entropy constant and kappa coefficient.

algorithm	# clusters	α_E	kappa (%)
KM	3	-	50.45
KME	13	1.	52.97
KM	13	-	52.72
KME	10	1.5	51.53
KME	4	6.6	52.48

posteriori select the number of clusters have revealed to be data dependent. In this paper, we have proposed to embed the clustering problem into a Bayesian inference. We use the entropy as a prior in order to overcome the problem of selecting the number of clusters. Our model depends on a parameter weighting the prior term and the goodness of fit term. The parameter allows us to define the coarseness of the clustering. Although related to the number of clusters, we claim that this parameter is data independent which is not the case of the number of clusters. We have tested this approach on our database containing 65 volumetric (3D) MR images and found an acceptable clustering with the same parameter for all images whereas the cluster count varied. Further work is directed towards extension for multispectral data and other applications, e.g. remote sensing.

6. Acknowledgements

The work was done at MPI of Cognitive Neuroscience, Leipzig during working visits of the first two authors. The authors wish to thank Prof. D.Y. von Cramon for supporting this work and PD Dr. D. Norris for providing the data sets, and Dr. R.P. Velthuizen and Prof. L.P. Clarke, Dept. of Radiology, College of Medicine, University of South Florida and the H. Lee Moffitt Cancer Center and Research Institute, Tampa, FL, for providing the manually segmented image. Thanks to colleagues Dr. R. Bamler and Prof. M. Datcu for reading a manuscript and valuable comments on it.

References

1. D. Hall and G. Ball, "ISODATA: A novel method of data analysis and pattern classification," tech. rep., Stanford Research Institute, Menlo Park, CA, 1965.
2. J. Dunn, "A fuzzy relative of the ISODATA process and its use in detecting compact well-separated clusters," *Journal of Cybernetics*, **3**, (3), pp. 32–57, 1974.
3. J. Bezdek and J. Dunn, "Optimal fuzzy partition: a heuristic for estimating the parameters in a mixture of normal distributions," *IEEE Trans. on Computers*, **24**, pp. 835–838, 1975.
4. Z. Liang, R. Jaszczak, and R. Coleman, "Parameter estimation of finite mixtures using the EM algorithm and information criteria with application to medical image processing," *IEEE Trans. on Nuclear Science*, **39**, pp. 1126–1133, 1992.
5. X. Descombes, "Application of stochastic techniques in image processing for automatic tissue classification in MRI and blood vessel restoration in MRA," Tech. Rep. KUL/ESAT/MI2/9603, Laboratory for Medical Imaging Research (ESAT-Radiology), K.U.Leuven, Belgium, 1996.
6. P. Sahoo, C. Wilkins, and J. Yeager, "Threshold selection using Renyi's entropy," *Pattern Recognition*, **30**, (1), pp. 71–84, 1997.
7. Y. Zimmer, R. Tepper, and S. Akselrod, "A two-dimensional extension of minimum cross entropy thresholding for the segmentation of ultrasound images," *Ultrasound in Med. and Biol.*, **22**, (9), pp. 1183–1190, 1996.
8. C. Shannon, "A mathematical theory of communication," *Bell Systems Technical Journal*, **27**, p. 379, 1948.
9. X. Descombes and F. Kruggel, "A markov pixon information approach for low level image description," *IEEE Trans. on Pattern Analysis and Machine Intelligence*, 1997. (under review).
10. J. Immerkaer, "Fast noise variance estimation," *Computer Vision and Image Understanding*, **64**, (2), pp. 300–302, 1996.
11. R. Velthuizen and L. Clarke, "An interface for validation of MR image segmentations," in *Proc. 16th Annual Conf. of the IEEE Engineering in Medicine and Biology Society, Baltimore Maryland, USA, vol. 16*, pp. 547–548, 1994.
12. J. Cohen, "A coefficient of agreement for nominal scale," *Educational and Psychological Measurement*, **20**, (1), pp. 1–38, 1960.

APPLICATIONS OF THE MAXIMUM ENTROPY METHOD IN CRYSTALLOGRAPHY

MARTIN SCHNEIDER AND SANDER VAN SMAALEN
Laboratory of Crystallography, University of Bayreuth,
D-95440 Bayreuth, Germany

Abstract. The application of the maximum entropy principle is briefly reviewed. Particular attention is given to the problem of deriving an appropriate constraint between the density map and the experimental data. Examples are presented pertaining to crystal structures with anharmonic displacements and disorder.

Key words: anharmonic displacements, disorder, crystal structures, maximum entropy

1. Introduction

The principle of maximum entropy (MaxEnt) has found application to a number of very different crystallographic problems [1]. It has been used in conjunction with direct methods to determine the phases of the structure factors from the experimental intensity data in x-ray diffraction [2]. MaxEnt has been used to derive accurate values for the integrated intensities of Bragg reflections [3]. Finally, it has been used to determine accurate electron density maps from phased reflections [4].

In a structural analysis of crystals, the final stage usually is to refine a structure model against the experimental diffraction data. Commonly, a model of independent spherical atoms with anisotropic temperature parameters is employed, corresponding to an electron density with maxima with an ellipsoidial shape. Deviations from this idealized model occur due to anharmonic displacements of the atoms, disorder, and the presence of chemical bonds between the atoms. Methods have been developed to describe these deviations [5]. However, they involve the introduction of many additional parameters, while the deviations are relatively small. Correlated and ill-determined parameters are the result. The maximum entropy method (MEM) is an attractive alternative, because it is a computational stable procedure, and it is of the same complexity for structures with spherical atoms and for structures which are more difficult to describe in a model.

The MEM requires that the map used to define the entropy also determines the data used in the constraints. For the electron density as used in crystallographic

W. von der Linden et al. (eds.), Maximum Entropy and Bayesian Methods, 335–340.

applications of the MEM this is not true. Corrections to the experimental data are required, in order to match experimental and calculated structure factors.

In section 2 the MaxEnt principle for crystallographic applications is introduced, and the various types of contraints are discussed. Particular attention is given to the phase problem and the corrections of the experimental data. In section 3 applications are reviewed to the disordered crystal structure of LiCp* [6], and to the structure of Cs_2HgCl_4, with anharmonic displacements [7].

2. The crystallographic MaxEnt problem

2.1. DEFINITION OF THE ENTROPY AND THE CONSTRAINTS

The unit cell of the crystal structure is divided in N pixels of equal volume, and the electron density is defined by its values ρ_j in the $j = 1, \cdots, N$ pixels. Then, the entropy is defined as

$$S = -\sum_{j=1}^{N} \rho_j \log \left(\frac{\rho_j}{r_j} \right) \tag{1}$$

where $\{r_j\}$ is the reference density. The crystallographic MaxEnt problem is stated as finding the electron density that maximizes the entropy (Eq. 1), subject to constraints provided by the data.

The structure factors $F_{MEM}(\vec{H})$ are defined as the Fourier transform of the electron density of the unit cell. In x-ray scattering only the magnitude of the structure factor is measured. Accordingly various types of contraints have been proposed [4]. The first constraint is always that $\sum_{j=1}^{N} \rho_j$ is equal to the number of electrons in the unit cell. Reflections with known magnitudes and phases are combined into a so-called F-constraint:

$$\left(\sum_{i=1}^{N_1} \frac{\left| F_{MEM}(\vec{H}_i) - F_{obs}(\vec{H}_i) \right|^2}{\sigma_i^2} \right) - N_1 = 0 \tag{2}$$

where N_1 is the number of reflections contributing to the sum. $F_{obs}(\vec{H}_i)$ is the experimental value for the structure factor with scattering vector $\vec{H}_i$, and σ_i is its standard deviation. Reflections for which only the magnitudes of the structure factors are known, are combined into a G-constraint:

$$\left(\sum_{j=1}^{N_2} \frac{\left| G^j_{MEM} - G^j_{obs} \right|^2}{\sigma_j^2} \right) - N_2 = 0 \qquad \text{with} \quad G^j = \sqrt{\frac{\sum_{i=1}^{n_j} \left(m_i \cdot |F(H_i)|^2 \right)}{\sum_{i=1}^{n_j} m_i}} \tag{3}$$

G^j corresponds to the square root of the sum of intensities of n_j reflections, weighted with their multiplicities m_i. Experimentally G^j_{obs} values often are obtained from x-ray powder diffraction.

If only magnitudes of the structure factors are know, the MaxEnt problem is ill-posed. However, Sakata has shown that if part of the reflections can be used in an F-constraint it is advantageous to incorporate the remaining information on intensities of reflections into a G-constraint [4]. Typically, this situation is encountered in x-ray powder diffraction.

2.2. FROM EXPERIMENTAL DATA TO CONSTRAINTS

The analysis requires that both the entropy and the constraints depend on the same electron density function (map). However, the integrated intensities of the Bragg reflections obtained by x-ray scattering do not correspond to the Fourier transform of the electron density [5]. It is possible to correct the experimental data, such that the corrected values $F_{electron}$ do correspond to the Fourier transform of the electron density. First, measured intensities are transformed into observed structure factors, F_{obs}, through the application of several geometric factors and the correction for the effects of absorption of the radiation. The absorption correction does not depend on the structure, but it does depend on the contents of the unit cell. This is the first point where a model enters the analysis, and the ideal situation cannot be achieved by the MEM providing a map independent of any model.

More serious are the effects of secondary extinction and anomalous scattering. They do depend on the details of the structure. In particular, anomalous scattering is not simply a function of the electron density, but it depends on the chemical elements present, and on the wavelength of the radiation used in the scattering experiment. Corrections for these effects must be made, in order to have a meaningful comparison of the experimental values with the Fourier transform of the electron density map.

Finally, the phases of the structure factors are not obtained from the experiment. The usual procedure is to use the phases of the calculated structure factors of an approximate model, and to assign them to the observed magnitudes. Then the resulting quantities are employed in the F-constraint (Eq. 2).

We have proposed a procedure, which applies several of the corrections, and transfers the calculated phase to the observed magnitudes in a single step [7]. $F^{obs}_{electron}(\vec{H})$ is obtained as

$$F^{obs}_{electron}(\vec{H}) = \frac{|F_{obs}(\vec{H})|}{|F_{calc}(\vec{H})|} F^{calc}_{electron}(\vec{H}) \tag{4}$$

where $F_{calc}(\vec{H})$ are the structure factors obtained for the best model, including all corrections. $F^{calc}_{electron}(\vec{H})$ corresponds to the Fourier transform of the electron density of the model.

Deviations between calculated structure factors and the real (unknown) structure factors are mainly in their magnitudes and not in their phases, if a model is used that fits the data well. This remains true even if the model does not capture all aspects of the structure. (e.g. average deviation between magnitudes of calculated and observed structure factors is less than 10 %), Similarly, it may be expected that the model provides a good estimate for the effects of anomalous scattering and extinction. It follows that MaxEnt methods can be used favorably to determine structural features beyond that of a reasonable fitting model.

3. Applications

The method of undetermined Lagrange multipliers is used to define the density map with the maximum entropy that fits the constraints. Likewise, algorithms that have been developed for various applications of the MaxEnt principle can be used for the crystallographic problem too [8]. We have used the computer program MEED, that employs a modified Gull-Daniell algorithm [9]. Sakata and coworkers used their program MEED to study the electron density in the chemical bond [4,10,11], and to determine structures from x-ray powder diffraction [12]. We have used MEED to determine the disorder in LiCp* [6], and to study the anharmonic displacements in Cs_2HgCl_4 [7].

3.1. ANHARMONIC DISPLACEMENTS IN CS_2HGCL_4

A_2BX_4 compounds have been studied extensively for their ferroelectric and ferroelastic properties. Usually, the room-temperature structure is centrosymmetric orthorhombic. On cooling a series of phase transitions occur towards phases that may exhibit ferroelectric or ferroeleastic behaviour. We have used the MEM to study the room-temperature structure of Cs_2HgCl_4 [7]. The analysis shows that in the high-symmetric phase, anharmonic displacements of the atoms exist, as a precursor to the static displacements in the low-temperature superstructures.

Cs_2HgCl_4 crystallizes in space group *Pnma* with lattice parameters a=9.8136 Å, b=7.6018 Å, and c=13.4201 Å. Refinement of the structure model with spherical atoms and harmonic temperature parameters lead to a final agreement between observed and calculated structure factors of wR = 0.031[1]. Application of the MEM with $F^{calc}_{electron}(\vec{H})$ as observed quantities in a F-constraint (Eq. 2) led to an electron density map that reproduced the model. The ellipsoidal shape of the maxima directly corresponded to the anisotropic, harmonic temperature factor (Fig. 1). Using $F^{obs}_{electron}(\vec{H})$ as observed quantities (Eq. 4) the MEM lead to a modified density map, that clearly shows the anharmonic contributions to the displacements of the atoms Cl(1) and Cs(2) (Fig. 2).

In MEED, convergence was decided when the constraint was less or equal to zero. This corresponded to wR = 0.02 for all calculations. Note that both the approximate model and ρ_{MEM} fit the data well, but that the MEM reveals much more details of structure. Confirmation of this result was obtained from a refinement with a model including parameters for anharmonic displacements. Despite correlations, and large standard deviations for these parameters, the principal modes of anharmonic displacements were the same in the model approach and as they were obtained by the MEM.

[1]wR corresponds to the squareroot of χ^2 in Eq. 2 divided by the weighted sum of the observed structure factors.

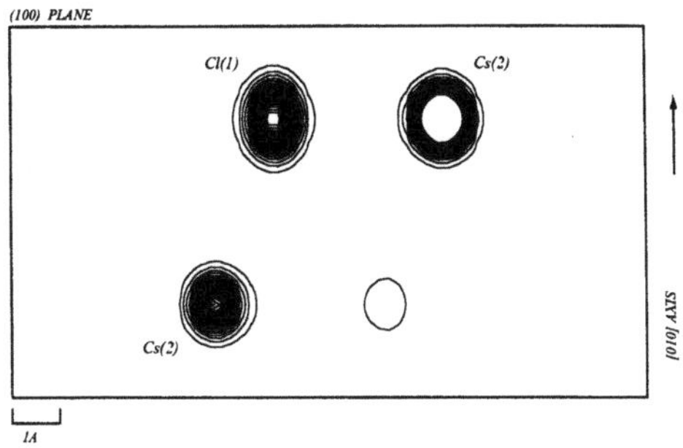

Figure 1. Section of the electron density of Cs_2HgCl_4 as obtained for a model with spherical atoms and harmonic temperature factors

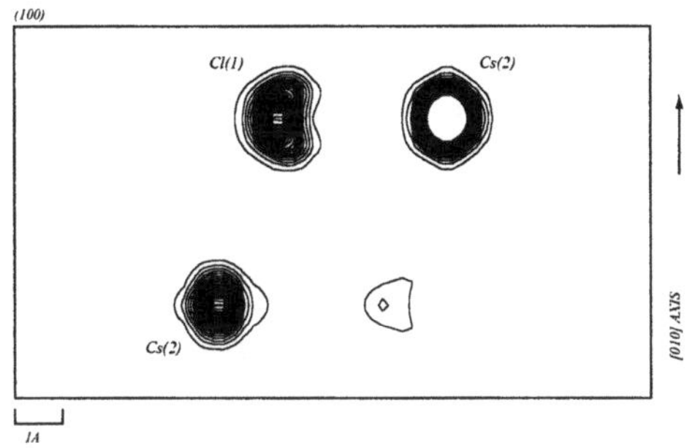

Figure 2. Section of the electron density obtained by the MEM, using the phases of the harmonic refinement and $F_{obs}(\vec{H})$

3.2. DISORDER IN THE STRUCTURE OF LICP*

LiCp* is an organometallic compound of lithium and the molecule $C_5(CH_3)_5$ (Cp*). It crystallizes in space group *R3m* with lattice parameters a=14.7711 Å and c=3.82206 Å. The structure consists of columns of alternatingly Li atoms and Cp* molecules. X-ray powder diffraction lead to the intensities of 33 reflections, and intensities of 60 groups of overlapping reflections [6]. Rietveld refinement did not lead to an entirely satisfactory fit. The model in space group *R3* (the molecules were free to rotate about their axis) gave a better fit than the model in *R3m* (the molecules were lined up with the mirror planes), but the deviation was only a rotation of 2° with a standard deviation of 3°. Obviously, the refinement cannot discriminate between the two symmetries.

The MEM reproduced the model in *R3* when $F^{calc}_{electron}$ for 157 reflections was used in a F-constraint. The experimental data correspond to a F-constraint with 33 reflections and a G-constraint with 60 groups of reflections. With a flat initial density, MEED led to an electron density map that obviously contained artifacts. Apparently, the number of reflections in the F-constraint was not sufficient to guide the iterations towards the true minimum. With the electron density of the

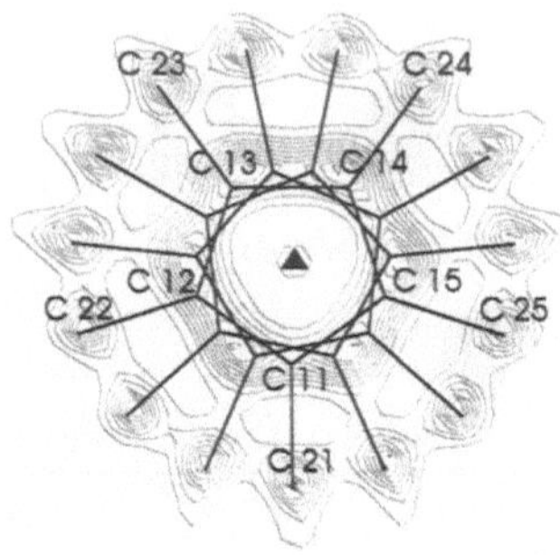

Figure 3. The electron density of LiCp* obtained by the MEM overlaid with a model of the Cp* molecule in 3 orientations.

model as initial value, MEED converged smoothly towards a density showing the three-fold disorder of the Cp* molecule. With phases from the Rietveld refinement in *R3*, the MEM led to a density that was much closer to *R3m* symmetry than the initial density. It was thus concluded that the MEM gave evidence for *R3m* symmetry contrary to *R3*. The result of a calculation with MEED with phases from a refinement in *R3m* is shown in Fig. 3. The three-fold disorder of the Cp* molecule is obvious, while major deviations between model and ρ_{MEM} are again found in anharmonic contributions to this disorder.

References

1. C. Gilmore, "Maximum entropy and bayesian statistics in crystallography: a review of practical applications," *Acta Cryst.*, **A52**, pp. 561–589, 1996.
2. G. Bricogne, "Direct phase determination by entropy maximization and likelyhood ranking: Status report and perspectives," *Acta Cryst.*, **D49**, pp. 37–60, 1993.
3. D. Sivia and W. David, "A bayesian approach to extracting structure-factor amplitudes from powder diffraction data," *Acta Cryst.*, **A50**, pp. 703–714, 1994.
4. M. Sakata and M. Sato, "Accurate structure analysis by the maximum-entropy method," *Acta Cryst.*, **A46**, pp. 263–270, 1990.
5. C. Giacovazzo, ed., *Fundamentals of Crystallography*, 1990.
6. R. Dinnebier, M. Schneider, S. van Smaalen, F. Olbrich, and U. Behrens, "Disorder determined by high resolution powder diffraction: The structure of pentamethylcyclopentadienyllithium," *Acta Cryst. B*, in press.
7. B. Bagautdinov, J. Luedecke, M. Schneider, and S. van Smaalen, "Disorder determined by high resolution powder diffraction: The structure of pentamethylcyclopentadienyllithium," *Acta Cryst. B*, in press.
8. N. Wu, *The Maximum Entropy Method*, Springer Berlin, 1997.
9. S. Kumuzawa, Y. Kubota, M. Takata, and M. Sakata, "Meed: a program package for electron-density-distribution calculation by the maximum-entropy method," *J. Appl. Cryst.*, **26**, pp. 453–457, 1993.
10. R. Y. de Vries, W. Briels, and D. Feil, "Critical analysis of non-nuclear electron-density maxima and the maximum entropy method," *Phys. Rev. Letters*, **77**, pp. 1719–1722, 1996.
11. M. Takata and M. Sakata, "The influence of the completeness of the data set on the charge density obtained with the maximum-entropy method. a re-examination of the electron-density distribution in si," *Acta Cryst.*, **A52**, pp. 287–290, 1996.
12. M. Takata, E. Nishibori, B. Umeda, M. Sakata, E. Yamamoto, and H. Shinohara, "Structure of endohedral dimetallofullerene," *Phys. Rev. Letters*, **78**, pp. 3330–3333, 1997.

TRACE METAL DETECTION IN NATURAL WATER

A. TAGLAUER AND R. NIESSNER
Institut für Wasserchemie, Technische Universität München
München, Germany †

AND

W. VON DER LINDEN, V. DOSE AND E. TAGLAUER
Max-Planck-Institut für Plasmaphysik, EURATOM Association
D-85748 Garching b. München, Germany

Abstract. Fluorescence measurements of metal ions in drinking water employing complex forming organic agents result in signals depending linearly on the metal ion concentration in the low concentration range. In this paper we present and analyze measurements with and without additives with potential influence on the complexation. We found that sulfite addition to neocuproine does not sensitize the fluorescence. Moreover we found that neocuproine can be used as an efficient masking agent for spurious copper in the analysis of nickel with Newport Green®.

Key words: Fluorescence, Ni - Cu - detection, model comparison, odds-Ratio

Human drinking water supplies have to be checked routinely for possible contaminants. In particular heavy metal contaminants are of interest since they are health hazardous. Among the various analytical methods for detection of contaminants in water, laser induced fluorescence has the best prospects for continuous real time analysis [1]. In this technique the metal ion is first bound to an organic molecule as complexing agent. The complexing agent is excited by e.g. nitrogen laser ultraviolet radiation. The excited state then decays by emission of visible radiation. Upon formation of the metal-ligand complex the intensity of the fluorescence is changed. It may be either amplified as in the case of Newport Green®which is employed for nickel detection or the fluorescence may be attenuated as in the case of neocuproine which is used for copper detection (see Fig. 1). Complex formation of a specific metal ion with the complexing agent is an equilibrium reaction. This means, that it depends on various parameters among which the oxidation state of the metal ion and concentrations of competing metal ions are of special importance in the context of this paper. Consider the influence of the oxidation

†Email: Andreas.Taglauer@ch.tum.de

W. von der Linden et al. (eds.), Maximum Entropy and Bayesian Methods, 341–344.

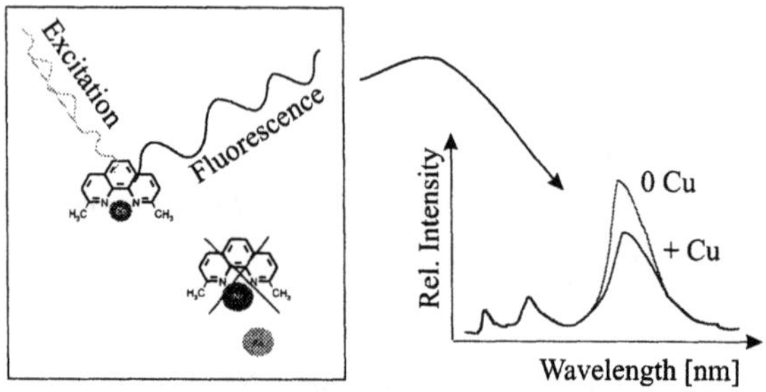

Figure 1. Schematic diagram of the reaction neocuproine + copper: Complexation of copper ions leads to a quenching of the neocproine-fluorescence. Nickel does not exhibit an effect on the neocuproine-fluorescence. In contrast thereto the Newport Green®fluorescence is strongly enhanced by nickel complexation and copper ions interfere by quenching this fluorescence.

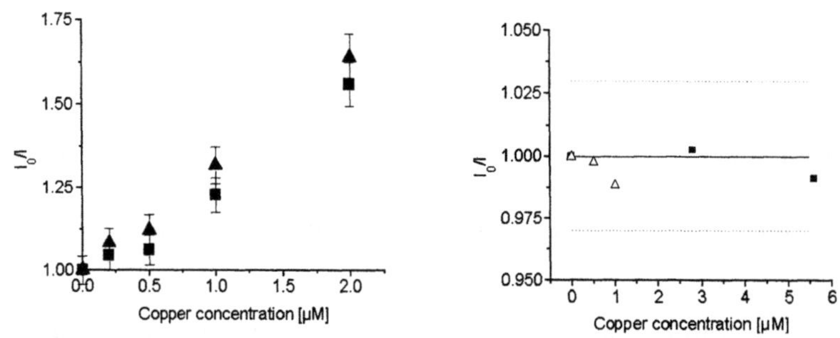

Figure 2. a) Influence of the copper concentration on the fluorescence intensity of neocuproine with (■) and without (▲) reducing SO_3^{2-} ions b) Distribution of the Ni^{2+}-Newport Green®fluorescence intensity with varying Cu^{2+} concentration. (■) 5.6 μM Ni^{2+}, 3.5 μM Newport Green®, 9.35 μM neocproine. (△) 1 μM Ni^{2+}, 3.5 μM Newport Green®, 1.6 μM neocuproine. The dotted line indicates the range of 3% error.

state first. Literature results [2] suggest that the complex formation reaction constant for reactions of copper with neocuproine increases, when Cu^{2+} is reduced to Cu^{+}. This raises the question whether copper ion detection in water by complex formation with neocuproine can be sensitized upon addition of a reducing agent. Experimental results on the (inverse) attenuation I_0/I of the neocuproine fluorescence as a function of copper concentration with and without addition of reducing SO_3^{2-} ions are shown in Fig. 2a. If the two measurements differ at all, then the difference is very small. This is not necessarily at variance with the observation that reduction of Cu^{2+} to Cu^{+} speeds up the metal organic compound formation. It may well be that the equilibrium of the reaction is strongly in favor of the metal organic compound, such that further increase of the reaction constant does not lead to a measurable modification of the equilibrium concentration of the product. However, let us return to the question whether the two measurements in Fig. 2a should be described by the same or by different linear relationships. This is a problem of Bayesian model comparison [3]. We denote by H_1 the hypothesis that a common model applies and by H_2 the complementary hypothesis that the distinction is justified. Our interest is then in the odds ratio [3].

$$O = \frac{p(H_1|\boldsymbol{d}_1, \boldsymbol{d}_2, \boldsymbol{x}_1, \boldsymbol{x}_2, \boldsymbol{\sigma}_1, \boldsymbol{\sigma}_2, I)}{p(H_2|\boldsymbol{d}_1, \boldsymbol{d}_2, \boldsymbol{x}_1, \boldsymbol{x}_2, \boldsymbol{\sigma}_1, \boldsymbol{\sigma}_2, I)} = \frac{p(H_1|I)\, p(\boldsymbol{d}_1, \boldsymbol{d}_2|\boldsymbol{x}_1, \boldsymbol{x}_2, \boldsymbol{\sigma}_1, \boldsymbol{\sigma}_2, H_1, I)}{p(H_2|I)\, p(\boldsymbol{d}_1, \boldsymbol{d}_2|\boldsymbol{x}_1, \boldsymbol{x}_2, \boldsymbol{\sigma}_1, \boldsymbol{\sigma}_2, H_2, I)} \ . \tag{1}$$

$\boldsymbol{d}_1, \boldsymbol{d}_2$ denote the two sets of fluorescence signals with known true errors $\boldsymbol{\sigma}_1, \boldsymbol{\sigma}_2$ measured at copper ion concentrations $\boldsymbol{x}_1, \boldsymbol{x}_2$. Since we have no prior information which favors H_1 over H_2 we put the prior odds equal to one which leaves us with the calculation of the Bayes factor [3]. The two measurements are of course uncorrelated. The total likelihood function $p(\boldsymbol{d}_1, \boldsymbol{d}_2|\boldsymbol{x}_1, \boldsymbol{x}_2, \boldsymbol{\sigma}_1, \boldsymbol{\sigma}_2, b_1, b_2, s_1, s_2, I)$ factors therefore into $p(\boldsymbol{d}_1|\boldsymbol{x}_1, \boldsymbol{\sigma}_1, b_1, s_1, I) \times p(\boldsymbol{d}_2|\boldsymbol{x}_2, \boldsymbol{\sigma}_2, b_2, s_2, I)$ where b_i and s_i are intercept and slope of the linear model function.

$$p(\boldsymbol{d}|\boldsymbol{x}, \boldsymbol{\sigma}, b, s, I) = Z_L^{-1} \exp\{-\frac{1}{2} \sum_i \left(d_i - b - s x_i\right)^2 / \sigma_i^2\} \ . \tag{2}$$

We introduce the definitions $\omega^{-2} = \sum_i \sigma_i^{-2}$, $\overline{y} = \omega^2 \sum_i \sigma_i^{-2} y_i$ and marginalize over b assuming a flat prior to obtain

$$p(\boldsymbol{d}|\boldsymbol{x}, \boldsymbol{\sigma}, s, I) = Z_L^{-1} \omega \sqrt{2\pi} \exp\{-\frac{1}{2\omega^2} \left[\overline{\Delta x^2}(s - s^*)^2 + \overline{\Delta d^2} - s^{*2}\overline{\Delta x^2}\right]\} \tag{3}$$

where $s^* = \overline{\Delta d \Delta x} / \overline{\Delta x^2}$ is the maximum likelihood value of the slopes. (3) may now be used to calculate the marginal likelihoods entering the Bayes factor. The numerator of (1) is then simply the product of two factors obtained by marginalizing (3) over s with a suitable prior. The denominator on the other hand is the marginal over a common s of two factors of the form (3) for datasets 1 and 2, respectively. Unlike the previous marginalization over b, where an improper prior was used without complication since it entered twice in the numerator and the denominator of the Bayes factor and therefore a proper limit existed, marginalization over s has to employ a proper prior since it enters twice in the numerator but only once in the denominator. If we assume a flat prior distribution of angles between the linear function and the x-axis, this transforms into

$$p(s|I) = (\pi(1 + s^2))^{-1} \ . \tag{4}$$

The integrations over s in either case (H_1, H_2) are not possible analytically. We employ the peak approximation, that is we assume that the curvature of the prior in the region where the likelihood differs appreciably from zero can be neglected. The prior can then be taken out of the integral with s replaced by s^*, the maximum of the likelihood function. This procedure yields for the odds ratio O

$$O = \sqrt{\frac{2}{\pi}} \sqrt{\frac{\mu_1 \mu_2}{\mu_1 + \mu_2}} \frac{1}{1 + s^{*2}} \exp\{\frac{\mu_1 \mu_2}{2(\mu_1 + \mu_2)} (s_1^* - s_2^*)^2\} \ , \tag{5}$$

with $\mu_\alpha = \overline{\Delta x_\alpha^2} / \omega_\alpha^2$ and the additional assumption $s_1^* \approx s_2^* \approx s^*$ which is mild because otherwise the data would already present clear evidence for H_2. For the data displayed in Fig. 2a we obtain an odds ratio $O \approx 19$ which clearly favors a description of all data by a common linear relationship. Sulfite addition to neocuproine does not sensitize the fluorescence. The second experiment which we want to present and interpret here is concerned with the competition of metal ions in the reaction with a complexing agent. We consider fluorescence detection of nickel reacted with Newport Green®(see Fig. 1). This fluorescence is normally quenched in the presence of copper ions. Copper on the other hand is quite selectively bound by neocuproine which in turn raises the question whether the Ni^{2+}-Newport Green®fluorescence can be made independent of competing copper traces when neocuproine is added to the solution as a masking agent. Data relating to two different Ni ion concentrations (5.6 μM Ni^{2+} + 9.35 μM Neocuproine and 1 μM Ni^{2+} + 1.5 μM Neocuproine) but with the same ratio Ni^{2+}/Neocuproine are shown in Fig. 2 b. Since it is not immediately obvious whether Cu^{2+} can be masked completely, we employ Bayesian model comparison to decide whether the Ni^{2+}-Newport Green®fluorescence changes linearly with or is independent of the copper concentration. The likelihood function for the constant model (H_1) is simply

$$p(\boldsymbol{d}|\boldsymbol{x},\boldsymbol{\sigma},b,I) = Z_L^{-1} \exp\{-\frac{1}{2}\sum_i (d_i - b)^2 / \sigma_i^2\} \tag{6}$$

and the likelihood function for the linear model (H_2) is the same as in (2). Proceeding with the marginalizations as before we obtain the odds ratio for H_1

$$O = \sqrt{\pi\overline{\Delta x^2}/(2\omega^2)}\left(1 + c^2\overline{\Delta d^2}/\overline{\Delta x^2}\right)\exp\left\{-c^2\overline{\Delta d^2}/(2\omega^2)\right\} \tag{7}$$

where we introduced the correlation coefficient $c = \overline{\Delta x \Delta d}/(\overline{\Delta x^2}\,\overline{\Delta d^2})^{\frac{1}{2}}$. For the optimal case for the constant model, i.e. zero correlation, we obtain an odds ratio of $\sqrt{\pi\overline{\Delta x^2}/(2\omega^2)}$ in favor of the constant model. For increasing c the increase in the prior factor is overcompensated by the decreasing exponential and the odds ratio decreases monotonically with c. For approximately constant errors $\sigma_i = \sigma$ we have $\omega^2 = \sigma^2/N$ and see the strong dependence of the odds ratio on the number of measurements given the signal to noise ratio $(\overline{\Delta d^2}/\sigma^2)^{\frac{1}{2}}$. For the data in Fig. 2b we obtain $O = 143$. The analysis therefore clearly favors constant nickel-Newport Green®fluorescence and we conclude that neocuproine is a perfect masking agent for the fluorimetric detection of nickel by complexation with Newport Green®.

This research was financially supported by the Deutsche Forschungsgemeinschaft.

References

1. E. Müller-Ackermann, U. Panne, and R. Niessner, "A fiber optic sensor array for the fluorimetric detection of heavy metals," *Anal. Meth. Instr.*, **2**, p. 182, 1995.
2. T. M. Florence, B. G. Lumsden, and J. J. Fardy, "Algae as indicators of copper speciation," in *Complexation of Trace Metals in Natural Water*, C. J. M. Kramer and J. C. Duinker, eds., p. 411, Nijhoff & Junk, The Hague, 1984.
3. P. Gregory and T. Loredo, "A new method for the detection of a periodic signal of unknown shape and period," *Astrophys. J.*, **398**, p. 146, 1992.

OSCILLATOR STABILITY ANALYSIS USING A BAYESIAN APPROACH

GILLES ZALAMANSKY
Université de Metz
Campus Bridoux, F-57000 Metz ‡

AND

FRANCOIS VERNOTTE
Observatoire de Besançon
41 bis av. de l'Observatoire, BP 1615, F-25010 Besançon Cedex §

Abstract. The power spectral density of frequency fluctuations of an oscillator is generally modeled as a sum of power laws with integer exponents (from -2 to +2). However, a power law with a fractional exponent may exist. We propose a method for measuring such a noise level and determining the probability density of the exponent. This yields a criterion for the compatibility with an integer exponent. This method is based upon a Bayesian approach described in the paper "BAYESIAN ANALYSIS WITH APPLICATION TO THE TIMING RESIDUALS OF A PULSAR" by G. Zalamansky and C. Robert, in these proceedings [1].

Key words: oscillator characterization, time stability, noise analysis, Bayesian analysis

1. Introduction : characterization of the time stability of oscillators

The output signal of an ideal oscillator may be modelled as :

$$V(t) = \sin(2\pi\nu_0 t) \tag{1}$$

where ν_0 is the nominal frequency of the oscillator.

A real oscillator is altered by a phase noise $\phi(t)$. We assume that this phase noise is a random process.

$$V(t) = \sin(2\pi\nu_0 t + \phi(t)). \tag{2}$$

‡Email: zala@sciences.univ-metz.fr
§Email: francois@obs-besancon.fr

W. von der Linden et al. (eds.), Maximum Entropy and Bayesian Methods, 345–352.

The instantaneous frequency is :

$$\nu(t) = \frac{1}{2\pi}\frac{d}{dt}\left(2\pi\nu_0 t + \phi(t)\right) = \nu_0 + \frac{1}{2\pi}\frac{d\phi(t)}{dt}. \tag{3}$$

Thus, the frequency of the oscillator is altered by a frequency noise $\Delta\nu(t)$ linked to the phase noise by the following relationship :

$$\Delta\nu(t) = \frac{1}{2\pi}\frac{d\phi(t)}{dt}. \tag{4}$$

The most often used quantity in the frequency and time domain is the instantaneous normalized frequency deviation $y(t)$ defined as :

$$y(t) = \frac{\Delta\nu(t)}{\nu_0}. \tag{5}$$

The autocorrelation function $\mathcal{R}_y(\tau)$ of $y(t)$ may be defined as :

$$\mathcal{R}_y(\tau) = \langle y(t)y(t+\tau)\rangle\,. \tag{6}$$

However, only the average of $y(t)$ over a time interval τ is observable. Generally, $\overline{y}_k$ samples are defined as :

$$\overline{y}_k = \int_{t_k}^{t_k+\tau} y(t)dt. \tag{7}$$

In the frequency domain, the power spectral density $S_y(f)$ of $y(t)$ is defined as the Fourier transform of $\mathcal{R}_y(\tau)$:

$$S_y(f) = \left\langle \lim_{T\to\infty}\frac{1}{T}\left|\int_{-T/2}^{+T/2} y(t)e^{-2i\pi ft}dt\right|^2\right\rangle. \tag{8}$$

Different behaviors of this spectral density have been experimentally observed. It is commonly assumed that $S_y(f)$ may be modelled as the sum of 5 power laws, defining 5 types of noise :

$$S_y(f) = \sum_{\alpha=-2}^{+2} h_\alpha f^\alpha \tag{9}$$

where h_α is the level of the f^α noise. It may be noticed that models with non-integer exponents are occasionally used.

The estimation of the noise levels is mainly achieved by using the Allan variance [2], which is defined versus the integration time τ as :

$$\sigma_y^2(\tau) = \frac{1}{2}\left\langle\left(\overline{y}_{k+1} - \overline{y}_k\right)^2\right\rangle. \tag{10}$$

In the frequency domain, the Allan variance may be considered as a filter. If the Allan variance versus the integration time τ is plotted, the graph exhibits different slopes, each slope corresponding to a type of noise :

$$\sigma_y^2(\tau) = C_\mu\tau^\mu \quad\Leftrightarrow\quad S_y(f) = h_\alpha f^\alpha \quad\text{and}\quad \alpha = -\mu - 1. \tag{11}$$

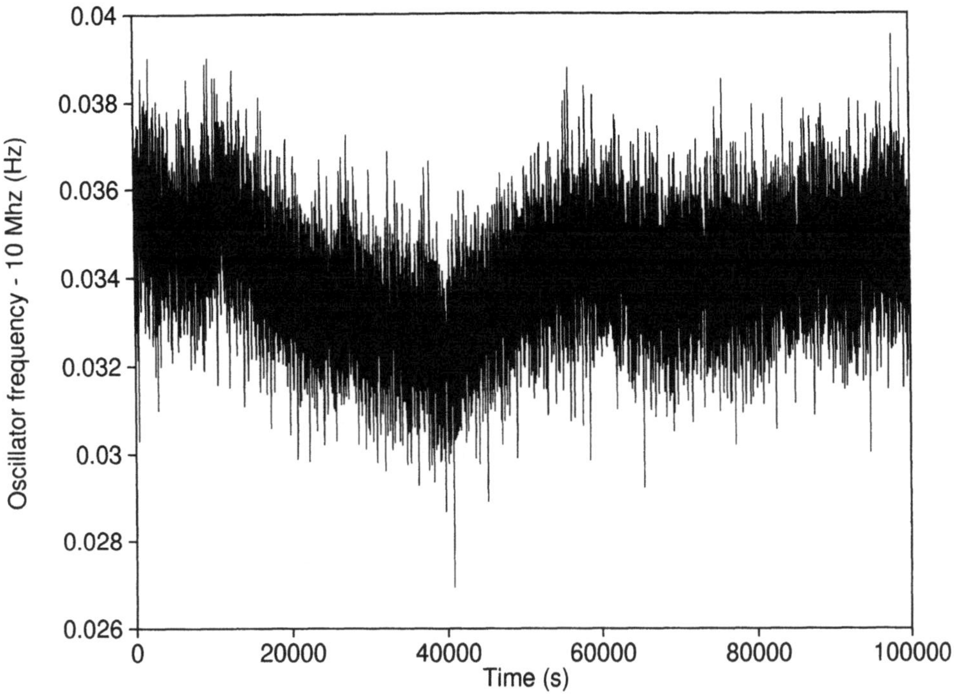

Figure 1. Sequence of frequency measurements

The estimation of C_μ yields an estimation of the noise level h_α.

2. Classical stability analysis of an oscillator

2.1. SEQUENCE OF FREQUENCY MEASUREMENTS

Figure 1 shows averaged frequency measurements $\overline{\nu}_k$ of a 10 MHz quartz oscillator compared to a Cesium clock. The sampling rate is 10 s and the integration time of each frequency measurement is also 10 s (sampling without dead time).

In order to obtain dimensionless $\overline{y}_k$ samples, we must substract the nominal frequency ν_0 (10 MHz) from the frequency measurements and normalized by ν_0 :

$$\overline{y}_k = \frac{\overline{\nu}_k - \nu_0}{\nu_0}. \tag{12}$$

2.2. VARIANCE ANALYSIS

Figure 2 is a log-log plot of the Allan variance of the quartz $\overline{y}_k$ samples versus the integration time τ. A least square fit of these variance measurements (solid line), weighted by their uncertainties, detects only two types of noise : a white noise and a f^{-2} noise. The corresponding noise level estimations are :

$$h_0 \quad = \quad (2.2 \pm 0.4) \cdot 10^{-5}\text{s} \qquad \text{at } 1\sigma \text{ (68\% confidence)}$$

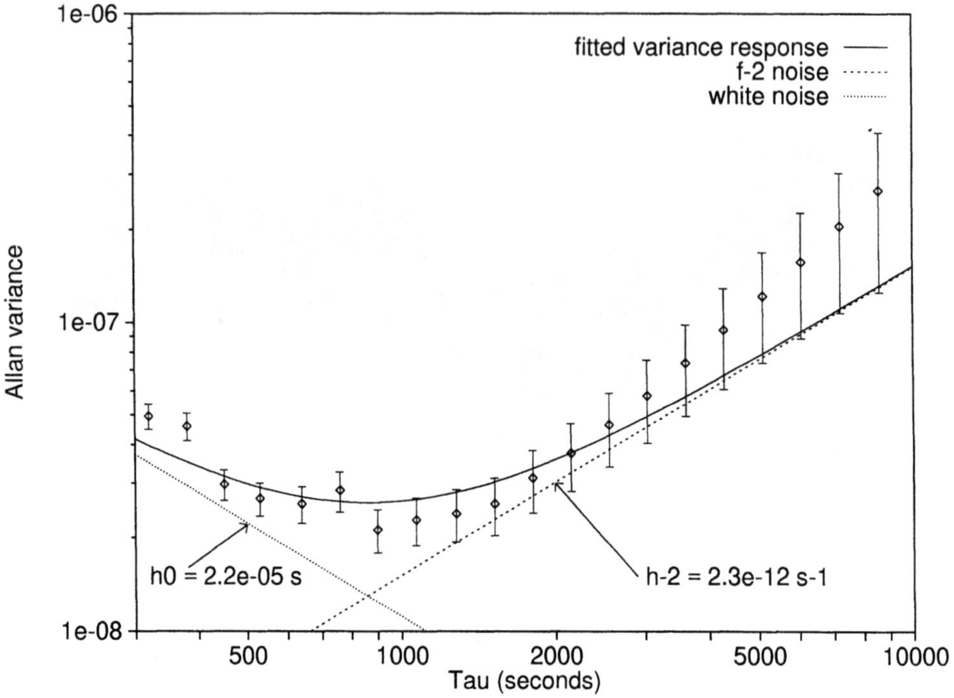

Figure 2. Allan variance of the sequence of frequency measurements

$$h_{-2} \quad = \quad (2.3 \pm 0.6) \cdot 10^{-12}\mathrm{s}^{-1} \qquad \text{at } 1\sigma \text{ (68\% confidence)}$$

(for the assessment of the h_α noise levels and their uncertainties, we used the multivariance method described in [3]).

However, for large τ values (corresponding to low frequencies), the variance measurements get away from the fitted curve. Two explanations are possible :

- instead of a f^{-2} noise, there is a noise whose non-integer exponent is contained between -2 and -3 ;
- since the uncertainty domains of the variance measurements contain the fitted curve, this apparent divergence may be due to a statistical effect.

In order to choose between these two explanations, we decided to estimate the probability density of the exponent with a Bayesian approach.

3. Bayesian approach

3.1. SPECTRAL DENSITY AND COVARIANCE MATRIX

Let us define the vector y which components are the N $\overline{y}_k$ samples normalized by the standard deviation of the white noise component (estimated from high sampling rate frequency measurement). We assume that y is a gaussian vector.

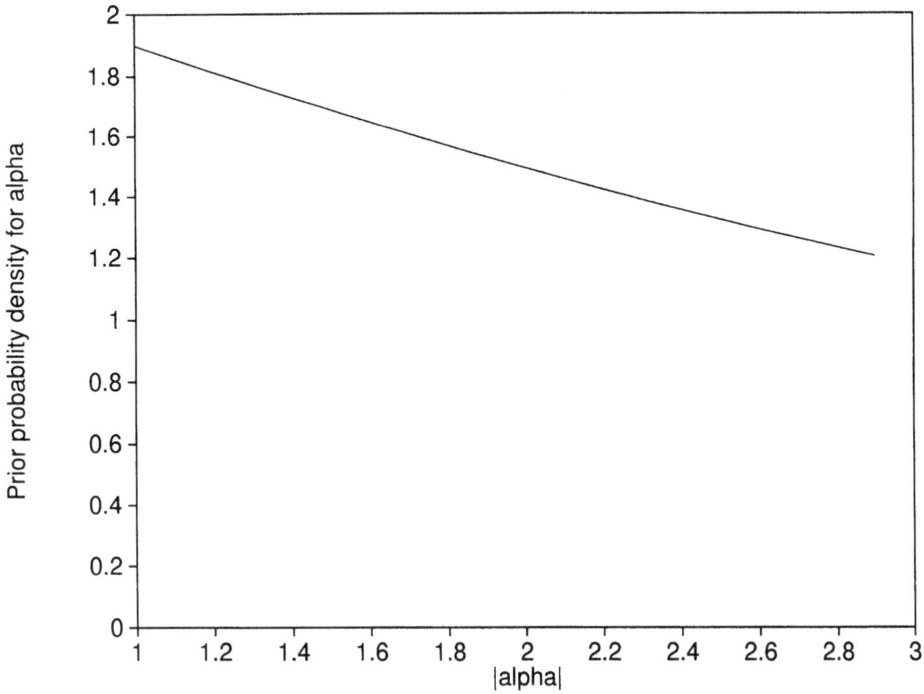

Figure 3. Reference prior for the power α

The probability distribution of y is

$$p(y) = \frac{\exp\left(-\frac{y^t C^{-1} y}{2}\right)}{(2\pi)^{\frac{N}{2}} \sqrt{|C|}} \tag{13}$$

where C is the covariance matrix. Since $S_y(f)$ is the Fourier transform of the autocorrelation function $\mathcal{R}_y(\tau)$, the general term of C is :

$$C_{ij} = 2 \int S_y(f) \cos\left(2\pi f (t_i - t_j)\right) df. \tag{14}$$

Equation (14) reveals the key role played by the spectral density of the noise in the expected fluctuation.

3.2. ASSUMED MODEL FOR THE SPECTRAL DENSITY

We assumed that the sequence of frequency measurement is composed of a white noise y_w, whose variance (level) is unitary, and of a red noise y_r whose level is unknown :

$$y = y_w + y_r. \tag{15}$$

This yields to the folowing model for $S_y(f)$:

$$S_y(f) = h_0 + h_\alpha \cdot f^\alpha \qquad \text{with } -3 \leq \alpha \leq -1 \tag{16}$$

where $h_0 = 2.2 \cdot 10^{-5}$s, h_α and α are the unknown parameters.

3.3. STATISTICAL MODEL

The part of the spectal density due to the red noise y_r may be written :

$$S_r(f) = H \cdot u_\alpha \cdot f^\alpha \tag{17}$$

where u_α is an amplitude factor chosen in such a way that, for $H = 1$, the averaged quadratic norms of the white noise and of the red noise are equal (see [1]).

We used the Bernardo-Berger analysis [4,1] for estimating the unknown parameter $\theta = (\alpha, H)$.

For a detailed description of this method (construction of the estimators, of the priors and of the posteriors) see [1]. The only differences with this paper are :

- we limited to 16 the number of estimators p_j (from degrees 0 to 15) for limiting the computation and because the high degrees, estimating the high frequencies, are less informative for a red noise ;
- we assumed $\alpha \leq -3$. In order to ensure the convergence for very low frequencies (moment condition, [5,6]) the first 2 estimators (p_0 and p_1) must been removed.

Thus we have $n = 14$ estimators $\{p_2, \ldots, p_{15}\}$ and $n = 14$ estimates $\{\xi_2, \ldots, \xi_{15}\}$.

The reference prior for the power α is shown in figure 3.

4. Results and discussion

4.1. COMPATIBILITY WITH AN INTEGER EXPONENT

Figure 4 shows the posterior probability distribution for the exponent α of the red noise using the Bernardo-Berger prior and the Jeffrey prior. As we expected, the curves are almost identical.

The exponent value obtained for the maximum of likelihood, just as for the maximum of the distribution, is $\alpha = -2.2$.

However, $\alpha = -2$ is fully compatible with this prior distribution. Thus, we may conclude that the apparent divergence between the variance measurements and the fitted curve in figure 2 is probably due to a statistic bias of the data. The spectral density $S_y(f)$ is then compatible with the following model :

$$S_y(f) = h_0 + h_{-2} f^{-2}. \tag{18}$$

4.2. NOISE LEVEL ESTIMATION

Selecting a exponent value $\alpha = -2$, we obtained the posterior probability distribution plotted in figure 5. As in the variance analysis, we chose a confidence interval of 68% (16% probability that h_{-2} is smaller than the low bound and 16% probability that h_{-2} is greater than the high bound) :

$$h_{-2} = \left(2.3 \begin{array}{c} + \\ - \end{array} \begin{array}{c} 2 \\ 0.8 \end{array}\right) \cdot 10^{-12}\text{s}^{-1} \qquad \text{at } 1\sigma \text{ (68\% confidence)}$$

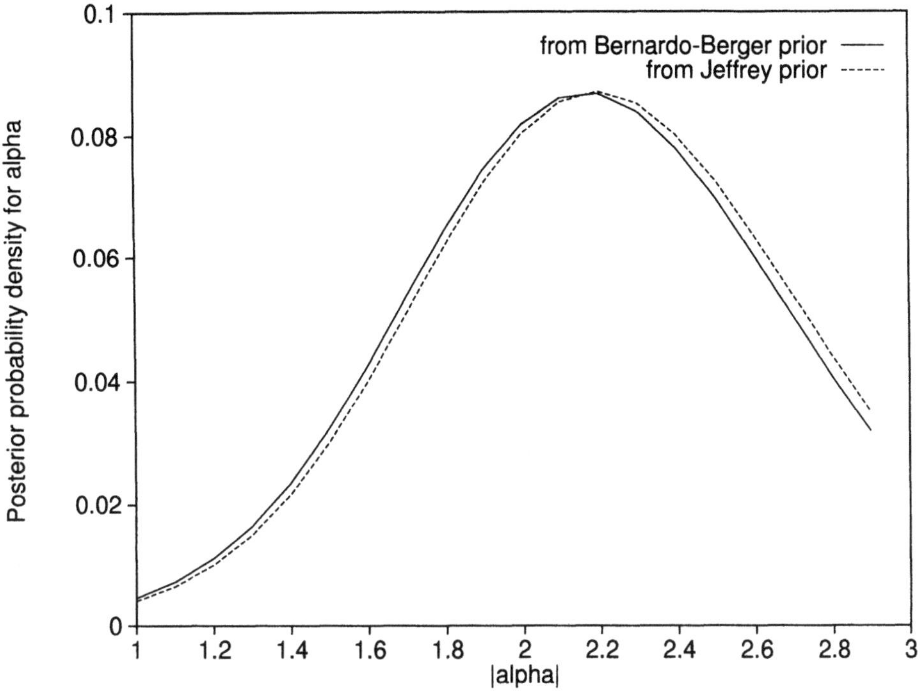

Figure 4. Posterior probability density for the power α

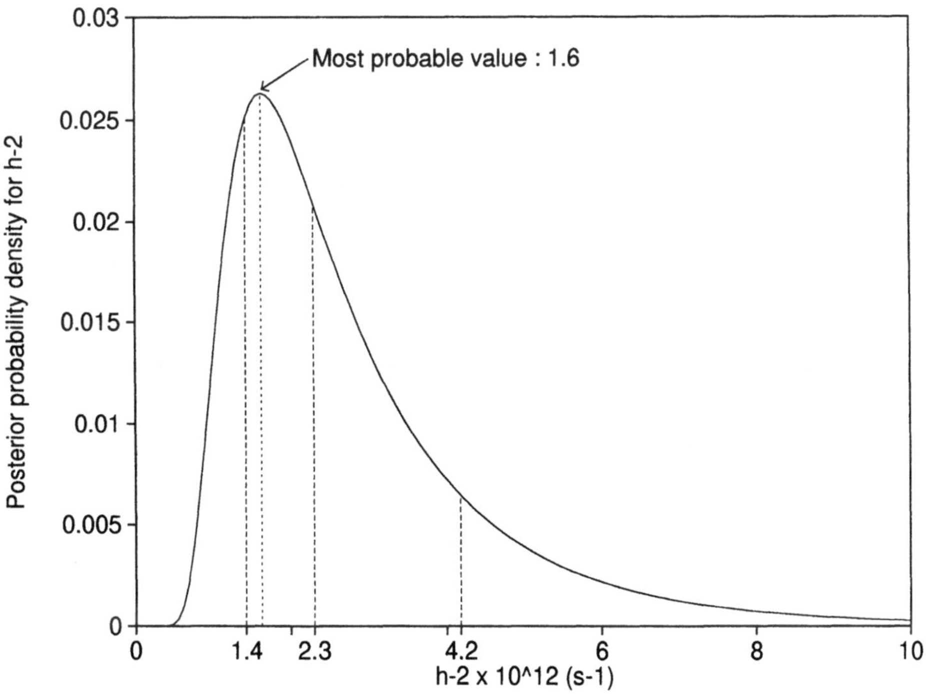

Figure 5. Posterior probability density for the noise level h_{-2}

The difference between the maximum likelihood value ($h_{-2} = 2.2991 \cdot 10^{-12}\mathrm{s}^{-1}$) and the variance analysis value ($h_{-2} = 2.2949 \cdot 10^{-12}\mathrm{s}^{-1}$) is only 0.18%.

However, the confidence intervals given by these two methods are quiet different. The main difference concerns the symmetry of the variance analysis interval : in this case, we don't take into account the fact that the noise levels are positive, whereas the prior of the bayesian approach is null for negative values of h_{-2}.

Moreover, the variance analysis interval seems to be a bit underestimated.

5. Conclusion

The variance analysis is an useful tool for a quick estimation of the noise levels in the output signal of oscillators. However, a negative estimation of a noise level may occur. Generally, in this case, this value is rejected and the noise level corresponding is assumed to be null. On the other hand, although the Bayesian method is a bit heavier, it takes into account properly the a priori information, and gives a more reliable estimation of these noise levels and especially of their confidence intervals.

But the main advantage of the Bayesian method concerns the verification of the validity of the power law model of spectral density. Each time the model is suspected, such an approach should be used in order to estimate the exponent of the power law. Particularly, this method should be very interesting for the study of the f^{-1} and f^{+1} noise, whose origin remains mysterious [7].

References

1. G. Zalamansky and C. Robert, "Bayesian analysis with application to the timing residuals of a pulsar," in *MaxEnt'98*, (Garching, Germany), July 1998.
2. J. A. Barnes, A. R. Chi, L. S. Cutler, D. J. Healey, D. B. Leesson, T. E. McCuningal, J. A. Mullen, W. L. Smith, R. L. Sydnor, R. Vessot, and G. M. R. Winkler, "Characterization of frequency stability," *IEEE Transactions on Instrumentation and Measurement*, **IM-20**, pp. 105–120, 1971.
3. F. Vernotte, E. Lantz, J. Groslambert, and J. Gagnepain, "Oscillator noise analysis : multi-variance measurement," *IEEE Transactions on Instrumentation and Measurement*, **IM-42**, pp. 342–350, Avril 1993.
4. J. Bernardo and A. Smith, *Bayesian theory*, Wiley & Son, 1993.
5. J. E. Deeter and P. E. Boynton, "Techniques for the estimation of red power spectra. i. context and methodology," *The Astrophysical Journal*, **261**, pp. 337–350, October 1982.
6. F. Vernotte, G. Zalamansky, M. McHugh, and E. Lantz, "Cut-off frequencies and noise power law model of spectral density : adaptation of the multi-variance method for irregularly spaced timing data using the lowest mode estimator approach.," *IEEE Transactions on Ultrasonics, ferroelectrics, and Frequency*, **43**, (3), pp. 403–409, 1996.
7. M. Planat, V. Giordano, G. Marianneau, F. Vernotte, M. Mourey, C. Eckert, and J. Miehe, "Is the frequency noise of an oscillator of deterministic origin ?," *IEEE Transactions on Ultrasonics, ferroelectrics, and Frequency*, **IM-43**, (2), pp. 326–330, 1996.

INDEX

adaptive kernels, 99, 112
adaptive procedure, 123
adaptive scale inference, 83
amplitudes, 23–25, 37–40, 43, 44, 46, 181, 192, 209, 218, 219, 222, 223, 237, 238, 240, 248, 325, 340
anharmonic displacements, 335, 338
array entropy, 238, 244, 245

Bayes factor, 174, 319, 324, 325, 343
Bayes' theorem, 24, 85, 87, 102, 139, 150, 162, 163, 167, 168, 170, 207, 212, 213, 218
Bayesian analysis, 7, 14, 56, 106, 174, 189, 193, 194, 197, 207, 208, 214, 309, 316, 345
Bayesian estimation, 125, 149, 304
Bayesian inference, 8, 14, 65–67, 83, 84, 115, 142, 162, 163, 201, 202, 207, 218, 305, 307, 309, 327, 328, 333
Bayesian model selection, 179
Bayesian networks, 199, 201, 300
Bayesian reconstruction, 73, 80, 318
Bayesian statistics, 7, 8, 150, 163, 170, 199, 340
Bayesian theory, subjective - , 157
binomial series, 153, 154
brain, 4, 8, 217, 223–225, 331, 332
BSS, 217–220, 222

Cauchy distribution, 149–151, 154
χ^2, 222
Clifford algebra, 247, 248
clustering, 120, 121, 199, 203, 204, 206, 327–334
clustering by melting, 199, 203, 205
constraints, linear - , 271, 288
contour integration, 151
correlation coefficient, 27, 145, 344
cost function, 123, 126, 127, 222, 328
Cox, 9, 16, 246–248, 271
cross-entropy, 69, 70, 206, 288
crystal structures, 335

decision theory, 289
deconvolution, 107, 163, 320
density estimation, 91–94, 97, 98, 102, 104
differentiation, 277, 278, 309, 312, 315, 318
digitization noise, 179, 183, 184
discrete field, 84
disorder, 57–60, 64, 91, 335, 338, 340
distributional inference, 289
duff data, 47

EEG, 217, 223–225
electroencephalography, 217
electromagnetic source estimation, 217, 218
electron energy, 99, 100
empirical data, 123, 124, 126, 224
entropic connections, 131
entropic priors, 131, 135–137
entropy, 4, 7–9, 11, 12, 14–16, 26–30, 46, 60, 61, 74, 75, 91–98, 104, 132, 135, 136, 139, 178, 188, 204, 212, 213, 215, 219, 227, 232, 237, 238, 243–245, 271, 273, 279, 280, 282, 283, 285–287, 293, 300, 304, 327–337, 340
ESE, 217, 218, 220–223, 225
evidence, 56, 65, 67, 70, 72, 74, 80, 103, 132, 162, 166, 179, 183,

184, 186, 187, 200–206, 272, 285, 320, 321, 340, 343

fiducial argument, 141, 142
fiducial inference, 141, 142, 146
field, 3–5, 18, 51, 66, 74, 83–85, 89, 101, 105, 110, 115, 117, 122, 124, 132, 137–139, 171, 250, 288, 302, 306, 307
fluorescence, 341–344
FMRI, 225
form-free reconstruction, 99, 100, 102, 104-106
foundations of statistics, 289
Fourier methods, 91
functional equation, 273, 276–278, 281, 282, 284, 288

Gauss markov random fields, 65
Gaussian distribution, 21, 22, 27–34, 37, 39, 41, 43, 44, 46, 49, 50, 53, 59, 60, 62–64, 67, 68, 73, 75, 78, 97, 102, 103, 109, 116, 137, 149, 157, 158, 162–164, 183–185, 189, 194, 220–224, 263, 264, 304, 313, 317, 320, 321, 323, 325, 327, 328, 330, 348
generalized Kalman filter, 83
Gibbs fields, 199
Gibbs prior, 73

height field, 85, 87
hierarchical modeling, 115
high energy physics, 157, 169

ICA, 219–223
image classification, 115, 199
image pyramid, 115
imaging, 66, 72, 225, 301, 318, 334
incomplete information., 300
independent component analysis, 219
inference, 4, 7, 9, 10, 17, 19, 21, 58, 61, 62, 65, 66, 83–85, 89, 91, 131, 132, 138, 141–146, 188, 194, 197, 199, 206, 212, 214, 215, 218, 227, 229, 236, 237, 245, 300, 309–311, 317
information geometry, 131, 132, 135, 138
information metric, 131
infrared detectors, 179
invariant measure, 171, 174
inverse problem, 99, 107, 217, 218, 224, 225
inverse problems, 8, 72, 74, 91–93, 98, 162
isodata, 327, 334

Jaynes, 1–13, 15–17, 19, 21, 22, 30, 38, 72, 91, 135, 170, 208, 211, 212, 214, 215, 238, 243, 245–247, 271–273, 288, 293, 300

k-means, 327, 328, 330–333
Kalman filter, 83, 227
knowledge representation, 83, 86
Kullback-Leibler distance, 88, 193

loss function, proper -, 141, 143, 145
loss function, quadratic -, 143, 146

magnetic resonance, 4, 6, 22, 224, 327, 332
magnetoencephalography, 217
manifold, 83, 133, 135, 136, 139, 193, 194
Markov chain monte carlo, 175, 176, 309, 316
Markov processes, 247
Markov random field, 65, 66, 68, 72, 84, 121, 201
Markov random fields, 116, 122, 199
maximally informative inference, 83
maximum entropy, 5, 7–9, 14, 16, 19, 22, 25–30, 46, 58, 59, 61, 75, 88, 91, 93, 98, 102, 104, 136, 178, 188, 201, 208, 214, 215, 219, 227, 228, 236, 271, 278, 279, 281, 288–290, 293–296, 298, 300, 304, 307, 317, 318, 320, 335, 338, 340

maximum likelihood, 75, 93, 121, 127, 128, 163, 219, 227, 327, 343, 352
MCMC, 68, 175, 211, 316, 317, 319, 321–323, 325, 326
measurement , 4, 6, 23, 47, 50, 51, 84, 85, 89, 99, 101, 102, 110, 159, 169, 170, 172, 177, 189, 192, 194, 211, 245, 309–311, 313, 347, 349
MEG, 217, 223–225
minimum description length, 83, 97, 328
model comparison, 171, 319, 341, 342
model parameter estimation, 65, 66
moment analysis, 149
moment generating function, 61, 150, 154
MRI, 224, 334
multi-scale markov, 115, 117
multi-scale markov random fields, 115, 116
multigrid, 83, 89

neural, 72, 217–219, 223–225
neural network, 219
Ni-Cu detection, 341
noise analysis, 345
noise scaling, 181, 185, 186
non-parametric function estimation, 91, 92, 96, 98
nonlinear dynamic models, 227, 234
nonlinear optimisation, 288
nonlinear programming, 288

observables, 160, 177, 238, 245
Occam's razor, 99, 103, 104, 106, 173, 176
odds-ratio, 341
oscillator characterization, 345
outlier, 47, 48, 52, 53, 55, 56

parameter estimation, 22, 47, 49, 65, 69, 72, 124, 149, 181, 199, 200, 334
parameter inference, 149
partial fractions, 151
PCA, 219
PET, 225
phaseless Fourier problem, 288
plasma, 99, 101, 104, 107, 110, 112–114, 171–173, 176, 178
posterior moments, 149, 150, 152
principal component analysis, 219
principle of insufficient reason, 242
principle of maximum entropy, 4, 14, 25–28, 46, 48, 335
prior-predictive value, 72, 171, 175, 178, 319, 325
probability theory, 4–7, 9, 18, 21–26, 33–35, 38, 45, 72, 92, 99, 100, 157, 160, 178, 212, 247, 251, 252, 256, 265–268, 300, 319
probability distribution, assignment 26, 271
properness, 143, 148

quantum entropy, 91–98, 104
quantum mechanics, 6, 18, 92, 132, 237, 240, 246, 247, 256, 268, 269
quantum theory, 2, 4, 8, 237, 239, 242, 243, 245, 248

reconstruction, 14, 65, 69, 70, 73–76, 78–80, 103, 105, 107, 123–130, 173, 305, 307, 309, 310, 313, 314, 318
remote sensing, 72, 122, 199, 200, 206, 301, 307, 334
residue theorem, 151
residues, 151, 302–306, 308
Riemannian geometry, 131, 197
Riemannian metric, 174
risk-unbiasedness, 141, 144
Rutherford backscattering, 107

sampling distributions, 22, 33, 36, 39
segmentation, 72, 77, 115, 116, 118, 119, 121, 122, 200, 201, 203–205, 332, 334

Shannon, 11, 12, 14, 26, 95, 96, 191, 212, 245, 246, 271–273, 287, 288, 329, 334
Shore–Johnson, 272, 274
singular value decomposition (SVD), 219, 224
smoothing, 70, 77, 91, 96, 98, 103, 104
source separation, 217–219, 223
speckle, 65-66, 70–72, 301
statistical inference, 7, 19, 91, 92, 132, 138, 139, 143, 149
stochastic process, 83, 268
sufficient statistics, 30, 31, 37, 46, 149
surface, 9, 59, 76, 83–89, 99–101, 107, 110, 112, 136, 139, 217, 218, 222–224, 303–306
synthetic aperture radar, 65, 72, 301

texture, 65, 66, 69, 70, 72, 118, 121, 122, 199, 200, 203
texture model, 66, 70, 115, 117, 120
time stability, 345
Tsallis entropy, 271

unbiasedness, strong -, 144, 146, 147
unbiasedness, risk -, 141, 144, 147
uncertainty, 12, 16, 26, 38, 40, 50, 66, 84, 126, 130, 132, 134, 135, 143, 157, 159–161, 163, 165, 167–169, 173, 200, 224, 244, 245, 272, 273, 300, 309, 316–318, 348
unitary time evolution, 245

wave function, 240, 241, 269

LIST OF PARTICIPANTS

Giulio D'Agostini
Dipartimento di Fisica
Università di Roma "La Sapienza"
Piazzale Aldo Moro 2
I-00185 Roma
Italy
Giulio.Dagostini@roma1.infn.it

Casper J. Albers
University of Groningen
Dep. of Mathematics
P.O.Box 800
9700 AV Groningen
The Netherlands
c.j.albers@wing.rug.nl

Mathias Anton
Max-Planck-Institut für Plasmaphysik
Postfach 1533
D-85740 Garching
Germany
Mathias.Anton@ipp.mpg.de

Valentin M. Artemiev
Institute of Applied Physics
Akademichiskaya 16, F. Scorina Street
Minsk 220072
Belarus
artemiev@iaph.bas-net.by

Rainer Behrisch
Max-Planck-Institut für Plasmaphysik
Postfach 1533
D-85740 Garching
Germany
Rainer.Behrisch@ipp.mpg.de

Tj. Romke Bontekoe
Bontekoe Data Consultancy
Herengracht 47
NL-2312 LC Leiden
The Netherlands
romke@astro.uva.nl

Larry Bretthorst
Dept. of Chemistry
Washington Univ.
St. Louis MO 63130
USA
larry@bayes.wustl.edu

Ariel Caticha
Department of Physics
University at Albany-SUNY
Albany, NY 12065
USA
ariel@cnsvax.albany.edu

Peter C. Cheeseman
NASA Ames Research Center
MS 269-2
Moffett Field, CA 94035
USA
cheeseman@pluto.arc.nasa.gov

Peter Coles
Astronomy Unit
Queen Mary & Westfield College
University of London
London E1 4NS
UK
P.Coles@qmw.ac.uk

David Coster
Max-Planck-Institut für Plasmaphysik
Postfach 1533
D-85740 Garching
Germany
David.Coster@ipp.mpg.de

Markus Donath
Max-Planck-Institut für Plasmaphysik
Postfach 1533
D-85740 Garching
Germany
Markus.Donath@ipp.mpg.de

Volker Dose
Max-Planck-Institut für Plasmaphysik
Postfach 1533
D-85740 Garching
Germany
Volker.Dose@ipp.mpg.de

Reinhard Drube
Max-Planck-Institut für Plasmaphysik
Postfach 1533
D-85740 Garching
Germany
Reinhard.Drube@ipp.mpg.de

Wolfgang Eckstein
Max-Planck-Institut für Plasmaphysik
Postfach 1533
D-85740 Garching
Germany
Wolfgang.Eckstein@ipp.mpg.de

Albrecht Elsner
Max-Planck-Institut für Plasmaphysik
Postfach 1533
D-85740 Garching
Germany
Albrecht.Elsner@ipp.mpg.de

Gary J. Erickson
Department of Electrical Engineering
Boise State Univ.
Boise, Idaho 83725
USA
gericks@bsu.idbsu.edu

Karl Ertl
Max-Planck-Institut für Plasmaphysik
Postfach 1533
D-85740 Garching
Germany
Karl.Ertl@ipp.mpg.de

Ozcan Esmer
MIDDLE EAST TECHNICAL UNIVERSITY
Department Of City & Regional Planning
O6531-ANKARA
TURKEY
esmer@vitruvius.arch.metu.edu.tr

Rainer Fischer
Max-Planck-Institut für Plasmaphysik
Postfach 1533
D-85740 Garching
Germany
Rainer.Fischer@ipp.mpg.de

Robert L. Fry
Johns Hopkins University
Applied Physics Laboratory
11100 Johns Hopkins Road
Laurel, MD 20723-6099
USA
robert_fry@jhuapl.edu

Robert K. Fung
Citicorp
New York, NY
USA
robert.fung@citicorp.com,
ca314159@bestweb.net

Anthony Garrett
63 High Street
Grantchester
Cambridge
UK
editor@scitext.com

Joachim Geiger
Max-Planck-Institut für Plasmaphysik
Postfach 1533
D-85740 Garching
Germany
Joachim.Geiger@ipp.mpg.de

Louis Giannone
Max-Planck-Institut für Plasmaphysik
Postfach 1533
D-85740 Garching
Germany
Louis.Giannone@ipp.mpg.de

Caio Görner
Max-Planck-Institut für Plasmaphysik
Postfach 1533
D-85740 Garching
Germany
Caio.Goerner@ipp.mpg.de

Amos Golan
Department of Economics
American University (Visiting 1997-1999)
4400 Massachusetts Ave.
Washington DC, 20016-8029
USA
agolan@american.edu, golan@are.berkeley.edu

Peter Goldstrass
Max-Planck-Institut für Plasmaphysik
Postfach 1533
D-85740 Garching
Germany
Peter.Goldstrass@ipp.mpg.de

Ales Gottvald
Institute of Scientific Instruments
Academy of Sciences of the Czech Republic
Kralovopolska 147
CZ-612 64 Brno
Czech Republic
gott@isibrno.cz

Amara Graps
Max-Planck-Institut für Kernphysik
Interplanetary Dust Group
Saupfercheckweg 1
D-69117 Heidelberg
Germany
graps@galileo.mpi-hd.mpg.de

Kenneth M. Hanson
Los Alamos National Laboratory
MS P940
Los Alamos, NM 87545
USA
kmh@lanl.gov

Friedrich Hertweck
Max-Planck-Institut für Plasmaphysik
Postfach 1533
D-85740 Garching
Germany
Friedrich.Hertweck@ipp.mpg.de

David Hestenes
Department of Physics and Astronomy
Arizona State University
PO Box 871504
Tempe, AZ 85287-1504
USA
hestenes@asu.edu

Dawn E. Holmes
School of Computing and Mathematics
University of Bradford
Bradford
BD7 1DP
UK
D.Holmes@scm.brad.ac.uk

Wolfgang Jacob
Max-Planck-Institut für Plasmaphysik
Postfach 1533
D-85740 Garching
Germany
Wolfgang.Jacob@ipp.mpg.de

Otto Kardaun
Max-Planck-Institut für Plasmaphysik
Postfach 1533
D-85740 Garching
Germany
Otto.Kardaun@ipp.mpg.de

Do Kester
Space Research Organization Netherlands (SRON)
PO Box 800
9700 AV GRONINGEN
The Netherlands
D.J.M.Kester@sron.rug.nl

Achim von Keudell
Max-Planck-Institut für Plasmaphysik
Postfach 1533
D-85740 Garching
Germany
Achim.von_Keudell@ipp.mpg.de

Kevin H. Knuth
Dynamic Brain Imaging Lab.
Department of Neuroscience
Albert Einstein College of Medicine
Rose F. Kennedy Center
Room 915
1410 Pelham Pkwy
Bronx NY 10461
USA
kknuth@balrog.aecom.yu.edu

Andreas Kohl
Max-Planck-Institut für Plasmaphysik
Postfach 1533
D-85740 Garching
Germany
Andreas.Kohl@ipp.mpg.de

Karl Krieger
Max-Planck-Institut für Plasmaphysik
Postfach 1533
D-85740 Garching
Germany
Karl.Krieger@ipp.mpg.de

Georg Kühner
Max-Planck-Institut für Plasmaphysik
Postfach 1533
D-85740 Garching
Germany
Georg.Kuehner@ipp.mpg.de

Karl Lackner
Max-Planck-Institut für Plasmaphysik
Postfach 1533
D-85740 Garching
Germany
Karl.Lackner@ipp.mpg.de

Bernhard Landkammer
Max-Planck-Institut für Plasmaphysik
Postfach 1533
D-85740 Garching
Germany
Bernhard.Landkammer@ipp.mpg.de

Wolfgang von der Linden
Max-Planck-Institut für Plasmaphysik
Postfach 1533
D-85740 Garching
Germany
Wolfgang.von_der_Linden@ipp.mpg.d

Christian Linsmeier
Max-Planck-Institut für Plasmaphysik
Postfach 1533
D-85740 Garching
Germany
Christian.Linsmeier@ipp.mpg.de

Jens Luthin
Max-Planck-Institut für Plasmaphysik
Postfach 1533
D-85740 Garching
Germany
Jens.Luthin@ipp.mpg.de

Friedrich Mast
Max-Planck-Institut für Plasmaphysik
Postfach 1533
D-85740 Garching
Germany
K.F.Mast@ipp.mpg.de

Christian Math
Max-Planck-Institut für
Plasmaphysik
Postfach 1533
D-85740 Garching
Germany
Christian.Math@ipp.mpg.de

Matthias Meier
Max-Planck-Institut für
Plasmaphysik
Postfach 1533
D-85740 Garching
Germany
Matthias Meier@ipp.mpg.de

Giovanni Nico
Deutsches Zentrum für Luft- und Raumfahrt (DLR) e.V.
Deutsches Fernerkundungsdatenzentrum (DFD)
Abteilung "Algorithmen und Verfahren" (AV)
Oberpfaffenhofen
82234 Wessling
Germany
nico@dfd.dlr.de

Gintautas Palubinskas
Deutsches Zentrum für Luft- und Raumfahrt (DLR) e.V.
Deutsches Fernerkundungsdatenzentrum (DFD)
Abteilung "Algorithmen und Verfahren" (AV)
Oberpfaffenhofen
82234 Wessling
Germany
Gintautas.Palubinskas@dlr.de

Peter Pecher
Max-Planck-Institut für
Plasmaphysik
Postfach 1533
D-85740 Garching
Germany
Peter.Pecher@ipp.mpg.de

Roland Preuss
Max-Planck-Institut für
Plasmaphysik
Postfach 1533
D-85740 Garching
Germany
Roland.Preuss@ipp.mpg.de

Georgi Rangelov
Max-Planck-Institut für
Plasmaphysik
Postfach 1533
D-85740 Garching
Germany
Georgi.Rangelov@ipp.mpg.de

Hubert Rehrauer
ETH Zuerich
Institut für Kommunikationstechnik
Gloriastr. 35
CH-8092 Zürich
Switzerland
Hubert.Rehrauer@vision.ee.ethz.ch

Nicole Reinecke
Max-Planck-Institut für
Plasmaphysik
Postfach 1533
D-85740 Garching
Germany
Nicole.Reinecke@ipp.mpg.de

Nicolas Rivier
LDFC
Universite Louis Pasteur
Institut de Physique
3, rue de l'Universite
F 67084 Strasbourg cedex
France
nick@ldfc.u-strasbg.fr

Carlos C. Rodriguez
Department of Mathematics
and Statistics
The University at Albany
1400 Washington Avenue
Albany, New York 12222
USA
carlos@math.albany.edu

Joachim Roth
Max-Planck-Institut für
Plasmaphysik
Postfach 1533
D-85740 Garching
Germany
Joachim.Roth@ipp.mpg.de

Josh Rychert
Department of Electrical
Engineering
Boise State Univ.
Boise, Idaho 83725
USA
jrychert@coe-ux2.idbsu.edu

Diemer Salomé
Department of Mathematics
University of Groningen
The Netherlands
D.Salome@math.rug.nl

Wolfgang Sandmann
Max-Planck-Institut für
Plasmaphysik
Postfach 1533
D-85740 Garching
Germany
Wolfgang.Sandmann@ipp.mpg.de

Martin Schneider
Laboratory of Crystallography
Faculty of Mathematics and
Physics
University of Bayreuth
D-95440 Bayreuth
Germany
martin.schneider@uni-bayreuth.de

Ralf Schneider
Max-Planck-Institut für
Plasmaphysik
Postfach 1533
D-85740 Garching
Germany
Ralf.Schneider@ipp.mpg.de

Ute Schneider-Maxon
Max-Planck-Institut für
Plasmaphysik
Postfach 1533
D-85740 Garching
Germany
Ute.Schneider@ipp.mpg.de

Michael Schröder
ETH Zürich
Institut für Kommunikationstechnik
Gloriastr. 35
CH-8092 Zürich
Switzerland
schro@vision.ee.ethz.ch

Thomas Schwarz-Selinger
Max-Planck-Institut für
Plasmaphysik
Postfach 1533
D-85740 Garching
Germany
Thomas.Schwarz-Selinger@ipp.mpg.de

Josef Schweinzer
Max-Planck-Institut für
Plasmaphysik
Postfach 1533
D-85740 Garching
Germany
Josef.Schweinzer@ipp.mpg.de

Richard N. Silver
T-11, MS-B262
Los Alamos National Lab.
Los Alamos, NM 87544
USA
rns@loke.lanl.gov

John Skilling
Cambridge University
Dept. of Applied Mathematics & Theoretical Physics
Silver Street
Cambridge CB3 9EW
UK
j.skilling@damtp.cam.ac.uk

Sander Van Smaalen
Laboratory of Crystallography
University of Bayreuth
D-95440 Bayreuth
Germany
smash@uni-bayreuth.de

Gerhard Staudenmaier
Max-Planck-Institut für Plasmaphysik
Postfach 1533
D-85740 Garching
Germany
Gerhard.Staudenmaier@ipp.mpg.de

Anton Steltenpohl
Max-Planck-Institut für Plasmaphysik
Postfach 1533
D-85740 Garching
Germany
Anton.Steltenpohl@ipp.mpg.de

Andrew W. Strong
Max-Planck-Institut für extraterrestrische Physik
D-85740 Garching
Germany
aws@mpe-garching.mpg.de

Ulrich Stroth
Max-Planck-Institut für Plasmaphysik
Postfach 1533
D-85740 Garching
Germany
Ulrich.Stroth@ipp.mpg.de

John C. Stutz
NASA Ames Research Center
MS 241-3
Moffett Field, CA 94035-1000
USA
j.stutz@mail.arc.nasa.gov

Andreas Taglauer
Institute of Hydrochemistry
Technical University
Marchioninistr. 17
81377 München
Germany

Gottfried Theimer
Max-Planck-Institut für Plasmaphysik
Postfach 1533
D-85740 Garching
Germany
Gottfried.Theimer@ipp.mpg.de

Christiane Tichmann
Max-Planck-Institut für Plasmaphysik
Postfach 1533
D-85740 Garching
Germany
Christiane.Tichmann@ipp.mpg.de

Udo von Toussaint
Max-Planck-Institut für Plasmaphysik
Postfach 1533
D-85740 Garching
Germany
Udo.v.Toussaint@ipp.mpg.de

Myron Tribus
350 Britto Terrace
Fremont, CA 94539
USA
mtribus@compuserve.com

Valery L. Vengrinovich
Institute of Applied Physics
Akademichiskaya 16, F. Scorina Street
Minsk 220072
Belarus
veng@iaph.bas-net.by

Gerhart Venus
Max-Planck-Institut für Plasmaphysik
Postfach 1533
D-85740 Garching
Germany
Gerhart.Venus@ipp.mpg.de

Svenja Vetter
Max-Planck-Institut für Plasmaphysik
Postfach 1533
D-85740 Garching
Germany
Svenja.Vetter@ipp.mpg.de

Marc Walessa
Deutsches Zentrum für Luft- und Raumfahrt (DLR) e.V.
Deutsches Fernerkundungsdatenzentrum (DFD)
Abteilung "Algorithmen und Verfahren" (AV)
Oberpfaffenhofen
82234 Wessling
Germany
Marc.Walessa@dlr.de

Hermann Walter
Max-Planck-Institut für Plasmaphysik
Postfach 1533
D-85740 Garching
Germany
Hermann.Walter@ipp.mpg.de

Arthur Weller
Max-Planck-Institut für Plasmaphysik
Postfach 1533
D-85740 Garching
Germany
Arthur.Weller@ipp.mpg.de

Christoph Wendland
Max-Planck-Institut für Plasmaphysik
Postfach 1533
D-85740 Garching
Germany
Christoph.Wendland@ipp.mpg.de

Andreas Werner
Max-Planck-Institut für Plasmaphysik
Postfach 1533
D-85740 Garching
Germany
Andreas.Werner@ipp.mpg.de

David R. Wolf
NASA Ames Research Center
MS 269-2
Moffett Field, CA 94035-1000
USA
greywolf@ptolemy.arc.nasa.gov

Gilles Zalamansky
University of Metz
France
zala@sciences.univ-metz.fr

Arnold Zellner
Graduate School of Business
University of Chicago
1101 East 58th Street
Chicago, Illinois 60637
USA
arnold.zellner@gsb.uchicago.edu

Fundamental Theories of Physics

Series Editor: Alwyn van der Merwe, *University of Denver, USA*

1. M. Sachs: *General Relativity and Matter.* A Spinor Field Theory from Fermis to Light-Years. With a Foreword by C. Kilmister. 1982 ISBN 90-277-1381-2
2. G.H. Duffey: *A Development of Quantum Mechanics.* Based on Symmetry Considerations. 1985 ISBN 90-277-1587-4
3. S. Diner, D. Fargue, G. Lochak and F. Selleri (eds.): *The Wave-Particle Dualism.* A Tribute to Louis de Broglie on his 90th Birthday. 1984 ISBN 90-277-1664-1
4. E. Prugovečki: *Stochastic Quantum Mechanics and Quantum Spacetime.* A Consistent Unification of Relativity and Quantum Theory based on Stochastic Spaces. 1984; 2nd printing 1986 ISBN 90-277-1617-X
5. D. Hestenes and G. Sobczyk: *Clifford Algebra to Geometric Calculus.* A Unified Language for Mathematics and Physics. 1984 ISBN 90-277-1673-0; Pb (1987) 90-277-2561-6
6. P. Exner: *Open Quantum Systems and Feynman Integrals.* 1985 ISBN 90-277-1678-1
7. L. Mayants: *The Enigma of Probability and Physics.* 1984 ISBN 90-277-1674-9
8. E. Tocaci: *Relativistic Mechanics, Time and Inertia.* Translated from Romanian. Edited and with a Foreword by C.W. Kilmister. 1985 ISBN 90-277-1769-9
9. B. Bertotti, F. de Felice and A. Pascolini (eds.): *General Relativity and Gravitation.* Proceedings of the 10th International Conference (Padova, Italy, 1983). 1984 ISBN 90-277-1819-9
10. G. Tarozzi and A. van der Merwe (eds.): *Open Questions in Quantum Physics.* 1985 ISBN 90-277-1853-9
11. J.V. Narlikar and T. Padmanabhan: *Gravity, Gauge Theories and Quantum Cosmology.* 1986 ISBN 90-277-1948-9
12. G.S. Asanov: *Finsler Geometry, Relativity and Gauge Theories.* 1985 ISBN 90-277-1960-8
13. K. Namsrai: *Nonlocal Quantum Field Theory and Stochastic Quantum Mechanics.* 1986 ISBN 90-277-2001-0
14. C. Ray Smith and W.T. Grandy, Jr. (eds.): *Maximum-Entropy and Bayesian Methods in Inverse Problems.* Proceedings of the 1st and 2nd International Workshop (Laramie, Wyoming, USA). 1985 ISBN 90-277-2074-6
15. D. Hestenes: *New Foundations for Classical Mechanics.* 1986 ISBN 90-277-2090-8; Pb (1987) 90-277-2526-8
16. S.J. Prokhovnik: *Light in Einstein's Universe.* The Role of Energy in Cosmology and Relativity. 1985 ISBN 90-277-2093-2
17. Y.S. Kim and M.E. Noz: *Theory and Applications of the Poincaré Group.* 1986 ISBN 90-277-2141-6
18. M. Sachs: *Quantum Mechanics from General Relativity.* An Approximation for a Theory of Inertia. 1986 ISBN 90-277-2247-1
19. W.T. Grandy, Jr.: *Foundations of Statistical Mechanics.* Vol. I: *Equilibrium Theory.* 1987 ISBN 90-277-2489-X
20. H.-H von Borzeszkowski and H.-J. Treder: *The Meaning of Quantum Gravity.* 1988 ISBN 90-277-2518-7
21. C. Ray Smith and G.J. Erickson (eds.): *Maximum-Entropy and Bayesian Spectral Analysis and Estimation Problems.* Proceedings of the 3rd International Workshop (Laramie, Wyoming, USA, 1983). 1987 ISBN 90-277-2579-9
22. A.O. Barut and A. van der Merwe (eds.): *Selected Scientific Papers of Alfred Landé. [1888-1975].* 1988 ISBN 90-277-2594-2

23. W.T. Grandy, Jr.: *Foundations of Statistical Mechanics.*
Vol. II: *Nonequilibrium Phenomena.* 1988 ISBN 90-277-2649-3
24. E.I. Bitsakis and C.A. Nicolaides (eds.): *The Concept of Probability.* Proceedings of the Delphi Conference (Delphi, Greece, 1987). 1989 ISBN 90-277-2679-5
25. A. van der Merwe, F. Selleri and G. Tarozzi (eds.): *Microphysical Reality and Quantum Formalism, Vol. 1.* Proceedings of the International Conference (Urbino, Italy, 1985). 1988 ISBN 90-277-2683-3
26. A. van der Merwe, F. Selleri and G. Tarozzi (eds.): *Microphysical Reality and Quantum Formalism, Vol. 2.* Proceedings of the International Conference (Urbino, Italy, 1985). 1988 ISBN 90-277-2684-1
27. I.D. Novikov and V.P. Frolov: *Physics of Black Holes.* 1989 ISBN 90-277-2685-X
28. G. Tarozzi and A. van der Merwe (eds.): *The Nature of Quantum Paradoxes.* Italian Studies in the Foundations and Philosophy of Modern Physics. 1988 ISBN 90-277-2703-1
29. B.R. Iyer, N. Mukunda and C.V. Vishveshwara (eds.): *Gravitation, Gauge Theories and the Early Universe.* 1989 ISBN 90-277-2710-4
30. H. Mark and L. Wood (eds.): *Energy in Physics, War and Peace.* A Festschrift celebrating Edward Teller's 80th Birthday. 1988 ISBN 90-277-2775-9
31. G.J. Erickson and C.R. Smith (eds.): *Maximum-Entropy and Bayesian Methods in Science and Engineering.*
Vol. I: *Foundations.* 1988 ISBN 90-277-2793-7
32. G.J. Erickson and C.R. Smith (eds.): *Maximum-Entropy and Bayesian Methods in Science and Engineering.*
Vol. II: *Applications.* 1988 ISBN 90-277-2794-5
33. M.E. Noz and Y.S. Kim (eds.): *Special Relativity and Quantum Theory.* A Collection of Papers on the Poincaré Group. 1988 ISBN 90-277-2799-6
34. I.Yu. Kobzarev and Yu.I. Manin: *Elementary Particles. Mathematics, Physics and Philosophy.* 1989 ISBN 0-7923-0098-X
35. F. Selleri: *Quantum Paradoxes and Physical Reality.* 1990 ISBN 0-7923-0253-2
36. J. Skilling (ed.): *Maximum-Entropy and Bayesian Methods.* Proceedings of the 8th International Workshop (Cambridge, UK, 1988). 1989 ISBN 0-7923-0224-9
37. M. Kafatos (ed.): *Bell's Theorem, Quantum Theory and Conceptions of the Universe.* 1989 ISBN 0-7923-0496-9
38. Yu.A. Izyumov and V.N. Syromyatnikov: *Phase Transitions and Crystal Symmetry.* 1990 ISBN 0-7923-0542-6
39. P.F. Fougère (ed.): *Maximum-Entropy and Bayesian Methods.* Proceedings of the 9th International Workshop (Dartmouth, Massachusetts, USA, 1989). 1990 ISBN 0-7923-0928-6
40. L. de Broglie: *Heisenberg's Uncertainties and the Probabilistic Interpretation of Wave Mechanics.* With Critical Notes of the Author. 1990 ISBN 0-7923-0929-4
41. W.T. Grandy, Jr.: *Relativistic Quantum Mechanics of Leptons and Fields.* 1991 ISBN 0-7923-1049-7
42. Yu.L. Klimontovich: *Turbulent Motion and the Structure of Chaos.* A New Approach to the Statistical Theory of Open Systems. 1991 ISBN 0-7923-1114-0
43. W.T. Grandy, Jr. and L.H. Schick (eds.): *Maximum-Entropy and Bayesian Methods.* Proceedings of the 10th International Workshop (Laramie, Wyoming, USA, 1990). 1991 ISBN 0-7923-1140-X

44. P.Pták and S. Pulmannová: *Orthomodular Structures as Quantum Logics*. Intrinsic Properties, State Space and Probabilistic Topics. 1991 ISBN 0-7923-1207-4
45. D. Hestenes and A. Weingartshofer (eds.): *The Electron*. New Theory and Experiment. 1991 ISBN 0-7923-1356-9
46. P.P.J.M. Schram: *Kinetic Theory of Gases and Plasmas*. 1991 ISBN 0-7923-1392-5
47. A. Micali, R. Boudet and J. Helmstetter (eds.): *Clifford Algebras and their Applications in Mathematical Physics*. 1992 ISBN 0-7923-1623-1
48. E. Prugovečki: *Quantum Geometry*. A Framework for Quantum General Relativity. 1992 ISBN 0-7923-1640-1
49. M.H. Mac Gregor: *The Enigmatic Electron*. 1992 ISBN 0-7923-1982-6
50. C.R. Smith, G.J. Erickson and P.O. Neudorfer (eds.): *Maximum Entropy and Bayesian Methods*. Proceedings of the 11th International Workshop (Seattle, 1991). 1993 ISBN 0-7923-2031-X
51. D.J. Hoekzema: *The Quantum Labyrinth*. 1993 ISBN 0-7923-2066-2
52. Z. Oziewicz, B. Jancewicz and A. Borowiec (eds.): *Spinors, Twistors, Clifford Algebras and Quantum Deformations*. Proceedings of the Second Max Born Symposium (Wrocław, Poland, 1992). 1993 ISBN 0-7923-2251-7
53. A. Mohammad-Djafari and G. Demoment (eds.): *Maximum Entropy and Bayesian Methods*. Proceedings of the 12th International Workshop (Paris, France, 1992). 1993 ISBN 0-7923-2280-0
54. M. Riesz: *Clifford Numbers and Spinors* with Riesz' Private Lectures to E. Folke Bolinder and a Historical Review by Pertti Lounesto. E.F. Bolinder and P. Lounesto (eds.). 1993 ISBN 0-7923-2299-1
55. F. Brackx, R. Delanghe and H. Serras (eds.): *Clifford Algebras and their Applications in Mathematical Physics*. Proceedings of the Third Conference (Deinze, 1993) 1993 ISBN 0-7923-2347-5
56. J.R. Fanchi: *Parametrized Relativistic Quantum Theory*. 1993 ISBN 0-7923-2376-9
57. A. Peres: *Quantum Theory: Concepts and Methods*. 1993 ISBN 0-7923-2549-4
58. P.L. Antonelli, R.S. Ingarden and M. Matsumoto: *The Theory of Sprays and Finsler Spaces with Applications in Physics and Biology*. 1993 ISBN 0-7923-2577-X
59. R. Miron and M. Anastasiei: *The Geometry of Lagrange Spaces: Theory and Applications*. 1994 ISBN 0-7923-2591-5
60. G. Adomian: *Solving Frontier Problems of Physics: The Decomposition Method*. 1994 ISBN 0-7923-2644-X
61 B.S. Kerner and V.V. Osipov: *Autosolitons*. A New Approach to Problems of Self-Organization and Turbulence. 1994 ISBN 0-7923-2816-7
62. G.R. Heidbreder (ed.): *Maximum Entropy and Bayesian Methods*. Proceedings of the 13th International Workshop (Santa Barbara, USA, 1993) 1996 ISBN 0-7923-2851-5
63. J. Peřina, Z. Hradil and B. Jurčo: *Quantum Optics and Fundamentals of Physics*. 1994 ISBN 0-7923-3000-5
64. M. Evans and J.-P. Vigier: *The Enigmatic Photon*. Volume 1: The Field $\boldsymbol{B}^{(3)}$. 1994 ISBN 0-7923-3049-8
65. C.K. Raju: *Time: Towards a Constistent Theory*. 1994 ISBN 0-7923-3103-6
66. A.K.T. Assis: *Weber's Electrodynamics*. 1994 ISBN 0-7923-3137-0
67. Yu. L. Klimontovich: *Statistical Theory of Open Systems*. Volume 1: A Unified Approach to Kinetic Description of Processes in Active Systems. 1995 ISBN 0-7923-3199-0; Pb: ISBN 0-7923-3242-3

68. M. Evans and J.-P. Vigier: *The Enigmatic Photon.* Volume 2: Non-Abelian Electrodynamics. 1995 ISBN 0-7923-3288-1
69. G. Esposito: *Complex General Relativity.* 1995 ISBN 0-7923-3340-3
70. J. Skilling and S. Sibisi (eds.): *Maximum Entropy and Bayesian Methods.* Proceedings of the Fourteenth International Workshop on Maximum Entropy and Bayesian Methods. 1996 ISBN 0-7923-3452-3
71. C. Garola and A. Rossi (eds.): *The Foundations of Quantum Mechanics – Historical Analysis and Open Questions.* 1995 ISBN 0-7923-3480-9
72. A. Peres: *Quantum Theory: Concepts and Methods.* 1995 (see for hardback edition, Vol. 57) ISBN Pb 0-7923-3632-1
73. M. Ferrero and A. van der Merwe (eds.): *Fundamental Problems in Quantum Physics.* 1995 ISBN 0-7923-3670-4
74. F.E. Schroeck, Jr.: *Quantum Mechanics on Phase Space.* 1996 ISBN 0-7923-3794-8
75. L. de la Peña and A.M. Cetto: *The Quantum Dice.* An Introduction to Stochastic Electrodynamics. 1996 ISBN 0-7923-3818-9
76. P.L. Antonelli and R. Miron (eds.): *Lagrange and Finsler Geometry.* Applications to Physics and Biology. 1996 ISBN 0-7923-3873-1
77. M.W. Evans, J.-P. Vigier, S. Roy and S. Jeffers: *The Enigmatic Photon.* Volume 3: Theory and Practice of the $\boldsymbol{B}^{(3)}$ Field. 1996 ISBN 0-7923-4044-2
78. W.G.V. Rosser: *Interpretation of Classical Electromagnetism.* 1996 ISBN 0-7923-4187-2
79. K.M. Hanson and R.N. Silver (eds.): *Maximum Entropy and Bayesian Methods.* 1996 ISBN 0-7923-4311-5
80. S. Jeffers, S. Roy, J.-P. Vigier and G. Hunter (eds.): *The Present Status of the Quantum Theory of Light.* Proceedings of a Symposium in Honour of Jean-Pierre Vigier. 1997 ISBN 0-7923-4337-9
81. M. Ferrero and A. van der Merwe (eds.): *New Developments on Fundamental Problems in Quantum Physics.* 1997 ISBN 0-7923-4374-3
82. R. Miron: *The Geometry of Higher-Order Lagrange Spaces.* Applications to Mechanics and Physics. 1997 ISBN 0-7923-4393-X
83. T. Hakioğlu and A.S. Shumovsky (eds.): *Quantum Optics and the Spectroscopy of Solids.* Concepts and Advances. 1997 ISBN 0-7923-4414-6
84. A. Sitenko and V. Tartakovskii: *Theory of Nucleus.* Nuclear Structure and Nuclear Interaction. 1997 ISBN 0-7923-4423-5
85. G. Esposito, A.Yu. Kamenshchik and G. Pollifrone: *Euclidean Quantum Gravity on Manifolds with Boundary.* 1997 ISBN 0-7923-4472-3
86. R.S. Ingarden, A. Kossakowski and M. Ohya: *Information Dynamics and Open Systems.* Classical and Quantum Approach. 1997 ISBN 0-7923-4473-1
87. K. Nakamura: *Quantum versus Chaos.* Questions Emerging from Mesoscopic Cosmos. 1997 ISBN 0-7923-4557-6
88. B.R. Iyer and C.V. Vishveshwara (eds.): *Geometry, Fields and Cosmology.* Techniques and Applications. 1997 ISBN 0-7923-4725-0
89. G.A. Martynov: *Classical Statistical Mechanics.* 1997 ISBN 0-7923-4774-9
90. M.W. Evans, J.-P. Vigier, S. Roy and G. Hunter (eds.): *The Enigmatic Photon.* Volume 4: New Directions. 1998 ISBN 0-7923-4826-5
91. M. Rédei: *Quantum Logic in Algebraic Approach.* 1998 ISBN 0-7923-4903-2
92. S. Roy: *Statistical Geometry and Applications to Microphysics and Cosmology.* 1998 ISBN 0-7923-4907-5

Fundamental Theories of Physics

93. B.C. Eu: *Nonequilibrium Statistical Mechanics.* Ensembled Method. 1998 ISBN 0-7923-4980-6
94. V. Dietrich, K. Habetha and G. Jank (eds.): *Clifford Algebras and Their Application in Mathematical Physics.* Aachen 1996. 1998 ISBN 0-7923-5037-5
95. *Not yet known*
96. V.P. Frolov and I.D. Novikov: *Black Hole Physics.* Basic Concepts and New Developments. 1998 ISBN 0-7923-5145-2; PB 0-7923-5146
97. G. Hunter, S. Jeffers and J-P. Vigier (eds.): *Causality and Locality in Modern Physics.* 1998 ISBN 0-7923-5227-0
98. G.J. Erickson, J.T. Rychert and C.R. Smith (eds.): *Maximum Entropy and Bayesian Methods.* 1998 ISBN 0-7923-5047-2
99. D. Hestenes: *New Foundations for Classical Mechanics (Second Edition).* 1999 ISBN 0-7923-5302-1; PB ISBN 0-7923-5514-8
100. B.R. Iyer and B. Bhawal: *Black Holes, Gravitational Radiation and the Universe.* Essays in Honor of C. V. Vishveshwara. 1999 ISBN 0-7923-5308-0
101. P.L. Antonelli and T.J. Zastawniak: *Fundamentals of Finslerian Diffusion with Applications.* 1999 ISBN 0-7923-5511-3
102. Reserved
103. Reserved
104. Reserved
105. W. von der Linden, V. Dose, R. Fisher and R. Preuss: *Maximum Entropy and Bayesian Methods.* Proceedings of the 18th International Workshop on Maximum Entropy and Bayesian Methods of Statistical Analysis. 1999 ISBN 0-7923-5766-3

KLUWER ACADEMIC PUBLISHERS – DORDRECHT / BOSTON / LONDON

MIX
Papier aus verantwortungsvollen Quellen
Paper from responsible sources
FSC® C105338

If you have any concerns about our products,
you can contact us on
ProductSafety@springernature.com

In case Publisher is established outside the EU,
the EU authorized representative is:
Springer Nature Customer Service Center GmbH
Europaplatz 3, 69115 Heidelberg, Germany

Printed by Libri Plureos GmbH
in Hamburg, Germany